WHERE TO WATCH BIRDS IN
FRANCE

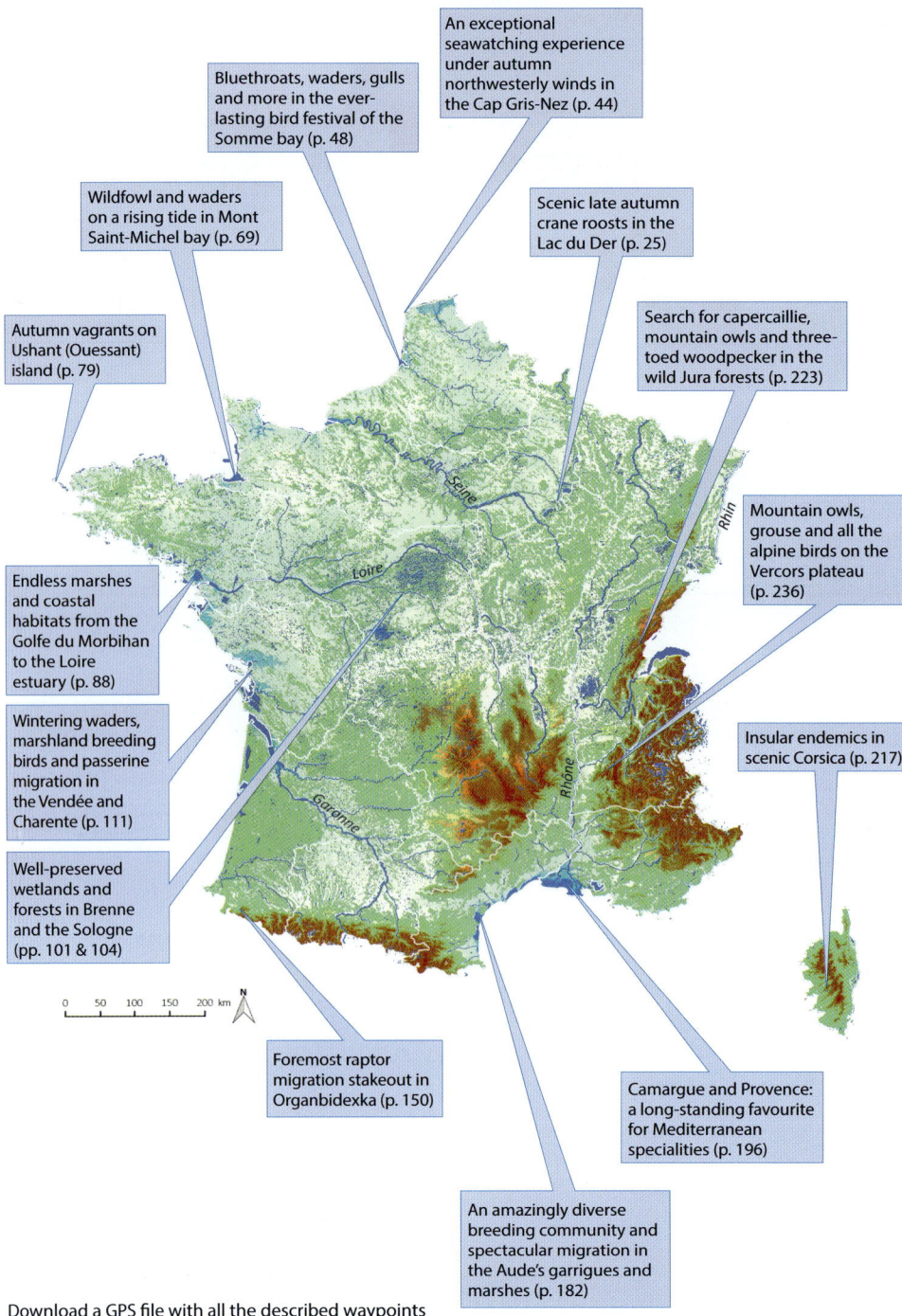

WHERE TO WATCH BIRDS IN
FRANCE

**JEAN-YVES BARNAGAUD,
NIDAL ISSA and
SÉBASTIEN DALLOYAU**

Published by Pelagic Publishing
www.pelagicpublishing.com
PO Box 874, Exeter, EX1 9YH, UK
www.pelagicpublishing.com

Where to Watch Birds in France

ISBN 978-1-78427-154-1 *Paperback*
ISBN 978-1-78427-155-8 *ePub*
ISBN 978-1-78427-156-5 *PDF*

© Jean-Yves Barnagaud, Nidal Issa & Sébastien Dalloyau 2019

The moral rights of the authors have been asserted.

All rights reserved. Apart from short excerpts for use in research or for reviews, no part of this document may be printed or reproduced, stored in a retrieval system, or transmitted in any form or by any means, electronic, mechanical, photocopying, recording, now known or hereafter invented or otherwise without prior permission from the publisher.

British Library Cataloguing in Publication Data
A catalogue record for this book is available from the British Library

Cover photograph: Common rock thrush, *Monticola saxatilis* (Photographer: Daniele Occhiato).

For any question or comment, please contact the authors at wwbf.authors@gmail.com

Typeset by D & N Publishing, Baydon, Wiltshire

Printed and bound in India by Replika Press Pvt. Ltd.

SYMBOL CHART

BACKGROUND
- Beach, dune
- Natural wetlands
- Open waters, ponds, marshes
- Forest, grass, meadows
- Built-up areas

LINEAR ELEMENTS
- Track or path
- Local road ('Départementale')
- Main road ('Nationale')
- Motorway ('Autoroute', usually with toll)
- Railway
- River, canal

FACILITIES
- Airport
- Passenger harbour

POINTS OF INTEREST
- Peak, hill, mountain
- Calvary (wayside cross)
- Contour line
- to St Brieuc — Direction to the nearest major urban centre
- PARIS, BREST: Administrative centre
- Reims, La Tranche sur Mer: Other inhabited locations

BIRDING INFORMATION
- Extent of a site
- Extent of a section
- Waypoint accessible by car
- Waypoint accessible on foot only
- no map — Section without site map

CONTENTS

Symbol chart	4
Foreword	7
Preface	8
Acknowledgements	9
Introduction	10
Regions at a glance	13
Practical information	18
Birdwatching in France	19
REGION 1 Paris and the Seine Reservoirs	22
REGION 2 From the North Sea to the Somme Bay and the Western Ardennes	38
REGION 3 Normandy	56
REGION 4 Brittany	73
REGION 5 The Loire Valley	92
REGION 6 Poitou-Charentes and the Vendée	111
REGION 7 Aquitaine	134
REGION 8 The Pyrenees	160
REGION 9 Western Mediterranean Coast and the Cévennes	174
REGION 10 Eastern Mediterranean Coast, Southern Alps and Corsica	195
REGION 11 Jura and the Alps	222
REGION 12 Massif Central	245
REGION 13 Burgundy	270
REGION 14 Northeast	288
Birdfinder	309
Checklist	317
Site index	332

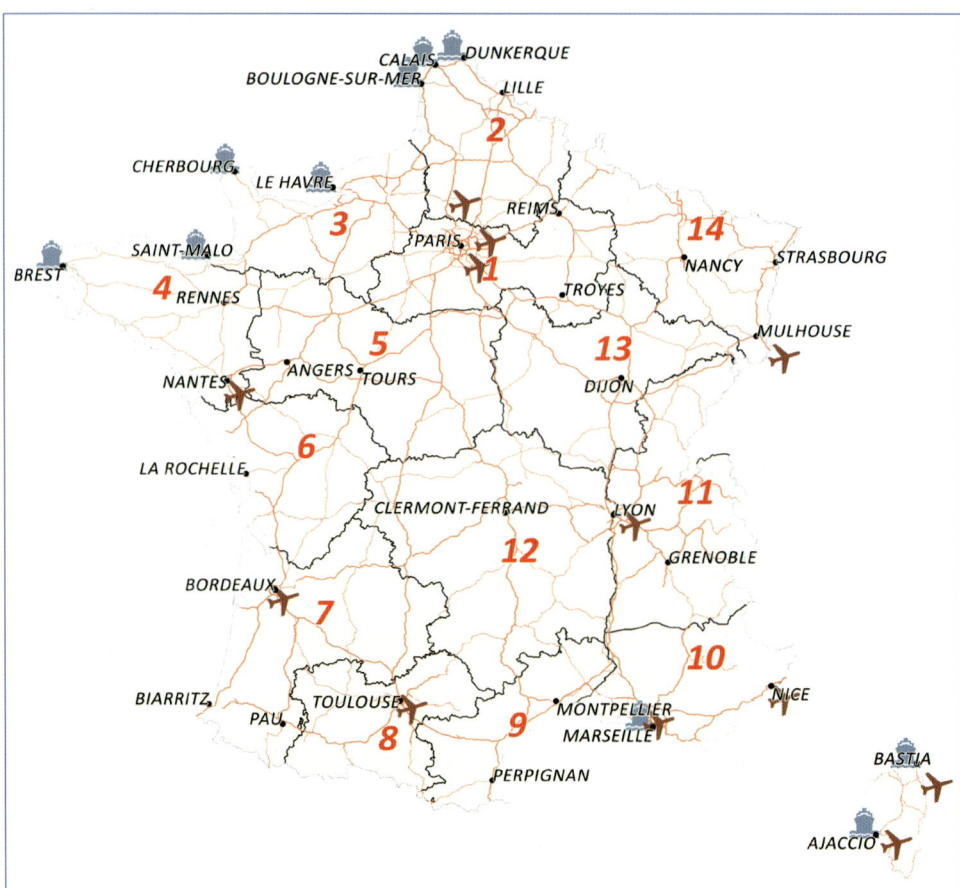

LEGEND	
REGION 1	Paris and the Seine reservoirs, p. 22
REGION 2	From the North Sea to the Somme bay and the western Ardennes, p. 38
REGION 3	Normandy, p. 56
REGION 4	Brittany, p. 73
REGION 5	The Loire valley, p. 92
REGION 6	Poitou-Charentes and the Vendée, p. 111
REGION 7	Aquitaine, p. 134
REGION 8	The Pyrenees, p. 160
REGION 9	Western Mediterranean coast and the Cévennes, p. 174
REGION 10	Eastern Mediterranean coast, Southern Alps and Corsica, p. 195
REGION 11	Jura and the Alps, p. 222
REGION 12	Massif Central, p. 245
REGION 13	Burgundy, p. 270
REGION 14	Northeast, p. 288

FOREWORD

When I started watching birds in the early 1970s, 'birdwatching' was the hobby of just a few. No more than a handful of professional ornithologists and a few tens of amateurs explored their regions, sometimes France, and even more rarely the world, searching for their cherished birds. I was feeding on their words and writings, mixing respect and desire. Only a handful of top birdwatching sites were well known, and as birding news did not travel at the speed of light, as they do today, you needed to rely on your own intuition to find 'hotspots' for yourself ... If only I had had, in those days, a book like the one you now hold in your hands, a veritable *vade mecum* for the travelling ornithologist.

In the last 20 years, the number of bird enthusiasts and field ornithologists has exploded, alongside a technological revolution that would have been unthinkable when I started out: cutting-edge observation devices, mobile phones, affordable digital cameras, online databases ... Birdwatchers have also increasingly become travellers, assisted by a greatly increased literature – not only identification books but also guides giving details of hotspots and good deals. Thus, birdwatching backpackers' guides multiplied across the world, usually dominated by English-language literature. Not until the 1990s would similar publications emerge in France, pioneered by the seminal *Où voir les Oiseaux en France* in 1991.

The present guide makes use of multiple innovations: for instance, it takes advantage of the technological advances mentioned above, indicating the GPS coordinates of every site described. This is a real comfort as I recall how, in the old days, we had to wander here and there in the hope of finding a spot roughly indicated on a paper map! While this book will be of interest to French '*birdwatcheurs*' and those of neighbouring countries, it is designed for anyone who wants to discover the extraordinary bird richness of our country. Being published in English, it is accessible to all. Although nowadays, many birders and specialised travel agencies favour better-known or more exotic destinations, this guide will make life so much easier for those who wish to explore France's unique natural richness, and especially its birds. France has a diversity of remarkable habitats hosting a diverse avifauna with few equivalents in Europe.

Ornithological tourism has become a commercial godsend that some of our neighbours have well understood. Countries such as Costa Rica have even gone so far as to focus their economy on 'green tourism'. We must not, however, conceal the contradictions that this implies for our environmental responsibility. Indeed, whether with respect to the disturbance of vulnerable habitats or species, or to greenhouse-gas emissions, this kind of tourism is sometimes labelled as non-sustainable. The counter-argument, however, is crystal-clear: green tourism undoubtedly contributes to the preservation of natural habitats across the world, and in many places offers a viable alternative to their destruction for unrestrained economic and urban development. France is not immune from this observation. And how can we not appreciate an increase in the number of birdwatchers, in a world where nature is under such threat? As the old adage has it, 'we conserve only what we love'.

All praise and thanks to the authors for having produced such a useful tool. They have themselves visited most of the sites they describe, and the practical advice they offer will enable visitors to make the most of their explorations. This new-generation guide will quickly become bedside reading for the world's travelling birders wishing to become acquainted with the birds of our beautiful France. If you are used to arranging your own trips, you will know that you live each journey three times: first as you prepare for it, then when you live it, and finally in your memories. This book will be an essential companion for the first two stages – and enjoyable reading even for someone who just stays at home.

'*Un voyage c'est une folie qui nous obsède, nous emporte dans le mythe*' (Sylvain Tesson)

Jean-Philippe Siblet, head of the Natural Heritage Service of the French Natural History Museum, secretary-general of the Société d'Etudes Ornithologique de France.

November 2018

PREFACE

This guide is designed as an introduction to France for anyone visiting with birds in mind, from casual birdwatchers checking a bird that flies over the terrace during a family holiday to addict birders who would sell their souls for a dream species or a record-breaking checklist. Some may have just a few spare hours to get their binoculars out between two business meetings or museum visits, others will be out in the field for two weeks or more, from sunrise to sunset. We hope this book will be helpful to all bird-lovers, birdwatchers and birders, whatever the duration of their stay, the number of kilometres they are prepared to travel and how they enjoy birds.

France is at the crossroads of several biogeographical regions. With over 400 regularly occurring species, of which 357 regularly breed or winter, it has arguably one of the most diverse avifaunas of all Europe, spanning an incredible variety from colourful Mediterranean flagship species such as roller, bee-eater or black-winged kite to secretive cold-climate or mountain specialists like three-toed woodpecker and Tengmalm's owl. Many warm-climate species reach their northern range limit in France, sometimes overlapping with the southernmost populations of some central-European species – a striking example being the local co-occurrence of melodious and icterine warblers in the northernmost quarter of the country. France is crossed by two major western Palearctic flyways and has several famous migration bottlenecks and stopover sites – Organbidexka, Ushant island, Gruissan, the Lac du Der and the Camargue being familiar names to most European birders.

Thus, we had to make choices. We gave up on the idea of an exhaustive description of all the possible birding sites found throughout the country – they are too numerous for a traveller's guidebook, and many of them have similar species compositions. Instead, we decided to describe, within each of the 14 covered regions, a selection of representative sites, chosen for their bird species composition, ease of access, complementarity, and admittedly with some level of subjectivity. We selected sites in order to enable the reader to see the widest possible species diversity and largest range of local specialities in a reasonable time and while respecting the basic ethical rules obvious to all birdwatchers. Whenever possible, sites are arranged in clusters or itineraries that can be covered in two to three days without hurrying. To supplement the use of the book in the field, all the sites we describe are geolocated in a file that can be downloaded from the publisher's website and downloaded onto any GPS device.

This book is a birder's companion, but it will not provide a last word on where to watch birds. Its job is to lead you to where the birds are, yours is to find the birds. Try to avoid merely hopping from site to site – perhaps your target species or the birding spot of your dreams waits for you at a random location along your way, not included in the book. Birdwatching can be about listing and ticking, but it is also about finding your own birds in your own places. Hence, it is probable – and desirable – that the book will soon be updated by readers' new discoveries: we'll welcome them warmly. There is still a lot to be explored and discovered in France: use the book to set the main path and let the birds lead you in the field.

ACKNOWLEDGEMENTS

The authors wish to thank all their colleagues, friends and fellow birdwatchers who have contributed, to any extent, to the completion of this guide. We owe particular thanks to the 38 chapter reviewers who have provided invaluable information, carefully inspected every single site location, access instruction and species list, performed field checks and proofread the text. This book is largely based on their in-depth experience of their respective regions: Aurélien Audevard, Luc Barbaro, François Bouzendorf, Yohan Brouillard, Sébastien Brunet, Michel Caupenne, Mikaël Champion, Jean-Michel Chartendrault, Thomas Chevallier, Valentin Condal, Marc Crouzier, Boris Delahaie, Quentin Dupriez, Jérémy Dupuis, Amine Flitti, Jean François, Julien Gonin, Laurence Guillosson, Tristan Guillosson, François Halligon, James Jean-Baptiste, Franck Jouandoudet, Thierry Joubert, François Legendre, Frédéric Malvaud, Vincent Palomares, Claude Parent, Julien Présent, Sébastien Provost, Sébastien Reeber, Antoine Rougeron, Karsten Schmale, David Simpson, Julien Vèque, Alain Verneau, Sylvain Vincent, Stanislas Wroza, Maxime Zucca.

We cannot name all the people who helped us at every point along the way. However, we wish to thank in particular those who tested preliminary versions of the text in the field or provided feedback on the maps: Volker Bahn, Luc Barbaro, Elsa Bugot, Jean Chevallier, Gilles Corsand, Aëlya Dalloyau, Thibault Daumal, Pierrick Devoucoux, Paul Doniol-Valcroze, Michel Doublet, Benoît Duchenne, Paul Dufour, Julie Fluhr, Charlotte Francesiaz, Nicolas Gendre, Julien Gernigon, Emmanuel Gfeller, Jean-Marc Guilpain, Marc Guyt, Christoph Haag, Nicolas Harter, Nicolas Macaire, Frédéric Malher, Sophie Meriotte, Jean-Philippe Paul, Julien Piette, Pierre Reveillaud, Serge Risser, Frédéric Signoret, Rune Tjørnløv.

We are also most grateful to the photographers who kindly provided the pictures that illustrate this book: Aurélien Audevard, Julien Boulanger, Cédric Caïn, Clément Caiveau, Michel Caupenne, Edouard Dansette, Armel Deniau, Aurélie De Seynes, Eric Didner, Paul Dufour, François Legendre, Hervé Michel, Corentin Morvan, Daniele Occhiato, Georges Olioso, Jean-Luc Pinaud, Louis-Marie Préau, Camille Reveillaud, Antoine Rougeron, Alain Verneau.

This book is supported by the Société d'Etudes Ornithologiques de France.

INTRODUCTION

Structure of the book
The book is divided into four main parts:

1. **Introductory chapters** (pages 10–21) explain how to use the book and provide the minimum practical information necessary to plan the outline of your birding trip.
2. **Regional accounts** (pages 22–308) make up the main body of the book. We have divided France into **14 regions** based on geographical and avifaunistic coherence. Hence, they do not match administrative divisions. Each regional chapter was written by one of the authors and thoroughly checked by one to five referees chosen from among the best birders of the region, who verified all the information provided, from the accuracy of site accounts to bird lists and phenologies. The regional accounts are designed to be used alongside the maps that are included in the text, and also with a GPS file that can be downloaded onto any GPS device (see below).
3. **The Birdfinder** section (pages 309–16) provides more targeted details for 30 species which often rank in the top wish-list of birders visiting France (the best times to look for them, the best sites, and a few tips). This selection is necessarily subjective. The aim is to help you focus your research effort – it will not replace a thorough check of the available identification and ecological literature for the species in question, but it may save you a bit of time.
4. **Species checklist and index of place names** (pages 317–36) will help your navigation through the book if you are looking for a specific bird or location.

Site accounts
We have divided France into 14 regions, representing a compromise between administrative boundaries, ornithological coherence and practical considerations. Each region is itself divided into sections that correspond to coherent birding areas. At the start of each section, a short introduction presents the highlights, the best times of year to visit, and some tips to help you make the most of your birding visit. Within these sections, **sites are the key divisions of the book. They are numbered continuously throughout the book from 1 to 312.** A site consists of a small area where you can concentrate your birding, from a single locality to a small valley, wetland or trail, or in a few instances an entire itinerary. **Within a site, one or more waypoints locate precisely the most suitable birding locations.** You can download all the waypoints from the publisher's website to help you navigate in the field with your smartphone or GPS device. In most cases, how to bird the site will become obvious once you are parked at the first waypoint – otherwise, the text and maps provide all necessary directions. **We do not list all the possible species for each site, but a selection of those that we believe are the most representative or sought-after.** Do not expect to find all the species mentioned: check their phenologies in your identification guidebook, and remember that surprises often show up.

Maps
The book includes four types of maps:

» **National maps** in the introductory chapter give a broad overview of what France looks like, how the 14 regions are organised, plus the main geophysical features and major road axes, airports and train stations.
» **Regional maps**, at the start of each chapter, provide a general view of the region covered, displaying the sections covered in the chapter.
» **Section maps** provide an overview of the sites within a section.
» **Site maps** display detailed views of the sites, waypoints, roads and tracks. These maps are not designed for road navigation: you should complement them with the use of a GPS device and/or a printed roadmap. Owing to limitations of space, we could not display some of the most isolated, least visited sites or parts of sites: the text or section maps indicate explicitly which sites are not mapped. All the access paths described in the text are displayed on maps, but there may be alternatives that are not mapped. Backgrounds on these maps represent forests, waterbodies, rivers, urban areas and topography. Each map was augmented and corrected by eye based on the most recent available satellite layers and verified with local reviewers or from personal checks in the field. The French names shown on the maps are those you are likely to encounter on a signpost or a local map.

INTRODUCTION | 11

Highlights report the main features of the region or area

Species are highlighted in blue type

Maps help you figure out the main features of described regions, areas and sites

Sites are numbered continuously throughout the text

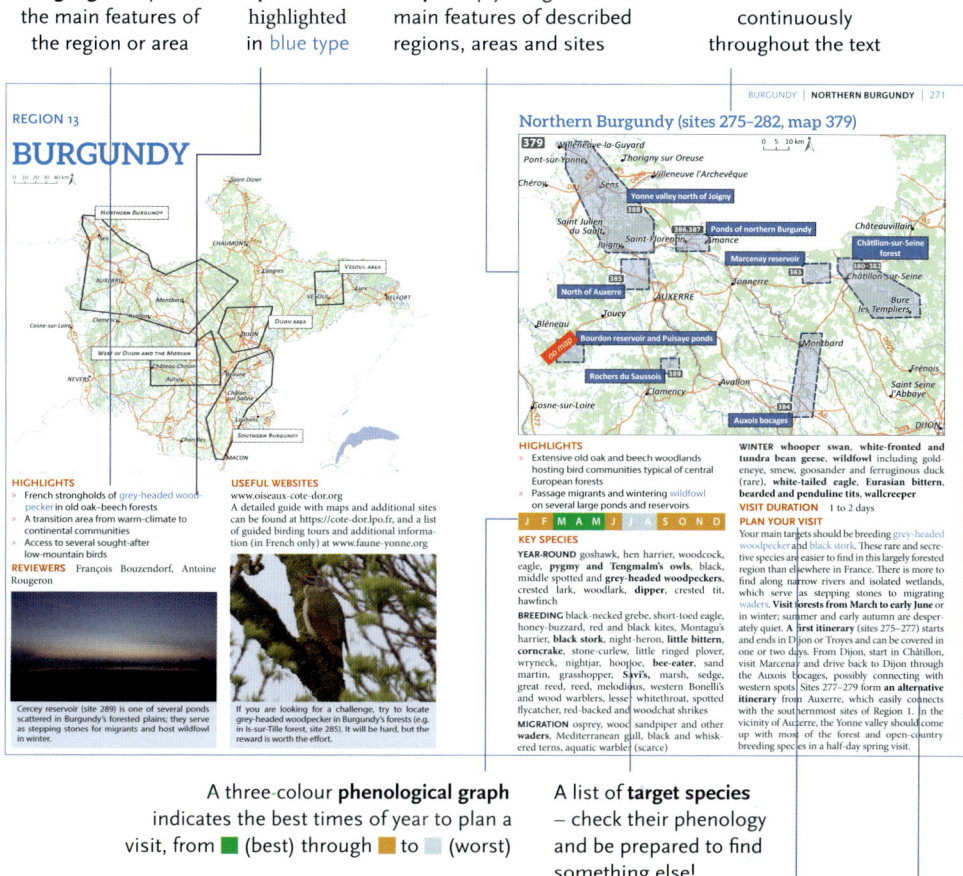

A three-colour **phenological graph** indicates the best times of year to plan a visit, from ■ (best) through ■ to ■ (worst)

A list of **target species** – check their phenology and be prepared to find something else!

Recommendations on **planning your visit** provide some basic guidelines to help you in setting your itinerary

Information not to be missed is highlighted in **bold** type; **Waypoints** are indicated in square brackets. For example, [2] in site 33 refers to waypoint 33-2 on the printed maps and in the downloadable GPS file

MAP SOURCES

Background: Open Street Map (http://openstreetmap.fr)
Topography: Digital Elevation Model from the Institut Géographique National (BD-Alti © 75m: http://professionnels.ign.fr/bdalti)
Road network: Institut Géographique National (Route 500 ©: http://professionnels.ign.fr/route500)
Wooded areas: CORINE-Landcover 2012 (www.eea.europa.eu/publications/COR0-landcover)

Waypoints

Throughout the text, numbers in square brackets refer to waypoints. All the waypoints were checked by the authors and the referees and are accurate and accessible at the time of publication. There are two ways to use the waypoints:

» **The printed site maps display most of the waypoints with small markers** (red for waypoints accessible by car, blue for those

INTRODUCTION

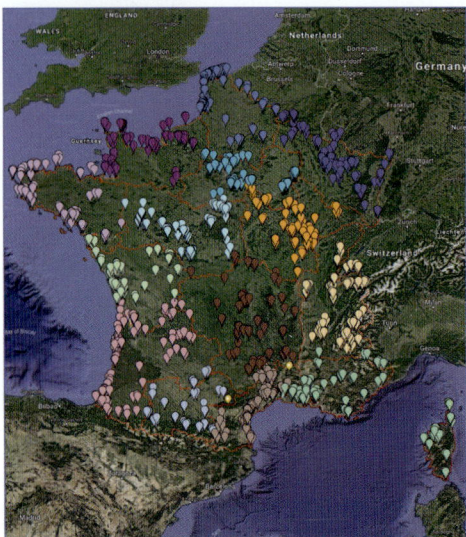

The GPS file as displayed by Google Maps. Each region is shown in a different color.

» **A GPS file with all the waypoints**, including those that are not displayed on printed maps, can be downloaded from the Pelagic Publishing website: https://pelagicpublishing.com/pages/wtwbf-gps. They can be used with any GPS device, on Google Maps, Google Earth or other mapping software (.kml and .gpx formats). In addition, on this website you will find a tutorial on downloading and using the file on a computer, a GPS or a smartphone. The waypoints are accurate to within a few metres wherever necessary, a few tens of metres when there is no precise viewpoint to reach.

DEVICES

You can open the GPS file in Google Maps, Google Earth as well as in any GIS software and it can be imported directly to most GPS devices. For smartphones, we recommend the free app Maps.me which can be used even in flight mode after downloading the suitable Open Street Maps layers. Other apps will work as well.

only accessible on foot). Associated with each marker are the number of the site and the number of the waypoint itself. For instance, waypoint 12-1 corresponds to the waypoint numbered [1] in the account of site 12. Note that some isolated or rarely visited waypoints are not displayed on printed maps. These waypoints are available in the GPS file (see below), and are identified in the text with the words 'see GPS file'.

USING THE FILE

The waypoints are numbered in the GPS file as they are on the maps. Hence, to reach the waypoint numbered [3] in site 111, type '111-3' in your GPS or Google Maps search engine.

If you open a waypoint, you will see five fields. They can be used to search all the waypoints within a region or a site, or according to access mode:

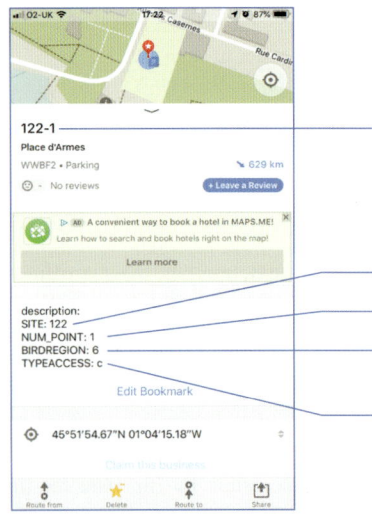

waypoint number: as displayed on printed maps

site number: from 1 to 312 as in the book

waypoint number: the number between brackets in the text

region number: from 1 to 14, as in the book

TYPEACCESS: 'c' corresponds to waypoints accessible by car (red on printed maps), 'f' to waypoints where you need to walk (blue on printed maps)

REGIONS AT A GLANCE

REGION 1

PARIS AND THE GREAT LAKES

REGION 2

FROM THE NORTH SEA TO THE SOMME BAY AND THE WESTERN ARDENNES

REGION 3

NORMANDY

ON THE FLYWAY. Visit the Lac du Der from November through to March and enjoy the scenic migration of hundreds of thousands of cranes and wildfowl.

URBAN BIRDING. Search for little bittern and peregrine falcon in a green escape from the city crowds, less than half an hour from the centre of Paris. Look for migrating passerines while sightseeing in the Père Lachaise or on the banks of the Seine.

STOP AND GO. Drive along the Seine and visit numerous wader and wildfowl stopover sites that host exciting surprises in both spring and autumn, starring osprey and black stork. Wander in Fontainebleau forest to fill your list with woodpeckers, warblers and raptors.

THE OTHER SIDE OF THE CHANNEL. Experience an autumn seawatching day on Cap Gris-Nez, and the French coast will never be that distant grey line facing Dungeness again.

GULL TIME. Do you need to challenge your ID skills? Search for yellow-legged and Caspian gulls in winter roosts near Dunkerque and Boulogne-sur-Mer. And maybe there will be a glaucous or Iceland gull as well.

A BIRD FESTIVAL. Bluethroats, icterine warblers, purple herons and other wetland specialities breed in the Baie de Somme in spring. It is also the time of the yearly Festival de l'Oiseau. In winter, countless flocks of waders and ducks fill the bay.

SETTING THE SEINE. The Seine estuary and its extensive reedbeds, marshes and meadows are a birding hotspot year-round.

THE SEA TWO HOURS FROM PARIS. Walk along the sandy D-Day beaches or sit on a rock at the Pointe du Hoc to scan flocks of scoters and auks from October to March. If you have a bit of time, try seawatching at Gatteville or La Hague.

SIGHTSEEING HALT. Search for pipits and buntings in saltmarshes around Mont Saint-Michel. Flocks of waders and wildfowl show up at high tide, and the Mont itself can be a great place to set your scope.

REGIONS AT A GLANCE

REGION 4

BRITTANY

REGION 5

THE LOIRE VALLEY

REGION 6

POITOU-CHARENTES AND THE VENDÉE

THE FAR WEST. Explore the scenic coast of western Finistère waiting for massive flocks of wintering waders in Goulven Bay or checking gulls after fishing boats coming back to harbours in Douarnenez and Le Guilvinec.

ISLAND RETREAT. Sail to Ushant (Ouessant) or Sein islands for a couple of days in search of an American or Siberian vagrant.

CELTIC SPRING. Kentish plovers, roseate terns and puffins breed in the Sept-Îles and Morlaix. Southern Brittany offers vast reedbeds, freshwater ponds and bays: search for bluethroats, breeding-plumage waders and herons.

RIVER SIDES. Watch birds along the dykes of the Loire from Orléans to Chinon during wader spring and autumn passage. In winter, look for goosander and goldeneye.

FOREST BLOOM. The Orléans and Loches forests are home to some of the last French breeding grey-headed woodpeckers. While searching for this ghost, enjoy ospreys, booted and short-toed eagles, black storks, and whatever forest passerines you may hope for.

REEDBEDS, MARSHES AND BOGS. The Brenne and Sologne are known as the northern counterpart to the Camargue, and they deserve their reputation. Herons, starring both bitterns, tern and gull colonies, waders, wildfowl, reedbed passerines and a mixture of northern and southern species breed in the area.

ATLANTIC FLYWAY. The Vendée and Charente seashores spread along the migration route of hundreds of thousands of shorebirds, seabirds and passerines. Watch their spectacular roosts in countless bays and coastal wetlands, and on several islands.

DEEP REED. From April to September, the vast Marais Breton, Marais Poitevin and Brouage marshes are home to a colourful and symphonic wetland community composed of herons, warblers, raptors and colonial waders.

IN THE FIELDS. France's last migratory little bustards share the Poitou agricultural plains with stone-curlew and several other emblematic species of European southern steppes.

REGIONS AT A GLANCE

REGION 7

AQUITAINE

REGION 8

THE PYRENEES

REGION 9

WESTERN MEDITERRANEAN COAST AND THE CÉVENNES

SEA, BIRDS AND SUN. There are countless options along the coastline, from the migration hotspot of the Pointe de Grave to the estuaries and ponds at the Spanish border. A high tide in the Parc du Teich in the Arcachon basin will probably remain high in your best French birding memories.

NEED BLACK-WINGED KITE? Search for it in the agricultural plains around Pau. No doubt you'll find it hovering over fields or sitting on a hedgerow tree.

ROCK IT. Griffon vulture, Egyptian vulture, white-winged woodpecker and Iberian chiffchaff reward those who brave the Pyrenean hiking trails. Lazy birders can have their share of mountains as well: just wait for passing-by raptors and the local lammergeier while enjoying some Etorki cheese on a sunny autumn day at Organbidexka.

MOUNTAIN GHOSTS. Two rare Pyrenean endemic subspecies will challenge the most determined birders. The *aquitanicus* capercaillie lives in remote beech forests, while *hispaniensis* grey partridge inhabits high-elevation rocky slopes above the tree line.

UPS AND DOWNS. Hike along high mountain trails from deep forested valleys to permanent snow. You will be rewarded with ptarmigans, soaring lammergeiers and Egyptian vultures, and a variety of highland passerines.

STEPPING STONE. A vast plain expands from Toulouse to the Pyrenean foothills. This is the home of the black-winged kite, and the last stopover before crossing the mountains for thousands of wildfowl, cranes and waders. You will find them in the appealing Mazères reserve and on many other gravel ponds throughout the region.

FILL YOUR BINS. Over 350 species have been recorded, from Pyrenean specialities to typical Mediterranean species, making this region one of the country's richest. If you're a list addict, this is the place for your big day. Will you break the 200 species barrier?

COLOURS AND SOUNDS. From late April, rollers, bee-eaters and shrikes are back in the garrigues. Wander among vineyards and groves around Montpellier or in the Aude valley for a photographic journey with Mediterranean specialities.

LET BIRDS FLY. Sit comfortably in a folding chair with your scope and and wait for migrants on a windy day at Gruissan or Eyne. If migration slackens off by midday, you'll have plenty of marshes and wetlands to check for flocks of waders and passerines.

REGIONS AT A GLANCE

REGION 10

EASTERN MEDITERRANEAN COAST, SOUTHERN ALPS AND CORSICA

REGION 11

JURA AND THE ALPS

REGION 12

MASSIF CENTRAL

FRENCH EVERGLADES. Get a taste of the Camargue by spending a morning in search of bittern and moustached warbler along the edges of vast reedbeds, or enjoying the arrival of migrating waders in the mythical Baisses de Cinq Cent Francs.

LOST IN THE DESERT. Search for pin-tailed sandgrouse as they flock to water in the Crau plain. Calandra lark, stone-curlew, little bustard, lesser kestrel and perhaps a red-footed falcon will detain you on your way.

BONAPARTE'S LAND. Look out for Corsican nuthatch, Corsican finch, Marmora's warbler, various endemic subspecies, Eleonora's falcon, Spanish sparrow and spectacular migration.

EASTERN TASTE. The old conifer forests of the Plateau du Risoux are home to three-toed woodpecker, capercaillie, pygmy and Tengmalm's owls. Finding just one of them is a challenge, but the reward is worth it.

TOP OF EUROPE. Wake up on a late spring morning with black grouse, rock partridge and wallcreeper at the Col de la Colombière, with Mont Blanc in the background. You'll probably cross paths with a lammergeier later in the morning.

UPS AND DOWNS. Start your day in the highest moraines of the Vercors to try for ptarmigan, stop for a lunch break in the shadow of coniferous forests with ring ouzels and lesser whitethroats, then descend to the plains to look for Mediterranean specialities.

VOLCANIC BIRDING. Search for Tengmalm's owl and rock bunting on the slopes of the Puys, a scenic range of volcanoes running on a north–south axis near Clermont-Ferrand.

INTO THE WILD. Most of central France is sparsely populated, and birders are scarce. Find new breeding populations, range expansions, unknown stopover sites in remote areas of the Limousin, Cantal and Ardèche.

THE LAND OF RAPTORS. Head to the prairie-like Grands Causses and the Gorges de la Jonte to feast your eyes on vulture flocks, golden and short-toed eagles, and perhaps a migrating lesser kestrel or a red-footed falcon.

REGIONS AT A GLANCE

REGION 13

BURGUNDY

REGION 14

NORTHEAST

FOREST LIFE. Old deciduous woodlands host middle spotted and grey-headed woodpeckers, wood warbler, nightjar and their accompanying cortege. Mid-March to early June give the best experience of Burgundy forests.

CÔTEAUX DE BOURGOGNE is a wine-producing district, but more importantly a hilly area with eagle owl, Alpine swift and wintering wallcreeper.

PONDS AND RIVERS. The Saône, the Seine and their tributaries are surrounded by well-conserved valleys and woodlands hosting a high species diversity reminiscent of southern species assemblages.

THE FOREST DWARF. From May, collared flycatchers breed in oak and beech forests all over the region. Learn their song and be patient.

ORIENT EXPRESS. Breathe the atmosphere of a continental climate in a day on the gentle slopes of the Vosges and nearby forests. Meet a troop of forest birds including mountain breeding species, with capercaillie and pygmy owl as local dreambirds.

THE WHITE-TAILED EAGLE'S KINGDOM. Will you feel the atmosphere of Fennoscandia while watching goldeneyes, goosanders, geese and the distant shadow of a large fish-eating raptor in these large forest reservoirs?

PRACTICAL INFORMATION

International access
The major international connections are Paris-Roissy Charles de Gaulle, Lyon Saint Exupéry, Toulouse Blagnac and Marseille Provence airports. Daily or weekly ferries from Great Britain serve Dunkerque, Calais, Le Havre and Roscoff. Express trains (TGV) link Paris to London, Brussels, Torino, Barcelona and several other cities in neighbouring countries. For lower-budget options, consider Euroline long-haul buses.

Travel within France
By air. Air France and Hop operate several connections per day to all the major regional airports. Flying is a good alternative for journeys taking more than four hours by train, such as between Paris and Brest, Toulouse, Bordeaux or Nice, between Lille and Marseille, or between Lyon and Brest. Several airlines serve Corsican cities on a daily basis.

By train. There are three main types of trains. Long-haul lines are served by express trains (TGV) that need to be booked in advance (From Paris: Marseille or Montpellier 3.5 h, Brest and Bordeaux 5 h, Toulouse 7 h, Lyon 2 h, Lille 1 h). Standard lines connect regional capitals without a transit through Paris. Numerous and relatively inexpensive local lines (TER) connect villages and smaller cities. Unlike the TGV, standard trains and TER usually accept cycles without any additional fee, making bike birding a reasonable option in most regions. All train trips can be booked from https://www.oui.sncf/ or in stations (prices decrease with time before trip for the TGV, but not for other lines).

By bus. Long-haul private bus lines connect most major cities and need to be booked in advance. Local bus services vary across regions and do not require booking. Buses can be a good option in mountain areas such as the Alps or the Cévennes.

By car. Driving is in many instances the best – or only – way to reach birding spots. France is a big country, so be prepared for long drives (check the table of distances on the inner back cover). There are essentially four types of roads. Motorways ('autoroutes', max 130 km/h, fee on some sections, paid at exit by credit card or cash) are useful for long-haul drives or quick connections but will rarely take you very close to birding spots. Rapid four-lane highways (max 110 km/h) replace motorways here and there, especially between major cities. Single-carriageway roads called Nationales (max 80 km/h) are the standard connection between bigger villages. They are completed by a network of smaller local roads (Départementales, max 80 km/h). Be aware that speed limits are strictly enforced. Unpaved roads are normally suitable for a standard car, but respect notices indicating private property and restricted access.

Accommodation
Most cities have several privately owned hotels in the centre and international chains in the suburbs near commercial centres. There is also a wide network of guesthouses ('chambres d'hôtes'), rentals ('gîtes'), collective housing ('gîtes d'étape') and youth hostels ('auberges de jeunesse'). All major urban centres and tourist areas have at least one camp site with basic service (washrooms, showers and drinking water). Wild camping is tolerated outside private land from sunset to sunrise, unless explicitly prohibited. Note that accommodation can be especially hard to find in remote areas and in tourist places during public or school holidays. Hotels can usually be found from €40, guesthouses from €70, collective accommodation and camp sites from €15.

Food
The easiest option is to buy food in supermarkets, which are easily found in the suburbs of all urban centres (usually open from 08.30 to 20.00). Better-quality fresh seasonal food can be found at outdoor markets ('marchés'). Restaurants are of varying quality, and even a modest one can sometimes be a nice surprise; they usually serve from 12.00 to 14.00 and from 19.00 to 22.00.

Fuel
In October 2018, diesel cost around €1.50 per litre, and unleaded petrol between €1.60 and €1.80 per litre. Prices are subject to some daily fluctuations. 'Total Access' and supermarket stations are cheaper, while prices are inflated at stations located along motorways and in city centres.

General literature

Numerous general tourist guides have been written about France, of which Lonely Planet, the Rough Guide and the Guide du Routard are the most useful as practical references for birdwatchers.

Maps

The best road maps are published by the IGN (Institut Géographique National) at 1:100,000 and 1:25,000 scales. They can be found in all major supermarkets and bookshops. The Geoportail portal (www.geoportail.fr) allows free online viewing of these maps together with numerous other layers including forest inventories (IFN), topography, protected areas, satellite views. We highly recommend visiting this website when preparing a birding trip, especially in underwatched areas or if you are in search of a particular species.

Language

French is the single common language. Most regional languages are only spoken in private contexts (virtually only in the Basque and Catalan areas). Do not expect most French people to speak more than a few words of English, or any other language, outside large cities and tourist places. Signposts and announcements are only in French in most locations. Younger birdwatchers usually read and speak decent English, but it is a good idea to have a phrasebook at hand.

Phone numbers

The European emergency number is 112.

BIRDWATCHING IN FRANCE

Apart from private areas, there is no strong obstacle to birding in most parts of the country. All paths and tracks can be used, and wandering off the paths is allowed unless explicitly forbidden by a fence or signposts. Remember that bird safety and tranquility as well as that of other wildlife, and habitat preservation, should always come first. We assume that users of this book are accustomed to birdwatching ethics, so we do not repeat them here. Bear in mind that all birders and naturalists will benefit from your own positive and respectful attitude towards local people and their properties.

Types of protected areas

The protected area network is complex, encompassing a wide range of protection levels, institutional support and names. Entry to protected areas is free unless explicitly stated otherwise, and the same rules usually apply to visitors whatever their status: all forms of camping, collection of animals and flowers, camp fires, use of motor vehicles and pets are forbidden.

Local reserves managed by private owners or public agencies are usually called 'réserve' and indicated by large signposts with an access maps and some basic information on wildlife. Birdwatching is usually done along tracks and hides where relevant. These reserves are watched over by private or public guards that one rarely ever sees. Do not be fooled by a notice saying 'réserve de chasse', which usually indicates private land managed for hunting.

National parks (Parcs Nationaux) are wide areas of protected land controlled by a scientific committee. They consist of a core area where most activities, including tourism, are strictly controlled, and a peripheral ring where some level of land use is allowed. There is always one to several park offices where rangers can provide useful information about paths, weather and sometimes wildlife. The full list of national parks can be found at www.parcsnationaux.fr.

Regional parks (Parcs Naturels Régionaux, PNR) are areas without formal protection status but where land use and economical activities are oriented towards environmental sustainability. PNR are usually of little visible consequence for visiting birdwatchers.

Private areas

Private property is a sensitive issue in France and it should be treated with care. Passage is in principle allowed on any public or private path not barred by a fence. In practice, it is inadvisable to wander where there are signposts such as 'Propriété privée' or 'Défense d'entrer'. Although, in law, such signs should not apply to roads and tracks, this is not the way most landowners understand it. Ask politely if you are in any doubt, and if there is no one to ask, refrain from entering. In some areas, even stopping in the vicinity of a private property makes landowners nervous, especially when facing someone with binoculars or a spotting scope. Be friendly and respectful, explain quietly that you are simply looking for birds, avoid stating any relationship with any environmentalist organisation and do not insist if you do not convince (in most cases, you will).

Other restricted areas

Driving is forbidden on many unpaved forest tracks; they are barred with signposts and/or fences. Do not ignore them. Using cameras or camera-like devices such as binoculars is forbidden around military areas, industrial harbours and airports. Expect to be quickly apprehended and excluded from the vicinity of such areas.

Hunting

The hunting season usually opens in early September and ends in late February, with some variability across years and regions. Hunting is generally not an issue inland as it is mostly practised in private areas or in cultivated fields where there is little to interest birdwatchers. It should be taken more seriously in major wetlands and coastal areas. Birding is generally slow in hunted areas, and birdwatchers are usually not welcome. In most instances, basic politeness with hunters and keeping your distance from hunting huts or hides should be enough for good cohabitation. However, birdwatchers should be especially cautious and avoid contact with hunters in areas of tension with local ornithological or environmental groups, especially in the northwest, Aquitaine and the Camargue.

French birdwatchers

French birdwatchers are far less numerous than in Great Britain or the US, so it should not come as a surprise that one can spend a full day birding without meeting any other observer, except maybe in critical spots such as the Lac du Der or the Camargue. Also bear in mind that although many birdwatchers are happy to help, they are likely to speak very little English (except younger people). The species index at the end of this book includes French names, to aid communication. Although twitching has become increasingly popular in the past decade, it is by no means as developed as in neighbouring countries. A vagrant or rarity will hardly attract more than a few dozen people at most. However, there has been huge improvement in communication among birdwatchers in recent years through the development of news web portals (see below). In general, you'll be welcome to exchange your mobile phone number with birders you meet in the field for bird news.

Bird news

Most of France is covered by the Biolovision framework, which has set a European standard. A few regions use other systems, as listed in the regional chapters.

Generic portal (www.faune-france.org): provides instant access to the most recent sightings, a rare bird species map, thematic maps, a search engine, general bird news and a list of regional sightings portals. The portal is currently only in French, but species' scientific names can be displayed instead of French names. It requires a login but is entirely free. All the regional portals under the Biolovision framework have a similar design, with a web address following the pattern 'faune-regionname.org', and use the same login.

NaturaList app (Android and iOS): free app for submitting your own sightings or checking others' records from a smartphone.

e-bird (http://ebird.org/content/ebird): few people use it in France, but it is still worth a check as foreign visitors sometimes report useful sightings.

Online birdwatching resources

Numerous personal or organisation websites provide information of variable accuracy about birds and birdwatching in France. Here are those we find the most useful for a traveller.

French Rarity Committee (www.chn-france.org): list of species subject to validation, record submission facility, reports and a search engine.

French twitchers' website (www.cocheurs.fr): all kinds of listings.

Cloudbirders (www.cloudbirders.com): 623 reports about France in October 2018.

Other online resources

Maps on the Geoportail website (www.geoportail.gouv.fr): this website displays various map layers including aerial photos, roadmaps, tracks and footpaths, topography, forest composition, protected areas and many others, down to a 1:25,000 scale.

Weather forecast (www.meteofrance.com): the website of the national weather forecast agency.
Tide tables (http://maree.shom.fr): the official tide tables for all French coasts.

Bird and wildlife organisations

Regional organisations are cited in the introduction of each regional chapter.

Société d'Etudes Ornithologiques de France (http://seofalauda.wixsite.com/seof): the SEOF has for a long time been established as the main authority for scientific ornithology. It publishes *Alauda*, a journal reporting studies covering all fields of ornithology.

Ligue pour la Protection des Oiseaux (www.lpo.fr): the LPO is is the major environmental organisation dedicated to bird conservation in France. Autonomous LPO subsections are active in most regions; their websites and contact details can be found on the main LPO page.

Office National de la Chasse et de la Faune Sauvage (www.oncfs.gouv.fr): the state agency dedicated to wildlife management and hunting control.

Inventaire National du Patrimoine Naturel (https://inpn.mnhn.fr/accueil/index): the INPN is a resource on habitats, protected areas and species, biodiversity indicators.

Global Biodiversity Information Facility (French section: www.gbif.fr): the standardised framework for biodiversity monitoring.

REGION 1
PARIS AND THE SEINE RESERVOIRS

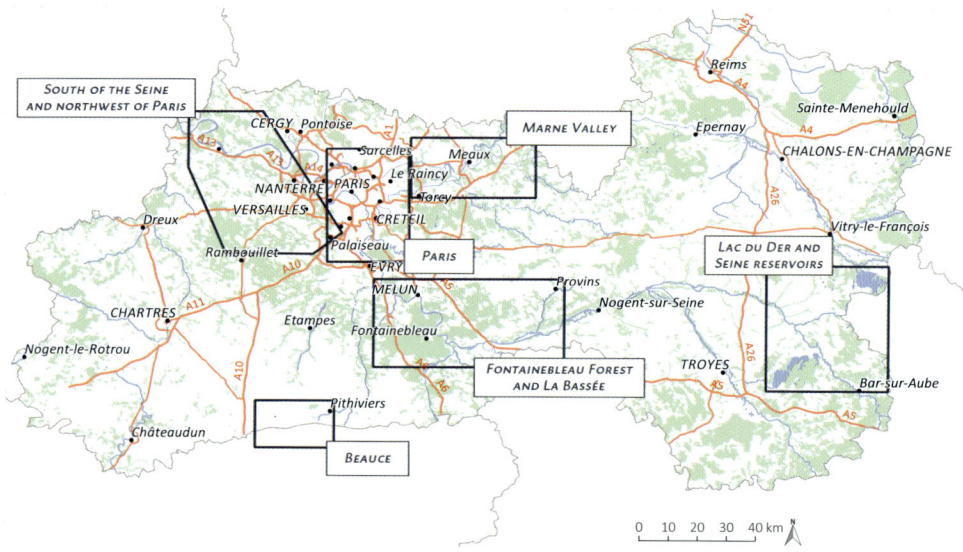

HIGHLIGHTS
» Europe's largest common crane roosts and large flocks of wildfowl on the Lac du Der from October to March
» Escape from the bustle of Paris to search for little bittern in the reedbeds at La Courneuve
» Search for autumn waders and wintering wildfowl and gulls along the Seine tributaries
» Raptors, woodpeckers and pied flycatcher breed in the majestic Fontainebleau forest

REVIEWERS Yohan Brouillard, Sylvain Vincent, Stanislas Wroza, Maxime Zucca

USEFUL WEBSITES
www.faune-iledefrance.org
www.faune-champagne-ardenne.org

The Lac du Der (site 4) is the place to be for wintering cranes and waterfowl from November to March.

An iconic species among wetland birds, the little bittern occurs in most reedbeds of the region, even in Paris!

The Lac du Der and the Seine reservoirs (sites 1–4, map 1)

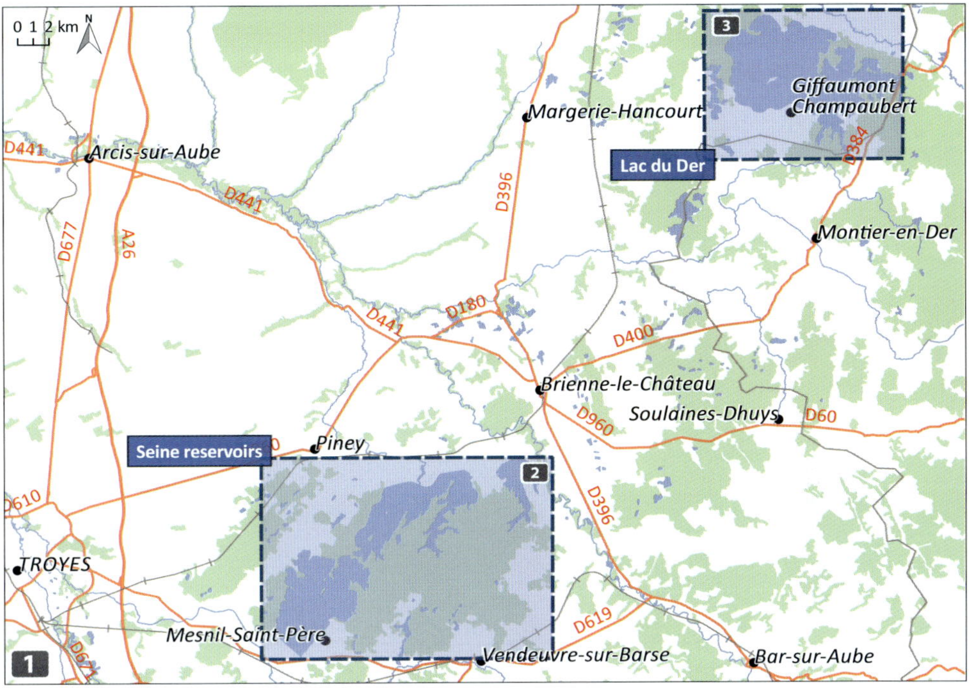

HIGHLIGHTS
» The largest stopover site for common cranes in Europe
» France's greatest diversity and highest counts of wildfowl
» White-tailed eagles year-round

KEY SPECIES
YEAR-ROUND common crane, great white, cattle and little egrets, goshawk, kingfisher, black, **grey-headed** and middle spotted woodpeckers, marsh and willow tits

BREEDING little bittern, purple heron, night-heron, squacco heron

MIGRATION black stork, waders

WINTER whooper and Bewick's swans, greylag, white-fronted and **tundra bean geese**, scaup, smew, goosander, goldeneye, red-throated, black-throated and great northern divers, black-necked, Slavonian, great crested, little and red-necked grebes, Eurasian bittern, **white-tailed eagle**, peregrine, water pipit, brambling, hawfinch, common redpoll

VISIT DURATION 0.5–2 days
PLAN YOUR VISIT
Four artificial lakes were built in the 1970s to prevent the Marne and Seine rivers from flooding Paris in late winter. They bear little resemblance to pristine wildlife reserves, being entirely artificial waterbodies surrounded by concrete dykes and dams. Still, the Lac du Der is France's hotspot for wintering wildfowl and geese and an internationally recognised stopover site for cranes. The sight of thousands of these birds coming in to roost on a November evening is enough in itself to justify a day trip. **A two-day visit between October and March is optimum, with one day for the Lac du Der and another for the three other reservoirs to the south**: the itinerary described here is optimised for this option.

Being smaller, the Orient, Temple and Amance reservoirs host less impressive bird numbers but they offer closer views of geese, swans and ducks. **Most of the birding is done from large dykes around the banks of the reservoirs; but do not neglect the surrounding fields and woods, which also host rare breeding species, including the**

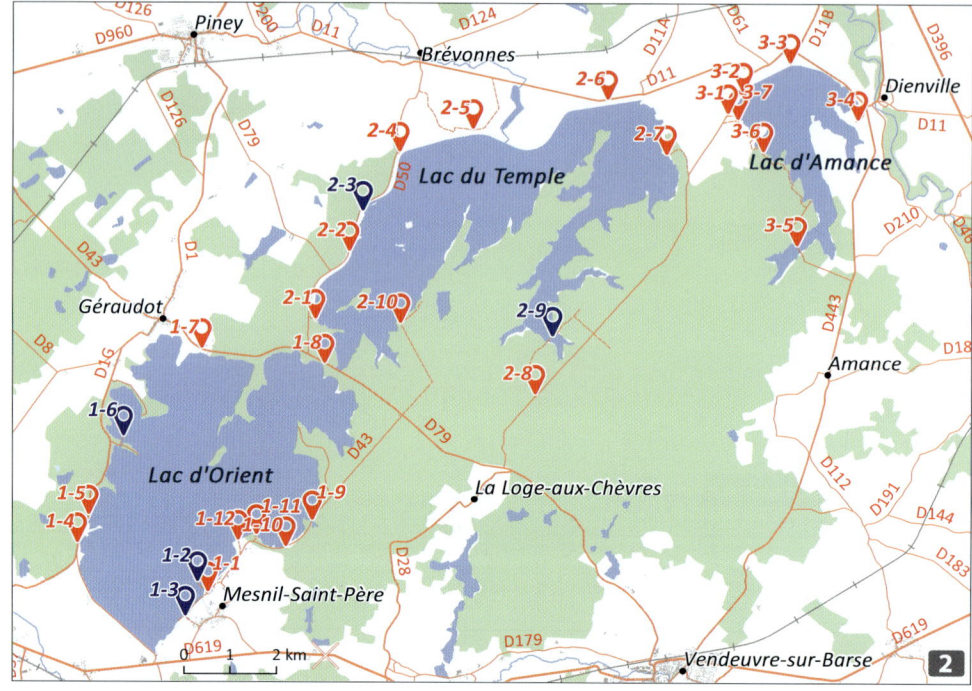

threatened grey-headed woodpecker. The yearly Montier-en-Der Nature Festival, in November, is a must for wildlife photographers: check www.festiphoto-montier.org.

1 Lac d'Orient

0.5 DAY; MAP 2 **Start your day in Mesnil-Saint-Père, on the southeastern bank of the Orient reservoir (Lac d'Orient)** [1] (27 km, 30 min from Troyes on the D619, follow signposts to 'Forêt d'Orient' and Bar sur Aube). Wildfowl can be sighted at sometimes close range along the dykes or even in the harbour [2]. Flocks of teal usually settle along the trail running westwards [3] with wigeon and other ducks. A pair of red-necked grebe even bred in the harbour itself a few years ago, and the species is still regular in this area of the reservoir. The place is also attractive for water pipit. Scanning open waters in mid-winter will likely yield large numbers of great crested grebes, possibly joined by black-necked grebes and the odd Slavonian grebe. In mid-winter, red-throated, black-throated or great northern divers, goosanders, goldeneyes and smew should frequent the deepest parts of the reservoir. Geese (mostly greylag and lower numbers of white-fronted, sometimes with bean) are easier to find on the other side of the reservoir with Bewick's swans (whooper are more common on the Lac du Der). There is always a white-tailed eagle somewhere around the reservoir: check dead trees, on which it often perches.

Drive around the reservoir to the west and north, stopping in the Anse des Terriers [4] or the Anse de Larivour [5], where up to several hundred pochard and tufted ducks may be found. Check the surrounding willows, especially in mid-winter, in the hope of hawfinches or a siskin flock in which to look for common redpoll.

The Presqu'île des Grands Sillons (car park at [6]) gives good views over the Anse de la Picarde, a reliable place for white-tailed eagle, peregrine, geese and wildfowl including smew, scaup or the rarer ferruginous duck.

The northern shore [7] has mudflats which can yield a few waders (most likely common sandpiper, snipe and green sandpiper) and water pipits. From there, either drive east to the Temple reservoir on the D50, or go back to Mesnil-Saint-Père. **There is a hide 400 m past the junction with the D50** [8] and several good viewpoints along the eastern bank, especially for cranes, wildfowl and swans [9,10,11,12]. Besides wildfowl, black and middle spotted woodpeckers are likely encounters in woods around

2 Lac du Temple

2–4 HOURS; MAP 2 **The southern shore of the Temple reservoir is located just 1 km north of the Orient reservoir along the D50** [1]. Two viewpoints on the dykes, located 1 km apart [2,3], allow good views of the open water where Bewick's swans, greylag and white-fronted geese and other wildfowl can be seen at a reasonable distance. Check dead trees on the nearby islands or on the distant southeastern forested bank in the hope of a white-tailed eagle. For another viewpoint, try the northern dyke, which is most easily walked from the **car parks located just before the D50 leaves the banks northwards** [4], **at the extreme north end of the reservoir** [5] (turn right just before the end of Brévonnes on the D11) **and near the barrage west of L'Etape** [6]. The reservoir is deep enough here to allow close views of all diving species among which goosander, smew, grebes (search Slavonian and red-necked), divers, kingfisher and common sandpiper.

Viewpoints on the eastern shore are harder to get to, but they are worth the effort for more wildfowl, waders and possibly white-tailed eagle or black stork (spring to autumn). From the north, head to Chantemerle on the D11 and follow signposts to 'Lac du Temple – Caron' [7] to visit the deep Anse du Caron, another stakeout for diving ducks. From the south, drive east along the D79 for 5 km past waypoint 1-8 to turn left on the 'Route Forestière du Temple'. Park at the end of the road [8] and walk 2 km to **the Digue de la Fontaine aux Oiseaux** [9], **which can be surrounded by extensive mudflats suitable for** waders or black stork **on stopover in spring and late summer.** Black and middle spotted woodpeckers as well as forest passerines are quite easy to see on the way. Grey-headed woodpeckers also occur in the vicinity but are far less likely to be seen (best chances in March). Alternatively, turn left from the D79 at the Maison du Parc (at the crossroad with the D43 to Mesnil-Saint-Père) to reach the **Valois hide** [10], a good stakeout for waders, white-tailed eagle and wildfowl in winter.

3 Lac d'Amance

1–2 HOURS; MAP 2 Amance is the smallest of the three reservoirs but its bird records are similar, including white-tailed eagle, all grebes, geese, goosander and other wildfowl. **Concentrate on the west and north banks** [1,2,3] **if daylight is limited** – this is a good area to end the day watching roosting cranes. If you have more time, try for white-tailed eagle, scaup and divers on the southern dyke [4]. Then, resume your loop around the southern tail [5] and the western shore [6,7], which eventually leads back to the D11. **If you follow the proposed itinerary you should be there at dusk, when large numbers of cranes come back to roost on the three reservoirs.** After dark, head to Saint-Dizier (55 km, 1 h) to be ready to resume birding on the shores of the Der lake at sunrise.

4 Lac du Der-Chantecoq

1 DAY; MAP 3 **Start your day early in front of Champaubert** church (20 km, 30 min from Saint-Dizier; head south on the D384 towards Giffaumont-Champaubert, turn right, following signposts before reaching the northeastern tail of the reservoir). Park at [1] and walk to [2] to scan the lake. **The sight of** wildfowl **and** cranes **slowly emerging from the fog with rising sun on your back can be magical, especially on a cold winter day.** Goosander, goldeneye, the three species of diver and smew should be around, together with large numbers of commoner species including all dabbling ducks, pochard and tufted ducks, sometimes with the odd scaup. **Check the islands to your north** for goose flocks (greylag dominates, with white-fronted and small numbers of bean geese) and Bewick's and whooper swans. Peregrine, goshawk and white-tailed eagle are regularly seen on trees or flying over the church.

When the sun rises, head to the Bassin Sud [3,4], a large bay isolated from the rest of the reservoir at its southern end. Large groups of pochard and tufted duck settle here with divers, other diving ducks and smaller numbers of goldeneye, goosander and smew. **Check the nearby Port de Giffaumont** for grebes (great crested, black-necked, and occasionally Slavonian or red-necked), three species of diver, smew and goosander. Also expect water and meadow pipits on the banks.

Head west, stopping here and there along the dykes, for instance around [5,6,7,8]. Park at Chantecoq [9] and walk west to the **hides on the banks of the Étang des Landres** [10] (take the path diverting to the right before entering the forest) **and the Étang du Grand-Coulon**

PARIS (SITES 5–12) | PARIS AND THE SEINE RESERVOIRS

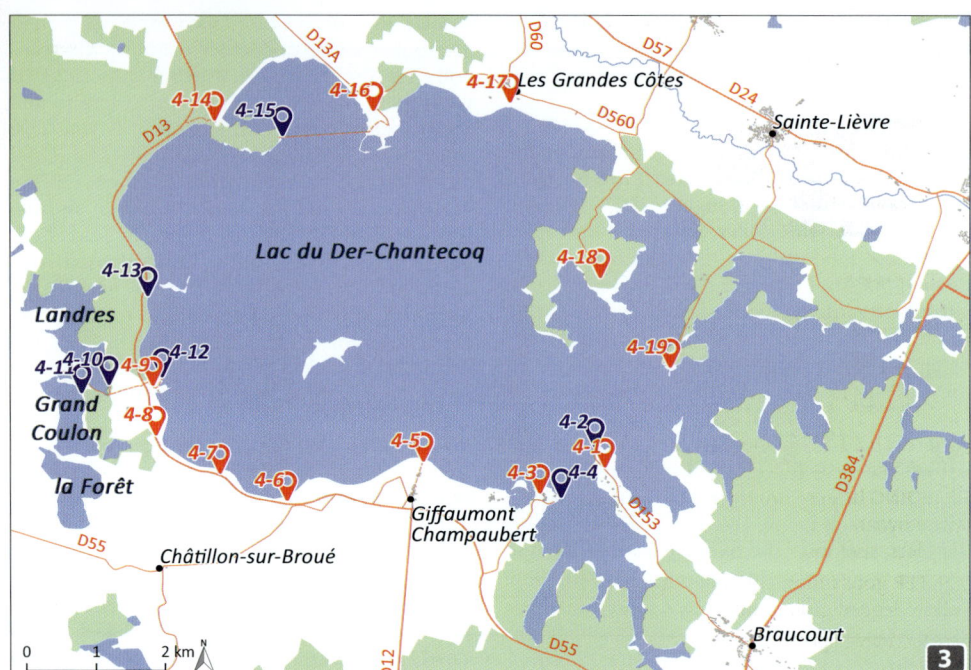

[11] (follow the main path until you find a hide on your left), two small ponds surrounded by reedbeds and woods. Beyond great white and little egret, **this is your best chance for bitten year-round and** little bittern **from May.** Purple heron, night-heron, squacco heron and cattle egret breed here as well.

Back on the Der lake, **a hide [12] offers among the best opportunities for** wildfowl **photography and chances for** bean **and** white-fronted geese. Resume your drive northwards; you can stop along the Digue Ouest [13], but it is a better idea to keep it for the evening. Instead, **head to the Bassin Nord, which is best watched from the Presqu'île de Larzicourt** (park near the camp site [14] and walk to the dyke [15]), **the Port de Nuisement [16] and La Petite Ville [17].**

In these deeper parts of the lake, you will likely find diving ducks (including smew), grebes and divers (a white-billed diver stayed here in winter 2013).

If you are done with ducks, walk on the woody **Cornée du Der** [18] for middle spotted woodpecker, marsh and possibly willow tits. Just to the south, the Presqu'île de Nemours will yield more views of wildfowl and geese [19]. **As soon as the sun goes down, go back to the Digue Ouest [13], the evening meeting point of photographers and birders waiting for thousands of** cranes **that converge to the lake to roost.** Do not miss it for any reason and stay until full darkness: this will likely be the highlight of your visit to the reservoirs.

Paris (sites 5–12, map 4)

HIGHLIGHTS

» Locate cirl bunting, middle spotted or black woodpeckers in the Bois de Vincennes
» Search for migrant passerines in the Jardin des Plantes
» Little bittern in La Courneuve, half an hour from Paris city centre

J F M A M J J A S O N D

KEY SPECIES

YEAR-ROUND lesser black-backed gull, black, lesser spotted and middle spotted woodpeckers, kingfisher, ring-necked parakeet, marsh and crested tits, firecrest, bullfinch, cirl bunting

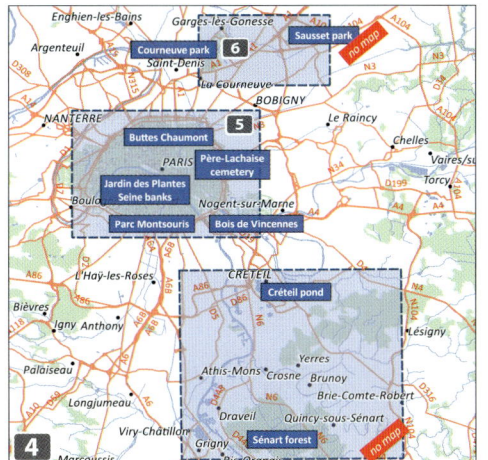

BREEDING little bittern, honey-buzzard, nightjar, whitethroat, reed, marsh, wood and melodious warblers, spotted flycatcher
WINTER goosander, goldeneye, Eurasian bittern, yellow-legged gull, brambling, siskin
MIGRATION spotted crake, jack snipe, pied flycatcher
VISIT DURATION 1 day
PLAN YOUR VISIT
Use a map of the Métro and local trains to plan your birding visit carefully. Travel can take time (e.g. 20 min from Châtelet to the Bois de Vincennes, add a 20% margin). All spots inside the city are busy every day, so try to arrive as early in the morning as possible. Most places are busy at the weekend.

5 Jardin des Plantes and the Seine banks

1 HOUR; MAP 5   The Jardin des Plantes (main entrances [1,2] 07.30–20.00, free access, Métro line 5, Gare d'Austerlitz; line 10, Jussieu; line 7, Censier-Daubenton) is among the most frequented parks of inner Paris. It also hosts an impressive list of bird records, starring the first kelp gull for Europe in 1995. **Wander along wide paths among typical French gardens, paying special attention to tree-shaded areas, bushes (visit the Jardin Alpin [3]) and animal enclosures around** [4]: expect gulls, ring-necked parakeet, grey and white wagtails (alba and yarrellii subspecies), robin, goldcrest and firecrest, crested tit and short-toed treecreeper or maybe a sparrowhawk passing through. During spring migration, warblers, flycatchers, thrushes or even a hoopoe may turn up randomly. In winter, check the nearby Seine: **the Pont d'Austerlitz [5] is a decent spot for** gulls (black-headed, herring, yellow-legged, lesser black-backed – Caspian has also been seen a couple of times), grey wagtail and kingfisher.

6 Père-Lachaise cemetery

1–2 HOURS; MAP 5   Cemeteries in Paris are just as green and wooded as any other park and their quiet atmosphere is much sought-after by morning walkers, tourists, and birds. The most famous, **Père-Lachaise [1], is also a hotspot of Parisian birding** (09.00–20.00, seasonal hours, free access, Métro lines 2 or 3, Père-Lachaise). Look for all common woodland birds such as great spotted woodpecker, nuthatch, tits and migrating passerines (pied flycatcher in late April and September, thrushes, bramblings and siskins in winter).

7 Buttes Chaumont and Parc Montsouris

1–2 HOURS; MAP 5   **The Buttes Chaumont [1] is the most famous birding spot within Paris**, with over 30 breeding species including sparrowhawk, green woodpecker, grey wagtail, spotted flycatcher, crested tit, nuthatch among more common species. Herring, yellow-legged and lesser black-backed gulls turn up daily in winter, as well as feral species such as ring-necked parakeet and Canada goose (07.00–20.00, seasonal, free access, Métro line 7b, Botzaris or Buttes Chaumont; or line 5, Laumière). Less regular species include breeding lesser spotted woodpecker and redwings in winter. **The Parc Montsouris [2] hosts the same species and can be a better choice for those staying in the southern part of the city** (09.00–21.00, seasonal, free access, Métro line 6, Glacière).

8 La Courneuve park

2 HOURS; MAP 6   Your main reason for visiting the **Parc Valbon in La Courneuve [1] are its several pairs of** little bittern (mid-May to mid-July, best chances being in late June when the adults are feeding their young and are most conspicuous) **and wintering** bittern. Both occur in the **small reedbeds surrounding the waterbodies northeast of the railway** [2,3]. Other breeding species include sparrowhawk, tufted duck, Canada goose and reed warbler,

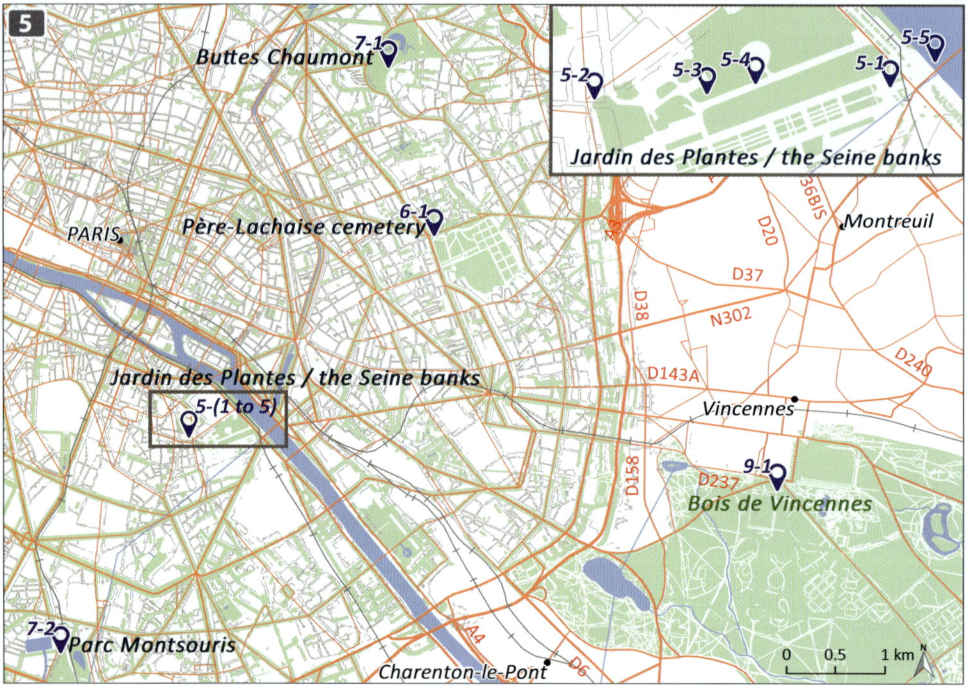

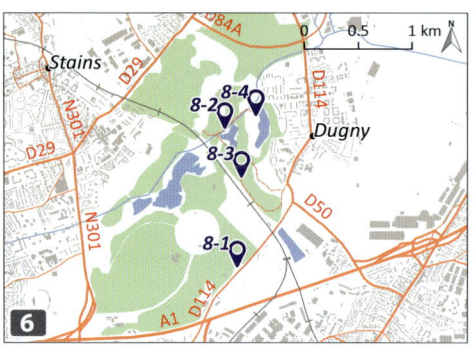

RER line A). Green, great spotted, lesser spotted, middle spotted and black woodpeckers occur here (in decreasing order of commonness): **walk among old broadleaved stands** far from the crowds. Other breeding species include ring-necked parakeet, tawny owl, cirl bunting (rare), bullfinch, common whitethroat, marsh and crested tits. Hobby could also breed. **The place is good year-round**, especially during spring and autumn migrations when passerines may turn up, including wood, melodious or reed warblers. Common pochard winters on the ponds at the western and eastern ends of the park.

and spotted crake occurs annually on migration. **The wasteland at the northeastern end of the park** [4] is also excellent for migrating passerines. The park is open 07.00–21.00 in spring (free access, RER line B, La Courneuve-Aubervilliers then bus line 249, Cimetière; or line D, Saint-Denis then tram T1, La Courneuve 6 Routes).

9 Bois de Vincennes

2–4 HOURS; MAP 5   **There is a large expanse of woodland** in Vincennes [1], at the eastern extremity of Paris (open 24 hours a day, Métro line 1, Bois de Vincennes, or same station via

10 Parc de Sausset

2 HOURS; MAP 4, NO DETAILED MAP   Sausset park [1] **can be a good place to spend a spare hour or two between connecting flights at Charles de Gaulle (Roissy) airport** (07.00–21.00, seasonal, free entrance, train line B, Gare de Villepinte – not all trains stop here on line B, check signs before boarding). Breeding species include little bittern, common tern, reed and marsh warblers, plus a wide range of common local species. In winter, common and jack snipes may turn up, plus bittern with a bit of luck and patience **in the marsh northwest of the main waterbody** [2].

11 Lac de Créteil

2 HOURS; MAP 4, NO DETAILED MAP   Créteil pond [1] is a large artificial waterbody southeast of Paris and **the best location for winter birding within reach of the Paris inner public transport network** (Métro line 8, Pointe Du Lac; RER line D, Créteil Pompadour, 35 min from Paris centre). Bittern is regular and wildfowl may turn up especially **in cold weather when other waterbodies are frozen**. Highlights may include goosander, goldeneye, and regional rarities such as Slavonian and red-necked grebes, ferruginous duck and smew. Check the local news before going. Breeding species are those typical of reedbeds, including reed warbler and possibly little bittern.

12 Sénart forest

0.5–1 DAY; MAP 4, NO DETAILED MAP   **Sénart is an extensive oak-dominated woodland in the outer suburbs of Paris** (follow signposts to 'Forêt de Sénart', main access from a large car park [1] located 10 km, 10 min north of Sénart or 8 km, 15 min south of Villeneuve-Saint-Georges on the N6; RER: line D, Quincy sous Sénart, 40 min from Paris centre). **Birding in the forest will probably be at its best from March** (for late wintering species and early migrants plus drumming woodpeckers) **to May** (for trans-Saharan species). If you are staying in the capital, this will be your closest chance of finding honey-buzzard (not before mid-May), nightjar (from May), black and middle spotted woodpeckers (unless you are lucky enough to find it in the Bois de Vincennes) and melodious warbler. Spring migration may yield local rarities such as wryneck, whinchat, spotted flycatcher or red-backed shrike. The full set of common woodland species is present, including willow and wood warblers. Be there as early as possible, visit the innermost parts of the forest, and try to avoid weekend crowds.

East and south of Paris: from the Marne valley to the Beauce through the Bassée and Fontainebleau (sites 13–16, map 7)

HIGHLIGHTS

» Search for Arctic and eastern migrants among the wildfowl and gull roosts in winter
» A day in the Bassée valley during spring migration will yield some of the largest bird lists of the region, starring little bittern, osprey, black-winged stilt and migrating waders
» Pied flycatcher, honey-buzzard and bee-eaters will fill a late spring day in Fontainebleau forest
» Make a stop on the Beauce plain for migrating waders and wintering short-eared owl

KEY SPECIES

YEAR-ROUND goshawk, hen and marsh harriers, woodcock, black and middle spotted woodpeckers, crested lark, Dartford warbler, crested and marsh tits
BREEDING red-crested pochard, garganey, black-necked grebe, night-heron and purple heron, **little bittern**, black kite, short-toed eagle, Montagu's harrier, stone-curlew, black-winged stilt, little ringed plover, Mediterranean gull, little tern, wryneck, hoopoe, nightjar, **bee-eater**, bluethroat, whitethroat, great reed, reed, sedge, melodious, wood, western Bonelli's and grasshopper warblers, pied flycatcher, red-backed shrike
MIGRATION spoonbill, **black stork**, osprey, spotted crake, waders (curlew sandpiper, knot, whimbrel, sanderling, little stint, black-tailed godwit, turnstone, avocet, wood sandpiper, **Temminck's stint**, jack snipe), whiskered and black terns
WINTER greylag goose, goosander, goldeneye, smew, scaup, ferruginous duck, great white egret, cattle egret, Eurasian bittern, **Caspian gull**, short-toed owl, woodlark, **great grey shrike**, siskin, redpoll, brambling, hawfinch

VISIT DURATION   0.5–2 days
PLAN YOUR VISIT
Try to devise a geographically coherent itinerary to avoid long drives, and be aware that the most urbanised parts of the area are affected by morning and evening traffic jams. Some sites, like Fontainebleau forest, are crowded at the weekend.

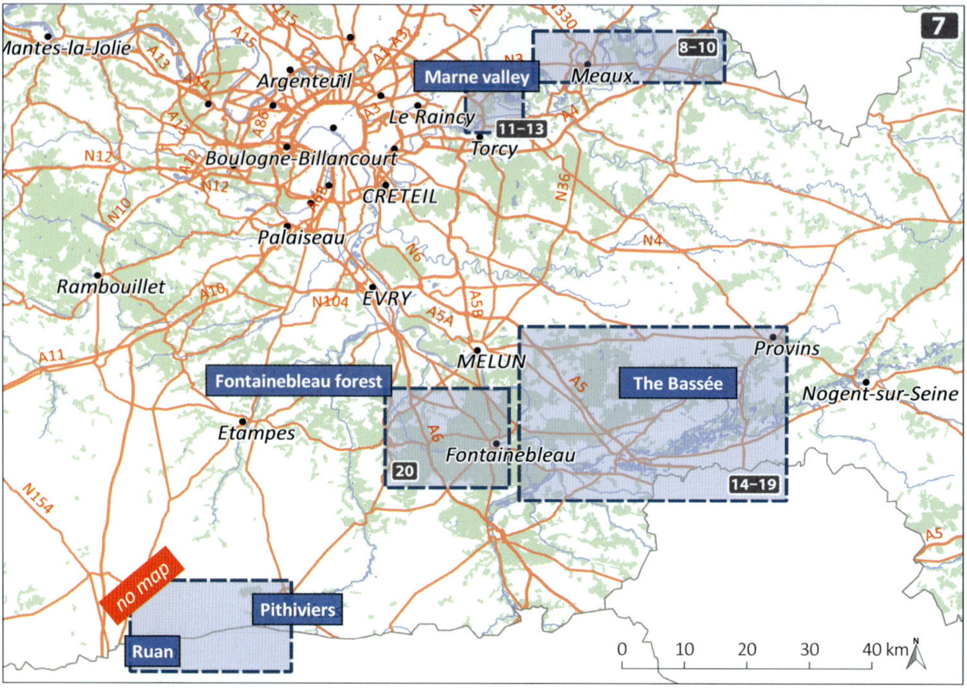

13 The Marne valley

0.5–1 DAY; MAPS 8-13 **Several quarries spread along the Marne river upstream and downstream of Meaux, 50 km, 40 min from Paris (drive east along the N3 or A4).** Some of them are still active, but others have been rehabilitated to form wildlife reserves with well-vegetated banks, mudflats and permanent waters suitable for waders stopping over on migration, wintering gulls or wildfowl, and wetland breeding birds.

Start birding in the Luzancy quarries [1], **the easternmost spot,** 25 km, 30 min from Meaux (From Luzancy, drive north on the rue de Messy, cross the railway and turn right after a concrete shed). Wildfowl and waders can be numerous at this site. The nearby meadows are excellent for short-eared owl and jack snipe, and there are recent records of great grey shrike.

In spring, visit the Grand Voyeux regional reserve [2] (12 km, 20 min northeast of Meaux, drive to Congis-sur-Thérouanne, then follow signposts along the D121E, check www.

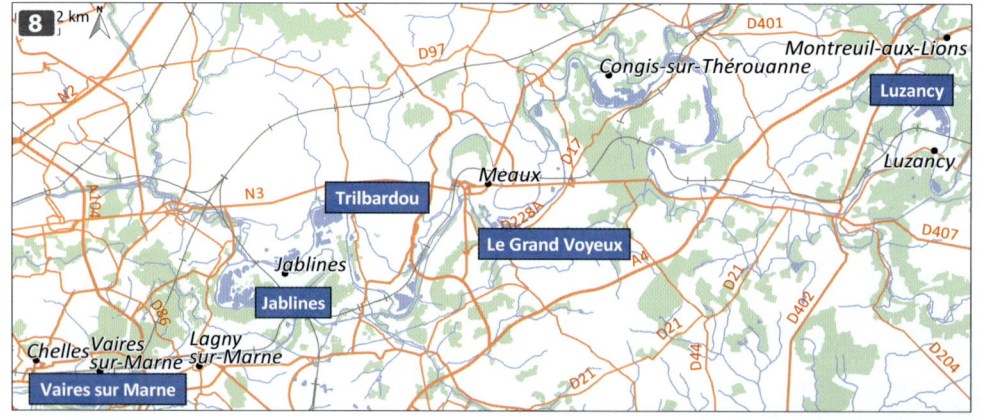

grandvoyeux.fr for opening times). This protected wetland is among the most reliable places close to Paris for breeding great white egret, purple heron, marsh harrier, great reed warbler and bluethroat, and is **arguably the best place to see many of these species in the area. If the gates are closed, there are still paths to the left and right with a few clearings that allow you to see inside.**

Drive 30 min west from the Grand Voyeux to visit Trilbardou [3] (park just after the D89 crosses the Marne river east of Trilbardou and walk around the ponds, stopping in the newly installed hide on the northern side [4]). Black kite and stone-curlew breed, and there is **a gull roost from October to March in which you should look for** Caspian, yellow-legged **and** common gulls **among larger numbers of** herring **and** lesser-black backed gulls.

Drive southeast for 9 km, 15 min to Jablines watersports centre (follow signposts from Jablines and park at [5]). This is the largest wetland complex of the area. It once held a major gull roost, but the birds have now been evicted by automatic scaring. It has remained an excellent winter wildfowl site, where the full range of diving and dabbling ducks are sometimes joined by regional rarities, especially after cold periods. **The best ponds are those at the northeast of the resort near the car parks** [6] **(for reedbeds) and those at the southeast end of the resort** [7,8]. Expect goosanders and goldeneye or even smew, scaup and ferruginous duck (rare). Regional rarities such as velvet scoter or long-tailed duck have occurred several times in recent winters. **A** bittern **sometimes winters in the reedbed of the pond just east of the main car park**. In May, you should look for breeding red-crested pochard, little bittern, marsh harrier, Mediterranean gull and great reed warbler.

In winter, **drive 15 km, 15 min southwest well before sunset to end the day with roosting gulls at the Vaires-sur-Marne watersports centre** [9] (cross the Marne and turn left to 'Base

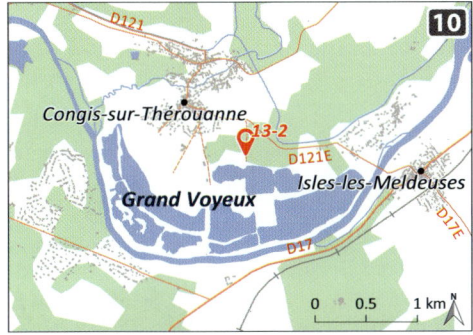

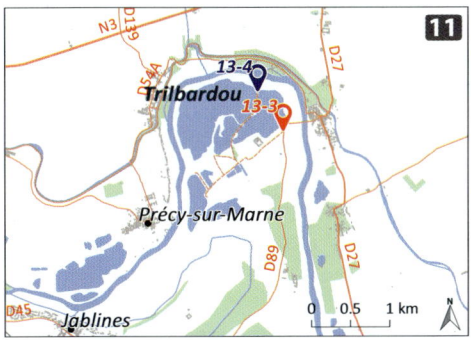

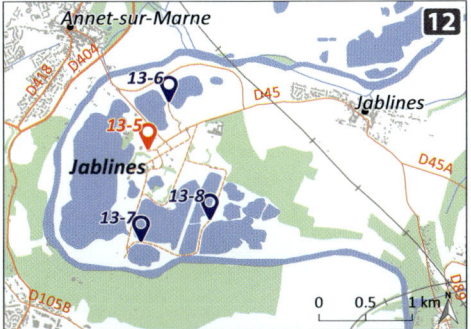

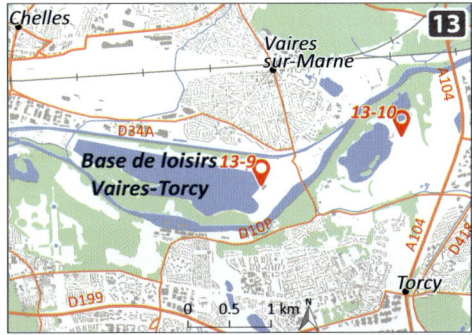

de Loisirs' at the roundabout just before entering Vaires from the south). Most of the birds are herring, black-headed and lesser black-backed gulls. Among them, search for Caspian gulls and rarities such as glaucous or Iceland gulls. In spring, Torcy resort can be a better choice [10] (cross the Marne south of Vaires and follow signposts to 'Parc de Loisirs de Torcy'). Singing bluethroats have been recorded here every year since 2011, along with more common wetland passerines.

14 La Bassée

0.5–1 DAY; MAPS 14–19 The Seine runs into the Bassée valley from Nogent-sur-Seine to Montereau, where it meets the Yonne valley and turns north towards Melun and Paris. This is arguably the birdiest area south of Paris during both spring and autumn migration, with the potential to yield over 100 species in an intensive birding day in May. **Most birding is done around quarries**, as the river itself is hard to access and quite birdless.

In May, first try Episy marshes [1] (car park 260 m south of Episy on the D40), where red-backed shrike and grasshopper warbler breed. There are also records of bluethroat and bittern. **Then, check out the ponds and quarries along the D606 in Ville Saint-Jacques** [2,3,4,5] between Moret-sur-Loing and Varennes-sur-Seine. Black-necked grebes, Mediterranean gulls, little terns and little ringed plovers breed here, and it is an excellent spot for migrating waders (including regular wood sandpiper and sometimes Temminck's stint). **Go back towards Montereau**, straight on at the first roundabout, turn right and right again on rue des Dormelles. This leads to a track (follow signposts for 'Espace Naturel du Grand Marais') and to another pond with a hide [6] which has yielded frequent sightings of night-heron and purple heron, cattle egret, wildfowl, and whiskered and black terns stopping over in April.

From there, you can either go back to the D606 and **try the ponds that lie on the south bank of the Yonne river, or head north for 10 km, 15 min to Marolles-sur-Seine**. An old quarry signposted as the 'Réserve ornithologique du Carreau-Franc' is renowned for migrating waders, breeding purple herons and night-herons, and ducks (turn right just after the A5 entrance/exit on the first roundabout after Marolles). A palissade just before the bridge over the A5 highway is the main viewpoint [7].

Resume your drive for 10 km, 10 min east along the D411, turn left in Balloy on the D77, **cross the Seine and visit the last pond before Vimpelles** [8]: goldeneye bred here in 2015 and 2016, and a Slavonian grebe stayed for several months in early 2016 with other wildfowl. Head back south, cross the Seine again and turn on the first road left, running along the river bank. Turn right just before the gate of a large quarry to reach **Champmorin hide** [9]: little bittern, Mediterranean gull, grasshopper, sedge and great reed warblers are fairly reliable here, and osprey is sometimes seen on migration. **Resume your drive to reach La Grande Bosse and park at the junction with the D109A [10]: this is one of

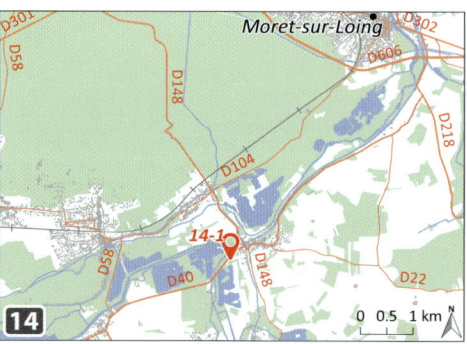

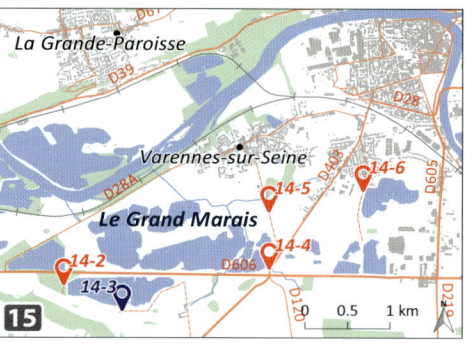

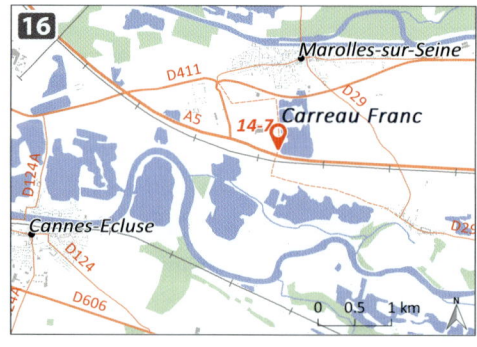

La Bassée's wader hotspots (including breeding black-winged stilt), and you can expect garganey, red-crested pochard, plus goldeneye, greylag goose and other wildfowl in winter. **Some regional rarities such as** spoonbill, Arctic tern **and** avocet **have also been recorded here**. Drive south to Bazoches-lès-Bray and turn left (east) on the D411 towards Jaulnes. On your way, the roundabout which marks the turning for Mousseaux [11] is a reliable place for crested lark.

Cross the Seine and **visit ponds around Neuvry sur Seine, 8 km, 15 min northeast of Mousseaux.** Park on the first track at the eastern end of the village [12] and walk to two hides on the northern side of the pond [13,14]: expect stilts, purple heron, cattle egrets, black-necked grebes, waders and black storks in spring. The surrounding fields are excellent for Montagu's and hen harriers, quail and stone-curlew, or even dotterel in late August. Drive 2 km further east to the Ferme d'Isle [15,16]. There can be high numbers of wigeon in winter, often accompanied by smew and goldeneye. Rarities such as Bewick's swan, great grey shrike and woodchat shrike (spring) have also been recorded here.

From there, drive north to Provins (20 km, 25 min), bypass it to the west and drive along the D619 to Nangis (25 km, 25 min). Turn left at the traffic lights as you exit the village and park after the railway [17]. **Walk 1.5 km south, then west, to two ponds** [18] **which have accumulated an impressive list of** waders **in the past** (wood sandpiper and black-winged stilt are regular, but curlew sandpiper, knot, whimbrel, sanderling, little stint, black-tailed godwit and turnstone have also occurred). There are also records of spotted crake, feldegg yellow wagtail and tawny pipit, all regional rarities.

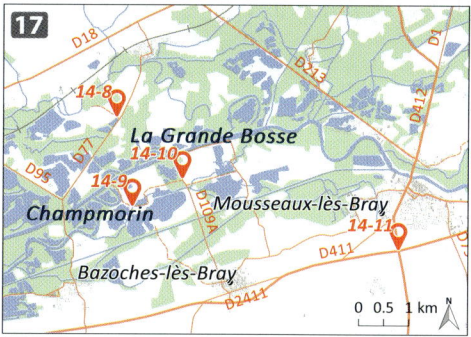

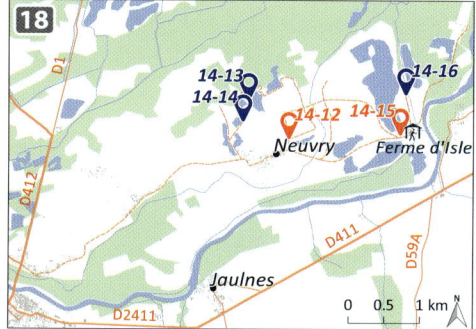

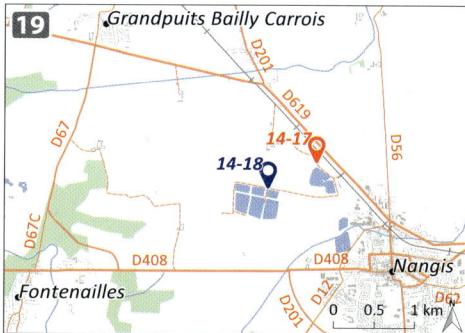

15 Forêt de Fontainebleau

0.5–1 DAY; MAP 20   **Fontainebleau forest is the best area to search for forest breeding birds in the region** (70 km, 1 h south of Paris on the A6), with a high diversity of raptors, fairly common pied flycatcher and Dartford warbler along with more common species. **Be there as early as possible in the morning, as the forest can becomes desperately silent even before 9 am on warm days.**

First visit Chanfroy plain (good year-round). Drive south from Arbonne-la-forêt on the D64 and turn left 1 km after leaving the village at the signpost 'Carrière des Fusillés' (car park at [1]). Walk east for 700 m to **a large clearing** [2] where wryneck, hoopoe, nightjar, woodlark, melodious warbler, whitethroat and red-backed shrike breed, with **peak activity in mid-May. Taking small paths north will lead you to the 'Tour de la Vierge'** [3], **a migration watchpoint on top of a rocky hill** (chances of short-toed eagle, goshawk, booted eagle and the rare black stork). Walk 500 m southeast on the Allée des Fusillés to another good viewpoint, which also has middle spotted woodpecker [4]. Go back to the car park through the south of the clearing and the **'Chemin des Sablières'** [5], **the best spot for** Dartford warbler **in the vicinity**.

The same breeding community occurs on the Macherin plain, 3 km, 3 min east of

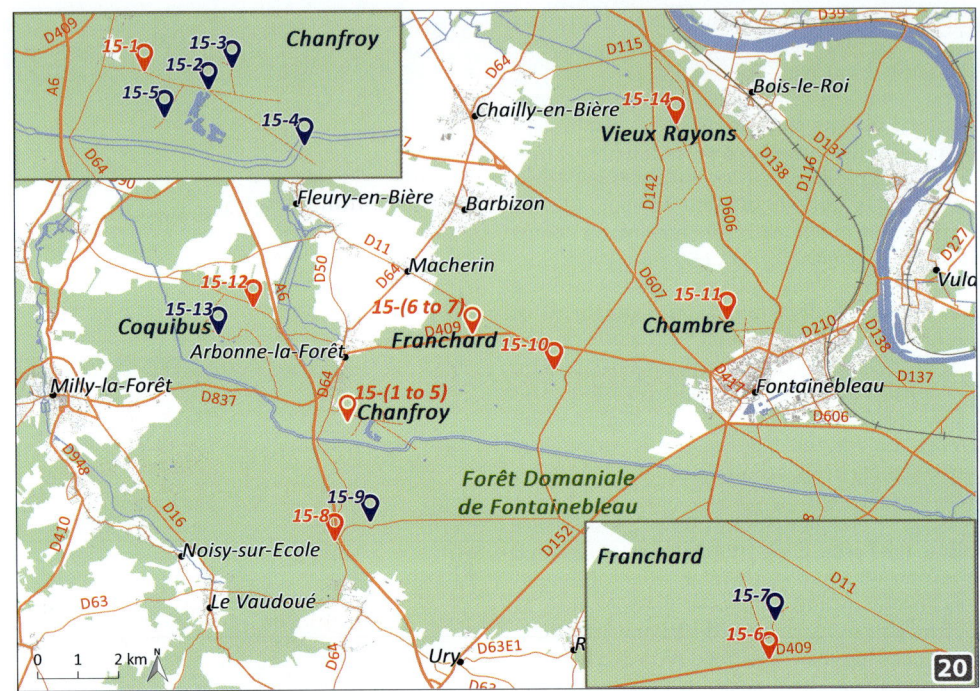

Arbonne-la-forêt on the D409 (park at [6] and walk north to [7]). Alternatively, drive further south on the D64 and park 700 m after the road passes under the motorway [8]. Walk east on the 'Route de la Haute Borne' for 1 km to the mares de Couleuvreux [9]. In winter, look for parties of siskins which may include redpoll, brambling and hawfinch. Check tit flocks for marsh, coal and crested along with great and blue tits and short-toed treecreepers. Middle spotted and black woodpeckers are uncommon breeders **most easily encountered in March when they are actively defending their territories.** Common Redstart is easy to find from late March. Warblers include chiffchaff (everywhere), willow (younger deciduous stands), wood (old oak stands) and western Bonelli's (which also occurs in mixed and conifer stands) – all of them are easy to find.

Pied flycatchers occur in small numbers in old stands around the forest reserve near **the car park of the Gorges de Franchard** [10] **and in the Chambre valley** [11] **or the plateau de Coquibus** (park at [12] and walk to [13], also good for Dartford warbler, woodcock and nightjar). In May, you might catch a glimpse of a bee-eater flying past, as there are colonies at undisclosed sites all around; **they are sometimes seen at the Carrefour des Vieux Rayons**, south of Bois-le-Roi on the 'Route de Solférino' (drive south from the junction of the D606 and the D138) [14].

16 The Beauce: Pithiviers and Ruan

0.5 HOUR – 1 DAY; NO MAP   These places are a long way south of Paris but are easily accessed via the A10, A5 or N20 (90 km, 1 h 20 towards Orléans or Etampes). **It can be a good idea to stop here on your way to sites in the Loire valley** (Region 5). **Pithiviers** [1] (from Pithiviers, follow signposts to 'Pithiviers-le-Vieil', take the D927 to Toury and turn left on the first road after the end of the village) **and Ruan** [2] (from Ruan village, follow signposts to Trinay and take the first track left after the last farm) **are excellent places for** waders, with yearly sightings of Temminck's stint, black tern and vagrants (including Baird's sandpiper and Caspian plover in recent years) – they will take less than half an hour each and can be rewarding. The fields from here to Orléans forest and west towards Chartres are good for dotterel in late April and early September, but it won't be an easy task finding them. Stone-curlew, short-eared owl (winter), whitethroat and quail are more reliable.

South of the Seine and northwest of Paris (sites 17–20, map 21)

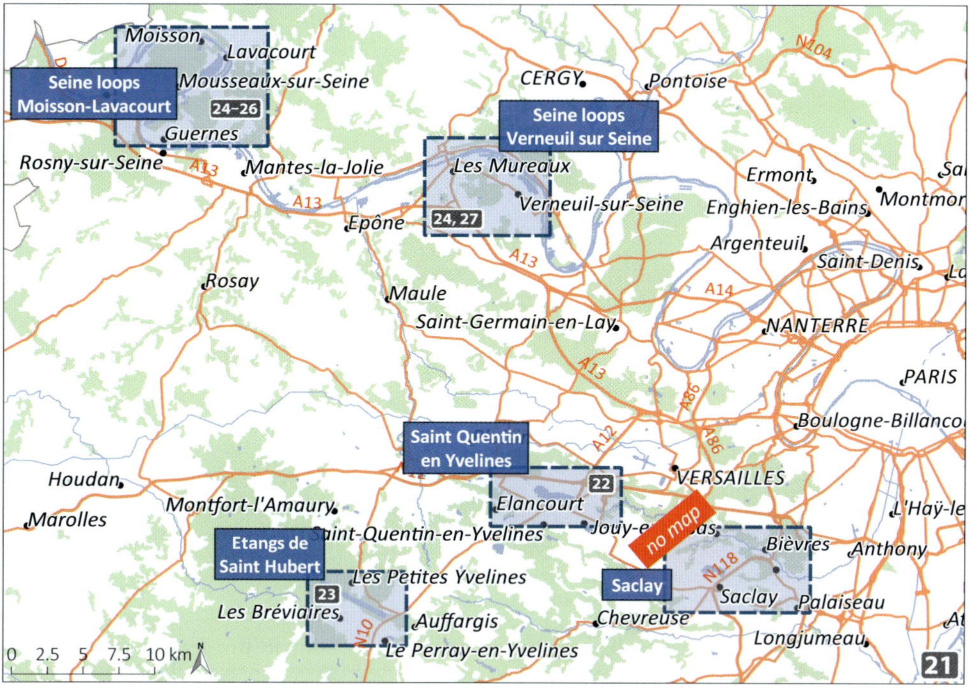

HIGHLIGHTS
» Little bittern and migrants in Saclay, just south of Paris
» The Seine and surrounding ponds for wildfowl and Eurasian bittern in winter

KEY SPECIES
YEAR-ROUND goshawk, black and middle spotted woodpeckers, Cetti's warbler
BREEDING little bittern, stone-curlew, Mediterranean gull, marsh warbler
MIGRATION osprey, **spoonbill**, crakes, waders, black tern
WINTER wildfowl including goosander and goldeneye, Eurasian bittern, great white egret, **Caspian gull, penduline tit**, siskin, redpoll, hawfinch

VISIT DURATION 1 day

PLAN YOUR VISIT
Plan your visit according to the season. Saint-Quentin-en-Yvelines is at its best during spring, but check the opening days before going. In winter, the Saint-Hubert ponds are a better choice. All the sites along the Seine can be birded in a single intensive day.

17 Saclay

1–3 HOURS; NO MAP   The artificial ponds in **Saclay** are a readily accessible site for wildfowl, **waders** and herons south of Paris if you rely on **public transport** (30 km, 35 min from Paris, take RER line C and stop at Jouy-en-Josas, then walk for 20 min; by car drive down the N118, and at the roundabout just south of Saclay turn onto the D446 to Jouy-en-Josas).

Walk along the dyke that separates the two **ponds** [1]. Little (spring) or Eurasian (winter) bitterns are reliable along the reedbeds that border both ponds. In winter, expect great white egret, all the common wildfowl and the odd goldeneye and goosander. Waders occur in good numbers on both passages. Regional rarities such as spoonbill, phalaropes or black tern may turn up in season: check the news on the internet beforehand.

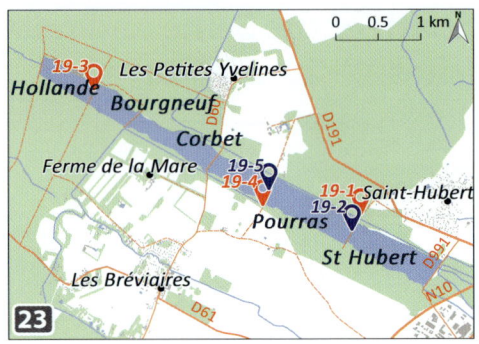

18 Saint-Quentin-en-Yvelines (Étang de Saint Quentin)

1–3 HOURS; MAP 22 The **Saint-Quentin-en-Yvelines watersports centre includes a small reserve** where little bittern breeds from May and Eurasian bittern winters (40 km, 50 min from Paris via the A13, A12 and N6, follow signs for 'Base de Loisirs', fee; opening days and times seasonal: check by calling 01 30 16 44 40 or reserve@basedeloisirs78.fr). Park at [1] and walk to a high point [2] which provides a general view over the reserve. There are marsh warblers in the surrounding bushes. The southernmost hide [3] faces a reedbed which is good for waders, crakes and penduline tit. A smaller hide 300 m further on [4] faces a black-headed and Mediterranean gull colony. **Outside the reserve, tracks surrounding the main pond** (e.g. [5,6]) can provide good views of wildfowl, with the possibility of goosander and goldeneye in winter. The site has regularly hosted regional rarities such as osprey, Arctic tern, Arctic skua, divers and Cetti's warbler.

19 Saint-Hubert ponds (Étangs de Saint-Hubert)

1 HOUR – 0.5 DAY; MAP 23 The ponds northwest of Le-Perray-en-Yvelines (Étang de Hollande, Étang de Corbet, Étang de Pourras and Étang de Saint-Hubert) **are a regional hotspot for** wildfowl **especially during cold winters** (30 km, 40 min from Versailles and 10 km, 20 min from Rambouillet: from both take the N10 to Le-Perray or Auffargis, then the D191 towards Montfort l'Amaury).

First stop between the Étang de Pourras and the Étang de Saint-Hubert (park in a small wood 2 km after leaving the N10 on the D191 [1] and walk to the dyke [2]). Check reedbeds for wintering Eurasian bitterns, which sometimes wander outside if the ponds are frozen. Parties of siskins may include a few common redpolls, and it is reasonable to expect single hawfinches. Black and middle spotted woodpecker and goshawk occur year-round in the area. Similar species can be found at the Étang de Hollande [3], 7 km, 10 min west past Les-Petites-Yvelines (follow signposts).

The Étang de Pourras and Étang de Corbet are reliable spots for little bittern **from May; the species is easier to see when adults are feeding their nestlings in late June.** Park at the base of the dyke in Les Bréviaires [4] (take the right-hand lane at the first junction north of the village) and walk until you find some clearings [5] among the trees from where you can see the reedbeds.

20 The Seine loops from Moissons-Lavacourt to Verneuil-sur-Seine

0.5–1 DAY; MAPS 24–27 **A number of spots can be birded in a day along the Seine as it flows from Paris westwards to Normandy.** All the common wildfowl species are present, plus rarer species including goldeneye, goosander, smew and divers. **This area is the northern counterpart to the Bassée and Marne valleys (sites 13 and 14),** and it hosts roughly the same bird diversity in similar abundance in winter (even better for geese and divers, because it is closer to the coast) – it is not as good in spring and autumn, however. Gulls roost at various locations: check for Caspian gull, which has become increasingly common in recent years.

Start at the Moissons-Mousseaux watersports centre (72 km, 1 h from Paris and 55 km, 1 h from Dreux; from either head to Mantes-La-Jolie, then Rosny-sur-Seine, Mousseaux-sur-Seine and Lavacourt, follow the watersports signs and park at [1], then walk on trails – you

can hide under a large concrete canopy [2] in case of rain). There are **poplar and birch stands interspersed with open areas on the nearby Moissons plain** [3]: visit them for goshawk, stone-curlew, wryneck (breeding), siskin and redpoll (winter). The site has occasionally hosted great grey shrikes in cold winters.

Next, drive to **the Guernes watersports centre** (40 km, 40 min from Moissons-Mousseaux, go back to Mantes-La-Jolie to cross the Seine and drive east along the D147 to Guernes) **and check the hide on the first pond** when entering the resort [4]. Apart from wildfowl and waders, **this is among the most reliable spots for stone-curlew in the region in spring**.

End your day at the **Verneuil watersports centre** (38 km, 40 min from Guernes; cross the Seine again in Mantes and exit the A13 at Les Mureaux, then turn right on the D154 towards Aérodrome des Mureaux and take the third exit at the first roundabout after the airport). From the car park [5], walk to **the westernmost pond** [6] ('étang Rouillard') to check for wildfowl.

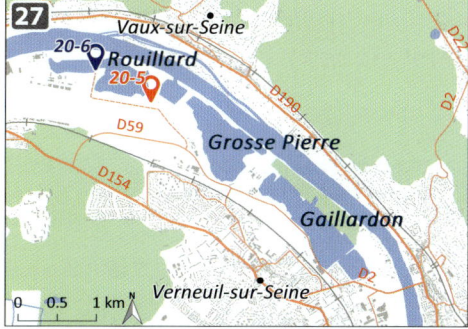

REGION 2
FROM THE NORTH SEA TO THE SOMME BAY AND THE WESTERN ARDENNES

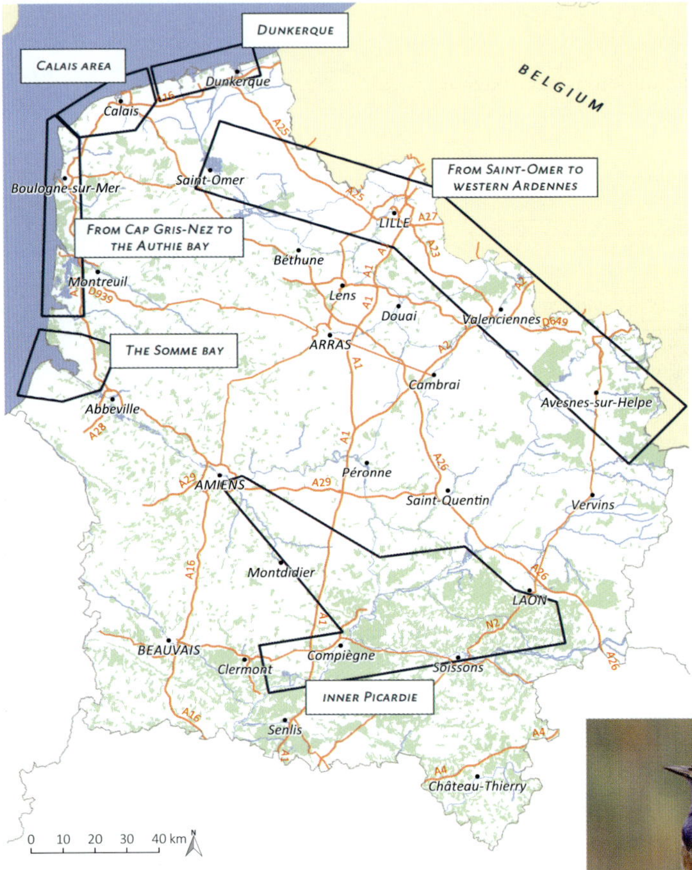

HIGHLIGHTS
» Spectacular seabird migration in spring and autumn
» Among the finest winter birding experiences in France along coastal marshes, beaches and harbours
» Rich communities of wetland and forest breeding species

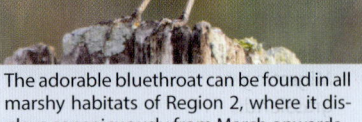

The adorable bluethroat can be found in all marshy habitats of Region 2, where it displays conspicuously from March onwards.

FROM THE NORTH SEA TO THE SOMME BAY AND THE WESTERN ARDENNES

Facing the British shore at the narrowest point of the Channel, the Cap Blanc-Nez (site 26) is an ideal watchpoint for seabirds and passage migrants.

REVIEWERS Valentin Condal, Quentin Dupriez

USEFUL WEBSITES
ornithonord.blogspot.fr
www.sirf.eu
www.nordpasdecalais.observado.org
chr-picardie.over-blog.com

For a more in-depth introduction to the region, refer to Dupriez, Q. (2015). *Où voir les oiseaux dans le Nord-Pas-de-Calais/Where to Watch Birds in Northern France*. Delachaux & Niestlé (205 pages with detailed site accounts and maps).

Dunkerque (sites 21–23, map 28)

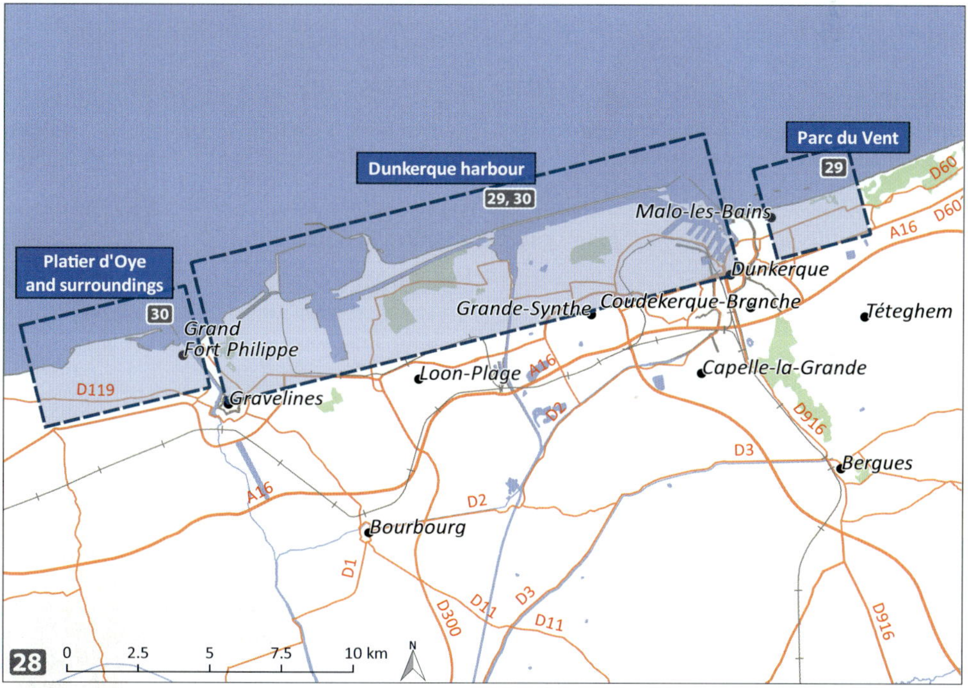

[Map of Dunkerque area showing sites 29, 30, Parc du Vent, Dunkerque harbour, Platier d'Oye and surroundings, Malo-les-Bains, Dunkerque, Grande-Synthe, Coudekerque-Branche, Téteghem, Grand Fort Philippe, Loon-Plage, Capelle-la-Grande, Gravelines, Bergues, Bourbourg]

HIGHLIGHTS
» Wintering and stopover area for large numbers of seabirds, waders and gulls
» Close views of sea ducks, divers and auks from the harbour walls
» Migrating passerines in willows, dunes and grasslands, with a good chance of finding a Siberian or American vagrant in autumn

KEY SPECIES
BREEDING black-winged stilt, lesser whitethroat, **icterine warbler, common rosefinch**

MIGRATION seabirds, waders, avocet, **little auk**, a wide range of passerines, **yellow-browed**, Hume's and Pallas's warblers (vagrants)

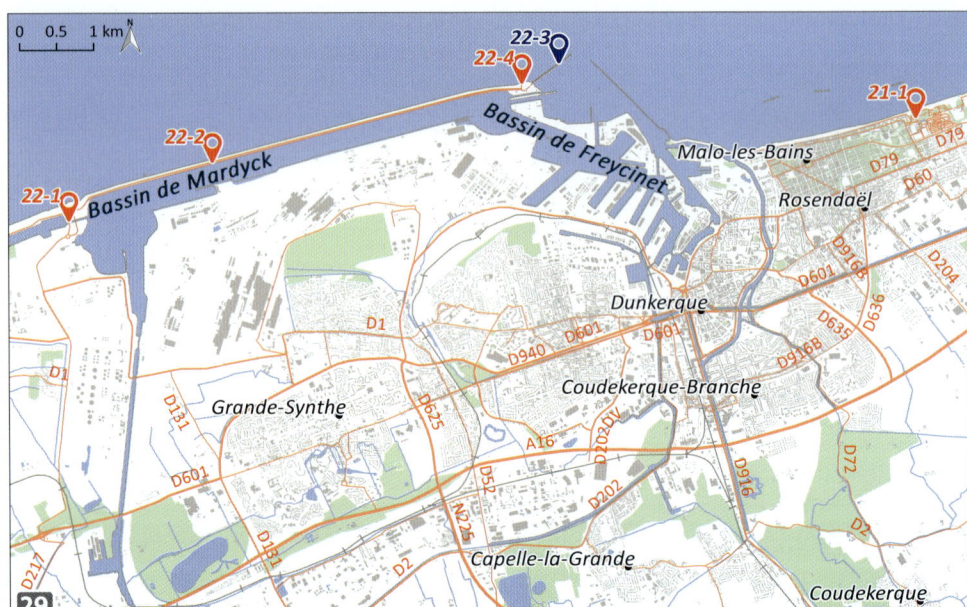

WINTER white-fronted goose, whooper swan, **long-tailed duck**, red-breasted merganser, Slavonian, red-necked and black-necked grebes, divers, herring, great black-backed, lesser black-backed, Mediterranean and **Caspian gulls**, **shore lark**, **twite**, snow bunting

VISIT DURATION   0.5–2 days

PLAN YOUR VISIT
When the wind is northwesterly in autumn, head straight to the seashore for seawatching. Seabird migration can also be excellent in spring in southwesterly winds. If the weather is not suitable for migration, stop-and-go from place to place in search of passerines, gulls and waterbirds. Be aware that the area is highly urbanised and driving may be slow.

21 Parc du Vent

1–3 HOURS; MAP 29 The **Parc du Vent is a magnet for migrating passerines from Belgium or Britain** (5 km, 15 min east of Dunkerque, follow signs to Malo-Les-Bains). Once in Malo follow signposts to the 'Camping de la Licorne' and 'Parc du Vent', and park near the tennis court [1]. The site consists of a park stuck between the seashore and the Dunkerque suburbs, in which several trails lead across wide stretches of bushes and grassy patches. Willows and brambles can be full of chiffchaffs, willow warblers, blackcaps and other insectivores from mid-August to October. In spring, icterine warbler and lesser whitethroat may also show up. Scarcer species (greenish, yellow-browed, Hume's and Pallas's warblers) have also been recorded in the park.

22 Dunkerque harbour

3 HOURS–1 DAY; MAPS 29 & 30 Dunkerque harbour may seem like an industrial hell with little room for nature, but leave this first impression behind you as its no-man's-land appearance makes it one of the regional birding hotspots, especially from September to March. You should concentrate on seabirds, gulls and waders, but it has good records of migratory passerines in autumn and a few snow buntings, sometimes with shore larks, winter in the sand dunes.

Exit Grande-Synthe to the west on the D601 (7 km, 11 min west of Dunkerque centre, signposted) and turn right at the first traffic lights after the bridge over the railway. Follow this road for 5 km until you reach the Ecluse des Dunes [1]. You can also reach this point by turning right towards Fort-Mardyck at the Hotel F1 Dunkerque in Grande-Synthe; take the third road at the roundabout (D1 towards 'port 3060 à 3100'). **Drive there and back along the 5 km of the Digue du Braek** [2], one way on the dyke itself and the other on the inner road just below. Scan the gulls on the beach, mostly herring and lesser black-backed mixed with small numbers of common, plus the odd yellow-legged or Caspian gull (beware of strange-looking herring gulls). There are small flocks of sanderlings and other waders, although most of them concentrate in the now inaccessible western harbour.

Check the inner basins and the harbour mouth [3] for divers, red-breasted mergansers, grebes (possibly including Slavonian, red-necked and black-necked), sea ducks and auks (little auk is possible after northwesterly winds in late autumn). The sand dunes on the dyke [4] are a good place for snow bunting, shore lark and twite after the first cold spells of the autumn, although the latter two have become irregular. Seawatching can be good from here as well, although nowhere near as good as from the nearby Le Clipon dyke, arguably the best seawatching spot in northern France in the 2000s, now closed to visitors.

Cross the Ecluse des Dunes again and go back, checking willows for migrating passerines on the way. Leave the D601 before Gravelines, turning right towards Institut Pasteur. Stop at the end of the road [5] (15 km, 20 min from the Ecluse). This site is excellent for waders as well as for sea ducks (often with one or two long-tailed ducks), great northern divers and grebes. Access to this area was restricted in 2016 but we still mention it here in the hope that it will open again in the near future. On your way back to Gravelines, **visit two ponds between the power plant and the town** [6] (3 km, 6 min from Gravelines centre, follow signposts to Sportica and turn right on the route des Enrochements) for black-necked grebe, ducks, waders (breeding stilt and avocet) and a Mediterranean gull colony.

23 The Platier d'Oye and its surroundings

3 HOURS – 1 DAY; MAP 30   **From here to Calais, the coastline consists of extensive beaches bordered inland by marshes** mainly used for wildfowl hunting, although parts of them are protected. From Gravelines, head to Grand-Fort-Philippe (4 km, 10 min, signposted) and park at the mouth of the Aa river [1]. **Walk across the large areas of saltmarsh embedded with small hidden ponds and canals connected with the beach and the sea**, in search of wintering snow buntings, twites (scarce) and shore larks which mix with linnets and greenfinches.

Drive westwards on the 'Route des Dunes' past the 'Camping de la Plage' and turn right after approximately 1.5 km towards Le Platier d'Oye. Stop at a first small hide [2], then park at the main entrance [3] and walk to the main ponds and hide [4] to check for wildfowl, herons and waders. **Bird numbers can be high at high tide, and** scarce species have been recorded,

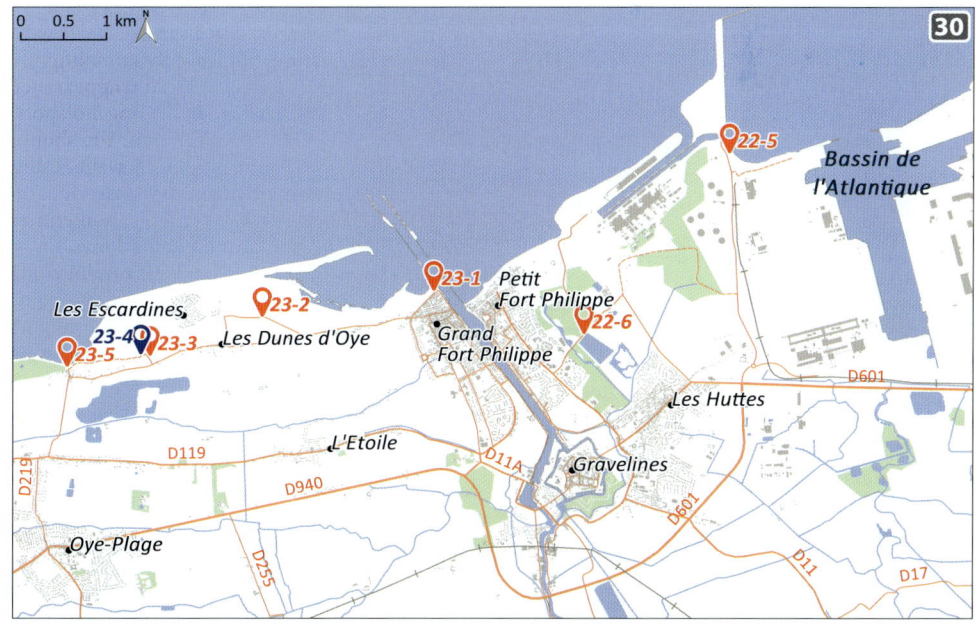

including whooper swan, white-fronted goose, and vagrants such as Baird's and pectoral sandpipers. Drive west for 1 km and park near L'Abri Côtier to **visit the large stretch of sandy beach and saltmarsh** [5] for wintering snow bunting, shore lark and waders that forage among saltworts – **avoid weekends as the area becomes crowded**. Territorial common rosefinches have occasionally been recorded in the vicinity in recent years (May–July).

Calais area (sites 24–27, map 31)

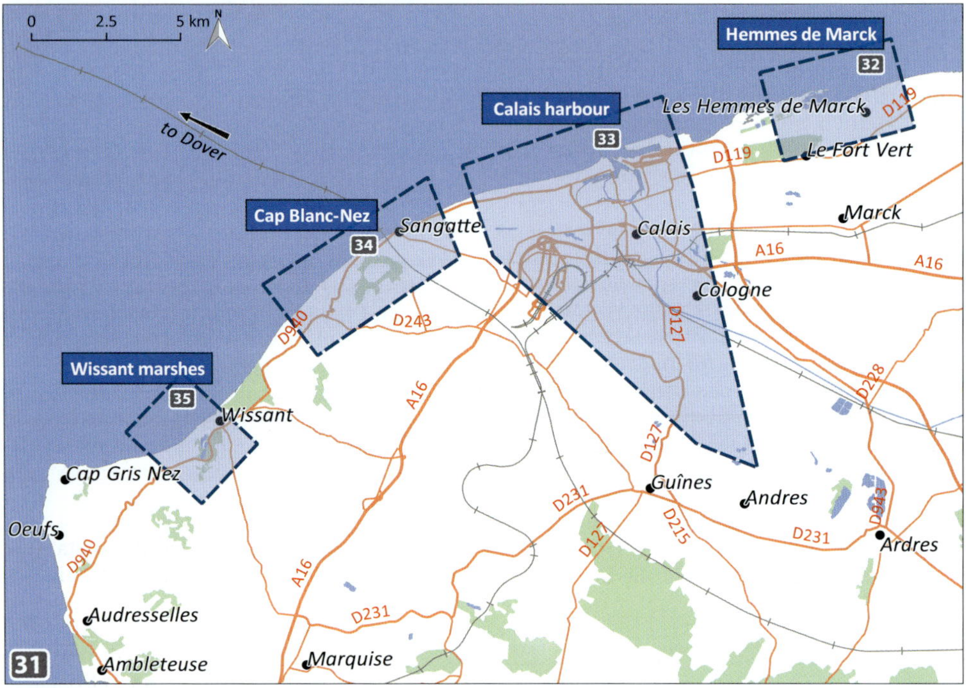

HIGHLIGHTS
» Migrating and breeding passerines along coastal wetlands and dunes
» Calais harbour for gulls, grebes and auks in winter
» Gull roosts, wildfowl and waders at various locations

KEY SPECIES
YEAR-ROUND cattle egret, marsh and hen harriers, peregrine, little owl, crested lark
BREEDING fulmar, kittiwake, bluethroat, lesser and common whitethroats, grasshopper, **Savi's**, marsh, reed, **icterine** and melodious warblers
MIGRATION seabirds, waders, passerines
WINTER red-breasted merganser, scaup, **rough-legged buzzard**, red-necked and Slavonian grebes, Eurasian bittern, **Caspian, glaucous** and **Iceland gulls**, shore lark, Richard's pipit, **twite**, snow bunting

VISIT DURATION > 1 day

PLAN YOUR VISIT
The area is good at all seasons, but spring and autumn passages offer the best birding. The sites are large and can take quite a long time to explore, so choose one and cover it thoroughly. Do not stick just to the sites described below; good birds can show up anywhere along the coastline and there are countless habitats to visit. **Hunting can be a serious issue** in autumn and winter, especially in the Hemmes de Marck.

24 Hemmes de Marck

1 HOUR – 0.5 DAY; MAP 32 The Hemmes de Marck are an extensive area of ponds and

marshes halfway between Calais and Gravelines (10 km, 20 min from Calais centre, take the D119 to Le Fort Vert and head north to the seashore). Park near the minigolf and the yacht rental on the Avenue de la Mer [1] and **explore the marshes, ponds and beach, which stretch several kilometres.** Although the area is heavily hunted in winter, it is much used by passage and wintering waders and passerines. Some ponds offer good views, while others are hidden or hard to see; in all cases, **keep to the tracks and do not trespass on private areas.** Walking towards the sea, explore the **grassy areas bordering the beach** for wintering passerines including regular snow buntings and shore larks (twite has become scarce in the last decade). Richard's pipit may occur here from time to time in late autumn. Once finished with the seashore, you can drive back to Calais, making stops here and there beside the ponds along the Digue Taaf [2].

25 Calais harbour

2 HOURS; MAP 33   First try the western side of Calais harbour [1] in search of wintering Caspian, glaucous and Iceland gulls among large numbers of herring, lesser black-backed and great black-backed gulls. **Check the basins for** red-breasted merganser **and occasional** red-necked **or** Slavonian grebes, **and the car parks for** crested lark. The Digue Blériot [2] is an excellent **seawatching spot** which can serve as an alternative to Cap Gris-Nez under strong northwesterly winds in autumn, with passage of shearwaters, skuas and auks that occasionally come within photographic range.

If you are leaving for the UK on the shuttle train, **stop at the nearby Eurotunnel ponds** [3] (from Calais harbour: 6 km, 10 min, follow signposts to the tunnel and Centre Commercial Cité Europe, park near the petrol station on the Avenue de France) for groups of tufted duck and pochard; they also sometimes host rarer species such as scaup. **If you still have a bit of time, visit the Guînes marshes**, especially in spring, for cattle egret, little owl, bluethroat, grasshopper, Savi's, marsh, reed and melodious warblers, lesser whitethroat and a generally good diversity of wetland species (13 km, 20 min from Calais centre, bypass the A16 and follow signposts to Guînes, then turn left on the D248E1 towards Andres and park after 2.2 km, on the right [4 – see GPS file]: a footpath with viewpoints and a hide starts from here). Bittern occurs in winter, and a harrier roost is regular from September to March.

26 Cap Blanc-Nez

0.5–1 HOUR; MAP 34   Drive along the coastal road (D940) south of Calais past Sangatte, turning on small tracks to check fields and ponds [1,2] for peregrine, lapwing, plovers, gulls and short-eared owl. Cap Blanc-Nez (car park at [3], then walk to the cape [4]) is worth a stop on your way from or to Calais (11 km, 20 min) for the scenery on sunny days. **It is also a good place for** peregrines**, which feed on breeding** kittiwakes **and** fulmars. The car park above the restaurant on the nearby hill [5] is an excellent viewpoint for raptors; rough-legged buzzards stay around virtually every winter. **All the surroundings are pastures with bushes that**

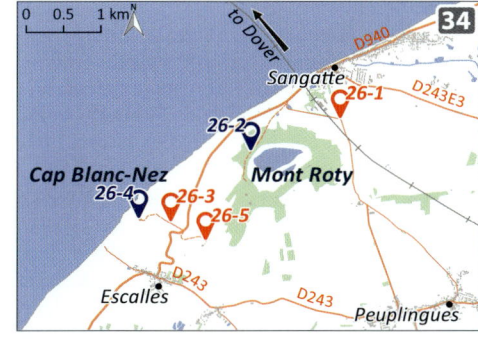

can be full of migrating passerines in late April and from September to November. **Skywatching can occasionally yield good passage** dominated by finches in winter and swallows in spring.

27 Wissant marshes

0.5–3 HOURS; MAP 35   **A wide belt of reeds and marshes separates the coast from inland hills between Cap Blanc-Nez and Cap Gris-Nez.** The easiest access is from a car park [1] on the right-hand side of the D948 1.3 km past Wissant westwards (10 km, 10 min from Cap Blanc-Nez along the coast road). **A track leads to the top of a hill where a hide [2] gives an excellent overview of the marshes.** Expect marsh and hen harrier, herons (with bittern in winter), ducks and waders (look for snipe). The reeds and bushes may be full of passerines in both spring and autumn, with good numbers of lesser and common whitethroats, reed warblers, melodious (rare) and icterine warblers

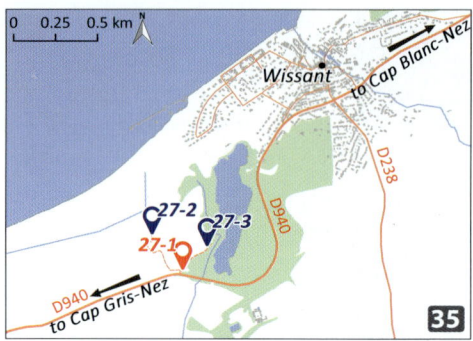

(the latter breeds in the vicinity), yellowhammer and reed bunting. Go back to the car park and walk downhill, then turn left onto a small path [3] which leads to the beach after 30 min. **There are more hides on the way (signposted) which offer closer views of** waders **and good opportunities to search for** passerines **in the bushes.**

From Cap Gris-Nez to the Authie bay (sites 28–30, map 36)

HIGHLIGHTS
» Exceptional autumn seawatching from Cap Gris-Nez when the wind is northwesterly
» Explore harbours for wintering gulls, grebes and auks
» Coastal bushes can be full of migrating passerines
» Waders and wildfowl in the large estuaries to the south

KEY SPECIES
YEAR-ROUND   black woodpecker, woodlark, crested tit

BREEDING   honey-buzzard, black-winged stilt, nightjar, sand martin, bluethroat, lesser whitethroat, **icterine** and melodious warblers, red-backed shrike

MIGRATION   brent goose, fulmar, Manx and Balearic shearwaters, **Leach's petrel**, waders, **grey phalarope**, Arctic, pomarine, great and **long-tailed skuas**, little and **Sabine's gulls**, common and Arctic terns, auks including **little auk, Richard's pipit, yellow-browed warbler**

WINTER   barnacle goose, common and velvet scoters, all divers, Slavonian and red-necked grebes, Eurasian bittern, **rough-legged buzzard, Caspian**, Mediterranean, **glaucous** and **Iceland gulls**, bearded tit, **twite**, snow and Lapland buntings

VISIT DURATION   > 1 day

PLAN YOUR VISIT

In spring and autumn, moderate to strong southwesterly or northwesterly winds should lead you to Cap Gris-Nez for seawatching in the morning. If the weather is against you, try Boulogne harbour. During passerine migration rushes, focus on areas of scrub and dunes. The large bays to the south of this area host large numbers of waders and wildfowl in autumn and winter, but be sure to be there on a rising tide.

28 Cap Gris-Nez

1 HOUR – 1 DAY; MAP 37   Cap Gris-Nez is arguably one of France's best seawatching spots, if not the very best (follow signposts from Boulogne-sur-Mer or Calais: 30 km, 30 min on the A16 or the coast road). Seabirds, waders and wildfowl concentrate here as the Channel narrows to its minimum width and fly along the coast, sometimes at close range. Incomparable migration scenes can be expected under **moderate to strong northwesterly winds in autumn (mid-August to mid-November) and southwesterly winds in spring (March to May), which tie birds to the coast**. Depending on the time of the year, brent geese, scoters and other ducks, shearwaters, common terns, little gulls, divers (three species) dominate bird movements, alongside sometimes

FROM THE NORTH SEA TO THE SOMME BAY... | 29 BOULOGNE-SUR-MER

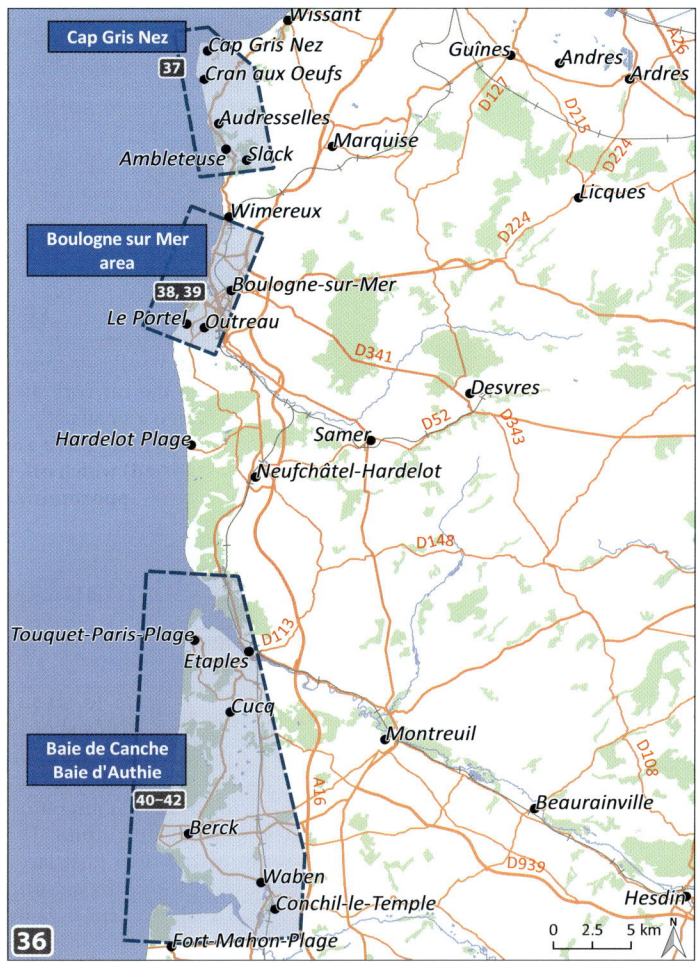

can be **full of migrating insectivorous** passerines including warblers and flycatchers with the possibility of a vagrant (yellow-browed and Pallas's warblers have been seen). In spring, icterine warbler and red-backed shrike can occur. The Bois d'Haringzelle [6] and surrounding fields down to Cran-aux-Oeufs [7] are the most famous area for migrating passerines in the vicinity. Also check fields for rough-legged buzzard, gull roosts, pipits and buntings (including snow and Lapland).

29 Boulogne-sur-Mer

2 HOURS – 1 DAY; MAPS 38 & 39 Boulogne harbour is an industrial hell (although less so since it has undergone extensive restructuring) and another hotspot for seabirds and gulls. From Cap Gris-Nez, drive south on the D940 towards Boulogne-sur-Mer. **Check the beaches at Audresselles** [1] **and Ambleteuse** [2], **as large** gull **roosts form at mid and low tides.** The Slack dunes are hard to bird but may be rewarding during passerine migration bursts, with flocks of passing swallows, larks or finches and resting warblers, thrushes or pipits in the bushes. Leaving Ambleteuse, cross the river and turn right just after the bridge to park [3], then walk along the fence facing the beach to **check the scrub and the grassy dunes** [4]. One or a few pairs of icterine warbler settle here from late May. The nearby Slack valley [5] can be viewed from the D237: **its wet meadows are excellent for** herons, white stork, waders and passerines including whinchat, yellow wagtail **and** water pipit.

Another good passerine area, albeit less well known, is the top of the Pointe de La Crèche cliffs, 2 km south of Wimereux (park at [6] and

high counts of Arctic, pomarine and great skuas, and Manx and Balearic shearwaters. Singles Leach's petrel, grey phalarope, long-tailed skua, Arctic tern and Sabine's gull invariably show up on good autumn days. Northwesterly winds in November yield one or two rushes of little auks per year. Spring migration is less popular, but breeding-plumage divers and waders that pass through in high numbers and at close range are just as spectacular.

There is only one car park [1]. **Walk to the right of the lighthouse and take up a position at the top of the cliff** [2] **or at its foot** (accessed from the Gris-Nez village [3]). The latter option is better but access might be restricted for safety reasons in future years. If the wind is not suitable for migration, visit the small woods around [4,5] as they

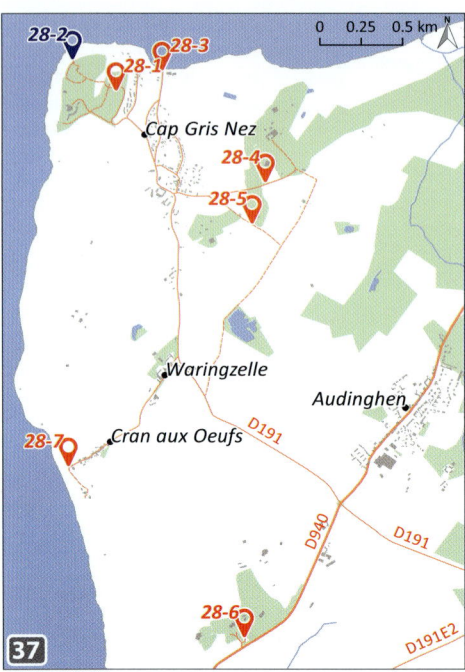

walk northwards through the fields [7] until you reach the woods around the camp site [8]). The cliffs also host a colony of fulmars. From the car park, the D940 descends steeply to Boulogne sur Mer harbour. Park near Nausicaa [9] and **check Boulogne harbour mouth** [10] for grebes (possibly including Slavonian or red-necked), gulls (chances of glaucous/Iceland/Caspian which show up every winter) and auks. **The same species can be searched for inside the harbour** at the Napoléon [11] and Loubet [12] basins. The surrounding roofs are often full of gulls, which can be spotted from **an elevated car park** [13] above the fishing harbour. Once you're finished with the harbour, you can **follow the river upstream from the Napoléon basin, parking intermittently along the road**: gull roosts can be viewed at close quarters at sunset and this also gives the chance of waders or kingfisher.

Carnot dyke [14], at the western end of the harbour, has long been one of the best areas for grebes and auks, but it is possible that access will be blocked. In general, the harbour is quickly changing and the viewpoints could change in the coming years. **For autumn passerine migration, visit the Parc de la Falaise in Outreau** (turn right off the main road on Rue de Reims, there is a car park at [15], then walk on paths for instance towards [16]). This extensive coastal park has become popular as rarities such as yellow-browed, Pallas's and Hume's warblers or least flycatcher have shown up regularly in October over the last few years. Search for them among chiffchaffs, willow warblers, blackcaps

and finches. **The dyke on Le Portel seashore** [17], 2 km, 5 min further south, hosts a Mediterranean gull roost.

30 Baie de Canche and Baie d'Authie

3 HOURS – 1 DAY EACH; MAPS 40–42 The Canche and Authie bays are two estuaries lying 30 km, 40 min and 50 km, 60 min south of Boulogne-sur-Mer (take the A16 south towards Rouen and exit at Etaples or Le Touquet). **They hold the largest concentrations of** waders and wildfowl **wintering on the French coast north of the Somme bay,** and form a handy alternative to the latter for those with only one day left or if you wish to connect with the Pas de Calais coast. **The two bays are somewhat similar, so chose one and work it extensively, preferably at high tide**.

The Baie de Canche has a reserve on its northern side (head to Etaples and take the D940 north to Hardelot-Plage; from Boulogne-sur-Mer, follow the D119 past Le Portel until you reach the D940). The reserve is divided into two parts **which can be accessed by walking on trails from car parks along the D940**. The north (park at [1] and walk towards [2]) is covered by sand dunes with pine trees in which honey-buzzard, nightjar (both from May), black woodpecker and woodlark breed in good densities. More common species like crested tit (year-round) and coal tit (winter), nightingale, lesser whitethroat, melodious warbler (from May) can easily be found in the pine woods or in surrounding bushes. **The southern part** [3] **is harder to access and needs to be watched from trails at its edges**. It consists of a wetland with reedbeds and ponds that are good for bittern (winter), bluethroat and bearded tit. **The estuary itself has a huge** gull **roost in winter and is an excellent area for** wildfowl, waders **and** terns **stopping over at both migrations**. The surrounding dunes and saltmarshes host snow buntings and loads of larks, meadow pipits and finches in winter. **The best spots are located along the south shore**, like at the Pointe du Touquet [4] (platform in [5]). The area is heavily hunted in autumn and winter so be discreet and cautious while walking in the surroundings.

The Baie d'Authie (35 km, 50 min from the Pointe du Touquet on the A16, take the D940/E1 to Conchil-le-Temple, then the D532 west to Fort-Mahon-Plage) is larger but **has roughly the same habitats and species**. Park at [6] and **walk in saltmarshes and sand banks** [7,8] to scan for waders and ducks foraging or roosting on the estuary (best at high tide). Meadow pipit is common around here, but snow and Lapland buntings, twite and the odd Richard's pipit may also occur in winter. In spring, check the ponds at Conchil-le-Temple [9] and Waben [10] for breeding avocets, oystercatchers, stilts and sand martins.

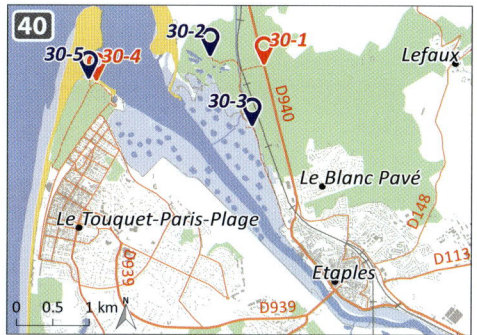

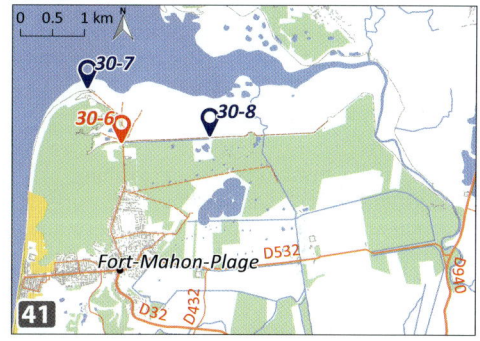

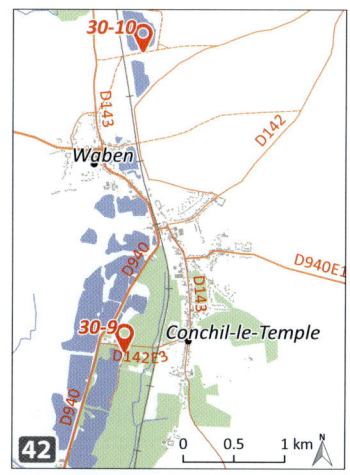

The Somme bay and inner Picardie (sites 31–33, map 43)

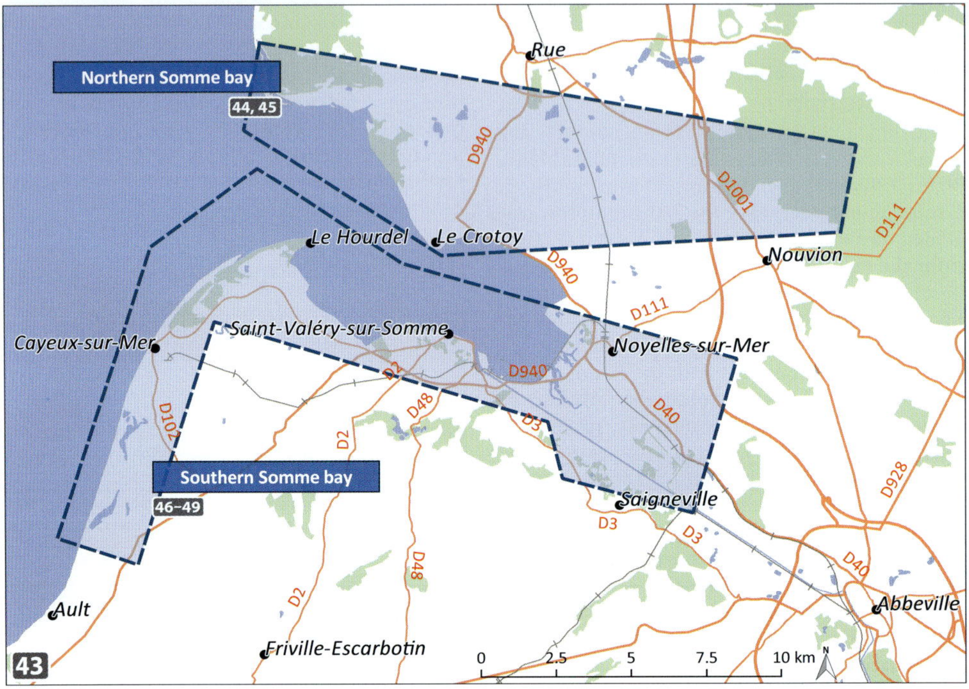

HIGHLIGHTS
» A definitive must for any birder visiting northern France at all seasons
» Huge wader and wildfowl roosts year-round
» Passerine migration and wintering shore lark and snow bunting from November through to March

| J | F | M | A | M | J | J | A | S | O | N | D |

KEY SPECIES
YEAR-ROUND little grebe, peregrine, great white egret, Eurasian bittern, avocet, black and middle spotted woodpeckers, fan-tailed and Cetti's warblers
BREEDING garganey, black-necked grebe, white stork, **little bittern**, quail, **corncrake**, black-winged stilt, **wryneck**, bluethroat, whinchat, lesser whitethroat, **Savi's**, grasshopper, reed, marsh and wood warblers, pied flycatcher, red-backed shrike
MIGRATION osprey, waders, black tern, sedge warbler
WINTER brent and barnacle (uncommon), white-fronted and greylag geese, goosander, common and velvet scoters, other wildfowl, divers and grebes, merlin, **shore lark**, **twite**, snow bunting

VISIT DURATION 1–3 days

PLAN YOUR VISIT
The Somme bay is rightly ranked among **France's most famous birding spots in all seasons**, although it is best from October to March. It deserves a bit of strategic planning, however, because the area is vast and driving times are long. **Be on the seashore at high tide** for wader roosts along La Maye estuary, in Le Hourdel or on the ponds of Le Hâble d'Ault. **Although the area is at its best in winter, spring birding can also be rewarding.** Photographers should allocate at least half a day to the hides at Le Marquenterre. There are more spots than can be described here, so we concentrate on a few of the most famous locations. **Do not forget to visit the banks of the River Somme** inland for wildfowl and the large stretches of bushes behind the coastline for passerines in autumn. **Hunting is a particularly serious issue** from November to March, so be cautious and behave politely in all situations.

31 Northern Somme bay

1 DAY; MAPS 44 & 45 The **Marquenterre reserve** [1] ('Parc du Marquenterre', 30 km, 40 min from Abbeville on the D40, follow signposts from Le Crotoy; €10.50 fee, open every day from 10.00 to 17.00 in winter or 19.00 in summer, guided visits) is a **200 ha hunting-free mosaic of marshes and ponds with several hides** spread along an easy foot trail. **Be there at high tide for the highest bird numbers.** Breeding species and spring migrants include white stork, great white egret, garganey, avocet and bluethroat. Autumn and winter bring high numbers of teal, wigeon and greylag geese. Waders occur at all seasons, including breeding lapwing, avocet and black-winged stilt.

Small parts of the Marquenterre can be viewed from outside the reserve, behind screens that can be reached by walking on the sand for several kilometres along La Maye river estuary (car park at [2] after the camp site, cross the river at [3], checking surrounding meadows for geese, the hides and good viewpoints are located around [4,5]). Most wildfowl, which sometimes include barnacle and white-fronted geese, and wader species can be seen from there at some distance. **The saltmarshes and meadows along the estuary** [6] are excellent for peregrine and merlin in winter, and small groups of greylag geese, ducks and waders gather here before roosting at high tide. **After about 4 km of a longish, yet worthy walk, a sandspit called the Banc de l'Ilette** [7] **is an autumn migration viewpoint.** Flocks of passerines converge here before crossing the Somme bay, with impressive numbers of swallows, pipits, larks, thrushes and finches from September to November. In winter, the Banc de l'Ilette is among the most reliable spots for snow bunting in France, and was formerly a classic place to see shore lark. This species has become less regular in recent years but still occurs from time to time, sometimes mixed with finch groups. Search for it among samphires or in the small sandy dunes that border the Marquenterre. Bear in mind that the walk from the car park at La Maye to the Banc de L'Ilette can be strenuous, especially in cold and windy conditions.

Once back at your car, drive to Le Crotoy (from La Maye car park: 6 km, 15 min on the D4) and **look for the hunting pond at the eastern end of the harbour** [8 – see GPS file] for gulls, grebes and waders. On your way back to Abbeville, a small diversion to Crécy forest to the north of Nouvion [9] can add several forest species, including middle spotted woodpecker

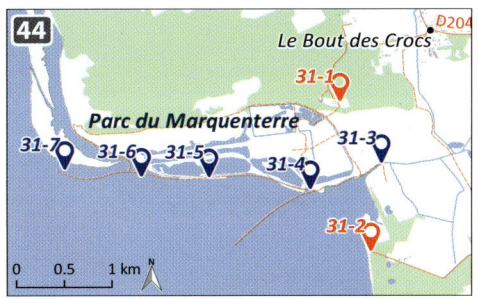

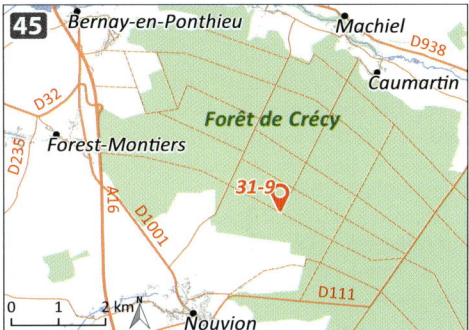

(year-round) and wood warbler (from April) (17 km, 20 min from Le Crotoy, enter the forest on tracks east of the village).

32 Southern Somme bay

1 DAY; MAPS 46–49 Stop here and there on your way to the bay from Abbeville on the D40 or from Le Crotoy on the D940 to **check the marshes and ponds that border the River Somme**, for instance around Port-le-Grand [1]. Expect great white egret, Eurasian bittern, bluethroat and Savi's warbler in spring. Fan-tailed warbler may also occur in reedbeds and grassy areas. Pairs of red-backed shrike breed along the D40 at locations near Saigneville [2] (park at the north end of the village and walk on paths towards the river) and Noyelles-sur-Mer [3].

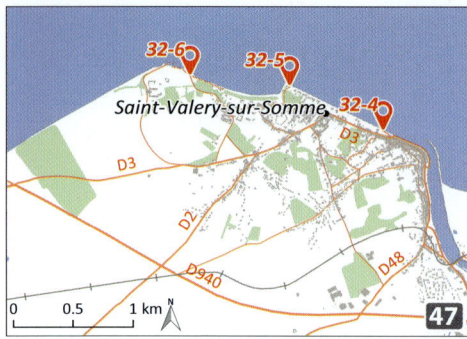

Drive to Saint-Valéry-sur-Somme (20 km, 25 min from Abbeville) **and stop anywhere along the seashore** [4,5,6] **to check for** wildfowl **and** waders **that forage on the estuary**.

Follow the D3 and D102 to Le Hourdel (signposted, 10 km, 10 min northwest of Saint-Valéry-sur-Somme). **Le Hourdel is a good alternative to the Banc de L'Ilette** (site 31) for those not willing to walk the 4 km trail along the Marquenterre, as **migrating** passerines **concentrate here before crossing the bay in spring under easterly winds**. Bushes and open areas behind the lighthouse [7,8,9] can be crowded with birds in spring and autumn. The sand and shingle spits [10,11,12] along the D102 between Le Hourdel and Cayeux-sur-Mer (8 km) are good places to find shore larks, snow buntings and possibly twites (scarce) among flocks of linnets and greenfinches in winter. Scan the sea for divers, grebes and scoters.

Immediately south of Cayeux, the southernmost spot in the Somme bay is a wide area of ponds known as Le Hâble d'Ault. This area hosts good numbers of wildfowl, waders and gulls with regular records of vagrants (including an oriental pratincole in 2014); do not hesitate to allocate several hours to birding here, especially if you are short of time to explore the bay more widely. Enter the area along some tracks that start from [13], and bird among ponds and meadows south of Ault, for instance at [14,15,16]. **Remember that the area is heavily hunted in winter**.

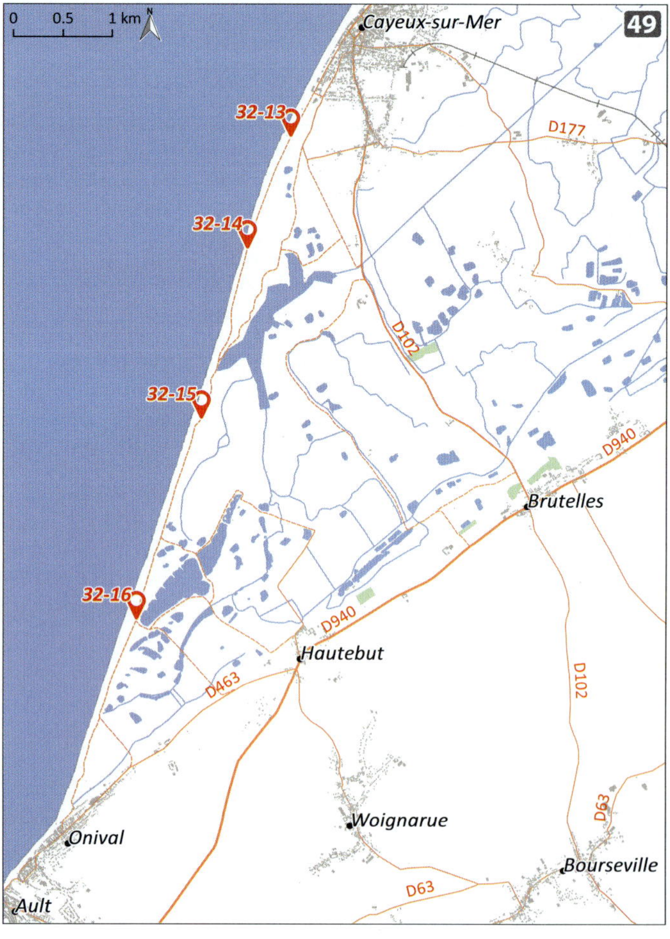

33 Inner Picardie

0.5 DAY; MAPS 50–56 Inland Picardie has few birding opportunities to offer, but several sites between Amiens and Laon are worth mentioning as they offer good opportunities for several sought-after species, including little bittern, bluethroat, Savi's warbler and red-backed shrike, barely one hour north of Paris.

The Étang de Saint Ladre nature reserve [1] is worth a visit in May–July for little bittern, reed warbler, bluethroat and a good diversity of wetland birds (15 km, 20 min from Amiens, take the A29 to Cagny, then the D116 to Boves). Further east on the A29, reach Saint-Quentin (86 km, 1 h), then take the A26 and exit south to Tergnier. The Oise valley between Tergnier and Amigny-Rouy [2] (26 km, 30 min from Saint-Quentin) is one of the last reliable spots for corncrake in northern France (calls are heard at dawn and dusk in May–June). Also search for black-necked grebe, garganey, quail, curlew, whinchat, lesser whitethroat and tree sparrow.

Less than 5 km south, the Saint-Gobain forest south of Servay [3] has black and middle spotted woodpeckers and wood warblers, and a few pairs of pied flycatchers might still hang on there. Further east, 16 km, 30 min southeast of Laon, lies the Lac de l'Ailette with its watersports centre

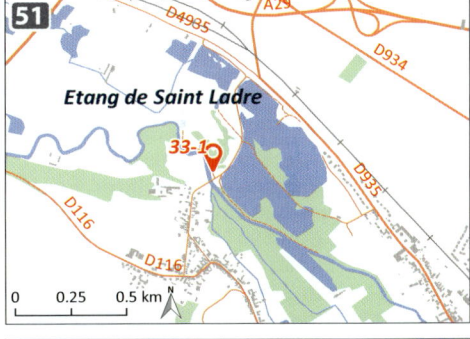

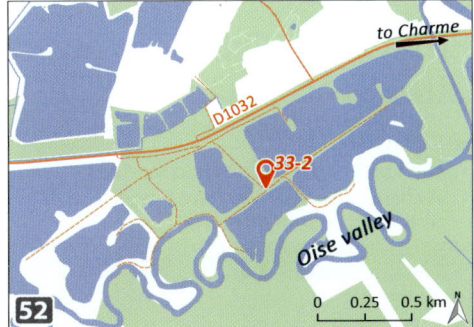

('Parc Nautique de L'Ailette': fee in summer; take the D967 past Chamouille to access the western

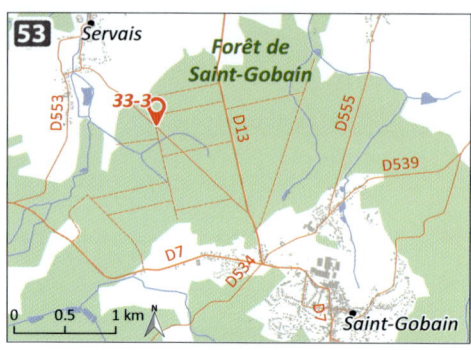
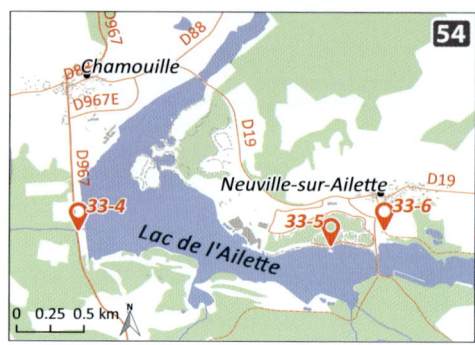
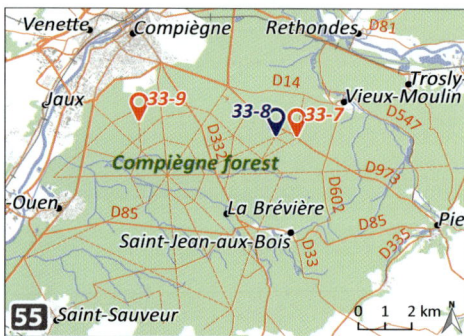

bank [4] or the D19 to Neuville-sur-Ailette for the northern bank and eastern tail [5,6]). Wildfowl and gulls roost in winter, with chances of goosander or divers. Black tern and osprey have been recorded as migrants, and waders may stop over if the water level is high enough. Little grebe, reed and Cetti's warblers and terns breed.

For a good diversity of forest birds, head to Compiègne forest (40 km, 40 min west of Soissons on the N31, and 90 km, 1 h north of Paris). Exit the N31 towards Pierrefonds on the D130 and the D973 to park at [7]. From here, walk southwest following signs to 'Mares Saint-Louis'

[8]. The oldest oak stands in the area and is good for middle spotted woodpecker (year-round, most conspicuous in March to May), pied flycatcher and wood warbler (the latter two from May). Clearings in the western side of the forest (e.g. around [9]) host wryneck, grasshopper warbler and red-backed shrike.

From Compiègne, head south towards Creil to reach **a marsh** [10] **along the D75 between Cinqueux and Sacy-le-Grand.** Breeding birds and spring migrants include osprey, grasshopper, marsh, Savi's and sedge warblers, lesser whitethroat and bluethroat.

From Saint-Omer to the western Ardennes (sites 34–40, map 57)

HIGHLIGHTS
» High bird diversity on inland marshes
» A chance to see to rare forest species such as black stork and goshawk

J F M **A M J** J A S O N D

KEY SPECIES

YEAR-ROUND great white egret, goshawk, Eurasian bittern, little owl, black and middle spotted woodpeckers, **dipper**, Cetti's warbler, willow, marsh and crested tits, hawfinch, common crossbill

BREEDING honey buzzard, white and **black storks**, **little bittern**, night-heron, purple heron, hoopoe, nightjar, bluethroat, whinchat, lesser whitethroat, grasshopper, **Savi's**, sedge, reed, marsh, melodious and wood warblers, red-backed shrike

MIGRATION garganey, gadwall, pintail, osprey, little ringed plover, Mediterranean gull

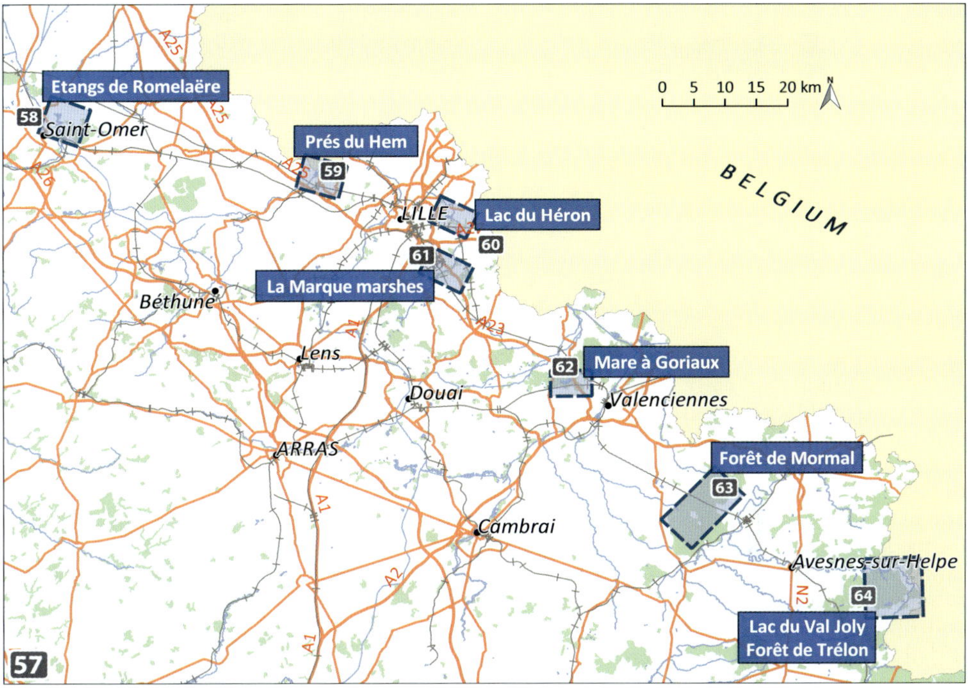

WINTER wigeon, goldeneye, goosander, woodcock, herring, great black-backed, lesser black-backed and **Caspian gulls**, **great grey shrike**, redpoll, brambling

VISIT DURATION 1 day

PLAN YOUR VISIT

This area is far less well known than the coastal part of the region. A few sites, however, are regularly monitored by local birdwatchers and have become regular birding haunts for those around Lille who do not wish to drive for one and a half hours to the coast. The area also offers a better chance of catching up with some forest species. The section is organised as a transect from Saint-Omer to the Ardennes. Most of the sites listed are inland wetlands or forest areas that can be covered in a couple of hours.

34 Romelaëre ponds (Étangs de Romelaëre)

0.5 DAY; MAP 58 The Romelaëre is a mosaic of protected swamps and ponds which hosts several breeding species that are hard to find elsewhere in the eastern part of the region (6 km, 12 min northeast of Saint-Omer on the D209 past Clairmarais, signposted car park after the camp site [1]). **Birding along the trails that cross the**

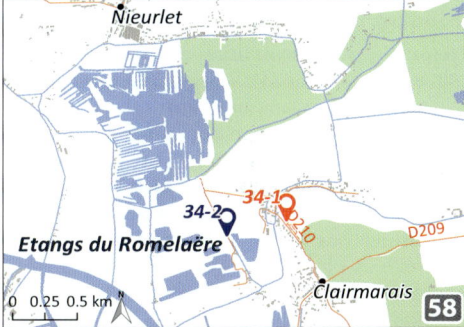

marshes [2] can yield an unusual diversity for the region, most wetland species being easier to see here than anywhere else. In spring your targets should be bittern, little bittern, night-heron, purple heron, white stork, bluethroat, common grasshoper, Savi's (rare), sedge, reed and marsh warblers. Commoner species such as hobby, turtle dove, Cetti's warbler and serin also breed here.

35 Prés du Hem

2–4 HOURS; MAP 59 The Prés du Hem is a lake with an outdoor centre located 25 km, 25 min west of Lille (exit the A25 in Armentières; follow signposts from the town centre to reach the car

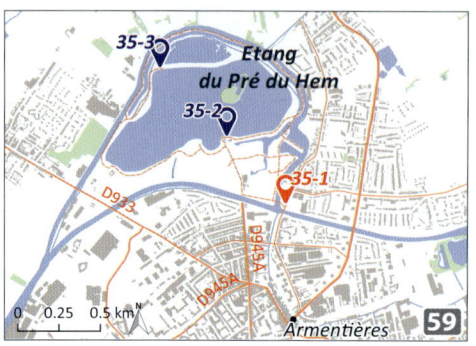

park [1]; closed from November to March, fee). **In late autumn and early March, Caspian, herring, great black-backed and lesser black-backed gulls roost at dusk on the main pond, best viewed from the southern bank** [2]. **A hide** [3] **gives closer views over the northern end of the ponds**, and there are two other hides on the same trail. Winter and early spring can bring smew, goosander, goldeneye, and there are records of great northern diver, Bewick's swan and long-tailed duck. More regular migrants include little gull, osprey and terns.

36 Lac du Héron
2 HOURS; MAP 60   The Héron pond (Lac du Héron) [1] is the most accessible birding spot in the vicinity of Lille (10 km east of Lille: drive to Villeneuve-d'Ascq and head north, bypassing the Musée d'Art Moderne on the Rue du 8 Mai 1945, and follow signposts to the 'Parc du Héron'). Although its location in a highly urbanised area may look uninviting, **it is a decent birding spot which has regularly hosted regional rarities** such as grey phalarope and Iceland gull. In winter, ducks and gulls roost here, sometimes in large numbers. Look for Mediterranean gulls among the flocks of black-headed gulls. The surrounding fields [2] host common snipe and water pipit in winter. Meadows [3] are good for herons and waders, as well as wheatear, whinchat and red-backed shrike (scarce), in both spring and autumn. Look for little owls in old willows and golden orioles in poplar groves; ring-necked parakeets have also recently become established. Ditches host bluethroat, marsh and reed warblers.

37 La Marque marshes
2 HOURS – 0.5 DAY; MAP 61   These **protected marshes** are located 17 km, 20 min south of Lille (exit the A1 motorway at Lille-Lesquin airport; bypass the airport to the east, cross the railway just north of Fretin and take the D54 towards Péronne-en-Mélantois and Templeuve to park beside the D19 just after Huvet [1]). Target species include great white egret, bluethroat, grasshopper and marsh warblers and golden oriole. Migrating common and green sandpipers also occur. **Several trails with hides depart from the car park. Early-morning birding in early May to mid-June will yield the best results.**

38 Mare à Goriaux
2–4 HOURS; MAP 62   The Mare à Goriaux is a forest pond located 6 km, 10 min west of Valenciennes (approaching from Lille, turn

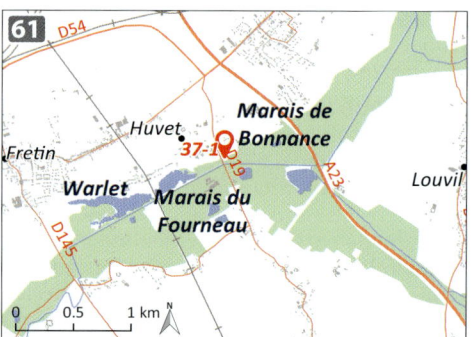

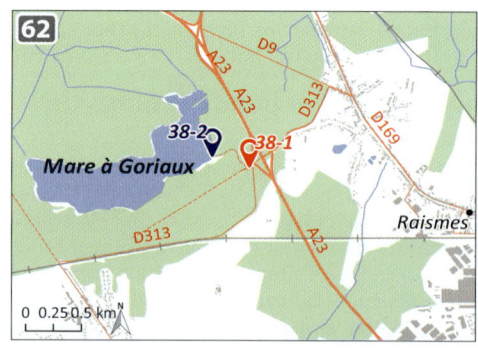

immediately right after the exit from the A23 in Raismes to find the car park [1]). Footpaths circle around the pond [2] through woods where common redpoll and woodcock can be found in winter with some effort, plus the usual breeding community featuring goshawk, black and middle spotted woodpeckers (rare), tree pipit and wood warbler. Winter brings great white egret and sometimes bittern, and garganey may occur from March.

39 Mormal forest
3 HOURS – 1 DAY; MAP 63   **The Mormal oak and beech forest** extends over 9000 ha, 80 km, 1 h southeast of Lille (exit the A23 in Valenciennes, bypass Le Quesnoy to reach Locquignol, in the middle of the forest) and 30 km, 30 min south-west of Maubeuge (from Bavais, take the D932 south). Park in one of several car parks [1,2,3] and **walk among pastures or forest stands from March to May depending on your target species**. Black, great spotted, middle spotted and lesser spotted woodpecker are easiest from mid-February to mid-April during their peak of drumming activity; playback is unnecessary. Tree pipit, common redstart and wood warbler are easy to find near the edges of old stands from April to June, while honey-buzzard, golden

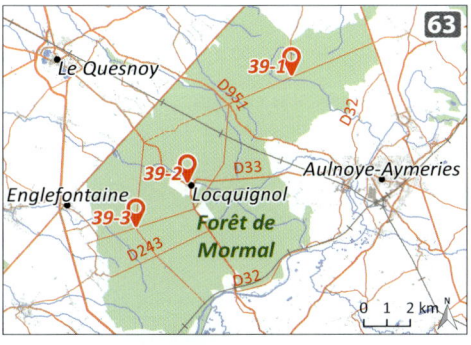

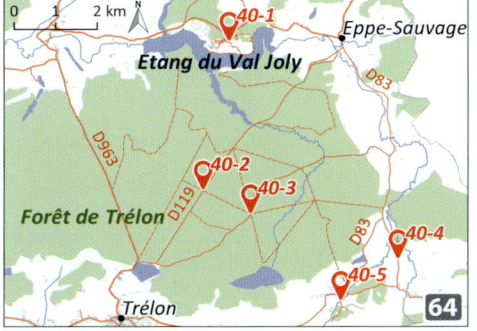

oriole, red-backed shrike and melodious warbler breed from early May. Spend an early summer evening near a clearing for woodcock, goshawk and nightjar. Your winter target should be great grey shrike (rare), which frequents **younger stands and pastures around the forest**. More common wintering species include hawfinch, common crossbill and brambling. At all seasons, look for willow (uncommon), marsh (common), coal and crested tits in mixed conifer stands.

40 Val-Joly ponds (Étang du Val-Joly) and Trélon forest
2 HOURS – 0.5 DAY; MAP 64   The artifical ponds forming the Étang du Val-Joly [1] are one of the stopover places for **migrating wetland species in the eastern part of the region** (28 km, 40 min southeast of Maubeuge). The wooded surroundings are worth some time as well, especially for great grey shrike and woodpeckers. Pintail, gadwall and wigeon are numerous on the ponds, together with small numbers of goosander and goldeneye. Caspian gull numbers can exceed those of herring gull. Black and middle spotted woodpeckers occur in the area year-round, although they are easiest when drumming in March and April. Osprey may stop over in the surroundings on both spring and autumn migration. Breeding species include little ringed plover, common tern (irregular), melodious warbler and golden oriole. Black stork and goshawk also breed in the surrounding woods, yet they are especially secretive.

South of Val-Joly lies the Trélon forest, among the largest in the region (10 km, 10 min from the Val-Joly, drive to Eppe-Sauvage and enter the forest by heading south on the D119). **Numerous public roads and tracks can be driven or walked across the forest**. Black and middle spotted woodpeckers, crested tit (conifer stands), marsh tit (deciduous stands) and hawfinch are common year-round. Great grey shrike can be found in open areas with some luck. **The Carrefour de Blois** (exit the D119 at [2] and park at [3]) is a good place to search for breeding species such as black stork (rare), honey-buzzard, woodcock, nightjar, hoopoe (rare), wood warbler and golden oriole. The bocage east of the forest around Baives and Wallers-en-Fagne (e.g. [4]) is also good for red-backed shrike, melodious warbler and lesser whitethroat. The river south of Wallers-en-Fagne [5] has dipper and grey wagtail.

REGION 3
NORMANDY

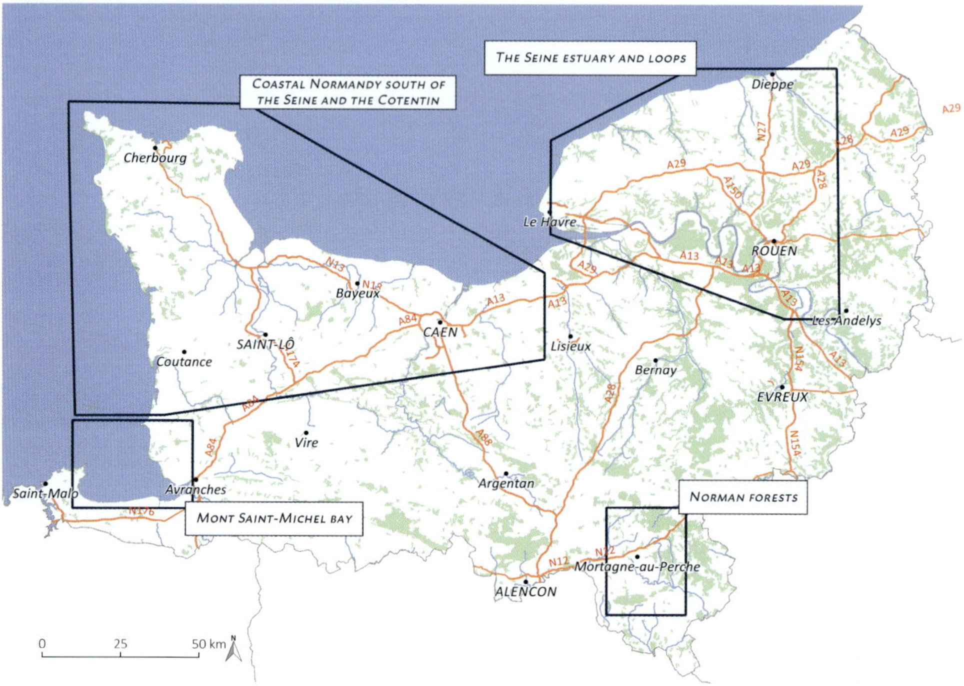

Colonies of fulmars and other seabirds are scattered along the Channel coast from Le Havre (site 41) to the Cotentin peninsula (site 51).

HIGHLIGHTS
» Numerous seabird and passerine migration hotspots along the ideally located Cotentin peninsula
» Some of France's largest reedbeds along the Seine estuary
» High diversity year-round, from pelagic seabird colonies on coastal cliffs to old-forest specialists, just two hours from Paris

REVIEWERS James Jean-Baptiste, Frédéric Malvaud, Sébastien Provost, Alain Verneau

USEFUL WEBSITES
www.faune-normandie.org/; www.migraction.net for migration in Carolles, Antifer and La Hague
http://gonm.org for monthly news
http://maisondelestuaire.org (Seine Estuary)
http://normandie.lpo.fr for supplementary sites and further details on those described here

NORMANDY | THE SEINE ESTUARY AND LOOPS | 57

The picturesque Vrasville marshes (site 50), at the northern end of the Cotentin peninsula, mark the first landing point for migrants from Britain.

The Seine estuary and loops (sites 41–46, map 65)

HIGHLIGHTS
» Northern France's largest reedbeds, surrounded by extensive areas of marsh and meadow
» A rich area featuring coastal seabirds and wetland specialists, as well as forest and meadow species

| J | F | M | A | M | J | J | A | S | O | N | D |

KEY SPECIES
YEAR-ROUND little and great white egrets, Eurasian bittern, goshawk, marsh harrier, peregrine, grey partridge, **spoonbill**, avocet, stock dove, little owl, middle spotted and black

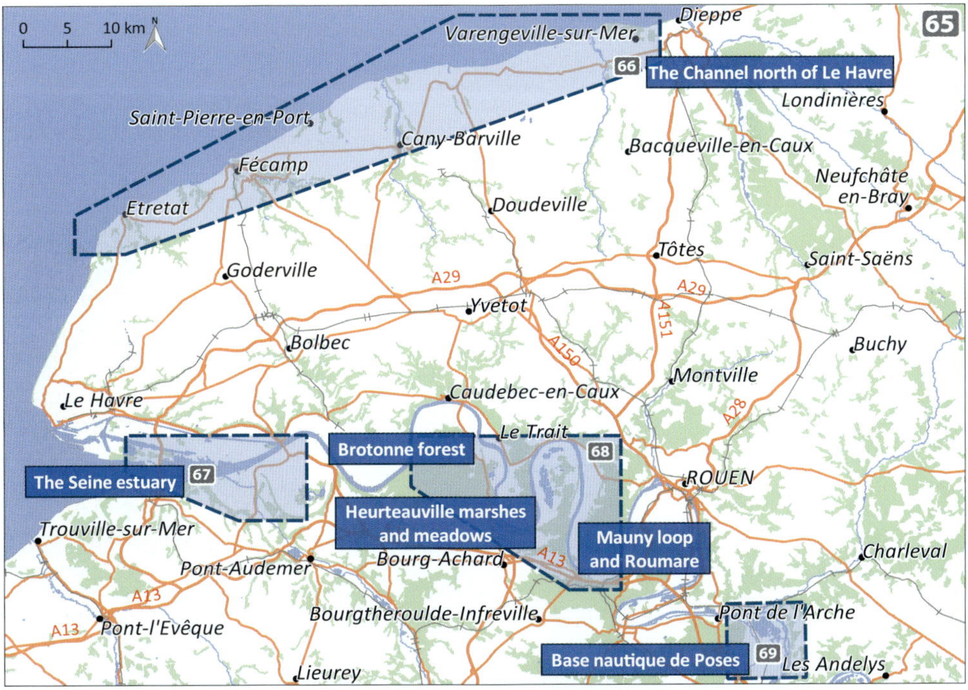

woodpeckers, fan-tailed and Cetti's warblers, crested, coal and willow tits, red-backed shrike, hawfinch

BREEDING fulmar, white stork, **little bittern**, honey buzzard, **corncrake**, stone-curlew, black-tailed godwit, Mediterranean gull, bluethroat, whinchat, lesser and common whitethroats, grasshopper, sedge, marsh, reed, melodious and wood warblers, bearded tit, red-backed shrike

MIGRATION pintail, wigeon, osprey, Balearic, sooty and Manx shearwaters, waders, skuas, auks

WINTER greylag, barnacle, white-fronted and **tundra bean geese**, goosander, goldeneye, smew, red-throated and great northern divers, merlin, **Caspian gull**

VISIT DURATION 1–2 days

PLAN YOUR VISIT

The Seine estuary has a European-level role for wildfowl migration, but it is not as accessible as its status deserves, owing to hunting and lack of suitable access roads. Plan your visit in advance according to the season. Sites 41 and 42 are organised along an itinerary which starts on the Channel coast, crosses the Seine at the Tancarville Bridge, runs through marshes and meadows and ends along the coast on the southern bank of the estuary. Marshes and reedbeds are at their best from April through to June; in winter, concentrate on coastal sites. A second itinerary covering sites 43–46 can make a perfect day trip from Paris or Le Havre to cover meadow and marsh species in late spring. In winter, concentrate on Poses (site 46) and surrounding ponds.

41 The Channel north of Le Havre

1 DAY; MAP 66 North of the Seine estuary, **coastal cliffs, beaches and small estuaries offer excellent opportunities for those looking for breeding seabird colonies and wetland species**. The Saâne estuary can serve as a good introduction to local species (100 km, 1 h 30 north of Le Havre, A29 to Yerville/Motteville, then D142; and 20 km, 30 min south of Dieppe, D75 or D925 to Longueil and Sainte-Marguerite-sur-Mer or Quiberville).

Although the area can be visited year-round, **it is mainly a spring and migration site**. Walk along the River Saâne on the 'GR212' footpath between the car park near the beach [1] and the D323 turn-off at Fond-de-Longueil [2]. Cetti's warbler can be heard along this path year-round. In spring, expect breeding red-backed shrike, sedge, marsh and reed warblers, serin, lesser and

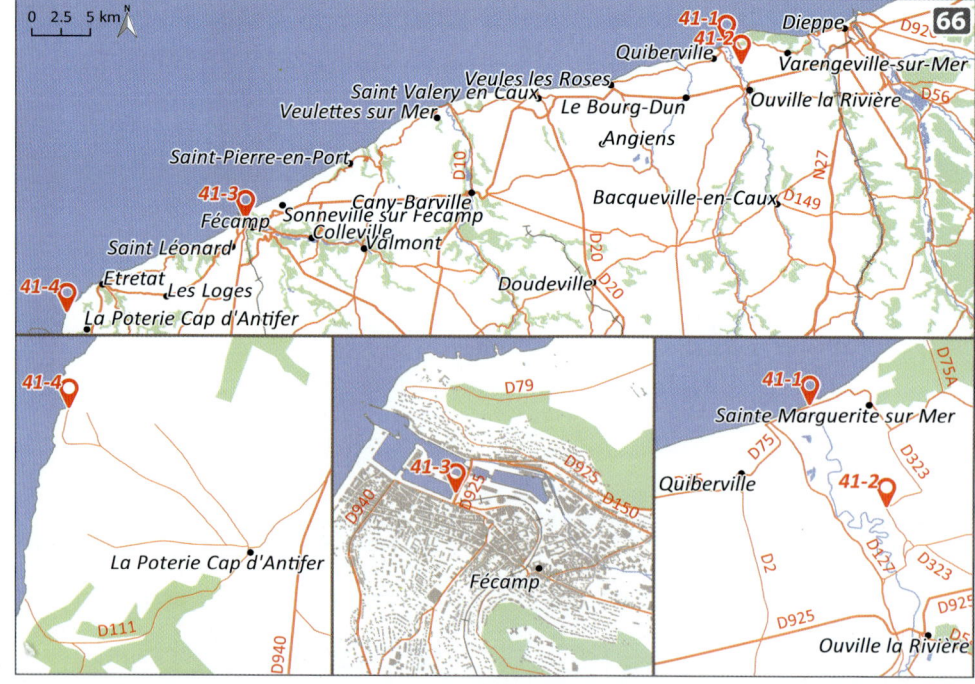

common whitethroats and grasshopper warbler. Among non-passerines, the most notable species are shelduck (one of the few breeding locations for this species in the region), grey partridge, hobby and peregrine. Winter and both migration periods bring small numbers of greenshank, snipe, common sandpiper and other waders. Watch the beach for fulmars, which breed on the nearby cliffs.

From there, drive south on local roads, stopping here and there along the coast to check the cliffs and the moorland. A stop at Fécamp harbour [3], for instance, can be rewarding for gulls (look for Iceland, glaucous and Caspian gulls in winter, all rare and irregular). **In autumn, try to keep a few hours for Antifer lighthouse** [4]. This is the northernmost of all Norman seawatching spots and one of the most renowned. From August to November, northwestern winds tie to the coast the usual suite of seabirds, among which divers, fulmar, Balearic shearwater, flocks of waders and wildfowl, skuas, gulls, terns and auks can be seen. Grey phalarope, Sabine's gull, Arctic tern and little auk pass through annually in very small numbers and will reward the most patient seawatchers.

42 The Seine estuary

0.5–1 DAY; MAP 67 The Seine estuary ranks among the most important wetland bird areas in northern France at all seasons and has a European-level importance for wetland birds. Besides an exceptional range of wetland species, it hosts several rare and declining breeding birds, including corncrake and black-tailed godwit. Unfortunately, access is difficult because of lack of roads and private areas. Heavy hunting from late summer to early spring also makes birding tricky, especially in the large Hode marshes [1], which are therefore better visited with a local birdwatcher (ask the local Ligue de Protection des Oiseaux; website at the start of this section).

The surroundings of the Pont de Normandie (20 km, 30 min from Le Havre on the A29) are the only easily accessible area. **Spending a few hours at dawn and dusk near the bridge should yield a good overview of local birds at all seasons**. A small car park and a path on the northern side of the bridge [2] allows a panoramic view over the surrounding marshes and wetlands. **Be there at first light, as bird activity declines quickly through the morning, and vehicle passage increases**.

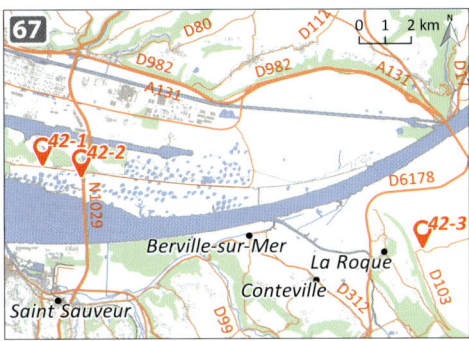

Likely species (season-dependent) include spoonbill, marsh harrier, avocet, bluethroat, Cetti's, sedge, marsh, reed and grasshopper warblers. Bearded tit also breeds but is trickier to find as it rarely emerges from the densest reedbeds. Bittern is a possibility but will remain an aspiration for most. There are also a few little owls in the area.

In winter, the site remains interesting for wildfowl (pintail, wigeon), waders, thrushes and sedentary reedbed species. **The Marais-Vernier is quieter and easier to visit** (30 km, 30 min inland on the A131, cross the Seine and leave the motorway immediately after the Tancarville Bridge). **Park somewhere along the D103 between Marais-Vernier and Quilleboeuf and check out both sides of the road along paths** [3]. A few red-backed shrikes breed here from May. The whole suite of open-field and meadow species occurs; depending on season, expect grey partridge, little owl, whinchat, fan-tailed, grasshopper, melodious, marsh and reed warblers. The area is not as good in winter but can still be worth a check for merlin, little owl (year-round) and flocks of lapwings.

43 Brotonne forest

0.5 DAY; MAP 68 **Upstream, the Seine forms a series of loops with well-preserved mosaics of forest and agriculture.** The Brotonne forest spreads over one of these loops, 60 km, 50 min from Le Havre and Rouen (from both cities, A131/A13 to Bourneville, then head to La-Mailleraye-sur-Seine). **Although the Perche forests (site 47), further south, are better, Brotonne offers an opportunity to catch most woodland species** (best March – early June). The forest is dominated by old beech stands here and there mixed with oak and pine, and the well-preserved understorey is dominated by hornbeam. From La Mailleraye-sur-Seine,

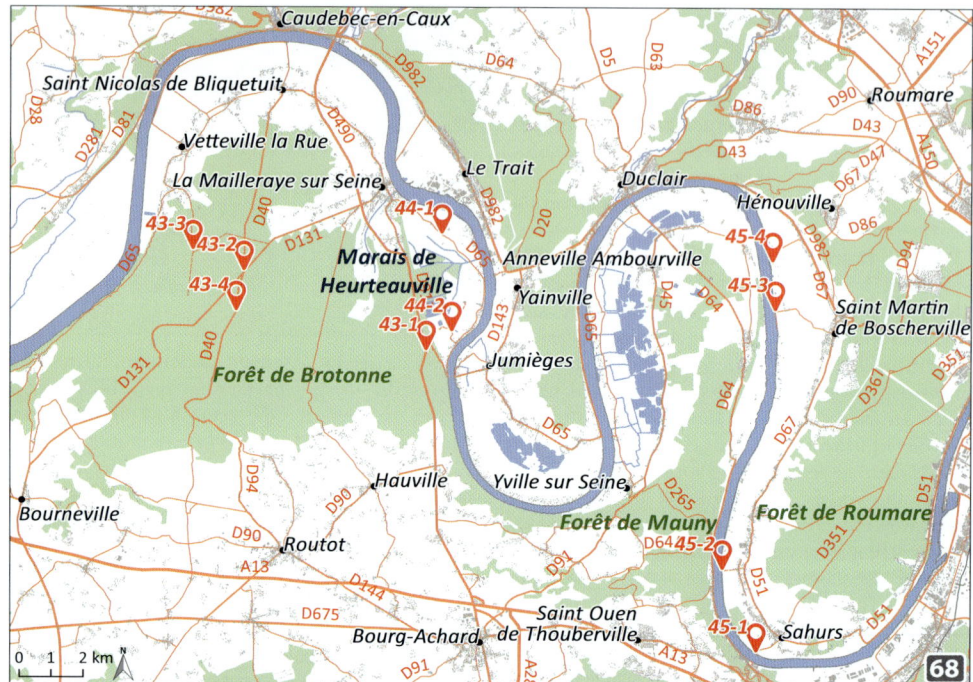

drive along the eastern edge on the D913 (car park at [1], just after the junction with the D143 which connects to site 44) or on the D131 and D40 between Bourneville and Routot (e.g. [2], clearings in [3,4] for edge and meadow species). Breeding birds include sedentary species (goshawk, stock dove, middle spotted and black woodpeckers, hawfinch, crested, coal and willow tit) and returning migrants such as turtle dove, honey-buzzard, tree pipit, wood and willow warblers.

44 Heurteauville marshes and meadows

2–4 HOURS; MAP 68 At the eastern edge of Brotonne forest, **the Heurteauville marshes and meadows host among the largest range of breeding** passerines **in the region in well-preserved habitats. This site offers easier birding than alternatives such as the Marais-Vernier and the Pont de Normandie.** From April to June, it can be a good idea to start birding here at dawn and shift to the forest when it gets brighter. **There are various places to park and walk along the D65 between La Mailleraye-sur-Seine and Heurteauville**, for instance just past La Douillère [1] or under the power-line by turning right on the rue Charretière just before the road joins the Seine bank [2]. White stork breeds and there are several heron colonies dominated by grey heron and little egret, sometimes joined by a few great white egrets on migration and in winter. The area is good for osprey (passage only), hobby, lapwing, little owl, stock dove (in the bogs), bluethroat, grasshopper, marsh and reed warblers. In winter, flocks of finches (including hawfinch) and tits from the Brotonne forest forage in the area, often joined by thrushes and stock doves.

45 Mauny loop and Roumare meadows

2–4 HOURS; MAP 68 On the northern bank of the Seine, **the Mauny loop and Roumare meadows are an alternative to Heurteauville marshes and Marais-Vernier** for those who wish to avoid the longish drive necessary to find a bridge over the Seine. They are also quicker to get to, being located only 15 km, 30 min from Rouen. **One good option is to park near the Seine south of Sahurs at the end of the D67** [1] **and walk along tracks for 3 km northwards, with the river on the left and meadows on the right** [2]. **An alternative is to stop-and-go by car in the vicinity of Saint-Martin-de-Boscherville** [3,4]. The bird community is quite similar to that of the south-bank sites, starring white stork, little owl and red-backed shrike (irregular) alongside a wide diversity of meadow and marsh species. Corncrakes may still stop over in these areas in April and May, but they have not bred for 20 years.

46 Poses

UP TO 0.5 DAY; MAP 69 **Poses is essentially a winter site where thousands of** ducks **stay from October to March, often joined by divers or sea ducks**. It ranks among Parisian birders' favourites, especially after frost or a storm, which can bring Arctic rarities, but it is also worth a visit during migrations for passing wildfowl and waders. The site, which operates as a watersports centre, consists of several large ponds that can be visited on a day trip from Paris (120 km, 1 h 30 via the A13) or Rouen (35 km, 35 min via the N338 and D321).

Several car parks and hides [1–7] offer good views, although birds are quite distant. Pochard, tufted duck and teal are among the most obvious species, together with good numbers of other ducks and a few dozen greylag geese sometimes joined by barnacle, white-fronted or the scarcer pink-footed and tundra bean geese (you might also see some more unusual duck species, which sometimes escape from a nearby bird farm). Black-headed,

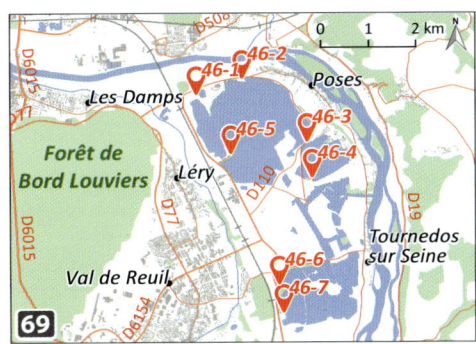

herring, yellow-legged and lesser black-backed gulls also come to roost in the evening, along with some Caspian gulls. Among rarer species, goosander, goldeneye, smew, red-throated and great northern divers are regular. Reed-beds towards the south of the site are good for wintering Eurasian bittern and breeding little bittern.

Poses is a stopover site for waders **from mid-August to October and in April–May.** Lapwing and golden plover can occur in large flocks, together with good numbers of shanks, godwits, dunlins and sandpipers. In spring, a colony of black-headed and Mediterranean gulls breeds on the ponds, and several good species breed in the surrounding woods and fields, including hobby and stone-curlew.

Norman forests (site 47, map 70)

HIGHLIGHTS
» Large extent of old oak forests with a rich bird community
» One of the few grey-headed woodpecker populations that survive in western France

KEY SPECIES
YEAR-ROUND woodcock, stock dove, **grey-headed**, black and middle spotted woodpeckers, crested and coal tits, Eurasian treecreeper, red crossbill, cirl bunting

BREEDING honey-buzzard, nightjar, grasshopper, melodious and wood warblers, red-backed shrike

VISIT DURATION 1 day

PLAN YOUR VISIT
All the target species except grey-headed woodpecker occur in all the forests, so choose one and cover it thoroughly. Start at first light; evening activity is never as good, and forests get desperately quiet by mid-morning. Woodpeckers will be easiest in March when they are drumming, while most other species are more conspicuous from mid-April to mid-June. If visiting in winter, find tit or finch flocks, which are often joined by woodpeckers and other species. Most tracks are closed to cars, so park your car in a strategic place and walk among the trees. Maps showing the composition of the stands are available at www.geoportail.gouv.fr, but be aware that they are regularly logged.

47 The Perche forests

1 DAY; MAP 71 **Several large oak-dominated forests can be found at the eastern edge of the region in the Perche bocage.** A number of species associated with agro-forest mosaics breed in the surroundings, including honey-buzzard,

47 THE PERCHE FORESTS | NORMANDY

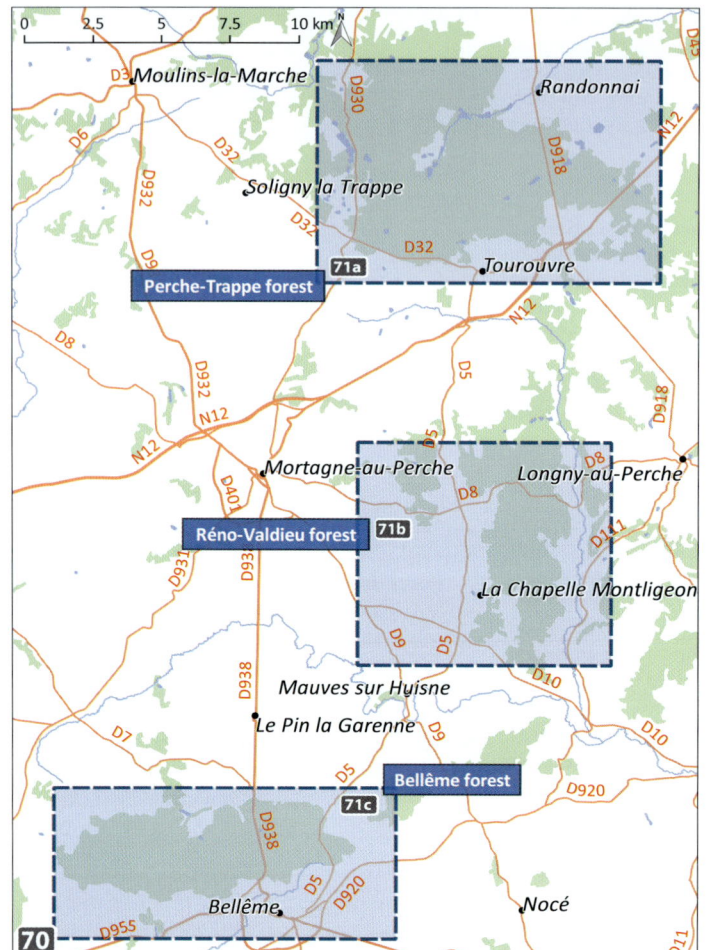

little owl, melodious warbler, red-backed shrike and cirl bunting: start in the forest at first light, then search the surrounding area later in the day.

The large Perche-Trappe forest is also the richest (75 km, 1 h 30 from Chartres; 135 km, 1 h 45 from Rouen on the A28, exit in L'Aigle). Park on the Étoile du Perche [1] (from L'Aigle take the D918 south to Randonnai and follow signposts) and walk along tracks and trails. Grey-headed woodpecker should be your main target. Expect it to be shy and tricky to find, but the reward is worth the effort and searching for it will lead you to most of the other specialities. Your best chances for the species are in old clear oak stands in March and April when birds are drumming. Another local rarity is Eurasian treecreeper, also found in old oak stands (beware: short-toed treecreeper also occurs). More common species include middle spotted and black woodpeckers, stock dove (year-round), common redstart (April), grasshopper warbler, wood warbler and golden oriole (May). Scattered pine stands host crested and coal tits and red crossbill (irregular). Flying woodcock and nightjar (from May) can be seen at dusk in scrubby areas and young stands. The smaller Bellême [2] and Réno-Valdieu [3] forests also host most of these species. The waypoints shown on the map are convenient car parks, but there is no specific spot; use the forest layers at www.geoportail.gouv.fr to make best use of your time.

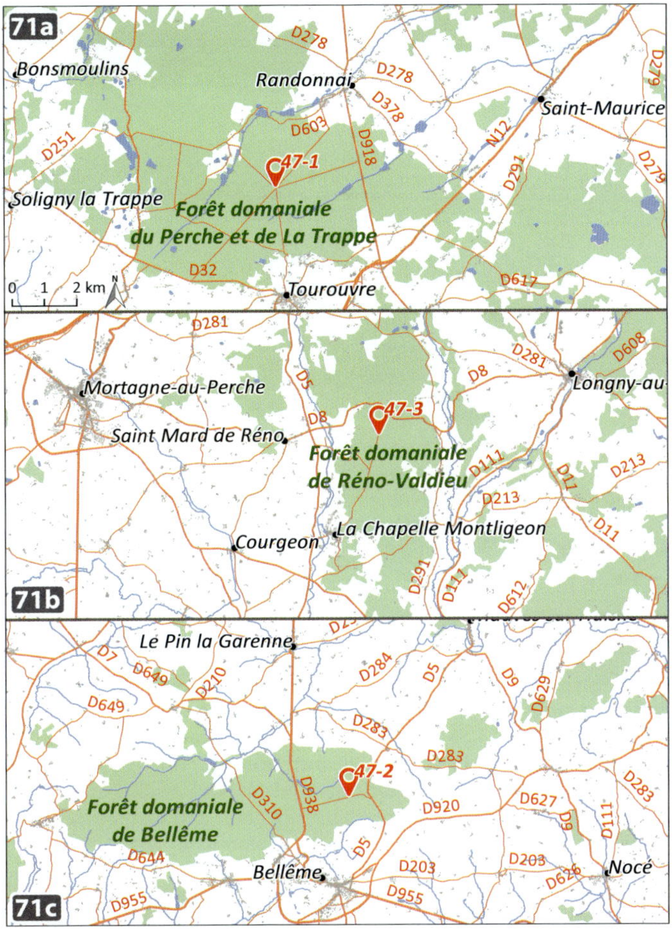

Coastal Normandy south of the Seine and the Cotentin (sites 48–52, map 72)

HIGHLIGHTS
» Gulls and waders on the D-Day beaches
» Seawatching and passerine migration on the main headlands
» Well-preserved wetlands and marshes for wildfowl and breeding passerines

| J | F | M | A | M | J | J | A | S | O | N | D |

KEY SPECIES

YEAR-ROUND fulmar, shag, grey partridge, marsh and hen harriers, cattle and great white egrets, avocet, woodcock, lesser and great black-backed gulls, long-eared, little and barn owls, black and middle spotted woodpeckers, fan-tailed warbler, Dartford and Cetti's warblers, crested tit, raven, cirl bunting

BREEDING honey-buzzard, white stork, quail, stone-curlew, stilt, Kentish and little ringed plovers, kittiwake, nightjar, whinchat, lesser whitethroat, marsh, sedge, grasshopper, melodious and wood warblers, red-backed shrike

MIGRATION Balearic and Manx shearwaters, storm petrel, Leach's petrel, osprey, waders including **dotterel**, black-tailed godwit and ruff, Sabine's gull (rare), all skuas, little and Arctic terns, auks, passerines, **Richard's pipit**, ring ouzel, **yellow-browed warbler**

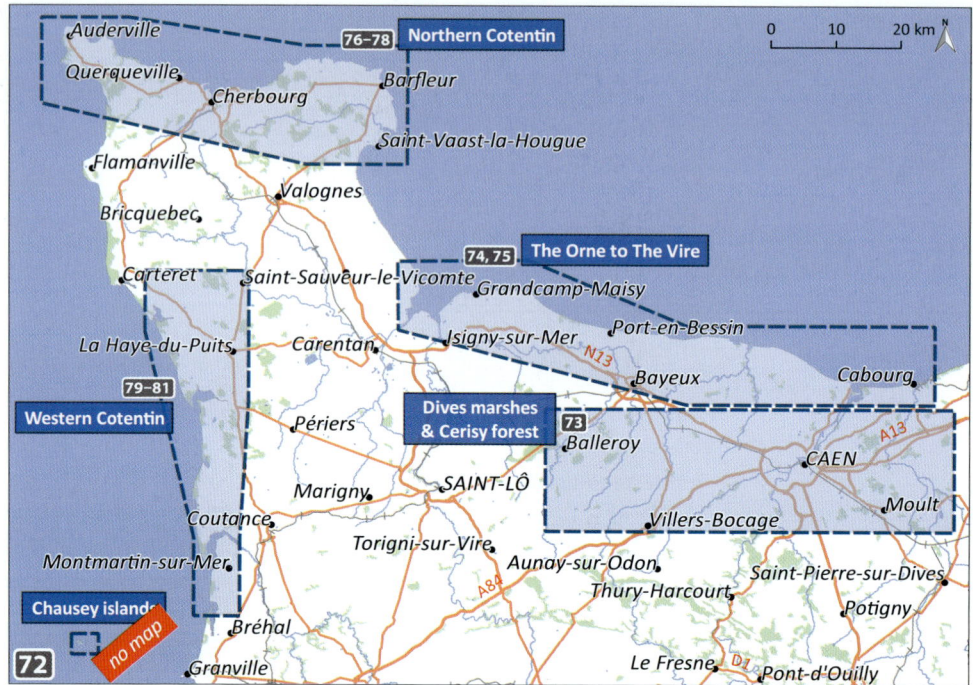

WINTER greylag, white-fronted, **tundra bean** and brent geese, velvet and common scoters, eider, red-breasted merganser, divers, black-necked, Slavonian and red-necked grebes, **spoonbill**, purple sandpiper, herring, yellow-legged, **Caspian**, Mediterranean and little gulls, short-eared owl, **shore lark**, Lapland and snow buntings

VISIT DURATION > 2 days

PLAN YOUR VISIT

This area covers a wide stretch of coast where the historic D-Day beaches are interspersed with marshy estuaries worth a visit year-round, although at their best from autumn to spring migrations. Choose your itinerary according to weather and tides. The area is rich and sites are large with long driving distances. Concentrate on a few of them. Northwestern winds from late August to November should lead you to seawatching spots, while sunny days with no wind in autumn, winter and early spring are ideal for birding on the beaches and marshes for sea ducks, passerines and waders.

48 From the Dives marshes to Cerisy forest

0.5 DAY; MAP 73   **If you are in the area in spring, the Dives marshes, 15 km, 15 min inland from Cabourg, is your best bet for meadow species. Stop-and-go on roads around Robehomme** [1]**, Brocottes** [2] **and Hotot** [3] **through pastures and meadows** which host white stork, hen harrier, quail, lapwing, little and barn owl, curlew, whinchat, lesser whitethroat, marsh warbler, red-backed shrike and yellowhammer. Check canals and ponds for waders, as well as breeding passerines such as marsh and sedge warblers. Poplar stands have honey-buzzard, hobby and golden oriole from May. **Two hours before high tide, head for the coastal sites. At low tide, visit the cultivated plains south of Caen** [4 – see GPS file] to search for hen harrier, grey partridge, stone-curlew and cirl bunting. **You can continue your trip by calling in on Cerisy forest** [5 – see GPS file] (15 km, 20 min southwest of Bayeux on the D572 to Saint-Lô; several car parks are located along the D13 to Cerisy). This old beech forest has woodcock (hard to find), honey-buzzard (from May), long-eared owl, black and middle spotted woodpeckers, tree pipit, wood warbler and crested tit.

49 From the Orne to the Vire

0.5–1 DAY; MAPS 74 & 75   **Dives-sur-Mer** [1] **and Ouistreham** [2] **are fishing harbours that can host good numbers of wintering gulls**, sometimes with yellow-legged and a few Caspian

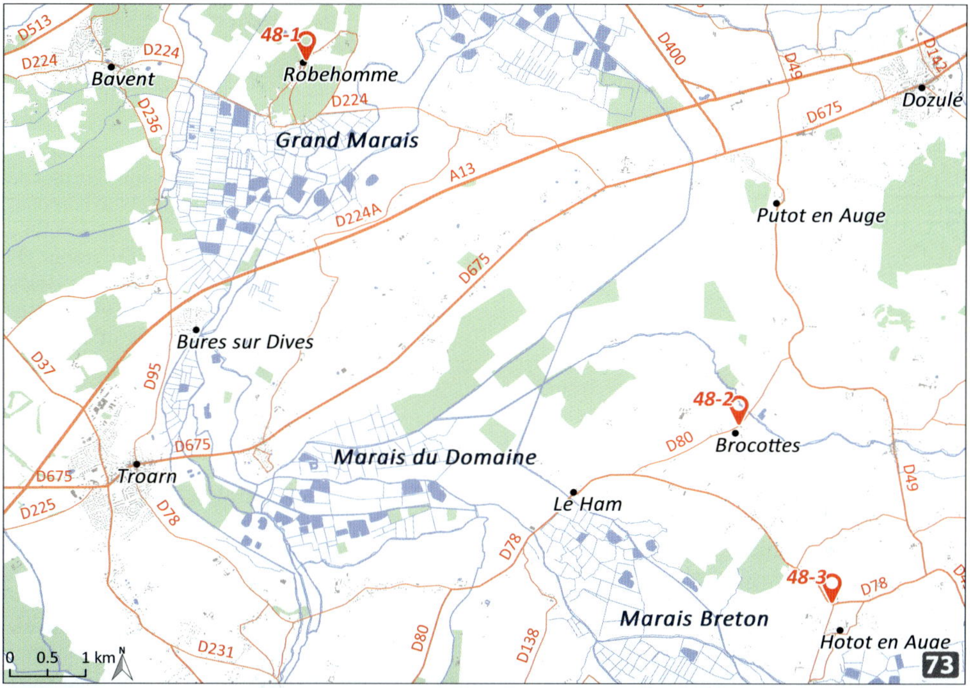

NORMANDY | **49 FROM THE ORNE TO THE VIRE** | 65

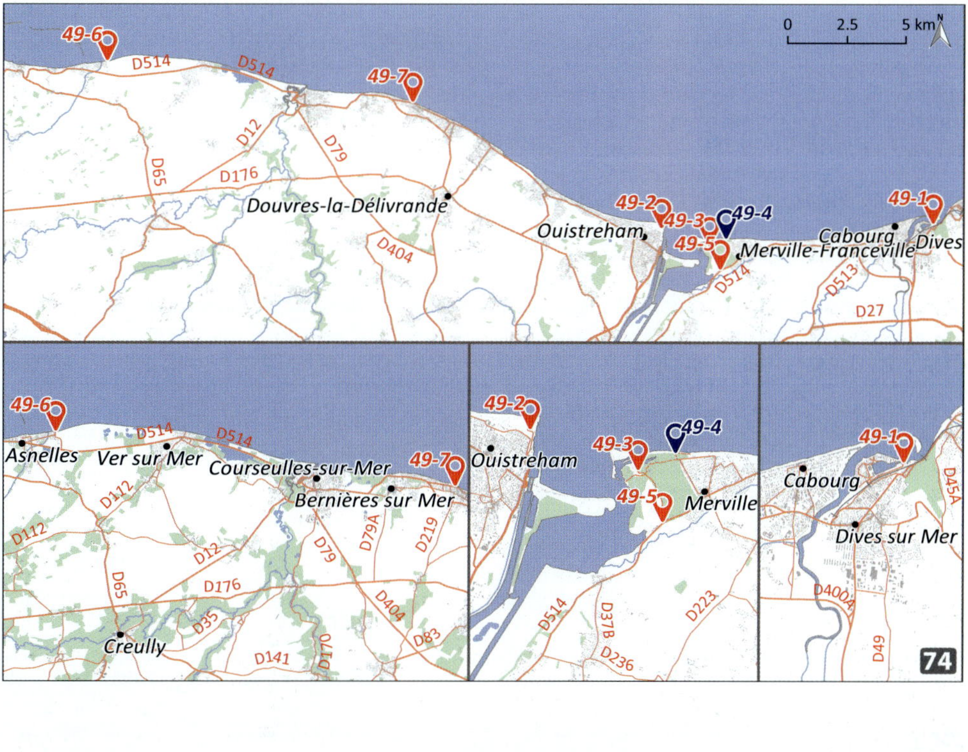

73

74

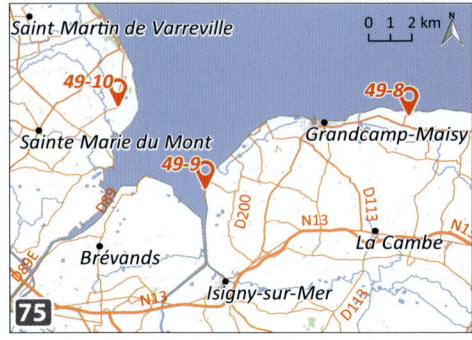

gulls from July to December (but beware of odd-looking Nordic herring gulls). Seawatchers congregate near the main rescue station on the beach 1 km west of the harbour; large flocks of velvet and common scoters can be spotted in winter from the eastern side of the main car park.

Merville-Franceville-Plage, on the eastern bank of the Orne estuary, is good year-round for waders, gulls and passerines at high tide (9 km, 15 min from Cabourg on the D514). Exit the village south on the D514, turn right at the signpost 'La Redoute', park after 1.1 km [3] and **walk on paths towards the nearby river, stopping at hides on your way**. Expect Kentish plover, yellow-legged gull among commoner herring, lesser and great black-backed gulls, turtle dove, marsh and melodious warblers in spring. Walk east for 1 km **for a view over the open sea** [4] **where** scoters, red-breasted mergansers, divers, skuas and auks **may rest in large numbers**.

Follow the coastline west of Ouistreham towards the D-Day beaches and stop-and-go along the D514. Stop in the Gros Banc reserve near the water treatment plan 700 m after exiting Franceville-Plage [5]. Cetti's and fan-tailed warblers are easy to find year-round here, and there are breeding avocet and stilt. The D-Day beaches are scenic but harder for birding, as wader and gull roosts tend not to allow close views; try to be there at high tide. Visit the marshes in Ver-sur-Mer for wildfowl and waders [6] (leave the D514 to the right when arriving in Asnelles from Ver, park near the sea and walk on the coastal path). **The Cap Romain reserve** [7] **and the Pointe du Hoc** [8] **are high points** from where you can scan the sea for auks (black guillemot has been recorded), divers, grebes (black-necked, Slavonian and the odd red-necked), red-breasted mergansers and scoters. The latter cliff hosts colonies of fulmars and kittiwakes. **In winter, the fields around the Pointe du Hoc** are a good place to search for Richard's pipit, Lapland and snow buntings among finches and roosts of long-eared and short-eared owls.

End your birding around the Vire estuary, where large numbers of wildfowl **roost in winter at high tide**. The estuary itself can be accessed by driving along the D514 between Grandcamp and Osmanville. One option is to turn right to Fontenay 350 m past Hameau-Penot, then drive west on a good track leading to the Pointe du Grouin [9]. **On the other side of the estuary, visit the Beauguillot reserve** (turn right on the D329 at the northern end of Vierville northwest of Carentan and park at a sharp corner after Pouppeville [10]). Greylag and brent geese can form large groups in marshes and meadows (best site in France for the latter), with smaller numbers of white-fronted and tundra bean geese (pink-footed is scarce). Also look for spoonbill, black-tailed godwit, ruff and other waders. Here again the seashore can be excellent for seabirds at close range.

50 Northern Cotentin

1 DAY; MAP 76–78 The Cotentin peninsula **deserves special attention from September to November for seawatching and migratory** passerines **and** waders. A visit in winter is still worth it for loads of divers, sea ducks and auks. In spring, concentrate on the western side, which hosts several good breeding species. **Start in Saint-Vaast-La-Hougue** (30 km, 30 min from Cherbourg on the D901 and D26, and 115 km, 1 h 30 from Caen on the N13). **Stop at Le Cul-du-Loup** [1] (park at the signpost 'sentier pédestre' and walk along the western edge of the camp site to reach the sea) **and the Tour de la Hougue** [2] **at high tide** to scan for grebes (mainly black-necked, with some Slavonian and red-necked), divers, red-breasted merganser and eider. Pelagics **can enter in the bay during strong northwesterly winds in autumn and winter.**

NORMANDY | **50 NORTHERN COTENTIN** | 67

At the northwestern end of the Cotentin, **Gatteville lighthouse** [3] **(15 km, 20 min from Saint-Vaast and 30 km, 30 min from Cherbourg) is a national-level seawatching hotspot** with good records of skuas, four tern species, shearwaters and auks (check www.migraction.net and www.trektellen.nl/maps/index/0/239). Northwesterly winds in autumn should lead you there without much hesitation, in the hope of good counts of common species and a chance of rarities such as Leach's petrel, long-tailed skua, Sabine's gull, Arctic tern and little auk. From late August to November, **find time to check the bushes around the lighthouse, south towards Barfleur harbour and along the coastal path to the marshes surrounding the Étang de Gattemare [4], as they can be crowded with passerines**. Rarity records include yellow-browed and Pallas's warblers, red-breasted flycatcher, Richard's pipit and little bunting.

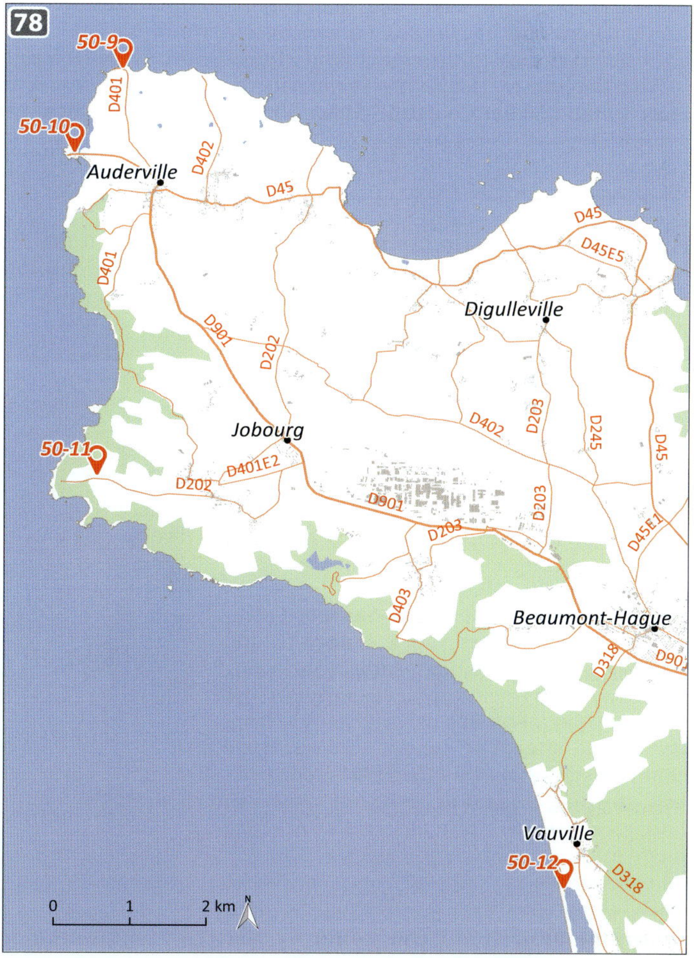

If you are short of time, take the D116, turn at the signpost 'Gouberville' and drive north until the road ends [5] to explore the surrounding dunes and bushes, or drive to the Pointe de Néville [6] (turn right towards 'La Mer' as you enter Néville-sur-Mer). You can also visit the marshes north of Vrasville (drive north past Réthoville church [7] or continue on the D116 to turn right on the D316 towards Les Mares past Vrasville [8]). **This area is the richest of all north Cotentin sites at all seasons.** In spring, expect migrants (ring ouzel, flava, flavissima and thunbergi yellow wagtail) and spoonbill, Kentish and breeding little ringed plovers, fan-tailed and Cetti's warblers. Cattle egret and great white egret occur in fields, where you should look for dotterels from mid-August to mid-September. Later in the autumn, yellow-browed warbler, Richard's pipit, shore lark, snow and Lapland buntings may show up. The coast is suitable for grebes, divers and auks throughout the winter.

The Cap de la Hague [9], the northernmost point of the Cotentin, offers similar seawatching conditions to those of Gatteville. Bushes, meadows and small stone walls between the semaphore and the Pointe de Goury [10] are especially good for migrating passerines. **From La Hague, drive south on the D901 to Jobourg and check out the cliffs of the Nez de Jobourg** [11], a small reserve with a few dozen shags, fulmars and great black-backed gulls. Stock dove, water pipit, Dartford and grasshopper warblers and raven are easy to find. Check the sea for razorbill and guillemot, which may occur in good numbers. **Further south, the Mare de Vauville reserve** [12] **(paths and hide) has marshes and dunes** that host great white egret, wintering wildfowl, waders (including breeding ringed and Kentish plovers), marsh harrier and wetland passerines.

51 Western Cotentin

1 DAY; MAPS 79–81 **Saint-Sauveur-Le-Vicomte forest** [1] **gives you a chance to add several forest breeding birds, including** middle spotted woodpecker **(year-round) and** nightjar **(May), without driving too far from the coastal sites. For coastal birding, drive south along the coast on the D901 from Portbail, stopping in several bays** [2–6] **which can** host good numbers of brent geese (including many pale-bellied brent), shelduck and waders (godwits, sanderling, Kentish plover in spring). **Look along the coastal stretches between these bays** for the two scoters, red-breasted

merganser, eider, divers and grebes, Lapland and snow buntings. Large gull roosts occur from September through to February, with regular Mediterranean and little gulls (late winter). **These areas can be surprisingly good under light southwesterly winds from mid-April to mid-May.** Osprey, dotterel and little tern occur occasionally, together with irregular bee-eaters (breeding in the area), red-throated pipits and more surprises to find.

NORMANDY | MONT SAINT-MICHEL BAY | 69

52 Chausey islands
1 DAY; NO MAP The small Chausey islands [1] attract attention from French birders as they have accumulated records of vagrants over the past few decades. There are still many rarities to add, owing to their location on the flyway of birds migrating south from Great Britain. A daily ferry service (www.vedettesjoliefrance.com) operates from Granville harbour, timetables vary with the tide. Be ready to board 30 min before departure. **Allocate a minimum of one full day to the island**, but there is accommodation for those who wish to stay for longer (book ahead).

Grande-Île is barely more than 1.2 km long and 500 m wide at high tide, so there is no specific spot. A scope is essential to check the surrounding islets, which are all protected. Red-breasted merganser can be seen year-round as it breeds on the island. Black-necked, Slavonian and great crested grebes, brent geese (with some pale-bellied brent) and great northern divers occur at sea in winter. Purple sandpiper may occur on the rocks, and there can be a few sanderling or plovers, but the main island is poor for waders, which tend to settle on surrounding islets. The island can be crowded with migrants (wryneck, wood warbler or golden oriole among high densities of wrens and blackcaps) during spring migration from mid-April to mid-May. Spring vagrants have included red-footed falcon, bee-eater, Iberian chiffchaff and trumpeter finch. The breeding season is quieter, but June to late August are ideal for storm petrel, Balearic shearwater and fulmar. Passerines turn up again from late August and migration peaks from September through to November with regular occurrences of ring ouzel, yellow-browed warbler, red-breasted flycatcher and scarcer species (red-flanked bluetail, Pallas's warbler, olive-backed pipit). Lapland bunting, crossbill and redpolls are often seen when the wind is northwesterly or northeasterly in winter, usually flying over.

Mont Saint-Michel bay (sites 53–54, map 82)

HIGHLIGHTS
» One of the major bird areas in northern France
» Huge flocks of waders and wildfowl in winter
» Wintering passerines in meadows and saltmarshes
» Passerine migration and seawatching from the main headlands

J	F	M	A	M	J	J	A	S	O	N	D

KEY SPECIES
YEAR-ROUND gannet, shag, marsh and hen harriers, peregrine, **spoonbill**, auks, long-eared owl, woodlark, Dartford and Cetti's warblers, cirl bunting

BREEDING Kentish and little ringed plovers, hoopoe, sand martin, bluethroat, lesser and common whitethroats, sedge, reed, grasshopper and melodious warblers, bearded tit, red-backed shrike

MIGRATION Balearic and Manx shearwaters, osprey, waders including snipe and jack snipe, **dotterel**, **wryneck**, passerines including **Richard's pipit**, **aquatic warbler (rare)**

WINTER whooper swan, brent, barnacle, greylag, white-fronted and tundra bean geese, other wildfowl, red-breasted merganser, common and velvet scoters, eider, red-throated and great northern divers, merlin, waders, short-eared owl, **shore lark**, Lapland and snow buntings

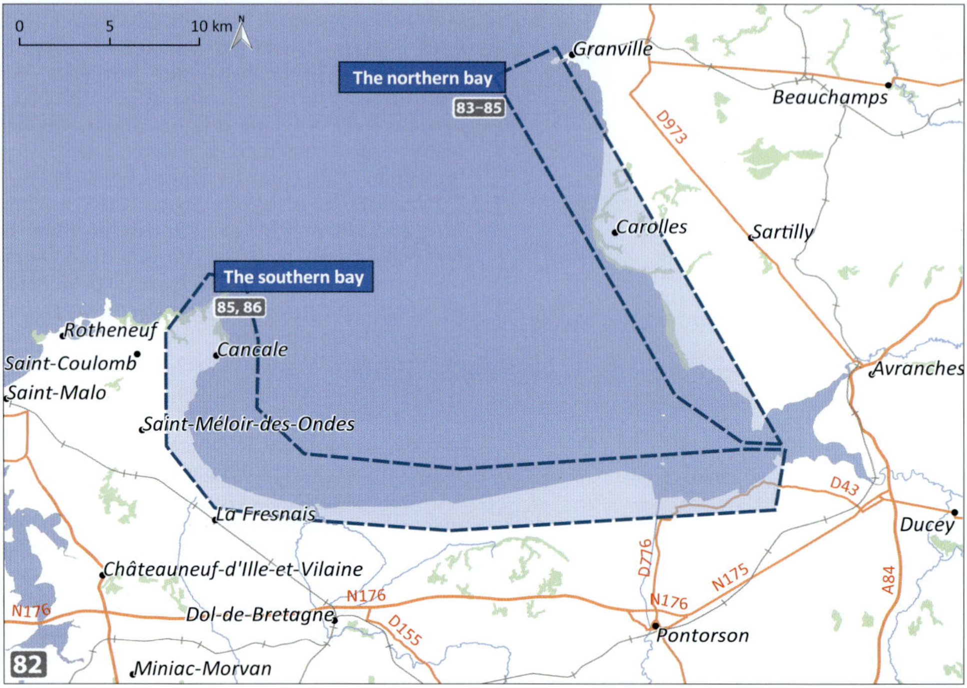

VISIT DURATION   > 2 days

PLAN YOUR VISIT

Mont Saint-Michel needs no introduction. Its extensive bay, inundated by the tide in record time twice a day, is **a European-level wintering and stopover site for** wildfowl, sea ducks **and** waders**, at its best from October to March, although a visit at any season will likely be a highlight of a birding trip to Normandy or Brittany.** Manage your visit carefully, as the bay is vast and your time is restricted by tides, **unless you want to spend your day at a migration spot.** The 2–3 hours before high tide offer the best sightings of waders and ducks in the bay itself. **Head to the seashore or the meadows at low tide, when the bay is empty.** Even then, there are always a few waders foraging along canals within saltmarshes and meadows. **Expect to spend significant time walking in meadows and across saltmarsh to approach birds.** We have chosen to split the bay into two parts, north and south of the Sélune river. The northern itinerary starts at the Carolles migration observatory and follows the coast, crossing the major roost sites. It also focuses on the best sites of the bay for the breeding season. **The southern part comprises a range of rather similar sites from Courtils to Cherrueix which are best at high tide**, plus the Pointe du Grouin, at the western end of the bay, which is also good at low tide.

53 North of Mont Saint-Michel

1–2 DAYS; MAPS 83–85   The northeastern side of the bay is marked by the Pointe du Roc (also called 'Cap Lihou') [1] and the Carolles cliffs, both good for seawatching and breeding water pipit. Carolles **is a major autumn migration observatory for passerines, with**

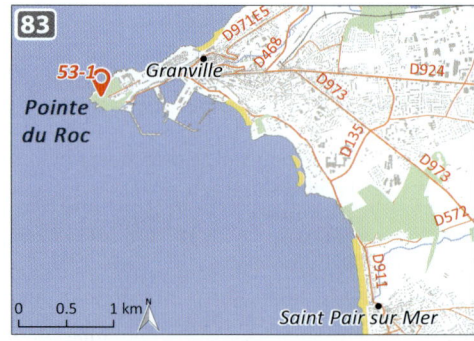

NORMANDY | **53 NORTH OF MONT SAINT-MICHEL** | 71

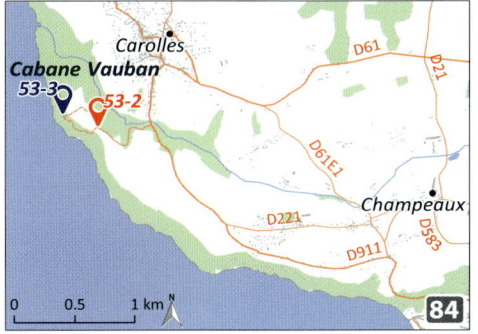

permanent monitoring from the end of August to November (from Granville, 15 km, 30 min on the D973 and D911; from Avranches, 22 km, 20 min via the D973 to Sartilly, then D61; check news by email: maisondeloiseaumigrateur@orange.fr or phone: 00 33 2 33 49 65 88). **The spotters position themselves a few hundred metres from the nearest car park** [2] **at the Cabane Vauban** [3] (signposted). Chaffinches form the majority of the yearly > 500,000 birds, with other finches, swallows, skylarks, woodlarks, pipits and the usual suite of autumn migrants. Irruptive species such as crossbills and coal tit have been counted in spectacular numbers, while scarce migrants, among them red-throated and Richard's pipits, show up annually. Seawatching is also good, with high numbers of red-throated and great northern divers, red-breasted merganser, common scoter from August to March, Balearic shearwater (Manx is scarcer), razorbill and guillemot. **The top of the cliffs is covered in a mosaic of moorland in which** lesser **and** common whitethroats **(from April),** Dartford **and** melodious warblers **(from May) and** cirl bunting **(year-round) occur.**

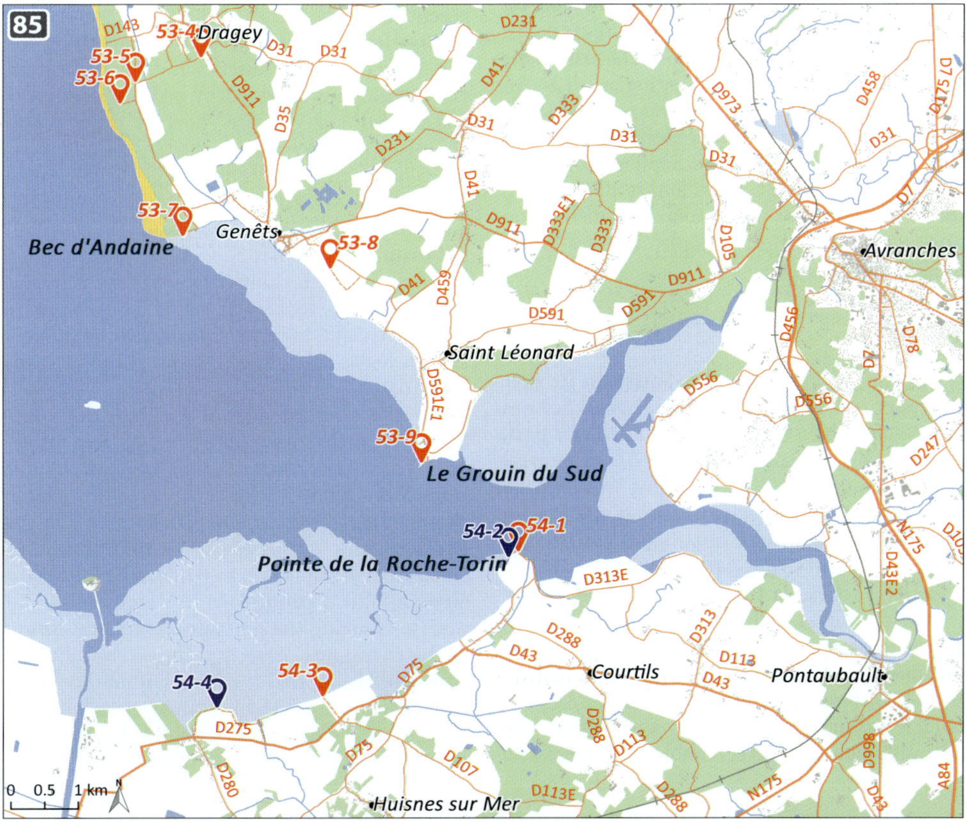

South of Carolles, visit the Claire-Douve marshes (8 km, 15 min from Carolles via the D911). Long-eared owls often perch in the conifer hedgerow bordering the camp site just past Dragey [4]. From here, turn right and drive among meadows [5], a prime roosting area for wildfowl, spoonbill and waders at high tide. In spring, the marshes host hobby, little ringed plover, sedge, reed and grasshopper warblers and cirl bunting; a few pairs of red-backed shrikes also breed along the hedgerows. **Visit the nearby dunes** [6] **as they may attract stopping-over migrants including** shore lark **(winter),** hoopoe **(April and May) and** wryneck **(spring and mid to late August).**

Drive 2 km further south to make a stop at the Bec d'Andaine [7], which marks the mouth of the Sélune river. This is a major roost site for wildfowl (including brent, barnacle, greylag, white-fronted and tundra bean geese and sometimes a few whooper swans), spoonbill, waders and terns at high tide. The wader flocks are dominated by ringed and grey plovers, knot, sanderling, dunlin and black-tailed godwit. Bar-tailed godwit occurs as well in smaller numbers. Breeding species include Kentish plover (from March), sand martin (from April), whitethroat (from April) and serin.

The same species also occur throughout the saltmarshes along the bay from Genêts [8] to the Groin Sud [9]; **walking along the coastal path can be rewarding at any season. In spring, the reedbeds near Genêts are home to key wetland species** including reed, sedge and Cetti's warblers, bluethroat and bearded tit. They are also an annual stopover place for aquatic warbler in May and September, but do not count on it. The Groin Sud itself is a good spot to search for migrating passerines and osprey in late August; it is also a brent goose roost in winter.

54 Southern Mont Saint-Michel Bay

1–2 DAYS; MAPS 85 & 86 **We insist on it: visit this part of the bay on a rising tide.** Species are roughly the same as in the northern part, although brent geese and wildfowl in general are less numerous. **A good point from which to watch** wader **and** duck **flocks is the Pointe de la Roche-Torin in Courtils** (10 km, 15 min from the Mont Saint-Michel; 20 km, 20 min from Avranches via the N175 and D43, follow signposts along small roads and tracks until they end in a car park [1]). The **surrounding meadows and saltmarshes** [2] host **common** and jack snipe, short-eared owl, Richard's pipit (which occurs on saltmarsh throughout the bay, although nowhere reliable), Lapland and

snow buntings from mid-October. Marsh and hen harrier, merlin and peregrine can easily be seen hunting over the area. Dotterel and shore lark are more or less annual, and rarities such as Blyth's and olive-backed pipits have shown up in recent years.

Similar viewpoints are spread all along the coastline between Courtils and the Couesnon river [3,4] (best accessed by walking along the coastal path; if you are short of time, driveable tracks head north to the bay all along the D275 between Courtils and La Caserne). Avoid Mont Saint-Michel itself. **Several viewpoints west of the Mont can be good alternatives, for instance the Chapelle Sainte Anne** [5 – see GPS file], **a regular wintering site for** Lapland bunting **and** jack snipe (drive to Saint-Broladre and turn right towards the coast immediately as you leave the village on the D797 towards Cherrueix).

At the western end of the bay the coast becomes rocky, and the tide has less influence when you reach Cancale (47 km, 45 min from Mont Saint-Michel). This area hosts large roosts of oystercatcher and turnstone (e.g. at the Pointe du Hoc [6 – see GPS file] or the Pointe de la Chaîne [7 – see GPS file]). **In winter, the Pointe du Grouin** [8] **is a good place to watch** ducks **and** waders **entering the bay during the first hours of the rising tide.** Also expect good numbers of divers, gannets, shags, eiders, common scoters (velvet scoter has become rarer), razorbills and guillemots. Manx and Balearic shearwaters can also be seen, especially in September.

REGION 4
BRITTANY

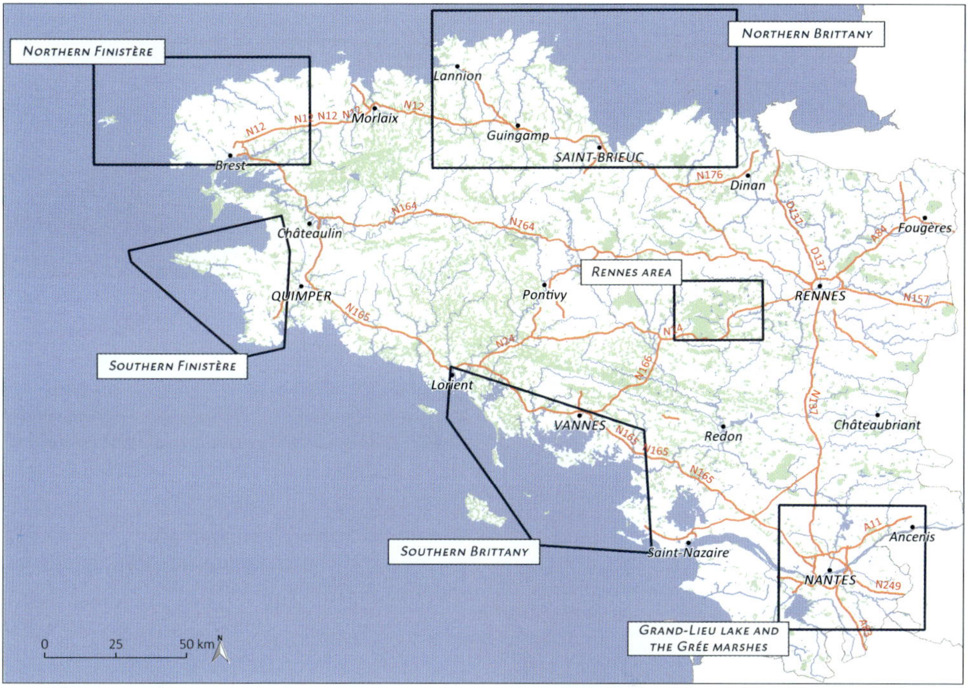

HIGHLIGHTS
- » A leading region for autumn migration of passerines, waterbirds and seabirds and a classic destination for birders in autumn and winter
- » Fascinating birding experience along a scenic rocky coast with wide sandy beaches, coves and bays and extensive reedbeds along river estuaries
- » Charismatic breeding seabirds, including gannet, roseate tern and puffin

REVIEWERS Mikaël Champion, Sébastien Reeber

USEFUL WEBSITES
www.faune-bretagne.org
www.faune-loire-atlantique.org
http://ano-ouessant.com
www.reserve-cap-sizun.org
www.snpn.com/rubrique.php3?id_rubrique=12
http://maisondulacdegrandlieu.com
www.sene.com/reserve-naturelle/reserve-naturelle-sene.php

As a French breeding species, puffins are confined mainly to the Sept-Îles (site 58), but individuals may be spotted anywhere at sea in autumn.

The wild and scenic Ushant island (site 63) remains the hottest French birding site in autumn, having hosted an incredible list of Siberian and Nearctic vagrants.

Northern Brittany (sites 55–58, map 87)

HIGHLIGHTS
» High seabird diversity during the breeding season
» The only roseate tern and puffin colonies of the French coast
» Leading coastal sites for migrating and wintering shorebirds, wildfowl and gulls

J F M A M J J A S O N D

KEY SPECIES
YEAR-ROUND shag, rock pipit, Dartford warbler
BREEDING gannet, fulmar, Manx shearwater, storm petrel, kittiwake, **roseate tern**, guillemot, razorbill, **puffin**, nightjar

MIGRATION wildfowl, Balearic shearwater and other pelagics, waders including knot, skuas, Arctic tern (uncommon)

WINTER brent goose, wigeon, pintail, common scoter, eider, red-throated, great northern and black-throated divers, black-necked grebe, short-eared owl

VISIT DURATION 2–3 days

PLAN YOUR VISIT
In May, the two must-see species of northern Brittany are roseate tern and puffin. This is the only area where they can be found breeding, apart from a few pairs of each in the westernmost part of the region. A visit to France's only gannet colony in the Sept-Îles is another must from April to June: book your trip ahead of time. From autumn to spring, Saint-Brieuc bay will be everythng you hoped for, with huge numbers of sea ducks, waders and other seabirds. There are various other sites interspersed among those described here: find a map and explore any seemingly suitable place. **Check tide tables first, then prepare an itinerary that will get you to a wader roost at high tide.**

55 Saint-Jacut-de-la-Mer

0.5 DAY; NO DETAILED MAP Saint-Jacut-de-la-Mer (24 km from Dinan on the D2) hosts **one of the two French** roseate tern **colonies** (from May to July). Park north of the village at the Pointe du Chevet [1] and scan the sea. You will probably not see the birds on their nests, as they are mixed in with larger numbers of breeding common and Sandwich terns on the distant Île de la Colombière, but you will probably catch them with your scope fishing or roosting on small rocky islands.

56 Cap Fréhel

0.5 DAY; NO DETAILED MAP The scenic cliffs of Cap Fréhel project north into the Channel 43 km west of Saint-Malo (D168 through Ploubalay, then D786 to Erquy and Fréhel; 50 km east of Saint-Brieuc on the N12 and D786 towards Erquy and Fréhel, signposted from both directions; you can either pay the fee at the car park near the cape or park 500 m before it, at the side of the road [1]). The landscape justifies a stop in itself, but you won't waste your time birdwise, as **the cliffs host a decently large** seabird **colony. Walk on the coastal path along the eastern side of the cape** and look for breeding fulmars, guillemots, razorbills, shags and kittiwakes (best from February to early July). Have a look offshore for passing Balearic or Manx shearwaters, gannets,

skuas and gulls. Rock pipit, Dartford warbler and stonechat are common year-round, and a few nightjars breed on the surrounding moorland.

57 Saint-Brieuc bay

0.5–1 DAY; MAP 88 **The Baie de Saint-Brieuc is a leading wintering and stopover area for** waterbirds **migrating through the Channel** (up to 40,000 birds). An extensive reserve covers most of the bay from Saint-Brieuc harbour to Yffiniac, 5 km inland, and 6 km west to the mouth of the Gouessant river. Most of the bay is exposed at low tide. **The vast majority of birds spread over large expanses of tidal flats, then gather into high-tide roosts along the coast on saltmarshes, mudflats and rocky islets.**

The reserve headquarters are located near Hillion at l'Étoile [1] (16 km, 20 min east of Saint-Brieuc, bypass Yffiniac on the N12 and exit north on the D80). **Walk on the coastal path south to Pointe d'Illemont hide** [2] or north towards the Grève de Saint-Guimont beach, which you can also reach by driving to the Pointe des Palus [3]. The most abundant species will likely be brent goose, wigeon, teal, pintail, shelduck, oystercatcher, ringed, grey and gloden plovers, knot, dunlin, curlew and godwits. Short-eared owls sometimes roost at the end of the bay between Pisse-Oison and les Grèves [4].

The Pointe des Guettes [5], 500 m north of Lermot, marks the eastern edge of the bay. **It is worth spending time here in autumn and winter, scanning the sea in search of** pelagic birds: common scoter, eider, red-throated, great northern and black-throated divers, great

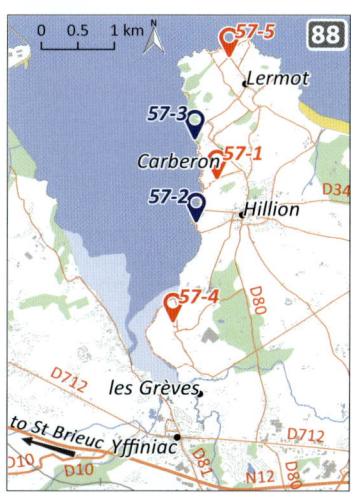

crested and black-necked grebes, up to several thousand Balearic shearwaters in late summer, gannet, skuas, Sandwich, common and Arctic (rare) terns, guillemot, razorbill.

58 The Sept-Îles

0.5–1 DAY; MAP 89 The Sept-Îles nature reserve [1], **5 km off Perros-Guirec, consists of five islands which host the largest French seabird colonies**. Seabirds have been protected here since 1912, making it France's oldest wildlife reserve. Apart from the conspicuous gannet colony (the only one in France), the islands are home to some of the last French puffins, which were decimated by hunting and oil spills during the twentieth century. Other breeding seabirds include Manx shearwater, shag, storm petrel, guillemot and razorbill.

Boats leave daily (April to September) from Perros-Guirec [2] (book ahead of time, especially in summer or at the weekend: www.armor-navigation.com, 00 33 2 96 91 10 00). **Landing is prohibited on the islands, so the boats sail**

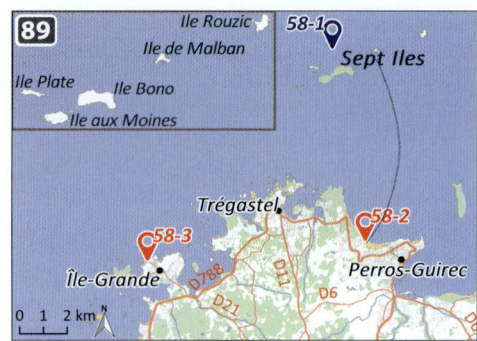

around them, which is perfectly adequate for good sightings and photos of most species. You can also visit the ornithological station at the Île Grande [3] (15 km west of Perros-Guirec, take the D788 along the coast towards Trébeurden, then follow signposts on the D21). **If you want to try for** roseate tern, make your way 70 km, 1 h 20 to Morlaix bay and scan the terns that are fishing or roosting at high tide from the shore, for instance at Carantec [4 – see GPS file].

Northern Finistère (sites 59–64, map 90)

HIGHLIGHTS
» Scenic bays and beaches with large numbers of migrating and wintering waders, gulls and wildfowl
» Hotspots for pelagic seabird migration
» Visit Ushant island in search of vagrant Siberian passerines and Nearctic shorebirds from August to November

KEY SPECIES
YEAR-ROUND shag, gannet, little grebe, marsh harrier, peregrine, **spoonbill**, rock pipit, fan-tailed, Cetti's and Dartford warblers, firecrest, **chough**, bullfinch
BREEDING Kentish plover, sedge and grasshopper warblers
MIGRATION Manx, Balearic, sooty, Cory's and **great shearwaters**, fulmar, storm petrel, **Leach's petrel, dotterel**, grey phalarope, great, **long-tailed**, arctic and pomarine skuas, kittiwake, **Sabine's gull** (rare), **Arctic tern** (uncommon), all the auks, whinchat, all the pipits including **Richard's**, warblers, pied flycatcher
WINTER whooper swan (rare), brent and **barnacle** (rare) geese, wildfowl including pintail, red-breasted merganser, great northern and red-throated divers, black-necked and Slavonian grebes, Eurasian bittern, waders including knot and **purple sandpiper**, Mediterranean gull, snow and Lapland buntings
VAGRANTS nearly yearly records of American wigeon, green-winged teal, lesser scaup, American golden plover, Iceland and glaucous gulls, yellow-browed warbler, red-breasted flycatcher, rosy starling, little bunting
VISIT DURATION 3–4 days
PLAN YOUR VISIT
France's westernmost stretch of coast is also arguably the most scenic, with incomparable rocky shores, moorlands and a multitude of bays and coves. For most birders, a trip to northern Finistère is as much an exciting experience as the Camargue or the Somme bay. **The area will yield top-level bird lists at any time of year, but it is at its best from late August to late November and remains surprisingly birdy throughout the winter.** We propose an east–west itinerary which starts in the scenic Goulven bay (be there two hours before high tide), most famous for its huge wildfowl and wader roosts, and ends in the coves of a battered rocky coast where you can search for divers, grebes, waders and seabirds as well as wintering Lapland and snow buntings. In

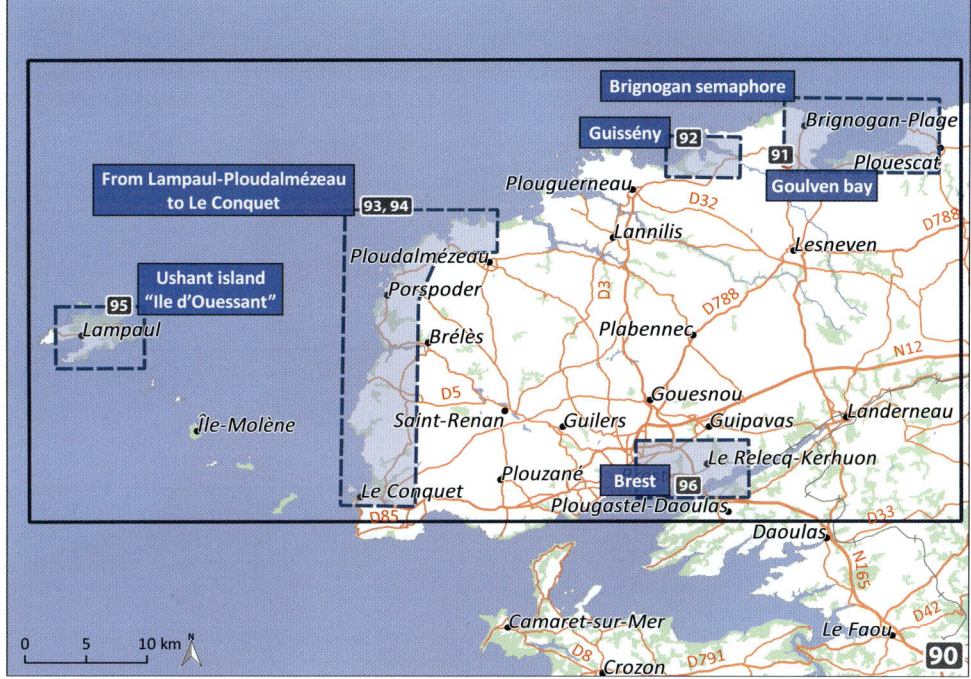

autumn westerly and northwesterly winds, try some seawatching at Brignogan. In October, you should try to plan a couple of days on the famous Ushant (Ouessant) island in search of Nearctic and Siberian vagrants.

59 Goulven bay

0.5–1 DAY; MAP 91 Goulven bay bird reserve lies 40 km north of Brest (bypass Lesneven on the D788 or N12, then head for Goulven). **The bay hosts impressive roosts of migrating and wintering** waterbirds **in a scenic landscape typical of western Brittany and is a must for anyone visiting the area.** Start with a stop in Kernic cove [1] for a first glance at waders, brent geese and gulls (west of Plouescat, park near the conspicuous casino) or in the Keremma dunes [2], where snow and Lapland bunting might be found wintering among the finches.

From the Goulven roundabout, turn towards Plouescat on the D10, then left after 370 m to park near the dyke which closes the bay [3]. Walk northwards on a small path leading to **a bridge where** waders **and** ducks **concentrate on a rising tide.** From here, expect sometimes large roosts of curlew, whimbrel, grey and golden plovers, dunlin, knot, sanderling, curlew sandpiper, snipe, godwits, redshank, spotted redshank,

greenshank, ringed and a few Kentish plovers, brent goose, shelduck, wigeon, teal and pintail. Singles or small flocks of spoonbill are sometimes seen, as well as little and black-necked grebes, barnacle goose and whooper swan (rare). It is a good place to find the occasional peregrine or merlin hunting over tidal marshes. With luck, bitterns can be seen flying over the reedbed on the eastern side. **The bridge area has repeatedly hosted vagrant** American wigeon, green-winged teal **and** American golden plover.

You may then resume your walk towards the dunes [4], where wheatears are abundant in March and September. A few Lapland and snow buntings sometimes winter from October to December in this area. **An arguably better viewpoint for roosting** waders **lies on the opposite side of the bay** (3 km north of Goulven, turn right 1 km before Plonéour-Trez in Trégueiller, park at the very end of the road [5]; **be there an hour and a half before high tide**). The spectacle of waders flying over the bay as the water rises is exceptional, especially if high tide occurs in the evening. Distances are suitable for great photographic shots with the scenic bay on the background. An equally good viewpoint is at Le Reun [6], at the end of 'rue de Pelleuz' in Plounéour-Trez.

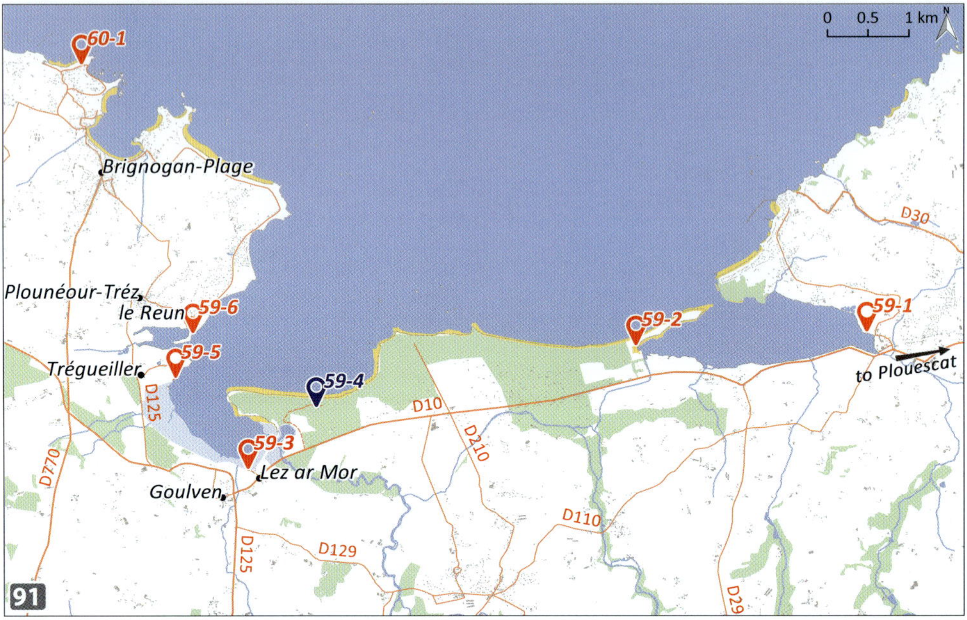

60 Brignogan semaphore

0.5–1 DAY; MAP 91 **Sighting conditions and large** seabird **counts make Brignogan semaphore** [1] **the most popular seawatching site in northern Finistère.** Once in Brignogan-Plage (6 km, 10 min north of Goulven, site 59) follow signs to 'Plage des Chardons Bleus' (2 km from the village centre) and settle at the foot of the semaphore. **Top migration days occur in northwesterly winds between mid-August and mid-November.** Expect good numbers of Manx, Balearic and sooty shearwaters (Cory's and great are scarce), great, arctic and pomarine skuas (long-tailed skua is possible but rare), common, Arctic and Sandwich terns, grey phalarope (uncommon), gannet, Sabine's gull (scarce), storm and Leach's petrels, red-breasted merganser, kittiwake, razorbill, guillemot and puffin. Divers also pass through in good numbers from late October.

61 Guisseny

0.5–1 DAY; MAP 92 The Guissény area features **several coves, ponds and beaches which can provide good views of** waders **and have regularly hosted Nearctic vagrants** (12 km, 20 min west of Goulven on the D10 through Kerlouan). Turn right past the church and park on the seashore along the 'route de la Croix' near an old cross [1]. **From here, you can either scan Tresseny bay or even walk on the coastal path or the beach at low or rising tide.** Expect good numbers of gulls, all the common waders, shelduck, wigeon (look for American wigeon), brent and barnacle geese (rare).

Drive 2 km further west to scan Le Curnic pond [2]. The pond itself is excellent for diving ducks (pochard, tufted duck, several records of lesser scaup), little grebe, marsh harrier and whooper swan (winter, rare). Breeding species include common tern, sedge, Cetti's and fan-tailed warblers. **A small mudflat often emerges at the western end** [3]. It is unconspicuous, but it has hosted Nearctic vagrants in the past, and there are always a couple of waders foraging on it. **On the other side of the dyke, Porz Olier bay is also worth a check.**

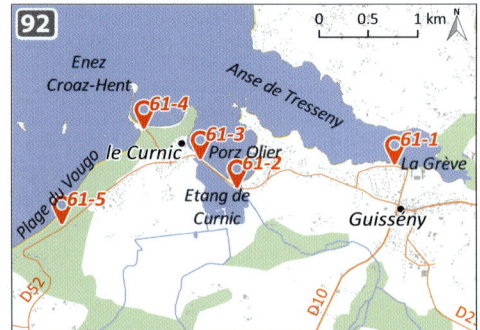

Le Curnic cove [4], 650 m further west past the hamlet, regularly hosts snow bunting and purple sandpiper. **Towards the southwest, walk along the long Vougo beach and scan the bay** for wintering black-necked (common) and Slavonian (rare) grebes, great northern (common) and red-throated (rare) divers. Wheatears and large numbers of rock pipits usually forage on tide wracks. **Park at the western end of the beach** [5] **for another viewpoint and to check the surrounding bushes** which may host good numbers of chiffchaffs, willow warblers, pied flycatchers and the odd yellow-browed warbler in October.

62 From Lampaul-Ploudalmézeau to Le Conquet

0.5–1 DAY; MAPS 93 & 94 The sandy beaches of Lampaul-Ploudalmézeau **are a hotspot which deserve a thorough visit for migrating/ wintering** shorebirds **and** gulls. You will find endless expanses of clean sand along bays protected from the open sea by reefs which offer ideal shelters for both birds and birders against northwesterly storms, frequent in winter. Drive to Ploudalmézeau (27 km, 30 min from Guisseny, site 61, through Plouguerneau and Lannilis or 25 km, 30 min from Brest through Saint-Renan).

First stop at the Plage des Trois Moutons beach, 4 km north of Ploudalmézeau past Lampaul-Ploudalmézeau (park near the camping resort [1]). Spend time scanning the bay for brent goose, divers, auks and grebes. Waders gather on the seashore, especially near the two river mouths. Expect large roosts of dunlin, sanderling, grey and ringed plover, curlew, oystercatcher and turnstone plus lower numbers of many other species. **The beach has several records of vagrants** (including semipalmated sandpiper, American golden

plover, cream-coloured courser to name but a few). There are also good numbers of gulls, among which glaucous, Iceland, Franklin's and Bonaparte's have been recorded in the past decade. Drive west for 2 km to visit the smaller Tréompan beach [2], which can be just as good.

All the stretch of coast from Lampaul-Ploudalmézeau to Le Conquet is worth stop-and-go in small coves and harbours along the incredibly scenic D27, for instance in the 'Presqu'île Saint-Laurent' in Porspoder [3 – see GPS file] (search for snow buntings in winter) or Lampaul Plouarzel [4 – see GPS file] (scan the roosting gulls). On a rising tide, check the sea beyond the 'Plage des Blancs Sablons' [5] for Balearic shearwater, divers, scoters and skuas and the river mouth and harbour [6,7] in Le Conquet for waders, gulls (including Mediterranean; Ross's gull was recorded here a decade ago) and terns.

63 Ushant island (Île d'Ouessant)

> 1 DAY; MAP 95 The Île d'Ouessant has a well-deserved reputation as **the most famous French playground for twitchers in search of autumn Siberian and Nearctic vagrants. Its location at the westernmost end of Europe, marking the entrance into the Channel, provides a logical stepping stone for migrants of the east Atlantic flyway and a landing point for lost birds arriving from the west.** Ouessant offers something more, and more endearing, however, not only because anything can turn up in any bush or on any stretch of moorland, but also because of its unique atmosphere and typical Celtic scenery.

Most birdwatchers visit the island in October, but it is always worth a couple of days at any season, even if nothing specific has been claimed. One or two ferries sail each day from Brest (3 h, departure around 08.00) or Le Conquet (1.5 h) and leave the island by 16.00 or 17.00 from a

small cove to the east [1]. They are often full at the weekend outside winter, so it is best to book ahead (+33 2 90 80 80 80; www.pennarbed.fr). The cheapest option for staying overnight on the island is the so-called 'CEMO' (+33 2 98 48 82 65) in the northwestern part of the island [2]. **The most efficient way to bird on the island is to rent a bike at the pier** (about €20/day).

Habitats include moorland, coastal grassland, open land with thick ferns and brambles, small patches of willow woodland and bogs, private gardens and rocky coast with beaches. Good breeding species include chough (mainly north coast and southwestern cape), shag, fulmar (rocky cliffs to the east), Dartford and grasshopper warblers, rock pipit, peregrine and a few pairs of puffins on the islet off the north coast – which, however, are rarely seen.

In the morning, start on the west side of the island around Le Créac'h lighthouse [3] and in the surrounding area: curlew, golden plover, wheatear, whinchat, pipits and wagtails usually forage in the pastures, while the bushes can be crowded with passerines arriving over the sea. Dotterel, snow and Lapland bunting are regular on coastal grasslands around the northeastern [4] and southwestern headlands [5]. **Migratory passerines may flock anywhere on the island, but the largest numbers of** warblers, flycatchers, thrushes **and** pipits **are found in the large central willow woods [6,7].** Waders roost on the southwestern rocky coves and on the large sandy beach near the village at high tide. Classical Siberian rarities include yellow-browed, Pallas's, dusky and barred warblers, Richard's pipit, red-breasted flycatcher, little bunting and rosy starling. Nearctic passerines are extremely irregular and tend to occur mainly in early October. The island is also good for seawatching, especially near the lighthouse and on the northeastern headland when the wind is in the northwest, with records similar to those from Brignogan (site 60).

64 Brest

0.5–1 DAY; MAP 96   Although Brest is a modern city organised around an extensive industrial harbour, **a number of birding spots are easily accessible from the city centre by bus or car.** They'll fill half a day before sailing to Ushant or heading to the northern Finistère coast.

Stang Alar [1], at the eastern edge of the city, is a wooded valley with a stream and pond which attract grey wagtail, kingfisher, moorhen, tufted duck in winter and common woodland birds such as firecrest and bullfinch year-round. The Moulin Blanc cove harbours gull roosts (mainly black-headed, Mediterranean, common and a selection of large gulls), Sandwich terns, waders on the beach, wintering great crested, little and black-necked grebes and sometimes auks. Park at La Cantine [2] or at the yacht club [3] and walk along the pier to a viewpoint.

Kerhuon pond in Le Relecq-Kerhuon [4] is also worth a check in winter (10 km east of Brest, cross the village along the seashore to park along the southern dyke). **The pond can only be viewed from the west**. Regular species include little grebe, water rail, tufted duck, teal and gulls, plus occasional bittern, goldeneye and scaup. Firecrest, goldcrest, Cetti's warbler and siskin are present in the surrounding trees and reedbeds.

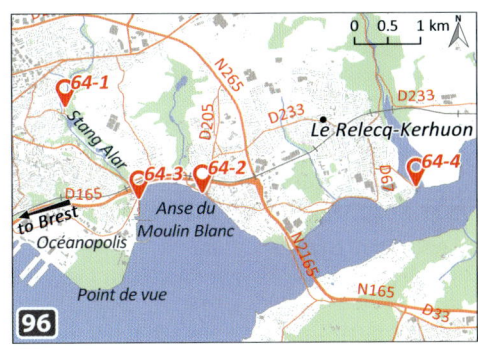

Southern Finistère (sites 65–71, map 97)

HIGHLIGHTS
» Countless bays, harbours and beaches with wintering seabirds and waders
» Sein island, another hotspot for vagrant Siberian passerines and Nearctic waders
» Gull roosts, lured in from the sea by fishing boats

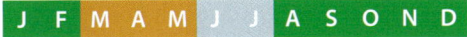

KEY SPECIES
YEAR-ROUND gannet, shag, great white egret, Eurasian bittern, marsh harrier, peregrine, rock pipit, fan-tailed, Dartford and Cetti's warblers, **chough**, raven

BREEDING fulmar, storm petrel, Kentish plover, kittiwake, razorbill, guillemot, **bee-eater**, sand martin, sedge and **Savi's warblers**

MIGRATION Cory's, **great**, Manx, sooty and Balearic shearwaters, **Leach's petrel**, **dotterel**, all skuas, **Richard's pipit**, all warblers, flycatchers

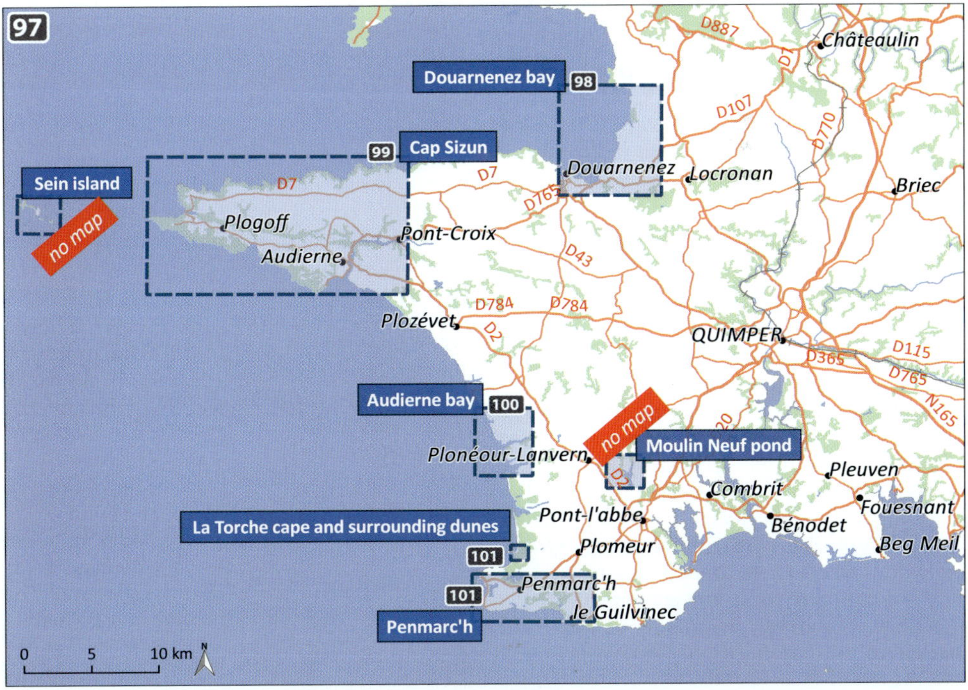

WINTER common scoter, **long-tailed duck**, eider, great northern diver, black-necked and Slavonian grebes, **purple sandpiper**, Mediterranean, **ring-billed, glaucous** and **Iceland gulls**, **bearded tit**, Lapland and snow buntings
VAGRANTS pectoral and buff-breasted sandpipers, yellow-browed warbler, red-breasted flycatcher
VISIT DURATION   3–4 days
PLAN YOUR VISIT

Some birders are addicted to northern Finistère, others to southern. Both are equally good in terms of bird diversity and abundance and the chances of finding national or local rarities, and both remain birdy year-round with a peak from late August to March. Beyond this, they do not look alike, and it is a good idea to try to visit both if you have enough time. In southern Finistère you won't find the numerous coves and bays of the north, but you'll get high cliffs, wide expanses of sandy beaches with ponds and reedbeds, and fishing harbours. The proposed itinerary runs along the coast from Douarnenez bay west to the Pointe du Raz and back along the southernmost coast of Finistère to the large fishing harbours around Le Guilvinec. **It is mainly an autumn and winter trip** focused on gulls and waders. **From mid-August to September, concentrate on the coastal ponds, beaches and reedbeds around Audierne,** where reasonable expectations include dotterel (September), bearded tit and breeding bee-eater (from May). **In September and October, plan a day trip to Sein island:** it has accumulated a long list of vagrant records over the years. **If the wind is in the northwest, head straight to Penmarc'h for seawatching** and keep other sites for later.

65 Douarnenez bay

0.5–1 DAY; MAP 98   Douarnenez fishing harbour lies at the very back of a wide west-facing bay 25 km northwest of Quimper. **It is rightly renowned for wintering** gulls **and** sea ducks. Large numbers of herring gulls roost in **the fishing harbour** [1]. Search for the odd glaucous or Iceland gull in January–February (the harbour has also hosted Ross's, American herring and ring-billed gulls in the past). Scan the bay from the end of the pier looking for divers, grebes (including black-necked and the rarer Slavonian), razorbill and guillemot and large groups of common scoter with the odd long-tailed duck or eider, both rare yet seen every winter.

Other viewpoints are along on a 15 km stretch of the bay north of Douarnenez. You

could for instance visit the Ry beach [2], Trezmalaouen beach [3], Pointe de Trefeuntec [4] and Sainte Anne La Palud beach [5]. Expect Mediterranean, black-headed, common and a few ring-billed gulls, together with sanderling, oystercatcher, small roosts of other waders and Sandwich terns. All these spots are equally good; try to visit all of them on a rising tide.

66 Cap-Sizun

0.5 DAY; MAP 99   The importance of the Cap-Sizun reserve derives mainly from its seabird colonies on spectacular cliffs [1] (25 km, 30 min from Douarnenez, follow signposts from Goulien or Cléden-Cap-Sizun). Walk the path to the left from the car park: **several viewpoints have been specifically designed to provide the best possible views of the colonies** where fulmars, kittiwakes, shags, guillemots and storm petrels breed. Watch the surrounding moorland for chough, peregrine, Dartford warbler, raven and all large gulls.

Drive west for 11 km, 12 min to reach the scenic Pointe du Van (car park at [2]). Walk along the coastal path between the cape and a small chapel [3]: it will not be long before you encounter Dartford warbler, chough and migrating chiffchaffs and blackcaps in autumn. **Settle somewhere on the trail to scan the sea**: in late August, it is an excellent viewpoint for storm petrel and Balearic shearwater. **In October, pay a visit to the pond at the nearby Baie des Trépassés** [4]. A few ducks, great white egret and snipe can usually be found here, and the gardens and willows up to the Lescleden camp site [5] may host stopping-over passerines, possibly including a yellow-browed warbler or other surprises. End your visit in the dramatic landscapes of the Pointe du Raz (car park st [6], fee). Walk as far as you can towards the cape: shags, gannets, Balearic shearwaters, fulmars and kittiwakes sometimes fly close enough to provide great shots.

67 Sein island

> 1 DAY; NO DETAILED MAP **Sein** [1] **is the southern counterpart of Ushant island** (site 63), 5 km off the Pointe du Raz. It has its own addicts, and the two islands engage every autumn in a friendly competition concerning the number and quality of vagrants that will show up – deciding which is best can only be a matter of opinion. Sein is barely 2 km long and 30–500 metres wide, with a maximum altitude of less than 10 metres. It mainly consists of moorland over former pastures surrounded by shingle beaches and rocky tidal flats, with a village on the eastern shore.

The only way to reach the island is by boat from Saint-Evette [2 – see site 66 map], 3 km south of Audierne (+33 2 98 70 70 70; www.pennarbed.fr; 1–2 return journeys per day, 1 h each way). **Sein is renowned for European, Siberian and Nearctic vagrants on autumn migration** (starring a chestnut-sided warbler in 2010), mainly in September and October. **Spend time around private gardens in the village in search of** red-breasted flycatcher, yellow-browed **and** Pallas's warbler **among** chiffchaffs, willow warblers, blackcaps **and** pied flycatchers. **The bushes around the lighthouse are a migrant magnet; they should be visited several times a day.** Look for wheatears, flocks of starlings (possibly with a rosy starling), rock, meadow (both abundant) and Richard's (rare) pipits and casual Lapland, snow or little buntings on shingle beaches and grassy stretches. Hundreds of waders roost on islets, reefs and rocks on the shore. Purple sandpiper occurs from October through the winter. Nearctic waders occur every year in August and September. The least unusual are pectoral and buff-breasted sandpipers, but rarer species, such as semipalmated and Baird's sandpiper or American golden plover, have been recorded in the past.

68 Moulin-Neuf pond

2 HOURS; NO DETAILED MAP **The Moulin-Neuf pond is worth a late-autumn or winter visit for wildfowl** (7 km, 15 min north of Pont l'Abbé; take the D2 towards Plonéour Lanvern and turn right 1.7 km after the last roundabout, then drive north on local roads). Park in Kervahut, and walk 200 m right past the hamlet to reach **a path leading to the northern bank of the pond** [1]. You'd better rely on your GPS device as the way is tricky, even after several visits. Expect varying numbers of wigeon, teal, pochard, tufted duck and gadwall, as well as waders passing through if the water level is low.

69 Audierne bay

0.5–1 DAY; MAP 100 **A series of shallow coastal ponds with reedbeds and mudflats** spreads along the flat Audierne bay between Tréguennec and Plovan (27 km, 30 min west of Quimper

through Plonéour-Lanvern, turn onto the D156 at the cemetery). **The largest of them is Trunvel pond. A ringing station is set up every autumn in the wide reedbed which occupies its northern bank** [1] (once at the western end of Tréguennec follow signposts to Tréogat, bypass the pond and turn left three times to reach Trunvel houses). Marsh harrier, bearded tit, Cetti's and fan-tailed warblers and reed bunting are common year-round. From April, look for sedge and Savi's warblers on the pond tail. A few aquatic warblers are ringed every year in late August, but seeing them is another matter.

To explore the seashore, park 2 km further on in Kerbinigou [2] **and walk 700 m south to the mouth of Trunvel pond** [3] **or 800 m north to Kergalan pond** [4]. The small bridge near the car park is a classic site for bearded tit and bittern. Look among roosting gulls and waders for breeding Kentish plover and possible Nearctic vagrants (late August).

The small Nérizellec pond, 7 km to the north, lies behind the beach 300 m from Stang ar Liou parking [5] (follow 'Plage de Ru Vein' from Plovan). Although it does not seem very appealing at a first glance, it is a regular stopover point for wildfowl, grebes and waders and has yielded one of the longest lists of Nearctic sandpipers in the area (semipalmated, Baird's, pectoral, spotted, mostly from late August to October).

70 Pointe de La Torche and surrounding dunes

3 HOURS; MAP 101 Pointe de La Torche, a flat sandy beach terminating in a rocky headland, marks the southern limit of Audierne bay (11 km, 15 min west of Pont L'Abbé; signposted from Plomeur). **The dunes north of the main car park** [1] **are a classic spot for** dotterel **(yearly) and** buff-breasted sandpiper **(vagrant)** in September. Wheatears are abundant on passage. Local breeding birds include a few pairs of bee-eaters (the only ones in Brittany), sand martin and Kentish plover on the beach.

71 Penmarc'h

0.5 DAY; MAP 101 **Eckmühl lighthouse** [1] **is the main seawatching spot of southern Finistère** (14 km, 20 min from Pont L'Abbé, signposted from Penmarc'h), at its best in southwesterly to northwesterly winds from mid-August to September. Under optimum conditions, expect Manx, Balearic and sooty shearwaters (Cory's and great shearwaters sometimes show up as well in late August), three skuas, good numbers of gannet, storm and Leach's petrels (there is at least one record of Wilson's), kittiwake, razorbill, guillemot and the odd Sabine's gull or grey phalarope. Look for purple sandpipers among the turnstones on rocky tidal flats, while good numbers of sanderling, dunlin, grey plover and turnstone roost all around the lighthouse.

In winter, visit the nearby Saint-Guénolé harbour [2]. It is a regular spot for Iceland and glaucous gulls among large roosts of herring, lesser black-backed and great black-backed gulls. **Gull numbers can be especially impressive when the fishing fleet returns to harbour in the late afternoon.** The large Le Guilvinec harbour [3], 10 km east, is just as good for Arctic gulls. Stop near the bridge over the river in the inner harbour [4] for waders (mostly ringed plover, redshank, dunlin, godwits, curlew and turnstone). Other species regularly found in the harbour in winter include great northern diver, little and black-necked grebes, auks and purple sandpiper (on tide wracks and rocks). Rock pipit is abundant all around.

Rennes area (sites 72–73, map 102)

HIGHLIGHTS
» A regionally high diversity of woodland species in large expanses of old-growth deciduous forest

KEY SPECIES
YEAR-ROUND black and middle spotted woodpeckers, woodlark, Dartford warbler, coal, marsh and crested tits, hawfinch, bullfinch
BREEDING black-winged stilt, nightjar, grasshopper, garden and wood warblers

VISIT DURATION 0.5–1 day

PLAN YOUR VISIT
Inland Brittany is a quiet mosaic of forests, meadows and fields. Bird activity is at its highest in spring but many target species are resident and can easily be found in mid-winter. Be in the forest at dawn as bird activity decreases quickly and will not increase much again in the evening. Start with old-growth stands and move towards moorlands and younger woods during the day. Keep Careil pond for the afternoon.

72 Paimpont forest

0.5 DAY; MAP 103 The old Paimpont forest (also known as Brocéliande forest) is located 35 km west of Rennes. Leave the N24 in Plélan-le-Grand and head north on the D59 towards Saint-Malon for 10 km until a left turn takes you back to the south towards Paimpont: stop after 1 km at the Rocher de Cadieu [1]. **A diverse habitat mosaic can be explored on multiple trails that run across the forest.** Young stands host nightjar (from May) and there are marsh tit, wood warbler and firecrest along the northern trail.

An alternative starting point is located at the western edge of the forest [2] (20 km, 25 min through Paimpont or Concoret, car park at the southern end of Tréhorenteuc). Take the GR37 trail and hike into the woodlands; there are several possible itineraries. Whichever route you follow, expect tree pipit, yellowhammer, woodlark, grasshopper, Dartford and garden warblers on moorland and coal and crested tits in coniferous plots, among other forest specialities.

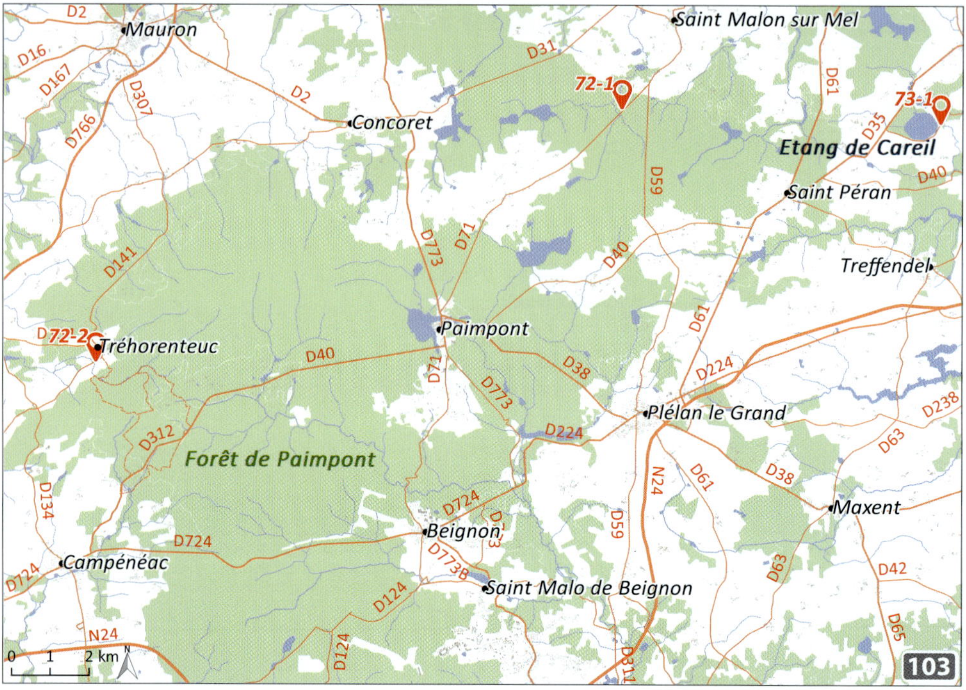

73 Careil pond

2 HOURS; MAP 103   Careil pond (Étang de Careil) [1] is located 30 km, 30 min west of Rennes and 15 km, 15 min east of waypoint 72-1 in the Paimpont forest (drive north on the D35 for 2.5 km from Saint-Péran, then turn right following signs for 'Domaine de Careil' – there are two hides on the banks). **The pond is a local reserve which attracts wintering and migrating** wildfowl (teal, wigeon, shoveler, gadwall, pochard, tufted duck), great white egret and waders (snipe, golden plover, lapwing, green sandpiper…). Black-winged stilt, hobby, golden oriole, little and great crested grebes breed on the pond.

Southern Brittany (sites 74–79, map 104)

HIGHLIGHTS

» A scenic coastal landscape, with good birds year-round
» Large number of wintering and migrating waders and wildfowls
» A wide diversity of marshland breeding birds

KEY SPECIES

YEAR-ROUND gannet, shag, little grebe, **spoonbill**, avocet, rock pipit, fan-tailed warbler, firecrest

BREEDING black-winged stilt, Kentish plover, bluethroat, grasshopper, sedge and reed warblers, **bearded tit**

MIGRATION Manx, sooty and Balearic shearwaters, storm petrel, waders, skuas, **roseate tern**, auks, passerines

WINTER brent goose, **long-tailed duck**, red-breasted merganser, other wildfowl, great northern diver, black-necked and Slavonian grebes, **purple sandpiper**, short-eared owl, Lapland bunting

VISIT DURATION   2–4 days

PLAN YOUR VISIT

Most of the sites covered in this area are wide bays at their best during both migrations. Checking the tide tables is critical when planning your itinerary. Waders roost on inland mudflats at high tide, then spread all across the gulf as the tide falls. You will therefore need at least two days to cover the area; if you have less time, cover one or two locations intensively. In autumn, a day trip to Hoëdic island can be rewarded by loads of seabirds, passerines and perhaps a vagrant sandpiper or warbler.

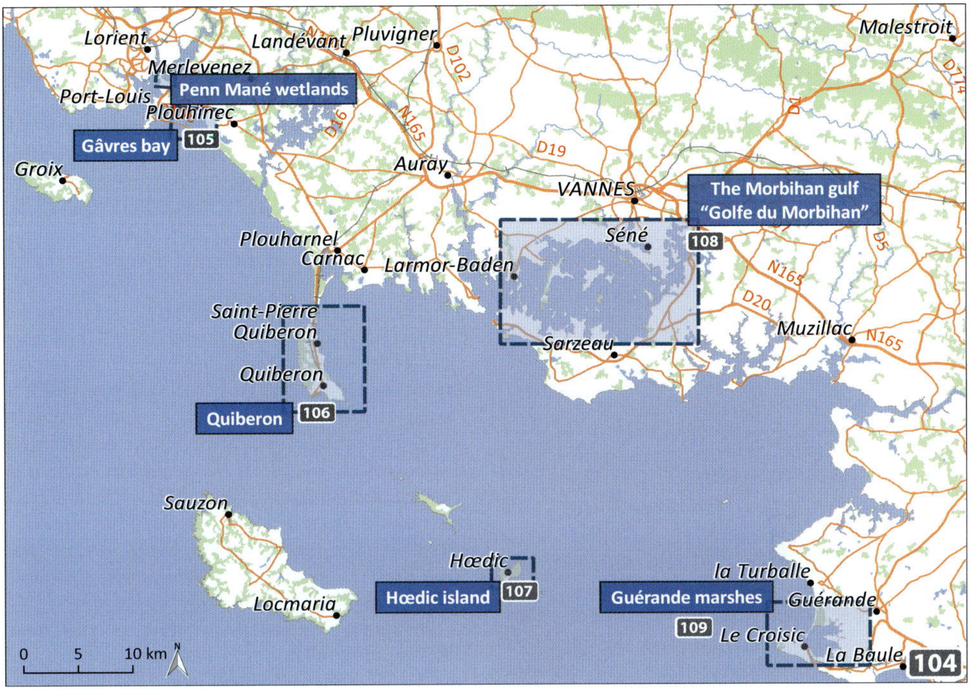

74 Penn Mané wetlands

< 0.5 DAY; MAP 105 The Penn Mané wetlands [1] extend along the eastern side of Lorient bay (exit the N165 in Hennebont and drive south to Locmiquélic on the D781, then follow signposts). **Walk carefully on the coastal path and between the ponds and marshes, where migrating and wintering waterbirds** include teal, redshank, greenshank, snipe, dunlin, black-tailed godwit, water rail, little grebe, kingfisher and Sandwich tern among other species. In spring, expect breeding common tern, black-winged stilt, shelduck, bearded tit, bluethroat, reed bunting, sedge and reed warblers.

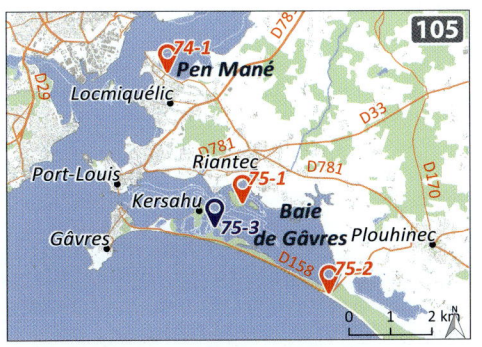

75 Gâvres bay

0.5–1 DAY; MAP 105 Gâvres bay is a logical follow-up to Penn Mané (site 74, 5 km to the northwest). **Be there on a rising tide** and stop at Île de Kerner at Riantec southern exit [1], at the eastern end of the bay southwest of Plouhinec [2] or near the water treatment plants 3 km east of Gâvres [3], depending on the time of day. Wintering and migrating waterbirds include a large diversity of waders (dominated by grey, golden, ringed and Kentish plovers, dunlin, knot, sanderling, curlew sandpiper, godwits, curlews), gulls, brent goose, black-necked grebe, spoonbill and wigeon. Long-tailed ducks sometimes stay a few days in the bay in winter, and there are regular records of short-eared owl, Richard's pipit, yellow-browed warbler and Lapland bunting, plus rarer species. In the breeding season, look for avocet, stilt and bluethroat.

76 Quiberon

0.5 DAY; MAP 106 The Quiberon peninsula is a narrow peninsula extending southwards towards Belle-Île, 30 km south of Auray. Start with Penthièvre beach [1] and the Portivy [2] coastline to look for sea ducks at sea. Long-tailed duck is rare but regular in winter, while black-necked and Slavonian grebes, great northern diver and

early November, and **Siberian vagrants** are seen every year (most usually yellow-browed and Pallas's warblers, little bunting, Richard's pipit, red-breasted flycatcher). The island is small and can be covered on foot in a few hours. Birds can show up anywhere, but **focus on gardens inside and around the village**, **open areas and grasslands** (Pointe du Vieux Château, Fort Anglais, Le Vieux Phare, Ancien Fort), the small marshland at Grand Marais and the water treatment plant.

78 Golfe du Morbihan

1–2 DAYS; MAP 108   The Morbihan gulf is **a European-level stopover site for** waterbirds **on the east Atlantic flyway**. It is a 15 × 8 km near-enclosed bay with several large wooded islands. The northernmost spot is the Pointe des Emigrés near Conleau island, in the southern suburb of Vannes. From the car park [1], a coastal path runs along the gulf to the Marle river mouth, a viewpoint for waders, gulls, black-necked and little grebes, red-breasted merganser, spoonbill and Sandwich tern.

red-breasted merganser are more usual. Small groups of purple sandpiper sometimes settle on Beg en Aud, while shag and rock pipit are common. **In a westerly wind, head 7 km south to Le Vivier** [3] **to scan the sea** for Balearic shearwater, gannet, auks, skuas, Sabine's gull (rare) and Manx shearwater (April). The Pointe du Conguel [4], at the southern extremity of the peninsula, hosts a few wader roosts **at high tide** (sanderling, oystercatcher, purple sandpiper, Kentish plover) **and can also yield** seabirds.

However, **the local hotspot is the Séné reserve**, a mosaic of mudflats, saltmarshes, coastal lagoons and meadows [2] (8 km southeast of Vannes). A wide diversity of local breeding birds includes avocet, black-winged stilt, common tern, redshank, shelduck, bluethroat, fan-tailed and grasshopper warblers. Waders and wildfowl forage and roost inside and around the reserve, especially at high tide. Depending on the season, expect wigeon, shoveler, pintail, plovers, sandpipers, spotted redshank, greenshank, curlews, godwits and rarer species in mixed flocks. To reach the east bank of the river, you will need to go back to the N165 in the vicinity of Vannes and exit south again towards Noyalo (11 km, 15 min). Park at La Métairie du Pont [3], a small harbour ideally located to scan waders feeding on mudflats or flying to their roosts at rising tide.

77 Hoëdic island

≥ 1 DAY; MAP 107   The Compagnie Océane (www.compagnie-oceane.fr) operates daily departures to Hoëdic island [1] from Quiberon harbour (1 h). Due to its location on a line between Quiberon and the Vendée, **Hoëdic is a stepping-stone for** passerines **migrating on the east Atlantic flyway**. The island can be crowded with warblers, flycatchers, firecrest, thrushes, redstarts, pipits and wheatear from August to

Resume your drive south for 9 km to park in Lasné near the causeway to the Île Tascon [4] or at the Pointe du Duer [5] in Saint-Colombier. This is another spot for **roosting** waterbirds; in winter, expect numerous wigeon, pintail, brent goose, grey plover, black-tailed godwit, dunlin and avocet. Short-eared owls sometimes hunt in this area. Sea ducks winter in the gulf and come near to the southern shore at high tide in Le Logéo [6]. Scan the bay in search of black-necked grebe, red-breasted merganser and the odd diver or auk.

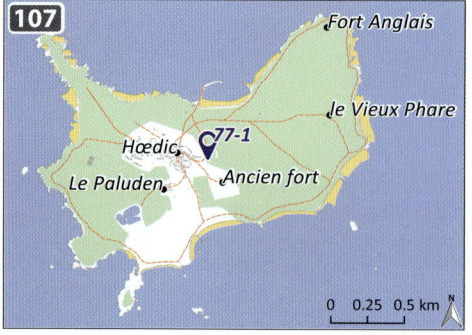

On the western shore of the gulf, the main viewpoint is Pen en Toul wetland in Larmor-Baden [7] (15 km, 20 min south of Vannes on the D101). In addition to the common species, expect spoonbill, water rail and bluethroat (spring). At the southern end of Larmor-Baden, **Le Passage de Berder [8] is the most reliable site in southern Brittany for** roseate tern. A few individuals stop over for a few weeks from mid-August to mid-September and are easy to see, especially (although not only) at high tide (they are normally absent at other seasons). Common and Sandwich terns are also regular at that time of the year.

79 Guérande marshes

1–2 DAYS; MAP 109 The Guérande saltmarshes are another hotspot for waterbirds, 30 km west of Saint-Nazaire. **First check Le Croisic harbour [1]** and look for wintering grebes (up to five species). Large flocks of waders and brent geese fly inland at rising and high tide and sometimes even form roosts inside the harbour. Walk along the pier [2] and scan the sea for divers, scoters, grebes, red-breasted merganser, auks and long-tailed duck (rare). From late summer to autumn, **seawatching from the pier may yield** kittiwake, Sabine's gull (rare), Balearic shearwater (Manx and sooty also occur but are scarce), terns and sometimes large numbers of storm petrels. Pelagic trips are sometimes organised from Le Croisic in August: check with the local LPO (http://loire-atlantique.lpo.fr).

The nearby Pointe du Croisic [3] is another possibility for seawatching, but it is more attractive for passerines in autumn. Park near the 'Fort de l'Océan' hotel and walk along the rocky coast and the surrounding golf course where scattered bushes and hedges may host warblers (including the odd yellow-browed warbler in October), redstarts, flycatchers, thrushes, pipits and wheatears.

Access the **Guérande saltmarshes themselves by driving towards Sissable** (15 km,

20 min from Le Croisic). First stop off at the junction between the 'Route des Marais' and the D92 [4] and walk onto the saltmarsh to look for breeding bluethroat, fan-tailed warbler, redshank and colonies of avocet and black-winged stilt. **At high tide, sandpipers and** plovers **roost on the saltmarshes** around the 'Saline de Sissable' and Grévin, 2 km west [5]. Closer to Guérande, look for bearded tit, reed warbler and water rail in the reedbeds, and waterbirds on ponds and canals around the '**Terre de Sel' complex** [6], **from where several paths run into saltmarshes for several hundred metres**.

To visit the northern part of the marshes, park 5 km further on at Pont de Lancly [7] (access through an easy track from the D92) or in front of Les Chardons Bleus camp site [8]. A further 2 km leads to the dunes around the 'Rade du Croisic' [8], where sea ducks, divers and grebes can easily be spotted. End at Pen Bron, the northern mouth of the saltmarshes [9], where gardens around the medical centre can be crowded with migrating passerines in April–May and August–October.

Lac de Grand-Lieu and the Grée marshes (sites 80–81, map 110)

HIGHLIGHTS
» A major stopover and wintering area for wildfowl, passerines and gulls
» A wide diversity of breeding waterbirds
» The closest birding hotspot to Nantes

KEY SPECIES
YEAR-ROUND spoonbill, cattle egret, great white egret

BREEDING purple heron, glossy ibis, garganey, Mediterranean gull, terns (black, whiskered and rare white-winged), bluethroat, sedge warbler

MIGRATION aquatic warbler (scarce)

VISIT DURATION 1 day

PLAN YOUR VISIT
These places are huge, but access restrictions mean that few sites are actually accessible. Spend time on each of them and wait for dawn and dusk bird movements.

80 Grée marshes

2 HOURS – 0.5 DAY; NO DETAILED MAP The Grée marshes are a 4.5 km² wetland dominated by **flooded meadows furrowed by a network of canals leading to the River Loire** (40 km, 30 min northeast of Nantes, leave the A11 on exit 20 towards Ancenis, turn right on the D923, then left after 500 m at the 'Observatoire ornithologique' signpost, which leads you to a car park with a hide [1]). **The site is a breeding and migration stopover area for** waterbirds. Expect large numbers of wildfowl (including pintail, shoveler, garganey and teal) and waders (breeding lapwing and migrating black-tailed godwit, snipe, shanks and sandpipers).

81 Lac de Grand-Lieu

1 DAY; MAP 111 **Grand-Lieu lake is one of France's largest natural waterbodies, surrounded by a vast expanse of marshes hosting an exceptionally rich wetland community** (20 km southwest of Nantes, drive past the airport towards Saint-Aignan-Grandlieu to the southwest). Although most of it is not accessible, there are a few viewing points on the western bank. **Start in Pierre-Aigüe** [1], west of Saint-Aignan-Grandlieu (turn left on the 'Route du lac' just after the church and park at the end of the road) for purple heron, cattle and great white egrets, black kite, grebes and terns.

In Bouaye, 6 km further west, visit the lake **headquarters** ('Maison du Lac', http://maison-dulacdegrandlieu.com) [2], which includes an exhibition centre, a nature trail bordering the lake and a museum in the 'Pavillon de chasse Guerlain'. **Another access point for the lakeside is at Les Prées Neuves** [3] (7 km south of Bouaye, bypass Saint-Mars-de-Coutais, turn left at the 'Sainte-Marie' signpost 1800 m past the end of the village and follow the road to the bank; the access may be slightly difficult to find).

There are two further viewing points on the southern bank near Saint-Lumine-de-Coutay: La Tuilerie [4] (signposted from the village) and Le Port [5], 200 m past the football field on the 'Rue du stade'. These places offer panoramic views over flooded meadows and swamps. **The area concentrates large numbers of migrating** wildfowl **from late winter to spring,** terns (including whiskered, black and the odd white-winged), **all possible** herons, glossy ibis, spoonbill **foraging or flying from/to colonies during the breeding season** and marshland passerines such as bluethroat and reed warbler.

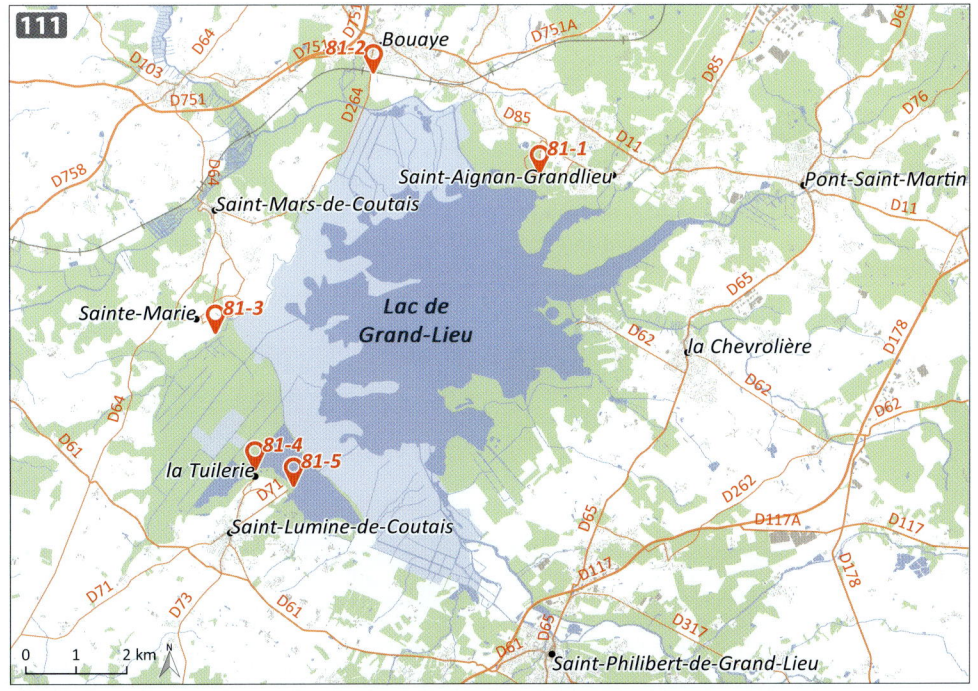

REGION 5
THE LOIRE VALLEY

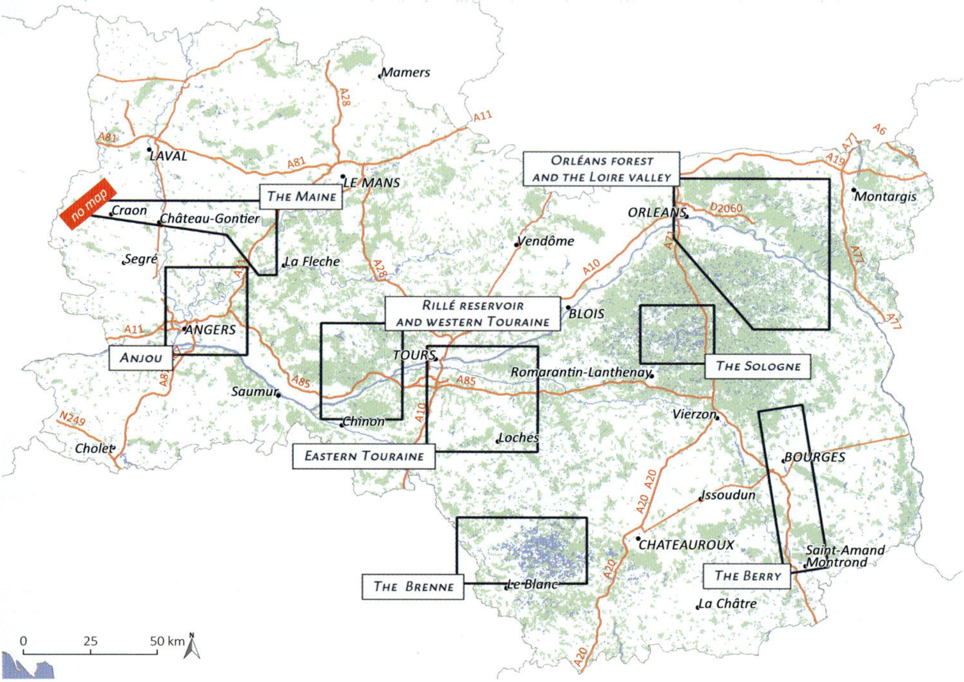

HIGHLIGHTS
» High species diversity at all seasons among the typical landscapes of France's longest river
» A rich and distinctive breeding community, with a mix of waterbirds, open-land, forest and reedbed species
» Wide expanses of old-growth forests with virtually all possible woodland species
» Two leading wetland areas with extensive marshes and meadows interspersed with countless ponds

REVIEWERS Sébastien Brunet, Jean-Michel Chartendrault, François Halligon, Julien Présent, Julien Vèque

Birding in the Loire valley (seen here upstream of Angers, site 87) offers a unique immersion in scenic landscapes along the river, where waders migrating north meet a rich breeding community in April and May.

The little tern is arguably the bird of the River Loire, found on most sandbanks as soon as the river level is low enough in May.

THE LOIRE VALLEY | **ANJOU** | 93

USEFUL WEBSITES
www.faune-anjou.org
www.faune-cher.org
www.faune-maine.org
www.faune-touraine.org
http://sirff.fne-centrevaldeloire.org
www.reserve-cherine.fr
https://obsindre.fr

The Maine (sites 82–84)

HIGHLIGHTS
» A few easy-to-bird spots along major road axes on your way to and from Brittany
» An quick introduction to typical bird communities of western France, year-round

KEY SPECIES
YEAR-ROUND little grebe, hen harrier, little, cattle and great white egrets, little owl, middle spotted woodpecker, reed and cirl buntings
BREEDING sedge, garden and wood warblers, golden oriole, red-backed shrike
MIGRATION waders
WINTER wildfowl, black-headed, common and lesser black-backed gulls
VISIT DURATION 1–2 days
PLAN YOUR VISIT
The covered sites are rather small and isolated but they form ideal birding stops on your way to or from Brittany.

82 Étang de la Rincerie
1 HOUR; NO MAP **La Rincerie watersports centre** [1] lies halfway between Angers and Rennes, 20 km, 20 min from the main road axis in Pouancé (signposted from Saint-Aignan sur Roë). It is worth a visit on your way to or from Brittany, as the only pond of the resort is hunting-free, and thus attracts good numbers of wildfowl in winter. Expect teal, shoveler, herons, gulls (black-headed, common and lesser black-backed) and migrating waders.

83 Curécy pond and Bellebranche forest
2 HOURS; NO MAP Bellebranche forest lies 13 km northwest of Sablé-sur-Sarthe on the way to Laval (signposted from the D21). **Park at** [1] **and walk 500 m back to check Curécy pond** for wildfowl, little and great white egrets, little and great crested grebes, lapwing, snipe and green sandpiper. Within the forest, expect **five** woodpecker **species** (middle spotted is reasonably common), wood and garden warblers, golden oriole (from May), hobby and hen harrier.

84 Cré-sur-Loir and La Flèche marshes
0.5 DAY; NO MAP **Cré-sur-Loir reserve is the largest alluvial marsh in the west of the region**, 8 km from La Flèche along the River Loir (not to be confused with the Loire, which is over 30 km to the south). Drive west on the D323 to Bazouges-sur-le-Loir, then head south on the D70 to reach Cré-sur-Loir. Park just outside the village [1] where **a trail with several viewpoints and a hide** leads to the river bank along the edge of the reserve. Local breeding species include red-backed shrike, little, cattle and great white egrets, reed and cirl buntings, hobby, golden oriole, sedge warbler, water rail, cuckoo and little owl.

Anjou (sites 85–87, map 112)

HIGHLIGHTS
» Stopover sites for waterbird migration
» One of France's handful of remaining corncrake populations
» The full range of riparian breeders typical of the lower Loire valley
» Winter gull roosts

KEY SPECIES
YEAR-ROUND peregrine, little and cattle egrets, kingfisher, fan-tailed warbler
BREEDING squacco heron, night-heron, **corncrake**, stone-curlew, little ringed plover, little and common terns, sand martin, bluethroat, whinchat, reed, sedge and grasshopper warblers
MIGRATION spoonbill, spotted crake, waders, black tern, **aquatic warbler** (scarce)
WINTER wildfowl and gulls (including **Caspian**)

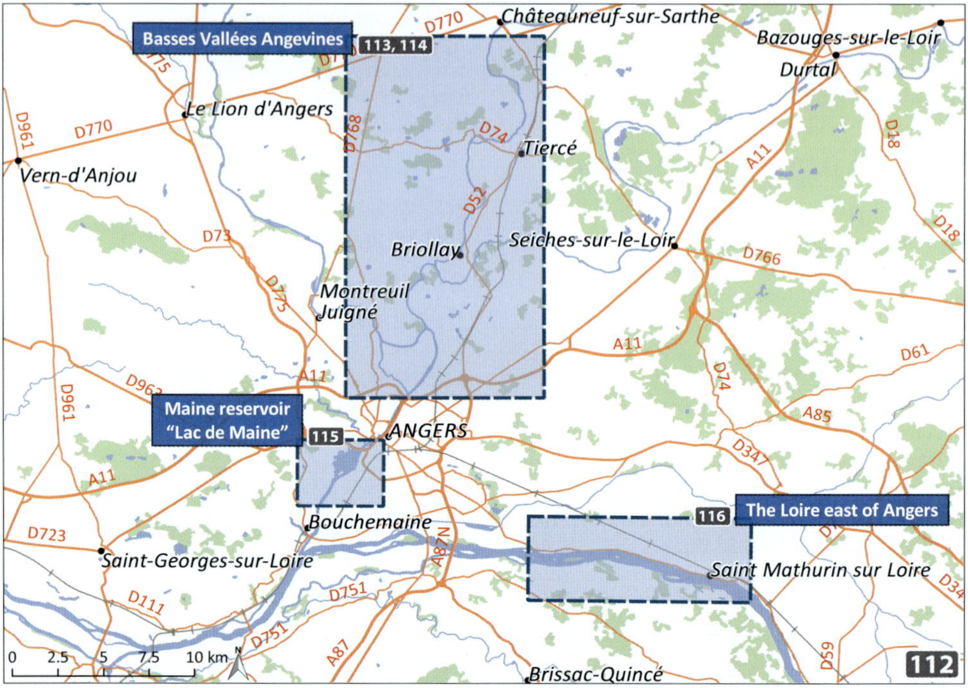

85 BASSES VALLÉES ANGEVINES | THE LOIRE VALLEY

VISIT DURATION 1–2 days
PLAN YOUR VISIT
The diversity of passage migrants and access to sites depends a lot on the Loire water levels, but can be high in both spring and autumn if sandpits are exposed. Wildfowl migration peaks from February to March, while waders reach their highest numbers and diversity from April to May and from mid-August to late September. Anjou is one of the last strongholds of corncrake in the country, and should be your main target in spring. Dusk is best to listen for singing males in April and May, but you may need to be patient and visit multiple spots. The Lac de Maine reservoir near Angers is worth a visit at any time of the year.

85 Basses Vallées Angevines

1 DAY; MAPS 113 & 114 The Anjou lowlands (Basses Vallées Angevines, sometimes abbreviated BVA) **are a wide expanse of meadows and marshes used as a stopover site by large numbers of waterbirds during the spring migration.** Thousands of ducks and geese in winter (dominated by pintail, greylag goose, teal, wigeon and pochard) and waders from March (black-tailed godwit, lapwing, golden plover, ruff, sandpipers) concentrate in the area, which also attracts sparrowhawk and peregrine. During the breeding season, the area is home to **high densities of grassland and meadow species**, starring corncrake, whinchat and commoner passerines including reed bunting, grasshopper and fan-tailed warblers and yellow wagtail. Singing spotted and Baillon's crakes are rarely but regularly recorded. The autumn migration brings reedbed species, among which the rare aquatic warbler sometimes features,

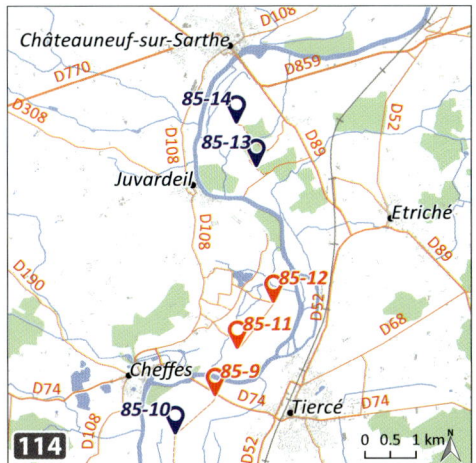
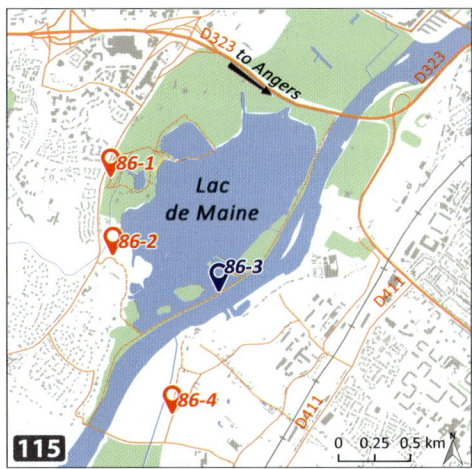

usually in ringers' nets; reed and sedge warblers and a few bluethroats are more often seen.

The valley can be visited by driving along a 20 km stretch of the River Sarthe from Angers northwards. The best viewpoints are located east [1] and west [2] of the Île Saint-Aubin, 300 m south of Cantenay-Epinard [3] and east of the village [4,5], 4 km further on, between Briollay and Soulaire-et-Bourg [6,7,8], east of Cheffes [9,10] and further north towards Châteauneuf-sur-Sarthe [11,12,13,14].

86 Maine reservoir (Lac de Maine)

3 HOURS; MAPS 115 The Maine reservoir is a 90 ha artificial lake in the southwestern part of Angers city. Exit the D523 towards 'Quartier du Lac de Maine', then follow 'Bouchemaine' and park near a roundabout [1]. Do your birding from the nearby watersports centre. Wildfowl numbers are at their highest in winter and a gull roost forms at dusk (black-headed, common, lesser black-backed, herring, yellow-legged and the odd Caspian). Common or black terns and waders may stop over on both migrations, and shrubs and trees in the surrounding park may host warblers and flycatchers. **A large heron colony is located on the southeastern side of the reservoir** (breeding little and cattle egrets, grey heron and night-heron, occasional spoonbill and squacco heron).

Park near the camp site [2] and follow the banks south for a better view of the colonies [3]. On the eastern bank of the River Maine, 800 m south of the reservoir, **the Prairies de la Baumette are flooded meadows where** corncrake **should**

be searched for at dusk in May. They also host whinchat, reed bunting, grasshopper and fan-tailed warblers. Follow 'Bouchemaine' from the centre of Angers and turn right towards Châteaubriant, then left to park near Les Tourelles and walk among the meadows [4].

87 The Loire east of Angers

0.5 DAY; MAP 116 **A number of viewpoints over the sandy banks and islets of the Loire** spread along a 10 km stretch of the D952 east of Angers. The river is wide, interspersed with wooded islands, and bordered by small roads or tracks that once served for boat towing. Stop for instance at La Daguenière [1], La Bohalle [2,3] and Saint-Mathurin-sur-Loire [4,5], where a bridge allows you to return to Angers along the southern bank. Breeding birds include little and common terns, little ringed plover, stone-curlew, kingfisher, hobby and sand martin. Sandpits emerge between April and October if the water level is sufficiently low. Scan them for common and green sandpiper, greenshank, redshank, ringed plover and stints.

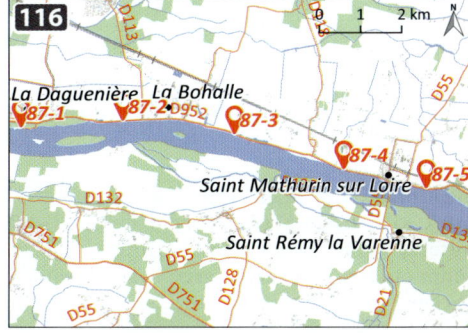

Rillé reservoir and western Touraine (sites 88–90, map 117)

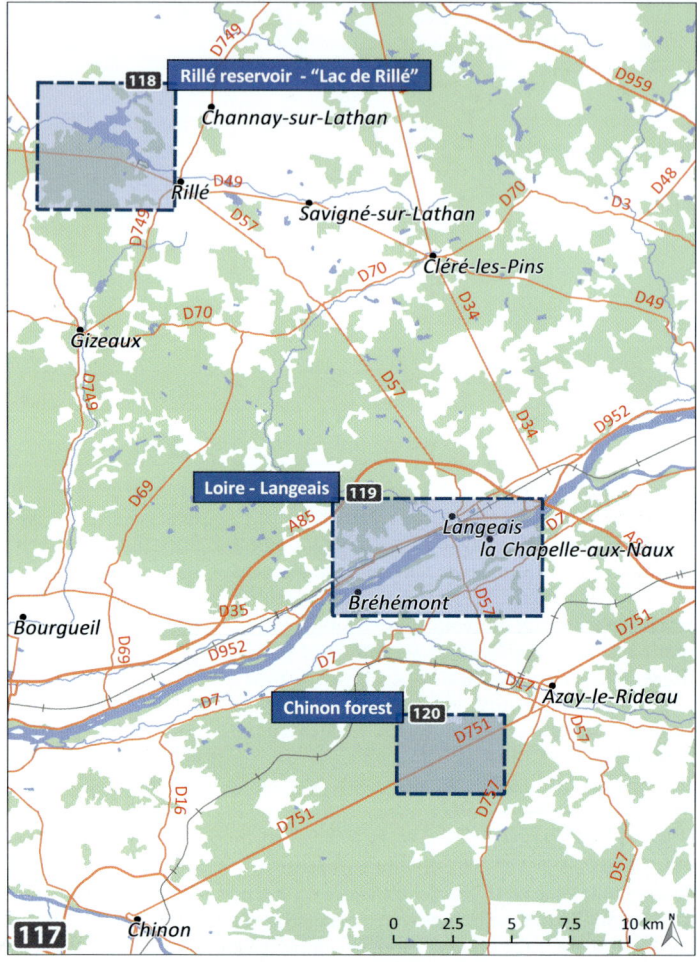

HIGHLIGHTS
» A mosaic of forest and farmland, crossed by the Loire valley and its tributaries
» Inland hotspots for rarities and vagrant migrants

| J | F | M | A | M | J | J | A | S | O | N | D |

KEY SPECIES

YEAR-ROUND goshawk, little grebe, great white and cattle egrets, black woodpecker, woodlark, Cetti's warbler, marsh and crested tits, firecrest, hawfinch

BREEDING short-toed eagle, honey-buzzard, black kite, **night-heron, purple heron, little bittern**, stone-curlew, little ringed plover, common tern, wryneck, **grey-headed woodpecker**, nightjar, sand martin, whinchat, whitethroat, Dartford, grasshopper, reed, melodious and western Bonelli's warblers, spotted flycatcher, red-backed shrike

MIGRATION black stork, spoonbill, osprey, waders including Temminck's stint, black tern

WINTER wildfowl

VISIT DURATION 1–2 days

PLAN YOUR VISIT

The region is more exciting than it looks, because the Loire hosts good numbers of waders migrating along the east Atlantic flyway. If you have just one day, spend your morning around Rillé reservoir, then stop-and-go along the Loire towards Tours. Stay around the reservoir if the Loire is too high for sandpits to be exposed. With two days, take more time to explore the grassy banks of Rillé in search of rare waders, or pay an early-morning visit to Chinon forest for woodpeckers (March to May). Do not stick to the viewpoints described here along the Loire: adjust your itinerary according to water levels, and check as many islets as you can. Do not forget to scan the sky regularly for raptors and storks.

88 Rillé reservoir (Lac de Rillé)

0.5 DAY; MAP 118 Rillé reservoir is a **major inland stopover site for migrating and wintering** waterbirds (45 km northwest of Tours, take the D952 to Langeais and follow signposts to Hommes then Rillé on the D57). **Either walk around the reservoir or spend time at each spot**, carefully checking bird movements. First stop at the main dam [1], where common and green sandpipers, snipe, kingfisher, grey wagtail and great white egret are always present. Little bittern occasionally turns up in June or July in the reedbeds east of the dam. **A hide 200 m north [2]** (first track on the left after the dam) allows closer sightings of reed warbler, purple heron, sandpipers (mostly common, green, wood and greenshank) and ducks (teal, gadwall, shoveler).

La Butte Noire [3] **and la Grande Maison** [4] **provide panoramic views over the open waters and surrounding fields and meadows.** Red-backed shrike, nightingale, yellowhammer, melodious warbler and whitethroat breed in hedgerows along the road and local cattle egrets, night-herons and purple herons feed on the most densely vegetated sections of the banks. Hobby, honey-buzzard, black kite and sometimes short-toed eagle are regularly seen here. **The surrounding bushes should be checked in August to September** for migrating flycatchers, warblers, redstarts, wheatears, whinchats and the odd wryneck. From **summer to late autumn, low water levels attract good numbers of** waders (black-tailed godwit, little stint, dunlin, sandpipers; Temminck's stint is regular and there have been a dozen pectoral sandpiper records in the past 20 years), wildfowl and gulls. Single ospreys, spoonbills, common and black terns could also show up.

A quick stop at the northern end of the reservoir [5] could yield little grebe and purple heron in May. **Further west, a bay surrounded by fields and grassy pastures** [6] hosts good numbers of greylag goose, wigeon, golden plover, lapwing and other waders from September to March and a breeding pair of stone-curlew in spring. If the reservoir is not so good, **walk in the forest to the south** [7] in search of honey-buzzard (from May), black woodpecker, woodlark, firecrest, western Bonelli's warbler and other woodland species.

89 Loire – Langeais

0.5 DAY; MAP 119 **The banks of the Loire from Tours to La Chapelle Sur Loire can be good for a full birding day during** wader **migration from mid-April to mid-May and from mid-August through September.** The landscape here is typical of the Loire, with wooded islands, sandpits and narrow towing roads along which large limestone houses and wine cellars are spread – in some ways, this is the ultimate French birding place.

If you are short of time, concentrate on a 7 km stretch of the river between La Chapelle-aux-Naux and Bréhémont (20 km downstream of Tours, exit 8 on the A85 or use the D7 towards Lignières de Touraine). **The D16 runs along the river bank** and offers a number of good viewpoints [1–4], some of which allow access to the water [2,3]. In late spring and early summer, little ringed plover, little and common terns breed on sandy islets and kingfisher and sand martin use holes in the banks (e.g. in front of [2]). Cattle egret, black kite and hobby are common on wooded islands, and there are Cetti's warbler, reed warbler and reed bunting in willows. In autumn and winter, little grebe, teal, cormorant

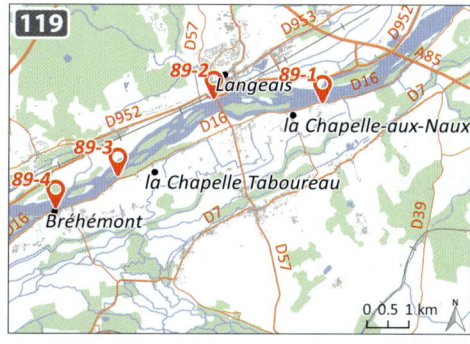

and great white egret commonly occur. **Wader numbers vary with water levels and tend to decrease towards winter** except for lapwing and golden plover. Expect green, common and wood sandpipers, shanks, ringed and grey plovers, little stint and the rare Temminck's stint. Osprey is seen daily in spring and autumn and black stork is less frequent yet possible with some effort and luck. In spring and late summer, also check the wooded banks for bluethroat (rare).

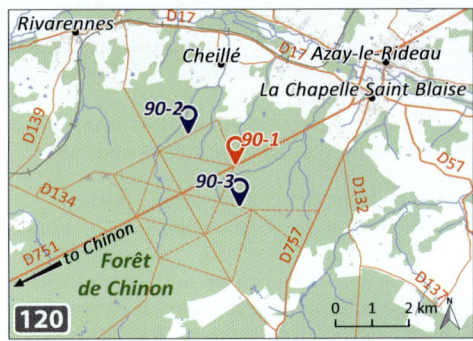

90 Chinon forest

0.5 DAY; MAP 120 Chinon forest hosts a **wide range of woodland birds** (30 km, 30 min southwest of Tours through Azay le Rideau). It is largely dominated by oak and beech, with interspersed pine stands and moorland – an ideal stop when days warm up in late spring and one of the most exciting forest birding sites of the region. There are car parks at all major junctions, from where **numerous tracks and trails run across the forest**. The 'Maison forestière d'Agnès Sorel' junction [1] is a good starting point for dawn birding. Walk along the forest trail north of the D751 towards [2]. Six woodpecker species breed around here, including the rare grey-headed woodpecker – which you should have as your main target during the drumming season in March. Later in spring, expect singing western Bonelli's and wood warblers (from April), wryneck, cuckoo, marsh and crested tits, firecrest, hawfinch, spotted flycatcher and golden oriole (the last two from May). Alternatively, head south of the main road to look for Dartford and grasshopper warblers in young stands and moorland (nightjar at dusk), for instance around the 'carrefour François 1er' [3]. Honey-buzzard, hobby, short-toed eagle, goshawk, sparrowhawk and osprey also breed in this area.

Eastern Touraine (sites 91–95, map 121)

HIGHLIGHTS
» An agriculture–forest mosaic with rare forest and steppe breeders
» All the breeding and migratory species of the River Loire, against a backdrop of scenic castles
» Several ponds suitable for a couple of productive birding hours at any time of year

KEY SPECIES
YEAR-ROUND little grebe, hen harrier, red-legged and grey partridges, cattle, little and great white egrets, **little bustard**, yellow-legged gull, little owl, **grey-headed**, black and middle spotted woodpeckers, crested lark, Cetti's warbler, marsh and crested tits, firecrest, hawfinch

BREEDING short-toed eagle, Montagu's harrier, **purple heron, night-heron, squacco heron,** stone-curlew, little ringed plover, Mediterranean gull, little tern, wryneck, sand martin, reed, melodious, western Bonelli's, and wood warblers

MIGRATION garganey, black-necked grebe, **black stork, spotted crake**, waders, whiskered and black terns

VISIT DURATION 2–3 days
PLAN YOUR VISIT
This part of Touraine has its share of waders and other waterbirds along the Loire. A birding day along the river can be combined with a visit to a winery and some of the most famous chateaus of the Loire – the Château d'Amboise would be a perfect choice. Slightly to the south, the wide-open cultivated fields of the Champeigne are unique in the region and host a suite of highlights such as little bustard and stone-curlew. Devote an early morning to Loches forest, and with luck and effort you might encounter a grey-headed woodpecker or a black stork. The Loire banks and ponds should be productive at any time of day.

91 Tours – La Bergeonnerie pond
1 HOUR; NO DETAILED MAP La Bergeonnerie pond [1] is located on the southern bank of the River Cher, 2 km south of central Tours in the 'Quartier des Deux Lions'. **A** heron **colony on the small central island** (scan it from the west bank) hosts night-herons, cattle and little egrets and one or two pairs of squacco heron (easy to

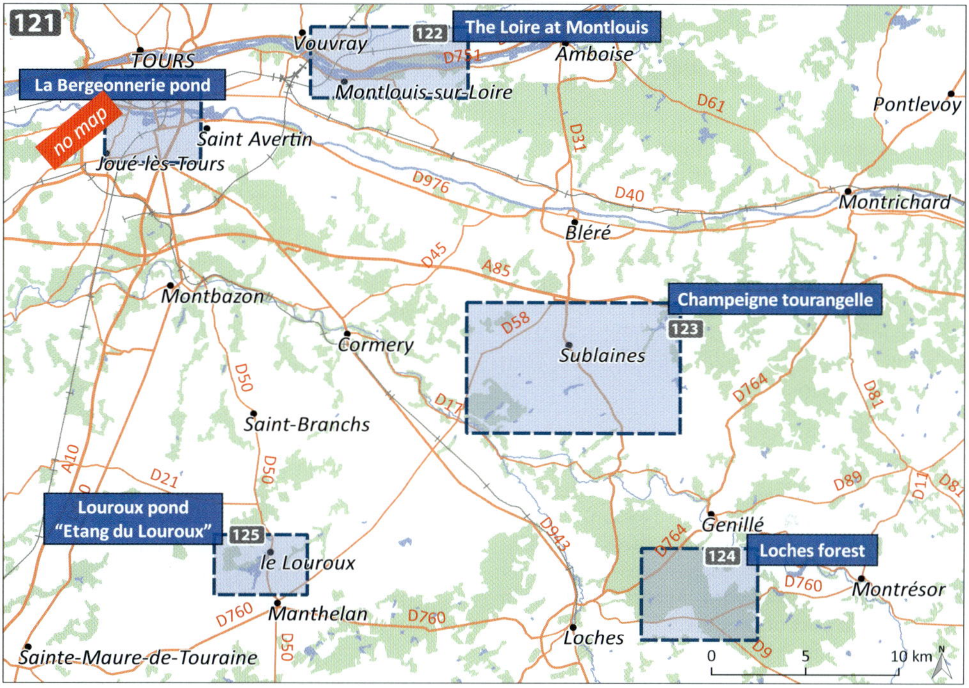

see in June–August on floating vegetation on the nearby river). Little and common terns, gadwall, cormorant, great crested and little grebes, yellow-legged gull and kingfisher usually occur on the pond and the River Cher [2].

92 The Loire at Montlouis
2 HOURS; MAP 122   Set out east from Tours and drive along the Loire towards Amboise for 13 km to reach Montlouis-sur-Loire. Terns, **Mediterranean** and black-headed gulls **breed in mixed colonies on sandy islands just past the railway bridge** [1], but their numbers vary strongly with early-season water levels. Nightingale and melodious warbler breed on the Île de Bondésir and at la Tuilerie [2] to the east of Montlouis, where you can reach the river bank by means of several paths. This stretch of the Loire is frequented by little ringed plover, common sandpiper, yellow-legged gull, little and common terns, kingfisher and sand martin during the breeding season, and by egrets, wildfowl and waders during migration. **On the northern bank, former quarries** past Le Gros Ormeau [3] and in La Frillière [4] have the same species.

93 Champeigne Tourangelle
0.5 DAY; MAP 123   A vast agricultural plain known as the Champeigne Tourangelle is home to a rich community of **farmland and steppe**

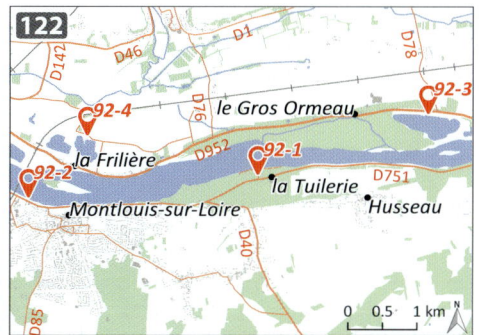
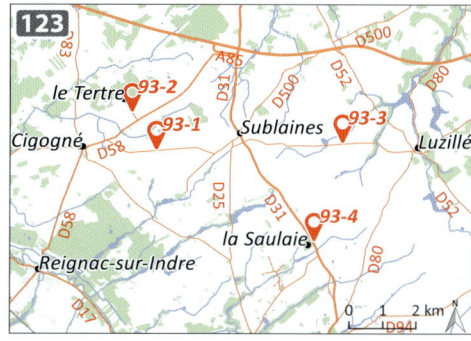

birds starring little bustard (40 km east of Tours, leave the A85 on exit 11 and head south toward Loches for 5 km). Stop-and-go on local roads between Cigogné, Sublaines, Luzillé and Chédigny, for instance at [1,2,3] in search of little bustard (especially around Le Tertre [2], east of Cigogné), stone-curlew, Montagu's and hen harriers, quail, crested lark, corn bunting, red-legged and grey partridge. The best spots for little bustard at dawn or dusk are located south of Sublaines (drive for approximately 4 km on the D31 towards Loches and park on the roadside near the entrance to the 'Village Vacance la Saulaie' [4]). Short-toed eagle is often seen hunting in the area.

94 Loches forest

0.5–1 DAY; MAP 124 **Loches forest hosts the full complement of** woodpeckers **and a particularly rich woodland community** (45 km, 50 min southeast of Tours on the D943 through Chambray-lès-Tours). From Loches, cross the River Indre and head east towards Montrésor or reach the northern part of the forest towards Genillé. Park near Pas aux Ânes pond [1] or at Beauchêne forest house [2] and walk along tracks and paths. Grey-headed woodpecker **breeds around the pond in old deciduous stands with beeches**. It will be a hard task to find it, but the reward is worth it – it has declined to the point that it is now one of French birders' most sought-after breeding species. Middle spotted woodpecker is common in the vicinity. Also look for wryneck, cuckoo, marsh tit, wood warbler, golden oriole, firecrest and hawfinch in deciduous stands,

crested tit in pine stands and western Bonelli's warbler in mixed deciduous–conifer stands.

95 Louroux pond (Étang du Louroux)

1–3 HOURS; MAP 125 Le Louroux is an isolated pond located 25 km, 30 min west of Loches and 35 km, 40 min from Tours. **It is worth a visit at all seasons but will be at its best from March to early May and from August to September**. Park at the northern end [1] and walk beside the pond looking for reed warbler, reed bunting and Cetti's warbler in the reedbeds. **A hide on the eastern bank** [2] **provides good views over the open water and southern section**. Migrating and wintering waterbirds include tufted duck, common pochard, garganey, teal, shoveler, little (year-round) and black-necked grebe (March–April), little and great white egrets, common, whiskered and black terns and kingfisher. Lapwing, snipe, shanks, sandpipers and stints should occur as soon as the water level is low enough, in addition to hundreds of wintering golden plover in the surrounding fields. Water rail and spotted crake (August–September) sometimes briefly forage on the mudflats.

The small and hidden Beaulieu pond [3], to the west of Le Louroux (access from the D50, turn right 800 m before Manthelan), attracts purple heron, egrets and sometimes black stork (August–September). Little owl breeds on buildings in the vicinity and should easily be heard at dusk in March. Hen and Montagu's harriers, stone-curlew, yellow wagtail, crested lark and corn bunting breed in cultivated fields.

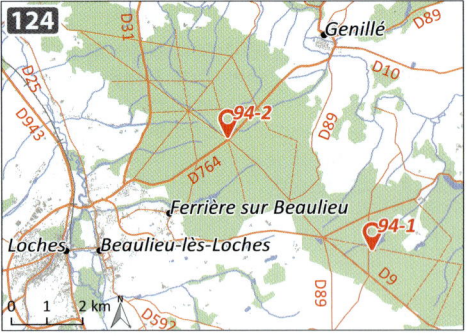

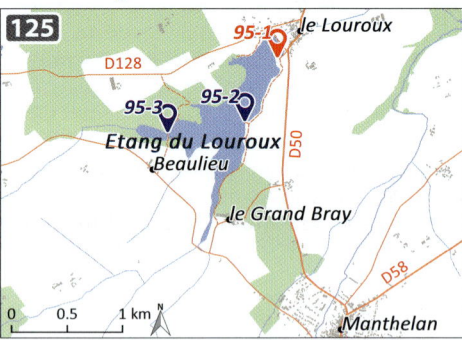

The Brenne (sites 96–97, map 126)

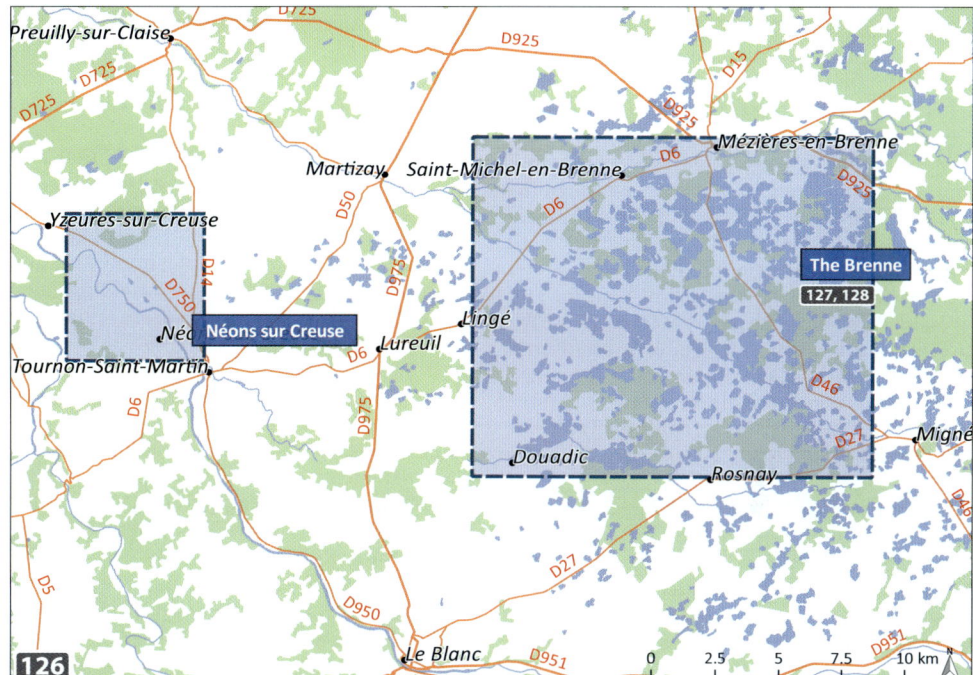

HIGHLIGHTS
- One of the most extensive inland wetland areas of the country with hundreds of ponds, reedbeds and marshes
- Worth a visit at any time of the year
- A stronghold for wintering and migrating waders and wildfowl

| J | F | M | A | M | J | J | A | S | O | N | D |

KEY SPECIES
- **YEAR-ROUND** marsh harrier, **Eurasian bittern**, great white and cattle egrets
- **BREEDING** red-crested pochard, black-necked grebe, **night-heron, purple** and **squacco herons, little bittern**, whiskered tern, **bee-eater**, sand martin, **Savi's** and great reed warblers, red-backed shrike
- **MIGRATION** osprey, spotted and little crakes, all **waders**
- **WINTER** all wildfowl, **ferruginous duck, white-tailed eagle**, common crane

VISIT DURATION 2–4 days

PLAN YOUR VISIT
The Brenne has something of an inland Camargue. It is good for birding year-round, peaking at both migrations, and wetland targets are easier here than anywhere else in central France. The marshes are hunted, but happily vast areas are protected year-round and serve as roosts for wildfowl and waders. Prepare your itinerary according to your target species and the season. Avoid wandering randomly in the area, as you are likely to waste your time in long drives with little reward: **some ponds are better for wintering waterbirds and some others are at their best during the breeding season.** Your best bet would be to plan a weekend or more in the area, staying in one of the numerous guest houses scattered everywhere, but bear in mind that you will find petrol and food only in Mézières en Brenne. Several hides in Chérine nature reserve need booking at the reserve headquarters. Bird occurrences on the various ponds vary between months and years according to water levels. **We have divided the area into two sites: the Brenne itself and three nearby localities worth a visit in May for** bee-eater **colonies.**

96 The Brenne

1–2 DAYS; MAPS 127 & 128   Start in Mézières-en-Brenne (40 km, 40 min east of Châteauroux through Vendoeuvres). As you drive south on the D15 from Mézières-en-Brenne towards Chérine nature reserve (signposted), make a first stop before the D17 junction at La Biénaise pond [1], which extends to both sides of the road. A ferruginous duck sometimes stays among pochard and tufted duck in winter, and waders can be seen foraging on the mudflats at both migrations if the water level is low.

You should spend significant time in and around Chérine nature reserve [2], which protects a mosaic of wet meadows, ponds and reedbeds typical of the region (rangers, guided visits, logbook and booking for the photographic hide at the headquarters). **Inside the reserve, the Étang Cistude hosts a large mixed colony of** purple heron, night-heron, cattle egret**, a few pairs of** little bitterns **and** squacco herons **and good numbers of** black-necked grebes **and** black-headed gulls. Still inside the reserve, **the Étang Ricot [3] has a hide facing a pond and reedbed which should be visited at dawn or dusk. It is one of the Brenne's best places to see** little bittern **(May–August).** You should also easily find reed and great reed warblers, marsh harrier, water rail, great crested and little grebes. Great white egret, bittern and wildfowl justify a winter or early spring visit. Before leaving the reserve, take a look from the road at the Étang de Montmélier and surrounding reedbeds for reed warblers.

Another hide is located at the Étang des Essards [4]**, 2 km further along the D44.** The pond is especially attractive for wintering and migrating wildfowl and waders, and in spring cirl bunting, yellowhammer, stonechat and red-backed shrike (May) breed in bordering hedges. Heading 2 km west again along the D44, the Étang de la Sous [5] is the western-most pond of Chérine reserve. **Its hide is one of the Brenne's nicest viewing points, especially for photographers**. Purple heron, little bittern, great reed warbler, whiskered tern and black-necked grebe all breed (April–July) and the pond is usually crowded with wildfowl in winter and early spring. From here, you can either visit southern sites or head northeast for 14 km, 15 min through Mézières en Brenne and Saint-Michel en Brenne to reach **the Étangs de Piégu and Étang Renard ponds** [6,7] (park at the junction past Subtray). **They host huge numbers**

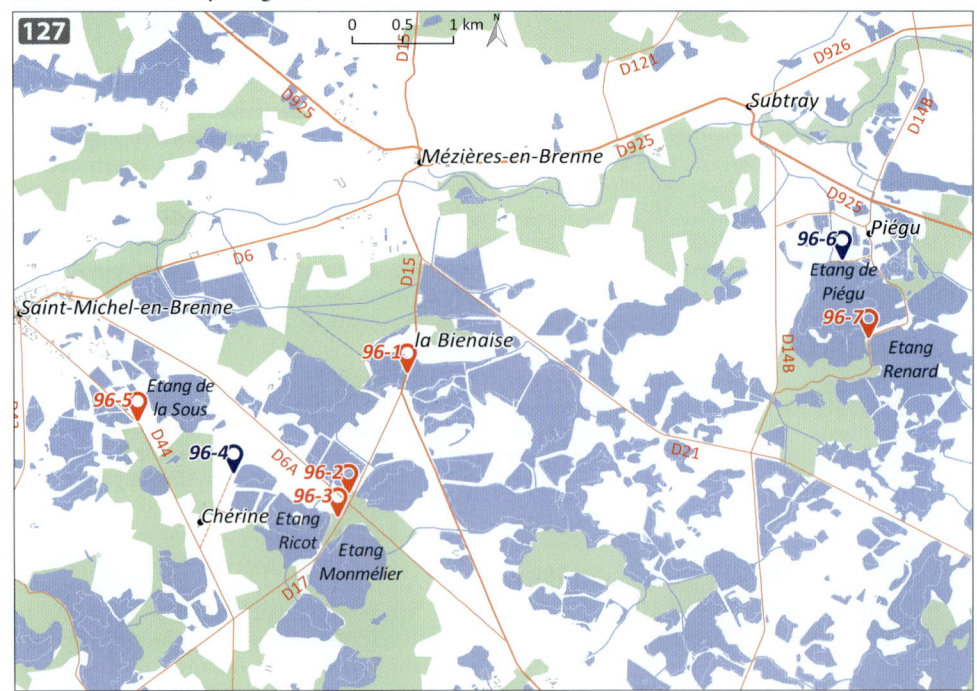

of ducks **in winter and on spring migration** and waders **when the water level is low.**

At the opposite corner, the Étang Purais [8] is best from late afternoon to dusk (2 km east of Lingé; 13 km, 15 min southwest of Mézières en Brenne). **Covered by waterlilies and surrounded by reedbeds, it hosts one of France's largest colonies of** whiskered tern together with numerous pairs of black-necked grebes, red-crested and common pochards and a pair of marsh harriers. Bitterns can occasionally be seen flying over the reedbeds at all seasons. **In winter, flocks of** cranes **fly over the pond heading to their night roosts in the surrounding fields and meadows.**

Head north for 6 km to reach the Étangs de Lérignon [9] and take the D78 for 5 km to visit Gabriau [10] and la Gabrière [11] ponds, then head north for 1 km on the D17 to check the Étang de Beauregard [12]. **These large ponds are mostly interesting for wintering** ducks **and** grebes **as well as for their spectacular** great white egret **roosts.** Back south along the D17, stop at the Étang de l'Ardouine and Étang de Rochefort and take a look on both sides of the road [13]. **Aquatic vegetation attracts** whiskered terns **and** black-necked grebes **during the breeding season,** as well as the usual wetland species including purple heron, night-heron and red-crested pochard, sometimes joined by bittern or a singing little crake at dusk.

A further 4 km south you come to the Étang de la Mer Rouge, the largest pond of the Brenne, which lies between Rosnay and Douadic [14]. **It is at its best in late winter and early spring** when greylag geese and other wildfowl concentrate here together with thousands of cranes. **It is worth spending a bit of time here, as there are regular movements of** waterbirds **from surrounding ponds.** Birds from the Étang de la Mer Rouge may also spread a few kilometres east to the Étang de Montiacre [15], Étang du Blizon [16] (a hide on the Étang Massé, on the south side of the road, looks towards a heron colony with all species) and Étang du Sault [17], which can be scanned from the D17a and D15. The whole area is a wintering ground for white-tailed eagle. In spring, look for breeding Savi's and great reed warblers in the reedbed belts surrounding some of these ponds. A last stop 1 km south on the D15 at the Étang Foucault/Étang Bénisme ponds [18] should yield more migrating/wintering ducks (mainly teal, wigeon, shoveler), waders, all breeding herons (including little bittern) and osprey in autumn.

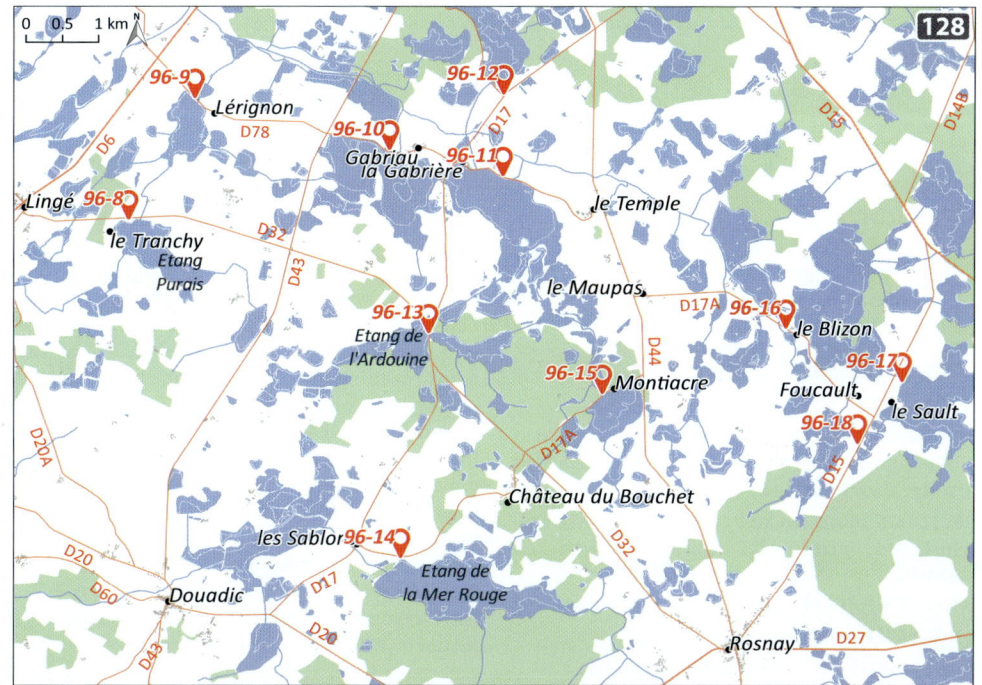

97 Néons-sur-Creuse

0.5 DAY; NO DETAILED MAP **The River Creuse, 30 km east of the Brenne, is worth a visit in May and June as it hosts several bee-eater colonies.** Drive from Mézières-en-Brenne to Tournon-Saint-Martin and turn on the D95 to Néons-sur-Creuse, where there are **several viewing points along the western bank of the river** [1,2,3]. Other breeding species include hobby, sand martin, kingfisher and grey wagtail.

The Sologne (site 98, map 129)

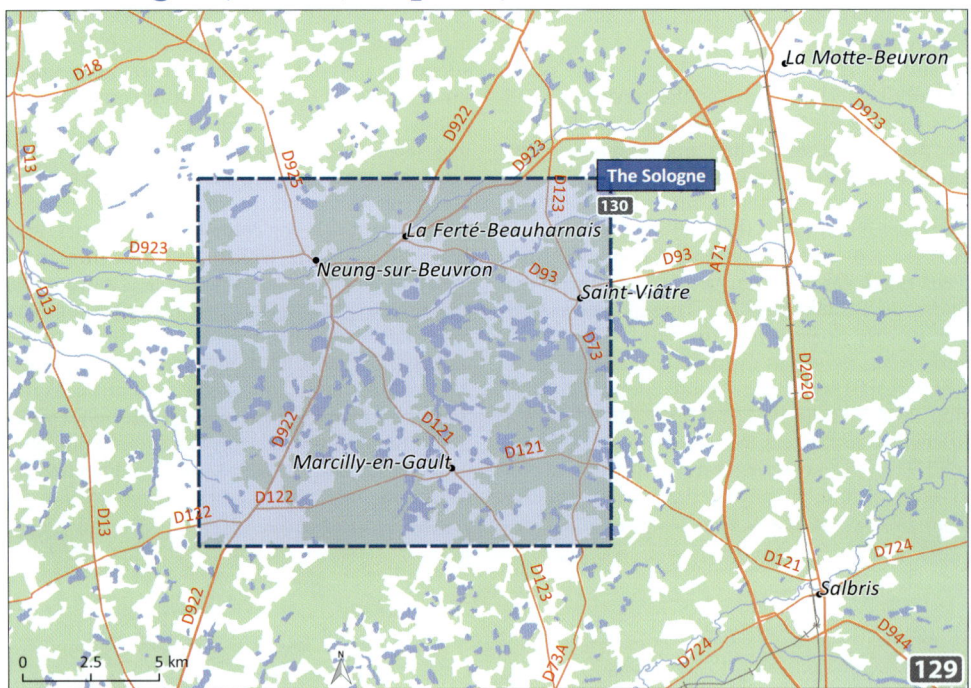

HIGHLIGHTS
» A unique mosaic of forested ponds with colonies of herons and waterbirds
» Reliable areas for several rare breeding raptors

| J | F | M | A | M | J | J | A | S | O | N | D |

KEY SPECIES
YEAR-ROUND Eurasian bittern, little and great white egrets, marsh harrier, middle spotted woodpecker, marsh tit

BREEDING black-necked grebe, honey-buzzard, **whiskered tern**, sedge warbler, red-backed shrike

WINTER wildfowl, ferruginous duck, wintering waders, finches

VISIT DURATION   1–2 days

PLAN YOUR VISIT
The Sologne wetland is a short drive from Blois and Orléans and can be visited as a day trip or on your way from Paris to the Brenne. Late winter to spring bring large numbers of migrating waders and wildfowl. Most of the main breeding specialities should have arrived by April. **Compared to the Brenne, the Sologne is more wooded and somewhat harder for the birder, most ponds and marshes being private.** Stay on the roadsides unless tracks are clearly public, and do not trespass past fences or signposts for any reason. As long as you respect these basic rules, you should enjoy some of the most exciting birding of central France, with large colonies of whiskered terns and endless reedbeds where you can search for black-necked grebes, bittern, reed warbler and many more. Mornings will yield the best of reedbeds and

ponds. But a midday visit on a sunny day is still worthwhile, as raptors will start soaring as soon as the temperature rises.

98 The Sologne

1 DAY; MAP 130 Concentrate on a triangle between Neung-sur-Neuvron (60 km, 40 min south of Orléans, exit 3 on the A71 then west; or D15 towards Romorantin-Lanthenay), Saint-Viâtre and Marcilly-en-Gault. **The Étang de Beaumont [1], an open pond surrounded by grasslands, is worth a visit at all seasons** (4 km west of Neung, signposted 600 m from the western exit of the village on the D923). A visit to its hide should yield breeding whiskered tern, black-necked grebe, marsh harrier and wildfowl (teal, pochard and tufted duck). Sedge warbler and reed bunting breed in the reedbed. Little egret, great white egret, little and great crested grebes are also present year-round.

Your next stop could be the Étang de Marcilly [2] (7 km from Neung, first south towards Marcilly then east to Saint-Viâtre on the D63), then the Étang de Favelle [3] and the Étang des Brosses [4], 7 km further south to Saint-Viâtre. These three ponds can be viewed from the roadside or from small trails. Expect breeding pochard (sometimes with a ferruginous duck in winter), tufted duck, gadwall, black-headed gull, great crested and black-necked grebes. Little and great white egret occur in the non-breeding season. Check hedgerows on your way between the ponds for warblers, linnet, yellowhammer and red-backed shrike.

Drive south for 7 km to Marcilly-en-Gault and park in the village or at the start of the Allée du Roi on the D122 near the Étang de Maintmont [5]. **A two-kilometre walk among woods leads to two other ponds; marsh tit and middle spotted woodpecker breed on the way**. The Étang de la Prée on your left [6] hosts a whiskered tern colony and numerous dabbling and diving ducks in winter and spring. A few hundred metres further, the Étang des Gâts [7] has wintering bittern, ducks, grebes and open areas where it is worth looking for honey-buzzard and hobby.

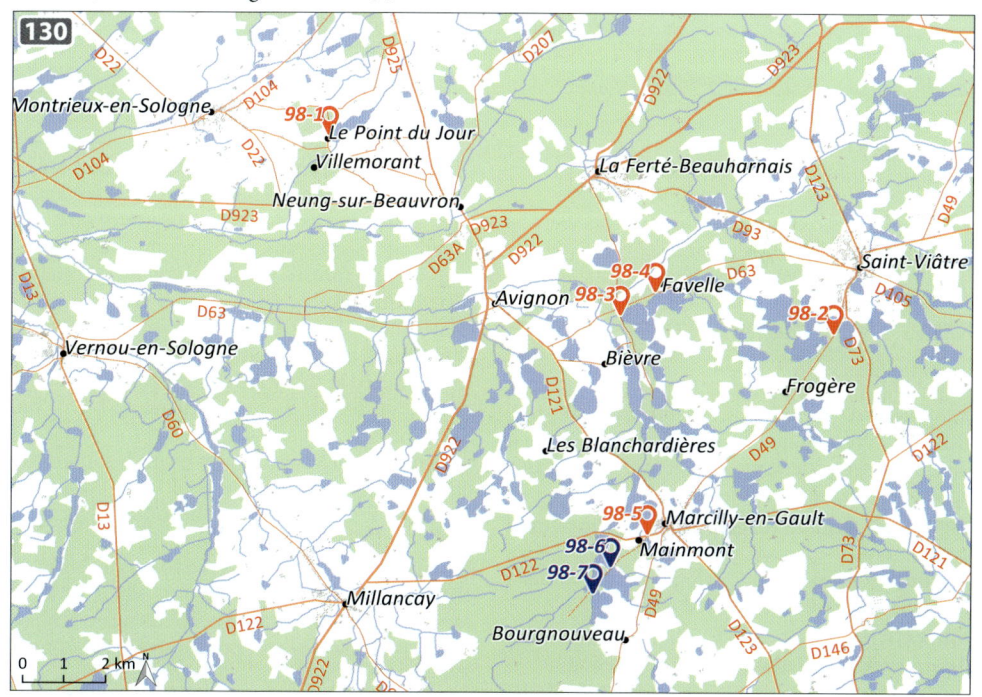

Orléans forest and the Loire valley (sites 99–101, map 131)

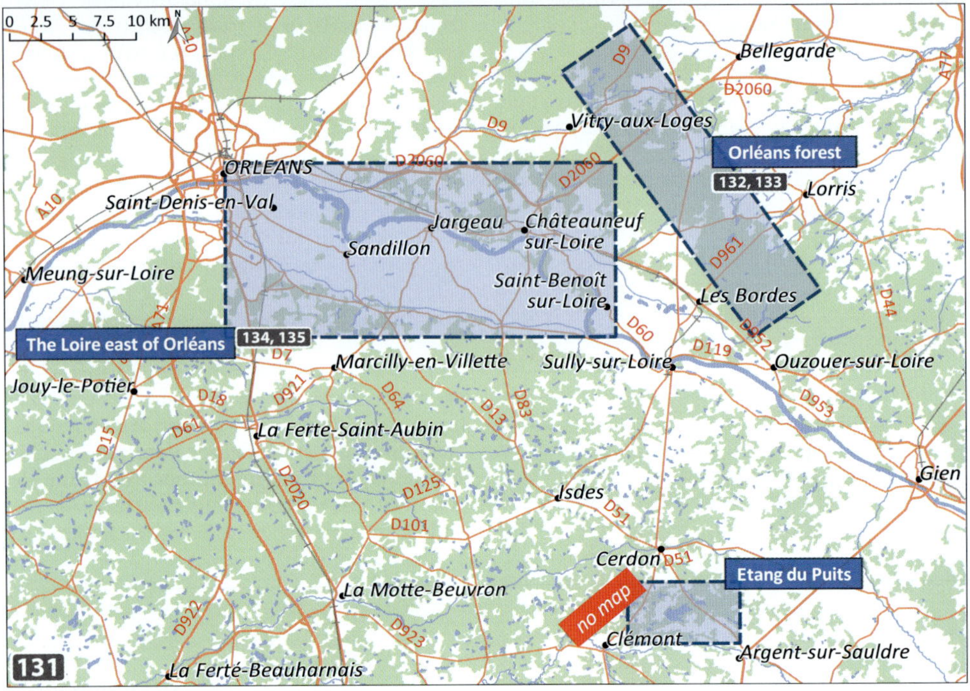

HIGHLIGHTS
- Locally high densities of rare breeding raptors
- Widespread old-growth mixed forests with a high diversity of woodland species
- Numerous stopover sites for migrants along the River Loire

| J | F | M | A | M | J | J | A | S | O | N | D |

KEY SPECIES

YEAR-ROUND woodcock, Mediterranean gull, little owl, nightjar, **grey-headed**, middle spotted and black woodpeckers, marsh and crested tits, spotted flycatcher, hawfinch

BREEDING booted eagle, osprey, short-toed eagle, honey-buzzard, goshawk, little tern, **wryneck**, grasshopper, wood and western Bonelli's warblers, red-backed shrike

MIGRATION black stork, waders including **Temminck's stint** and wood sandpiper, gulls

WINTER goldeneye, goosander, other wildfowl and geese, cranes, **Eurasian bittern**, siskin, redpoll

VISIT DURATION 1–2 days

PLAN YOUR VISIT

This is one of France's most extensive and oldest forests, and its bird community is accordingly rich, featuring a unique set of raptors and rare breeding species such as grey-headed woodpecker. The forest's flagship is osprey, which is near-unmistakable from March to July; other charismatic forest-breeding raptors, such as short-toed and booted eagles, are also likely encountered at that time of the year. All the sites can be visited in a day trip from Orléans, but their interest is seasonal. Orléans forest itself is at its best from March to May. A mid-winter visit could still be rewarded by tit or finch flocks and with luck a short glimpse of a woodcock, a great grey shrike or even a white-tailed eagle. The Étang du Puits can be covered in a half-day visit, especially in frosty conditions when bitterns tend to emerge from the reedbeds more readily. The Loire banks are productive year-round. Winter frosts will bring loads of wildfowl, and waders appear as soon as the water level falls in April and May, then in August and September. Strong westerly winds in autumn sometimes bring a lost diver, a kittiwake, or other surprises.

99 Orléans forest

2 HOURS – 1 DAY; MAPS 132 & 133 Orléans forest covers a 500 km² area delimited by the Loire to the south and the Beauce agricultural plain to the north and roughly divided into two parts, east and west of Lorris (45 km east of Orléans as the crow flies). **It is dominated by a mosaic of oaks, pines and mixed stands, with smaller proportions of beech and other trees.**

The **Belvédère des Caillettes** [1]**, a 24 m tower, provides a 360° overview over the forest** (45 km, 1 h east of Orléans, head to Châteauneuf-sur-Loire on the D2060 then north to Vitry-aux-Loges on the D137; turn right on a good forest track 2 km past Seichebrières and drive 2 km further to the car park). Visit the Belvédère on sunny spring days, late in the morning or even during the afternoon: raptors **breeding in the surroundings or flying over the forest will soar within telescope range or sometimes close enough for photography.** Expect booted eagle (from April), osprey and short-toed eagle from March, honey-buzzard (from May), common buzzard, sparrowhawk and goshawk (the latter rare and secretive) year-round. **The Belvédère ranks among a handful of good inland migration sites** (March to early May and August to November). Obviously, it does not equal the records of coastal sites or mountain passes, but you could still easily spend a day scanning passing storks, raptors, swallows, pipits, larks and finches (depending on season). A Bonelli's eagle was even seen flying by in late October 2005.

Several trails leave from the Belvédère; choose one of them at random in search of wood, western Bonelli's and willow warblers (from April), garden warbler (from May), middle spotted and black woodpeckers, marsh and crested tits, spotted flycatcher (May) and

hawfinch. Heading back south towards Vitry aux Loges, check hedgerows for little owl.

In the eastern part of the forest, a large pond known as the Étang du Ravoir [2] **is equipped with a hide which provides a direct view of an** osprey **nest from March** (50 km, 1 h east of Orléans or 20 km, 30 min west of Gien, access track to the northeast halfway between Les Bordes and Ouzouer-sur-Loire, marked with a small signpost). A pair of booted eagle breeds in the vicinity of the pond and is often seen soaring around, and migrating black stork are not unusual. A white-tailed eagle has wintered in the past nearby and others will likely show up in the future. **The** osprey **nest is monitored by webcams** which can be viewed at www.balbucam.fr or at the Carrefour de la Résistance [3] (turn left on a large forest track 6 km north of Ouzouer sur Loire).

Walk to the southeast and listen carefully for grey-headed woodpecker which breeds in surrounding old oak stands. Middle spotted woodpecker is far more common and should be found without too much effort in a wide range of habitats from inner forest to clear stands and edges, as long as there are old trees. Stay late in the forest. Nightjar is a usual inhabitant of young stands and moorlands from May, as well as grasshopper warbler. Wrynecks (rare) sing here and there on isolated trees in clear stands. An evening visit could also be rewarded by a woodcock. Western Bonelli's, wood **and** willow warblers, redstart **and** crested tit **are common to abundant**, together with the whole community of forest breeding birds. **In winter, walk along forest tracks in search of** siskin **flocks**, which can include the odd redpoll. Tit flocks are often followed by middle spotted woodpecker, nuthatch, treecreepers and other species. Check low stands for great grey shrike: it has become scarcer in recent years but single birds may still show up.

100 The Loire east of Orléans

0.5 DAY; MAPS 134 & 135 The 50 km stretch of the Loire between Orléans and Sully-sur-Loire is worth a half-day trip at any season, especially during frost episodes or when sanbanks are exposed in April–May and August–September. Start at the upstream bridge in Orléans [1] (leave the D951 at the southern end of the bridge and park just below the railway bridge). Common and little terns usually fish near the dykes and gulls (black-headed, yellow-legged and sometimes herring, common or Caspian) roost in winter.

The next spots are located along the south bank. Bypass Saint-Denis-En Val on the D951 towards Sandillon and turn left towards 'Zone Horticole de Melleray', then 'Base canoë-kayak' until you reach the river bank [2]. **Sandbeds and mudflats are often exposed, offering foraging grounds for passage** waders **from mid-April to May and in August and September.** Most of them will be green sandpiper, dunlin, greenshank and redshank, but wood and Temminck's sandpipers are seen yearly and true rarities have occurred in the recent past (including white-rumped sandpiper). **Further south, the Île aux Oiseaux near Sandillon** [3] hosts a black-headed and Mediterranean gull colony with common and little terns (irregular numbers).

The meander 12 km further east past Jargeau [4] is another good spot for waders, black stork, osprey and terns. In winter, cross the Loire at Châteauneuf-sur-Loire and check the surroundings of the bridge [5], the first meander past Germigny-des-prés [6] and Saint-Benoît-sur-Loire harbour [7]. **This section is often visited by small groups of** goldeneye, goosander **and occasionally a** great northern **or** red-throated diver. The opposite bank is better in spring and autumn for waders, especially in Bouteille, 1.7 km east of Guilly [8] (red-backed shrike and little owl are sometimes recorded around the mill) and near Sully-sur-Loire bridge [9]. From there, either go back to Orléans (50 km, 50 min), visit the Étang du Ravoir in spring (site 99, 13 km, 15 min) or the Étang du Puits (site 101, 20 km) in winter.

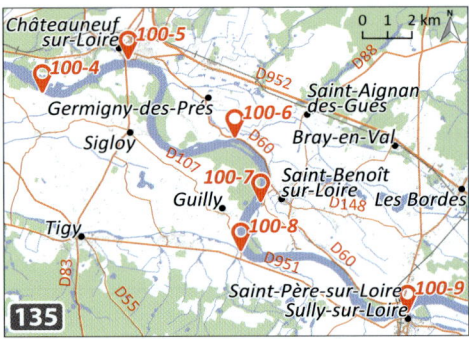

101 Étang du Puits

2 HOURS; NO DETAILED MAP The Étang du Puits pond is far from any other spot but still worth a visit for wintering wildfowl, especially during cold spells (60 km from Bourges on the D940 or 60 km from Orléans on the D13, signposted along the D948, 20 km south of Sully-sur-Loire between Cerdon and Argent-sur-Sauldre). Bitterns **winter annually** in reedbeds bordering the northern side of the pond and sometimes can be seen close to **the hide** [1] **located 800 m east of the main car park**. A good range of wildfowl can be found here: expect pochard, wigeon, gadwall, pintail, teal and small numbers of goldeneye and goosander. On your way to the hide, check siskin flocks for the occasional redpoll. Middle spotted woodpeckers breed in the surrounding woods and are usually conspicuous. **The southern dyke** [2] provides another view over the pond for wildfowl.

The Berry (sites 102–105, map 136)

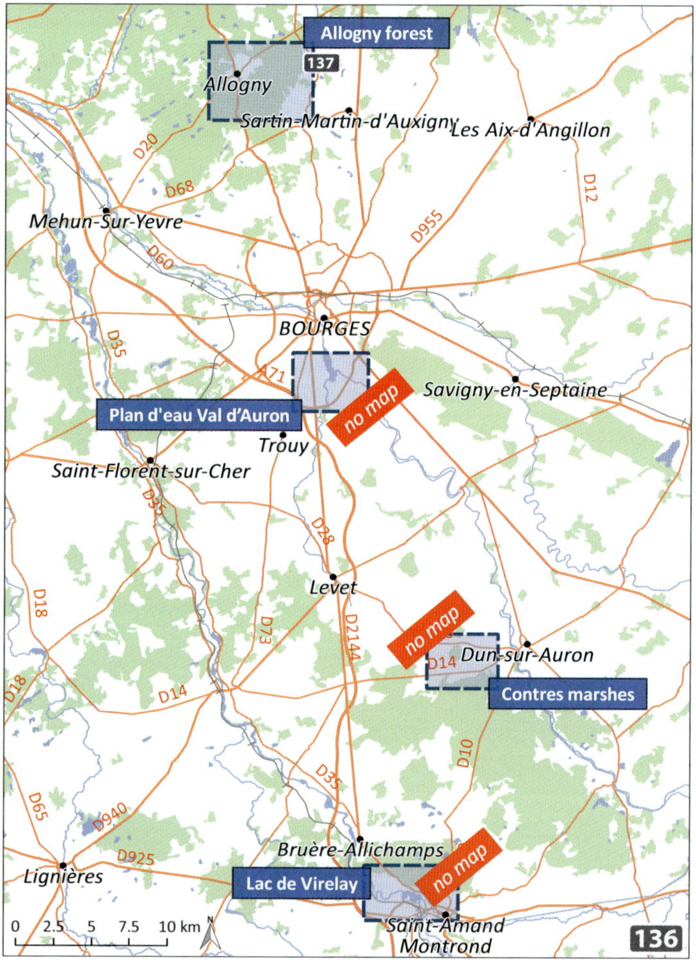

HIGHLIGHTS
- High diversity of woodland and farmland birds in remote agricultural landscapes
- Several regional breeding rarities

KEY SPECIES
YEAR-ROUND hen harrier, cattle and little egrets, **grey-headed**, black and middle spotted woodpeckers, marsh and crested tits, **great grey shrike**, hawfinch

BREEDING night-heron, short-toed eagle, black kite, **bee-eater**, sand martin, garden, sedge and wood warblers, red-backed and **woodchat shrikes**

MIGRATION black stork, jack snipe
WINTER wildfowl, grebes and gulls
VISIT DURATION 1–2 days
PLAN YOUR VISIT
The region is remote but a few sites south of Bourges deserve short visits at all seasons and could make a suitable stop on your way to Clermont-Ferrand and the south.

102 Allogny forest
2–4 HOURS; MAP 137   Allogny forest lies 20 km north of Bourges (take the D944 towards Salbris). Park near the Allogny pond south of the village or along nearby trails [1,2]. Look for black and

middle spotted woodpeckers, wood and garden warblers (from April), golden oriole (from May), hawfinch, marsh and crested tits, which are all common throughout the forest. Grey-headed woodpecker **is easier to find in the eastern part of the forest in old beech–oak mixtures.**

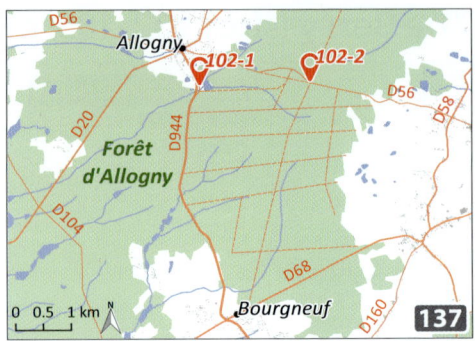

103 Val d'Auron
1 HOUR; NO DETAILED MAP   The **Val d'Auron [1] is a large pond used for watersports at the southern edge of Bourges, near the airfield.** A heron colony (night-heron, cattle and little egrets) and wintering wildfowl, grebes and gulls (mostly black-headed and common) justify a short visit. **Park towards the southern end and walk on the banks.**

104 Contres marshes
1–3 HOURS; NO DETAILED MAP   Contres marshes are a mosaic of marshes, fields and meadows located 30 km, 30 min southeast of Bourges (drive to Dun-sur-Auron on the D953, turn right on the D28 towards Levet and bird along the narrow road which crosses the marshes [1]). **Concentrate on flooded fields and hedgerows** to search for foraging black stork, hen harrier, short-toed eagle, wintering great grey shrike (rare), snipe and jack snipe, breeding sedge warbler and red-backed shrike.

105 Virelay
1–3 HOURS; NO DETAILED MAP   **The Lac de Virelay, with its boat club** [1]**, lies at the southern edge of the region near Saint-Amand-Montrond (42 km, 40 min south of Bourges on the D2144).** A mixed heron colony on an islet features grey heron, night-heron and cattle and little egrets. **Bee-eaters and sand martins are often seen hunting over the lake in May.** The nearby Noirlac Abbey [2] (4 km west) is surrounded by a bocage where black kite, hobby, red-backed and sometimes woodchat shrike, buntings and warblers breed.

REGION 6
POITOU-CHARENTES AND THE VENDÉE

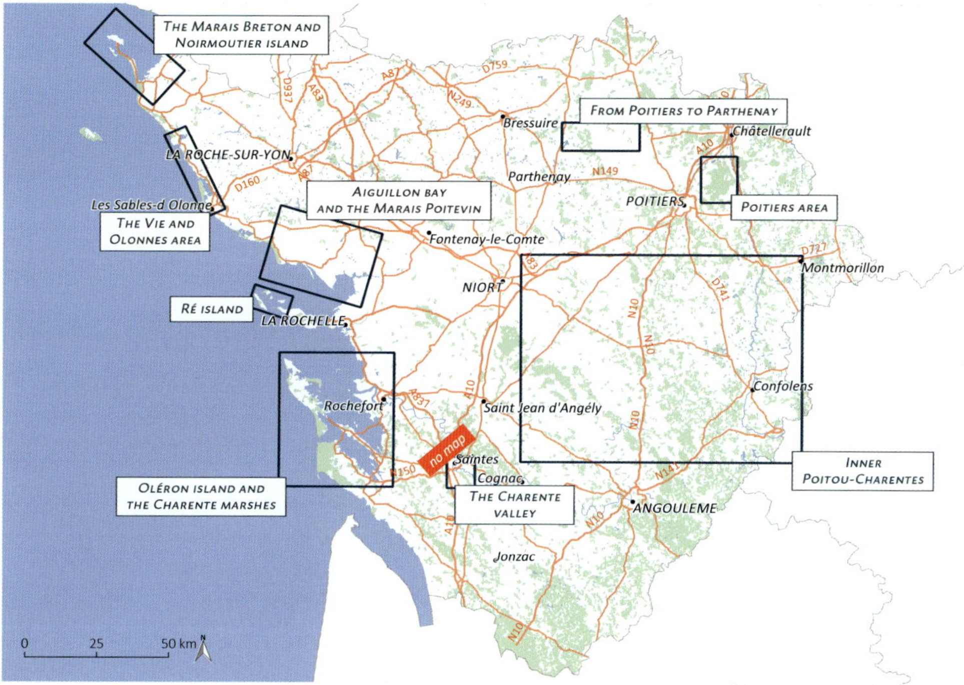

HIGHLIGHTS
» Countless freshwaters and coastal bays where waders and wildfowl stop over or winter from August to March
» One of France's most spectacular migration bottlenecks in the Aiguillon bay
» The highest densities of breeding little bustard and ortolan bunting in the northern half of France, sharing wide expanses of cultivated plains with other farmland specialities

REVIEWERS Michel Caupenne, Thomas Chevallier, Julien Gonin

USEFUL WEBSITES
www.faune-charente-maritime.org
www.faune-vendee.org
www.faune-charente.org
www.vienne.lpo.fr
www.ornitho79.org

The little bustard is an icon of farmland bird conservation in France. Although its western populations have dramatically declined, it can still be found locally between Poitiers and Parthenay (sites 130 and 131).

The Brouage marshes (site 122) offer some of the Vendée's best opportunities for breeding reedbed species and migrating shorebirds.

The Marais Breton and Noirmoutier island (sites 106–108, map 138)

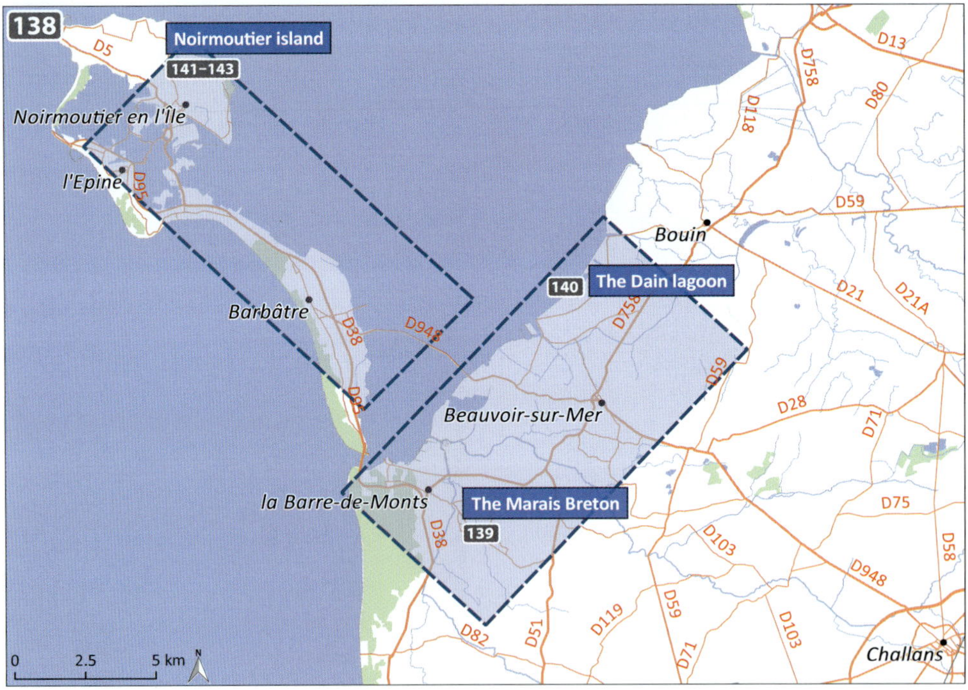

HIGHLIGHTS

» A mixed gull and tern colony, sometimes hosting an elegant tern or a slender-billed gull
» Huge numbers of wintering and migrating waders and wildfowl
» Breeding waders in coastal marshes and lagoons

KEY SPECIES

YEAR-ROUND avocet, yellow-legged and herring gulls
BREEDING garganey, Montagu's harrier, black-winged stilt, black-tailed godwit, Kentish plover, Mediterranean gull, **elegant tern** (one returning

individual), Sandwich tern, **slender-billed gull** (one returning individual), bluethroat
MIGRATION wildfowl, waders
WINTER brent goose, short-eared owl
VISIT DURATION   2 days
PLAN YOUR VISIT
There are extensive areas of flooded meadows, coastal marshes and reedbeds in this area – a year-round birders' playground where the highlights are the huge numbers and vast diversity of wildfowl and waders. Choose either to visit the mainland or Noirmoutier island, as it is unlikely that you will manage both in a single day. As at any coastal site, tides are critical when setting your itinerary: a rising tide concentrates wader and wildfowl roosts.

106   The Marais Breton

2 HOURS – 0.5 DAY; MAP 139   The Marais Breton wetland extends across a 45 km² coastal plain between Challans and Noirmoutier island (60 km, 1 h south of Nantes or 40 km, 50 min northwest of La Roche-sur-Yon). This vast expanse of flooded meadows interspersed with canals and polders is home to **one of France's most diverse breeding** waterbird **communities**. Start at La Barre-de-Monts church (20 km, 30 min west of Challans) and drive southeast for 2 km to the 'Écomusée Le Daviaud' [1] (follow signposts). From there, either walk along trails into Le Daviaud protected area or stop-and-go on surrounding roads looking for breeding avocet, black-winged stilt, lapwing, black-tailed godwit, redshank, shelduck, shoveler, short-eared owl (at dusk) and migrating sandpipers, pintail, garganey among other wildfowl. Alternatively, start 8 km north in Beauvoir-sur-Mer and drive south on the D22, then turn left as you leave the village at a roadside cross. Migrating (spring) and breeding waders and ducks roost

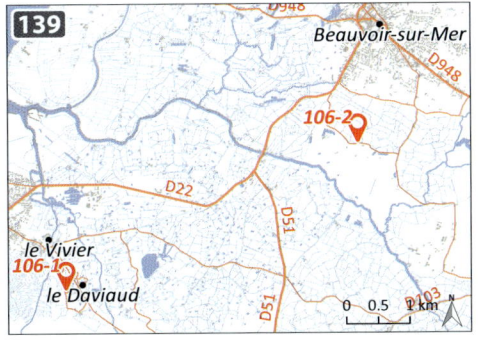

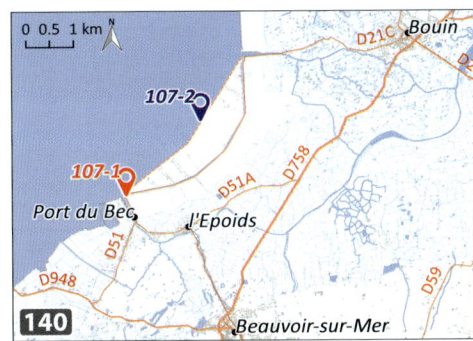

on surrounding ponds and flooded meadows and are easily seen from the road that loops through the marshes [2].

107   The Dain lagoon

2 HOURS – 0.5 DAY; MAP 140   Facing Noirmoutier island, the Dain polder and brackish lagoon is **a major high-tide roost site for migrating/wintering** waders **and a breeding area for** waterbirds. Drive from Beauvoir-sur-Mer to l'Epoids and park near the lock in Le Port du Bec [1]. **Cross the canal and walk along the dyke for 1 km to find a suitable viewpoint over the lagoon** [2]. Local breeding species include avocet, Kentish plover, redshank, lapwing, black-headed gull, common and Sandwich terns and Montagu's harrier on the polder.

108   Noirmoutier island

0.5–1 DAY; MAPS 141–143   Noirmoutier is an elongated flat island lying northwest of La-Barre-de-Monts, to which it is connected by a bridge. The island consists of a rocky islet connected to a 6 km sand bar by wide expanses of saltmarsh. **Although it is largely urbanised, it is one of the Vendée's hotspots at all seasons**. Starting from the south, there is **a large** gull **and** tern **colony is east of Barbâtre in the Polder de Sébastopol nature reserve** [1] (follow the main road from the bridge for 5 km and turn right just past a roundabout on 'Rue du polder'; park at the end of the road). **The best sightings are obtained by walking on the main dyke**. Avocet, common and Sandwich terns, Mediterranean and black-headed gulls account for most of the locally breeding birds, together with smaller numbers of other waders. An elegant tern **bred with** Sandwich terns **until 2016**, and a returning slender-billed gull has been seen annually in the colony.

114 | 108 NOIRMOUTIER ISLAND | POITOU-CHARENTES AND THE VENDÉE

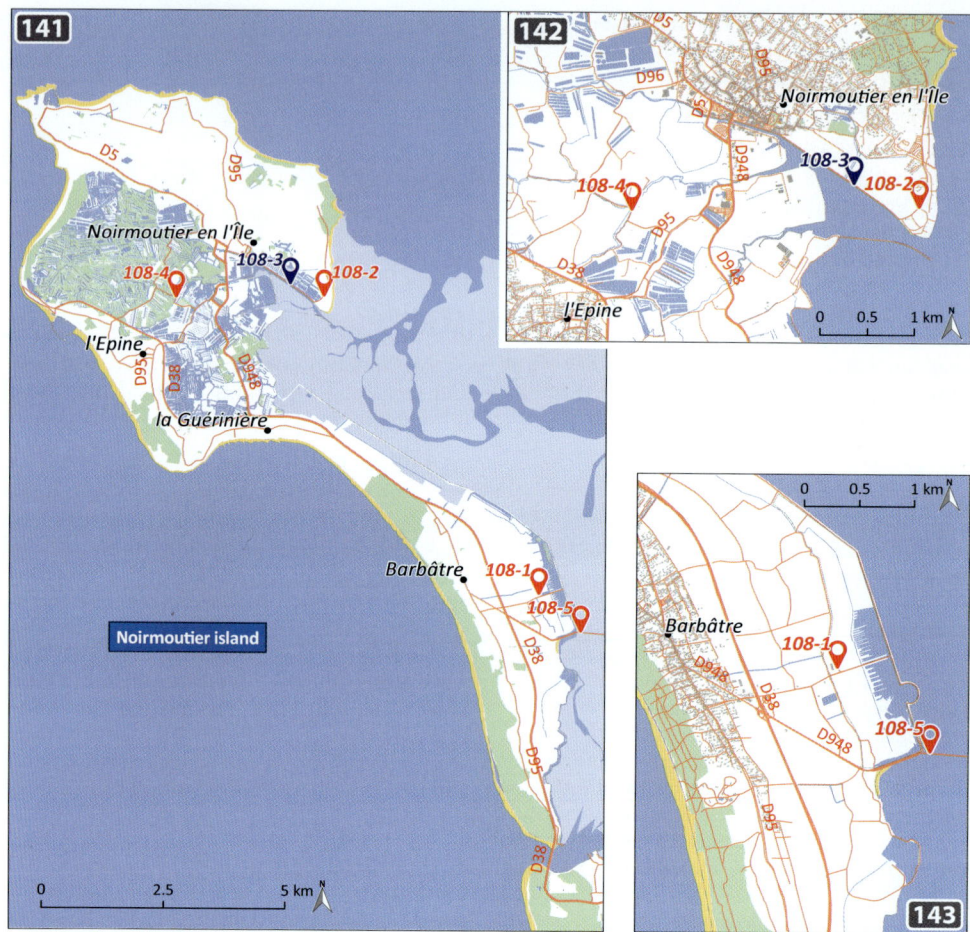

Another nature reserve lies on the northern islet. Follow signs to 'Marais de Müllembourg' in Noirmoutier-en-l'Île (15 km, 25 min from Barbâtre) and park at the entrance to the reserve [2] to **walk towards the 'Jetée Jacobsen'** [3] where avocet, stilt, redshank, Kentish plover, common and Sandwich terns breed. The site hosts good numbers of brent goose, greenshank, redshank, avocet, sandpipers and plovers roosting at high tide in winter and at both migrations. **A walk or drive inside L'Épine saltmarshes** (3 km from Noirmoutier-en-L'Île, turn right past the supermarket) through Pont de l'Arceau [4] should yield breeding shoveler, redshank, stilt, yellow-legged gull, shelduck, bluethroat and migrating waterbirds (mainly sandpipers, stints and dabbling ducks).

For more waders, take the road known as 'Le Gois' [5], a causeway that connects the southern tail of the island to Beauvoir-sur-Mer. Plovers, sandpipers, godwits and curlews usually feed on the mudflats along the road on a rising tide (but be aware that the road is completely flooded at high tide).

The Vie and Olonnes area (sites 109–110, map 144)

HIGHLIGHTS
» Numerous spots for wintering and migrating wildfowl, waders, spoonbill and gulls
» Breeding waterbirds in the marshes
» Morning seawatching at L'Armandèche lighthouse for Cory's shearwater

KEY SPECIES
YEAR-ROUND gannet, **spoonbill**
BREEDING black-winged stilt, bluethroat
MIGRATION Balearic, Manx, sooty and **Cory's shearwaters**, storm petrel, **Leach's petrel**, all skuas, terns (including Arctic, rare), auks
WINTER wildfowl, all divers, **purple sandpiper**
VISIT DURATION 1 day
PLAN YOUR VISIT
Waders, terns and gulls concentrate on restricted coastal areas to roost on a rising tide; find them and stay put for a while. Westerly winds from late August to November should take you to seawatching spots.

109 Sion-sur-l'Océan
2 HOURS; NO DETAILED MAP The coastline facing the Île d'Yeu has several major gull and wader roosts and offers seawatching opportunities around Sion-sur-l'Océan in Saint-Hilaire-de-Riez

(19 km, 20 min south of Challans in the northern vicinity of Saint-Gilles-Croix-de-Vie). Park along the coast road [1] wherever you find a suitable viewpoint. At high tide, look for wintering purple sandpiper among flocks of sanderling, dunlin, turnstone, plovers and large gulls on the rocky shore. **Check roosts of** common, little **and** Sandwich terns **from late August through September** for Arctic tern. **In autumn, westerly winds force** seabirds **to the coast.** Expect shearwaters (Balearic, sooty and Manx are frequent; Cory's is scarcer), storm petrel and possibly the odd Leach's petrel in October–November.

110 Olonne marshes (Marais d'Olonne)

2 HOURS – 1 DAY; MAPS 145 & 146 The Marais d'Olonne is a large expanse of saltmarshes 5 km north of Les Sables-d'Olonne which hosts most of the local wetland breeding species, including lesser black-backed and yellow-legged gulls, avocet, black-winged stilt, redshank, common tern, shelduck, bluethroat and large numbers of migrating waders, wildfowl and spoonbills. From Les Sables, drive north towards Saint-Gilles-Croix-de-Vie on the D38 and turn left at the 'Observatoire ornithologique' signpost as you enter L'Île d'Olonne to find a car park [1]. **From there, a path leads to a hide overlooking the saltmarshes; try to be there early in the morning, especially if the tide is rising.** Back in Les Sables-d'Olonne, Armandèche lighthouse [2] **ranks among the best seawatching sites of**

the region in a westerly wind, especially during autumn migration. Expect good numbers of gannet, skuas, terns, Balearic, sooty and Cory's shearwaters and storm petrel. In winter, look for gulls, auks, grebes, divers and wildfowl from the piers at the entrance and inside the fishing harbour [3].

The Aiguillon bay and the Marais Poitevin (sites 111–116, map 147)

HIGHLIGHTS
» European-level stopover and wintering sites for waders and wildfowl
» Spectacular late-summer migration in the Aiguillon bay

KEY SPECIES
YEAR-ROUND wildfowl, spoonbill, marsh and hen harriers, peregrine, avocet, both godwits, other waders, little owl

BREEDING garganey, white stork, black kite, Montagu's harrier, black-winged stilt, little ringed and Kentish plovers, tawny pipit, bluethroat, melodious, sedge and reed warblers

MIGRATION black stork, spotted crake, waders, great skua, black and Caspian terns, swallows, martins, larks, pipits, wheatears, thrushes, warblers, flycatchers, finches, icterine and aquatic warblers (scarce), ortolan bunting (rare)

WINTER brent, greylag and white-fronted geese, surf scoter (rare), long-tailed duck (rare), red-throated diver, merlin, common crane, auks, short-eared owl

VISIT DURATION 2–3 days

PLAN YOUR VISIT
The bay is one of the country's most impressive migration spots for wildfowl, waders and passerines, and a late-summer favourite of French birders, worth several birding days at all seasons.

To make the most of your experience, visit during autumn passage between late August and late November when loads of passerines concentrate at the Pointe de l'Aiguillon on their way south while massive roosts of waders and wildfowl gather at high tide. The breeding season will yield bluethroat, tawny pipit and breeding-plumage waders. Choose either the north or the south side of the bay and visit the mouth of the bay on a rising tide, keeping inner sites for high and falling tides.

111 The Aiguillon bay

1–2 DAYS; MAP 148 The Aiguillon bay links the Marais Poitevin to the sea 20 km north of La Rochelle. Its mouth is protected from the open sea by Île de Ré to the south and by the Pointe de l'Aiguillon to the north, resulting in **a near-closed bay in which low tide uncovers wide expanses of mudflat and saltmarsh twice a day**. Being located on the east Atlantic flyway at the northern extremity of the Bay of Biscay, it is a logical **stopover and wintering site for tens of thousands of waterbirds from late August to May**. Expect spectacular roosts of avocets, godwits (both species), plovers, dunlin, sanderling, other sandpipers and shanks, with large numbers of the more common species. Wildfowl numbers peak from November to March with among the largest wintering numbers of brent and greylag geese, a few white-fronted geese and teal, shoveler, wigeon, shelduck and pintail (the last of these especially in late February and March). All gulls and terns (with always a few Caspian terns) are also abundant at both passages. In late August, aquatic warblers stop over in very low numbers in grassy meadows (check the news). **Visit the spots around the seashore on a rising tide to scan the flocks inside the bay**. At high and receding tide, visit saltmarshes, mudflats and canals in the inner bay.

From the south, start at the Pointe-Saint-Clément [1] (drive to Esnandes, 13 km, 20 min north of La Rochelle, and turn left at the traffic lights; park after 2 km past Les Misottes). Short-eared owls winter in saltmarshes all around; there is no particular viewpoint, as roost sites change from time to time, but they are worth looking for, as the birds hunt at dusk or even during the day. Black kites, marsh and Montagu's harriers breed in the surrounding fields.

At high tide, waders **roost on salt meadows and ponds 4 km to the north** (bypass Esnandes and park near the 'Canal du Curé' [2], scan the area from the locks), **the 'Port du Pavé'** [3] (5 km further north past Charron, straight on the Rue du Pavé in Les Groies) **and 'La Prée Mizotière'**

[4] (turn left just after crossing the Sèvre Niortaise river on the D10A to Puyravault, park at 'La Petite Prée' and go through the wooden gate).

Be careful when entering the site: birds will see you arriving at a distance and are likely to fly away if you do not take the necessary precautions. This area is a regular spot for Temminck's stint and black tern on both spring and autumn passage within groups of sandpipers, curlews and geese. In spring, the site hosts colonies of avocet, black-winged stilt and a few pairs of little ringed plover, lapwing, redshank and shelduck. The habitat is suitable for passage aquatic warbler; search for this species in late summer among high grasses (beware of young sedge warblers).

An observation platform offers a view over the bay from its northern coast and is a suitable spot from which to scan the flocks on a rising tide [5] (drive to Saint-Michel-en-l'Herm, then take the D60 to La Dive; leave it to the south towards a grain silo and park after 4.5 km). Here as well, thistles and grassy patches could yield an aquatic warbler in late summer. There are Montagu's harriers in the surrounding fields in spring and little owls breed on La Dive cliff.

If you wish to visit the northern edge of the bay, drive to L'Aiguillon-sur-Mer and follow signs to 'Pointe de l'Aiguillon'. They will lead you to a dyke on which you can stop here and there (for instance at [6] or [7]) to check the gulls on sandpits and ponds. **The main viewpoint for the northern bay is the Pointe de l'Aiguillon itself at the end of the road** [8]. Kentish plover, bluethroat and tawny pipit breed in good numbers on sand dunes from April. Waders, wildfowl, gulls and terns get closer as the tide rises; expect shelduck, sanderling and Sandwich tern together with passage or wintering wildfowl and waders virtually year-round (lowest activity in June and July). **The Pointe is at its best from late August through November when flocks of passerines gather before crossing the bay on their way south;** a team of spotters is present most of the time in autumn. Thousands of swallows and martins (late August), pipits, wheatears and warblers (September), larks, thrushes and finches (from October) gather at the Pointe before crossing the strait to La Rochelle. **The tamarisk hedgerow north of the car park along the road can be crowded with** willow, garden and melodious warblers, blackcaps, common redstarts, flycatchers and other insectivorous passerines, sometimes joined by rarer species such as icterine warbler or ortolan bunting.

Another good place for migrating waders **and** passerines **is the Pointe d'Arçay, 7 km**

to the north. Cross the bridge from l'Aiguillon-sur-Mer to La-Faute-sur-Mer and follow signposts to park as far as possible along the road [9]. From here, a 3.5 km path leads to the Pointe d'Arçay through a network of ponds and lagoons which are worth a check for waterbirds.

112 The Belle Henriette lagoon

1–2 HOURS; MAP 148   La Belle Henriette reserve covers a natural coastal lagoon halfway between La-Faute-sur-Mer and La-Tranche-sur-Mer, 14 km north of the Pointe de l'Aiguillon. Several paths cross the lagoon from a car park at the southern edge of the 'Camping Les Violettes' [1]. **Bird numbers and diversity vary constantly because of changing water levels and habitat instability**; you will quickly see what to expect. Breeding species include Kentish plover, tawny pipit and bluethroat, and there are wader roosts and fishing terns year-round. If mudflats are exposed, expect stints, sandpipers and shanks at both passages, and scan the edges of reedbeds for spotted crake. Look for sedge and reed warblers in reedbeds, and flycatchers and warblers on the wooded edges of the camp site and in bushes. Aquatic warbler could stop in tall grasses in late August, but do not count on it!

113 La Terrière

2 HOURS; NO DETAILED MAP   La Terrière beach [1], 4 km northwest of La Tranche-sur-Mer, **is a viewpoint for wintering** common scoters (up to several thousand) joined by one to a few surf scoters almost annually. Red-throated divers, auks, long-tailed ducks (rare), Sandwich terns and great skuas should be easily found with a scope on clear days with light winds.

114 Saint-Denis-du-Payré

2 HOURS; MAP 148   Saint-Denis-du-Payré reserve is a well-preserved marsh typical of Marais Poitevin wetlands. The site consists of **flooded meadows and shallow ponds with mudflats used as a roost by migrating and wintering** waterbirds – the birdiest of all inland sites in the region. First visit the reserve headquarters in Saint-Denis-du-Payré [1] to buy a ticket (€5, open 09.30–18.00, midday break, seasonal changes, www.reservenaturelle-saint-denisdupayre.fr), then enter the reserve by driving along the D25 for 1 km eastwards [2]. The only viewpoint is **a large hide with scopes, guidebooks and one or two volunteers**, but you can also find additional species by driving

on the surrounding local roads. In mid-winter, expect dabbling ducks (teal, garganey, shoveler, pintail, wigeon, gadwall), greylag geese, cranes and common waders (lapwing, plovers, dunlin, godwits and curlews), raptors (peregrine, merlin, hen harrier) and short-eared owl. Wader diversity peaks in early May and late August (expect black-tailed godwit, ruff, curlew, black-winged stilt, sandpipers, plovers, stints), and black stork is commonly seen on mudflats at these periods. Spoonbill is present year-round and surprises may arise any time (long-billed dowitcher, Pacific golden plover and more commonly marsh sandpiper and Temminck's stint have been recorded from the hide).

115 La Vacherie nature reserve

1 HOUR; MAP 148   La Vacherie consists of 180 ha of **flooded marshes that attract breeding and migrating** waterbirds (10 km northeast of Aiguillon bay, inland, signposted from the D125, to the right 800 m after entering Champagné-les-Marais from the east). The main viewing point is a hide near the reserve headquarters [1]. The site hosts a black tern colony (from April), breeding redshank and black-tailed godwit. Winter and spring migration bring the usual wildfowl (including shoveler, pintail, garganey and teal) and waders. In winter, peregrine may be seen hunting over flocks of hundreds of golden plover and lapwing that often include a few ruff.

116 Lairoux-Curzon marshes

0.5 DAY; MAP 149   The Lairoux marshes are another **stopover site for** waterbirds **during spring migration**. Drive west for 8 km from

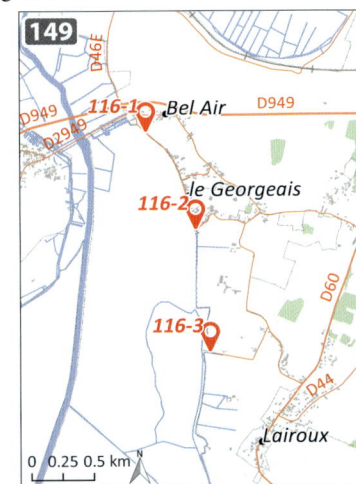

Luçon on the D949 towards Avrillé and turn left on the D2949 at Bel Air, then follow 'Communal' signposts. Drive for 100 m to reach a first car park, which offers a view over the flooded meadows [1], which may be crowded with **wildfowl** (shoveler, teal, garganey, pintail, wigeon) and **waders** (black-tailed godwit, golden plover, curlew, whimbrel, redshank, lapwing) plus greylag goose, herons and white stork. **There are two other viewpoints, 1 km south in Le Georgeais [2] (hide) and 2 km further south again in Les Terres Gâchées [3].**

Île de Ré (site 117, map 150)

HIGHLIGHTS

» Spectacular high-tide wader roosts in the Fiers d'Ars cove
» Wintering seabirds from the seashore and one of the best seawatching sites of the region
» A regional stronghold for scops owl

KEY SPECIES

YEAR-ROUND spoonbill, avocet
BREEDING black-winged stilt, both godwits, bluethroat, **scops owl**
MIGRATION Balearic shearwater, storm petrel, **waders**, skuas, terns, **yellow-browed warbler**
WINTER brent goose, red-breasted merganser, other **wildfowl**, black-necked and Slavonian grebes, black-throated, great northern and red-throated divers, **purple sandpiper**
VISIT DURATION 1–2 days

PLAN YOUR VISIT

The Île de Ré is vast, crowded with birds year-round, and a nice place for a family holiday as well as for a hard-core bird race. Whatever your style of birding, choose your itinerary according to the tides. Try seabirds and the Lilleau des Niges reserve at high tide, while the cove is better on a rising tide. In autumn, westerly winds should lead you to the Phare des Baleines lighthouse for seawatching.

117 Île de Ré

1–2 DAYS; MAPS 151–154 The Île de Ré is accessed via a well-signposted bridge from La Rochelle. **The best birding areas are located at the northwest extremity of the island**, the southeast being highly urbanised and covered with fields and woods. Turn left on a track between the 'Écomusée des Marais Salans' and Loix (22 km, 30 min from the bridge), bypass the equestrian centre and park in La Lasse, on the sea front [1]. **Be there at rising tide and walk along the dykes** to watch flocks of shorebirds, brent geese and gulls as they enter the Fier d'Ars cove to roost on beaches and rocky shores. In winter, scan the sea for black-necked and Slavonian grebes, red-breasted merganser, great northern diver and sometimes black-throated and red-throated divers. **Another viewpoint for wintering seabirds is the nearby Pointe du Grouin** [2], 3 km east of Loix (signposted).

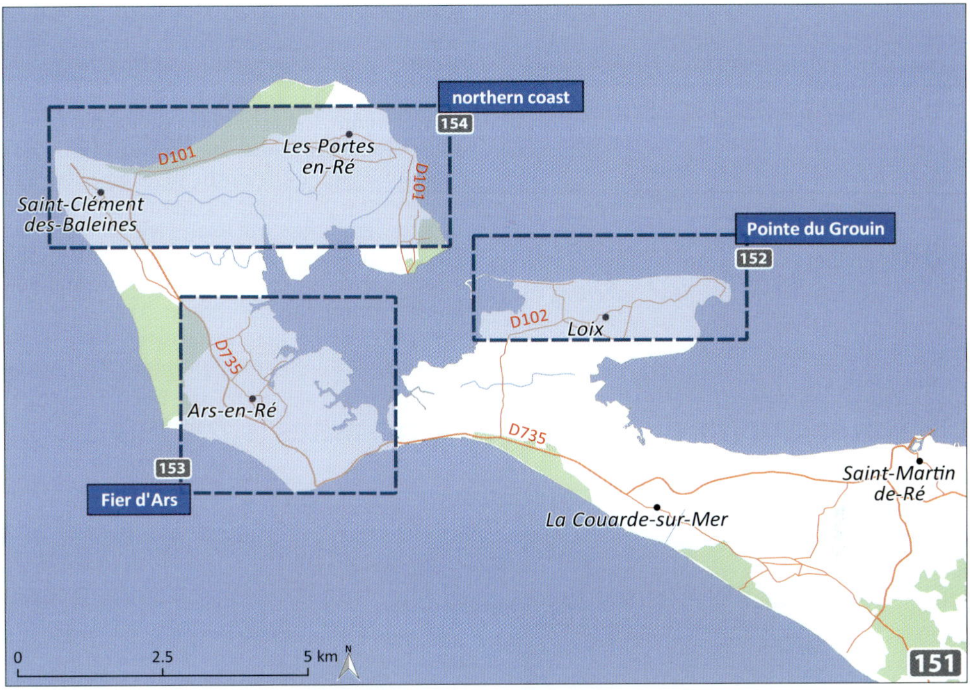

Resume your drive towards Ars-en-Ré on the D735 and make a stop at a car park where the road makes a sharp turn [3]. **Walk on the dyke at high tide for further seabirds**, including red-breasted merganser, black-necked

and Slavonian grebes. **Close views of roosting waders** at high tide can be obtained from the Fiers d'Ars cove [4] (cross Ars-en-Ré and turn right after the harbour, following 'Zone Aquacole', follow the road to its end). The roosts are dominated by ringed and grey plovers, bar-tailed and black-tailed godwits, redshank, greenshank, curlew, whimbrel, knot, dunlin, sanderling, oystercatcher and turnstone.

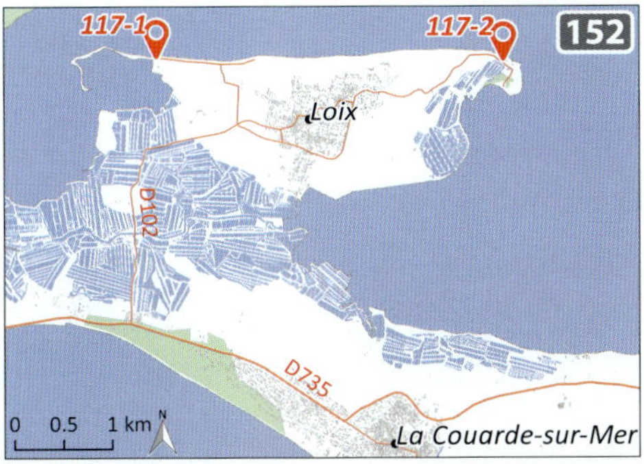

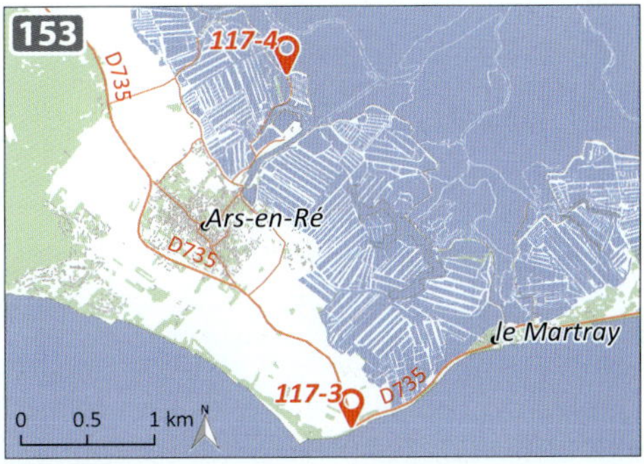

Other roosts form in the Lilleau des Niges nature reserve (drive to Les Portes en Ré and leave the D101 right in front of a small supermarket towards 'La Maison du Fier' [5]). An extensive network of walking trails spreads across the reserve, and you can expect close-range views

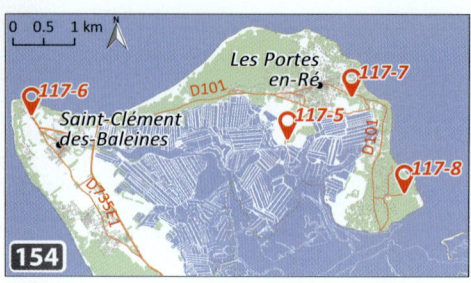

of wildfowl and waders, spoonbill, breeding stilt, avocet, common tern, bluethroat and yellow wagtail. From April to August, woodlands around Les Portes en Ré host **a population of scops owl**. They can be found singing at dusk in La Prée camp site [6] or in the Trousse-Chemise wood around the car park [7].

In autumn, southwesterly or northwesterly winds should lead you to the **Phare des Baleines lighthouse** [8], 2 km north of Saint-Clément-des-Baleines. A couple of hours' seawatching from the lighthouse will likely yield skuas, terns, Balearic shearwater and storm petrel. If passage is poor, search for **passerines in the woodlands that surround the lighthouse**. Yellow-browed warbler is regular here in October. Also check the shore for purple sandpiper and gulls.

Oléron island and the Charente marshes (sites 118–127, map 155)

HIGHLIGHTS

» An internationally important area for passage and wintering waders and wildfowl along the east Atlantic flyway
» A wide range of breeding marsh species, including herons, waders and passerines
» An autumn or winter day trip to Oléron for seabirds, waders and gulls

KEY SPECIES

YEAR-ROUND little grebe, great white egret, **spoonbill**, marsh and hen harriers, little owl, crested lark, reed bunting

BREEDING white stork, **night-heron, purple heron, squacco heron**, cattle egret, black kite, black-winged stilt, Kentish plover, **scops** and little owls, turtle dove, hoopoe, tawny pipit, bluethroat, fan-tailed, sedge, **Savi's**, reed, great reed, Cetti's, melodious and western Bonelli's warblers, whitethroat, **bearded tit**, red-backed shrike

MIGRATION garganey, Balearic shearwater, glossy ibis, osprey, peregrine, **waders** including curlew sandpiper, knot, ruff, **Temminck's stint**, little gull, black, whiskered and white-winged terns, auks

WINTER pale-bellied brent goose and black brant, common and velvet scoters, eider, **long-tailed duck**, red-breasted merganser, other wildfowl, black necked and Slavonian grebes, Eurasian bittern, **purple sandpiper**, other waders, yellow-legged, **Caspian** and Mediterranean gulls, short-eared owl, **penduline tit**

VISIT DURATION 3–4 days

PLAN YOUR VISIT

This area encompasses vast wetlands with extensive reedbeds and tidal flats that attract waterbirds at all seasons and the full breeding community of brackish-water marshes. As with anywhere in the region, tides are an essential parameter in planning your visit. Some sites can be covered in barely more than an hour, while others will require more time. Visit reedbeds at dawn to maximise your chances with singing warblers; moderate use of playback can help for some species. Keep tidal mudflats for the rising tide as waders flock to roost. Beware that some roads in Brouage may be hard to drive in a standard car. Plan a day or two in Oléron, especially from late August to May: it has more waders and seabirds to offer than you can actually count.

118 Yves bay

0.5 DAY; MAP 156   The Yves bay nature reserve (Réserve Naturelle du Marais d'Yves) is located halfway between La Rochelle and Rochefort along the D137, south of Châtelaillon-Plage. **This wide bay combines a mosaic of muddy foreshore, dry dunes, marshes with reedbeds and tamarisk hedges.** It compares to the Aiguillon bay bird-wise, but you will miss passerine migration. On the other hand, it can be even better for waders owing to its smaller extent. **Visit the northern part of the bay on a rising tide as** waders **roost along the shore, allowing short-distance views** (leave the D137 towards Yves, then follow 'Les Boucholeurs' and turn left 250 m after crossing the railway towards 'colonie de vacances', park at [1]). **Walk south along the coastal path until you reach a concrete dam, from which you can scan thousands of** ringed **and** grey plovers, bar-tailed **and** black-tailed godwits, redshank, curlew, whimbrel, knot, dunlin, sanderling, oystercatcher, turnstone, brent geese **and** shelduck, **with smaller numbers of various other species.** The reserve itself is worth a visit (you need to apply for a permit, which is only available for fixed times: check http://marais.yves.reserves-naturelles.org). Report at the 'Pôle nature' located on the 'Aire de la Baie d'Yves' on the D137 heading south towards Rochefort [2]. In May, the reserve hosts all local breeding species, including black-winged stilt, little grebe, shelduck, bluethroat and tawny pipit.

119 The Cabane de Moins

2–5 HOURS; MAP 156   The Cabane de Moins gives you access to a network of ponds and meadows that cover most of the area north of Rochefort (pass Breuil-Magné on the D116 northwards and turn left at the 'Cabane de Moins' signpost). On your way, you should find a little owl on the roof of La Baudette farm. Also check the meadows around Pré du Fond; they host farmland and

waterbird species, including wintering great white egrets, short-eared owls (at dusk), marsh and hen harriers. Once at the car park [1], **you can either visit the Cabane de Moins reserve itself or scan the area from the hill** [2]. Bird numbers are highest from late winter to spring when geese (greylag – and look for the odd pink-footed), dabbling ducks (shoveler, teal, pintail, garganey, gadwall) and waders (black-tailed godwit, sandpipers and shanks) roost in surrounding fields. In spring, focus on the road leading to the Marais du Roy [3] where red-backed shrike, turtle dove, white stork and black kite breed.

120 The Charente estuary

3 HOURS – 0.5 DAY; MAP 157 The Charente river mouth is surrounded by marshes with numerous wet pans where wildfowl and waders forage in numbers. **Start at the Pointe de la Parpagnole on the northern bank** [1] (15 km, 20 min from Rochefort via the D137, leave it in Saint-Laurent-de-la-Prée and turn left past the church on the 'Route de la grande levée'). In spring, black-winged stilt and lapwing breed on the ponds, while the meadows are home to fan-tailed warbler, corn bunting, yellow wagtail, shelduck and migrating snipe, harriers, purple heron, common and green sandpipers, kingfisher, great white egret and wintering short-eared owl. Look for singing bluethroat in the vegetation that covers the banks; high tide should also bring waders (avocet, redshank, grey, little and ringed plovers, dunlin, curlew sandpiper, knot, black-tailed and bar-tailed godwits).

For birding on the estuary itself, go back to Saint-Laurent-de-la-Prée and head south towards Soumard until you find the 'Route du Marais' (4 km). Park near the gate [2] and walk on the road towards Fort la Pointe where waders **concentrate on the many wet pans and coastal mudflats along the dyke.**

A further 6 km northwest, past Fouras, the rocky coast around Pointe de la Fumée [3] is a **roosting site at high tide** for migrating Sandwich, common and little terns and wintering Mediterranean, common and herring gulls. Scops owl **breeds in the small wood** at the base of the headland [4].

121 Rochefort water treatment plant

1–3 HOURS; MAP 158 The Rochefort water treatment plant (station de lagunage) and the surrounding marshes provide an easy access to well-preserved meadows that border the Charente banks (2 km south of Rochefort, drive the D733 south and turn right on Route de Soubise in front of the Cap-Vert supermarket, just before crossing the bridge). **These wetlands are hunting-free and thus serve as roosts for numerous migrating and wintering** ducks, waders **and** gulls, among which rarities are found every year (lesser scaup, black-winged kite, broad-billed sandpiper, Wilson's phalarope, Franklin's gull and white-winged tern, to name just a few).

Access to the hides inside the plant is restricted to visits organised by Espace Nature (phone: +33 5 46 82 12 44). Breeding specialities are sedge, Savi's, reed and great reed warblers, avocet, black-winged stilt and white stork. If you cannot join a visit, you can still bird from the car park [1] by **walking on the road in front of the external lagoons and the little trail that leads to a vast hide** [2]. These ponds concentrate shoveler, teal, a few pochard, little and black-necked grebes, lapwing on dykes, little gulls (up to hundreds in March–April), black and whiskered terns at both migrations (white-winged tern is regular in August–September). Back north along the D733, check La Beaune meadows and Sainte-Sophie lagoon for more wildfowl, lapwing, golden plover, ruff and black-tailed godwit (early spring) [3].

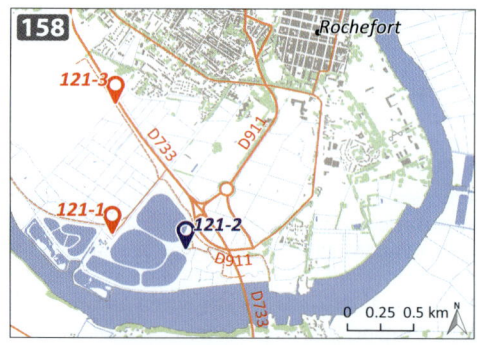

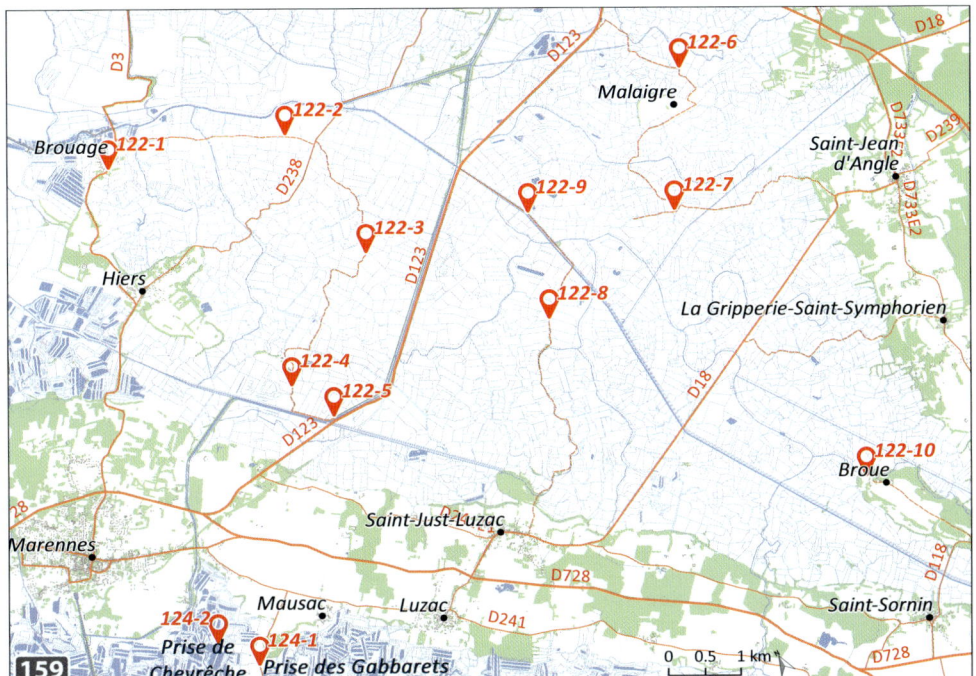

122 Brouage marshes

0.5–1 DAY; MAP 159 The Brouage marshes extend for over 160 km² between Rochefort and Oléron island. **The site is an extensive mosaic of reedbeds and meadows which is best visited by car or bike along local roads**. Bird spectacles at dawn and dusk are unique, offering countless photographic opportunities. **This is a key site for reedbed and freshwater species**. The most conspicuous species include marsh harrier, black kite, white stork, shelduck, yellow-legged gull, Cetti's, fan-tailed and melodious warblers, whitethroat, yellow wagtail and hoopoe.

Start at Brouage citadel [1] (18 km south of Rochefort towards Marennes), a viewpoint over wildfowl roosts in winter and spring (mostly teal, wigeon, shoveler, greylag goose and great white egret). **On your way to Beaugeay, make a first stop at Préveil reedbeds before the junction with the D238** [2]; it is a regular spot for wintering and migrating penduline tit and bittern. Local breeding species include reed bunting, reed, sedge and Savi's warblers and sometimes bearded tit. Stop-and-go across La Craie, Petit Matton, le Grand Sauvaget and Fousil [3] and **check the wet pans on both sides of the road** for wildfowl in spring (shoveler, pintail, teal, garganey) and waders on mudflats in both spring and autumn (sandpipers, snipe, curlew, ruff, avocet, stints including Temminck's, shanks, plovers, hundreds of black-tailed godwits). Spoonbill and herons (purple, grey, night-heron, possibly squacco heron and all the egrets) can be expected anywhere; also keep an eye open for the odd glossy ibis. Black-winged stilt and lapwing are also widespread throughout the marsh.

L'île d'Erablais [4] offers an overview of the southwestern end of the marshes. Red-backed shrike breeds in the hedge downhill from May onwards. Bird occurrences at the nearby Prise de l'Epée [5] depend on water levels, but a check can be rewarding for sandpipers. **An alternative itinerary follows a 10 km stretch of the D123 between Saint-Agnant and Marennes**. Take the path that runs across the marshes from 'Pont de la Roberte', 1.4 km after the start of the D123. From there you can circulate among wet pans to Malaigre [6] looking for herons, wildfowl and waders. A further 3 km south, reach the Cabane à Ballan [7] and walk on the trail for 500 m for further views over the marshes.

A third option starts from the south in Saint-Just-Luzac (7 km south of Marennes). Take the D18 towards St-Jean-d'Angle and turn into the last street on the left before leaving the village ('Rue de la Tonnellé'). This street leads to the marshes towards Reux, Gabaud, Cisière and Bellevue [8,9]. All the above-mentioned species breed in this area. **You might end with a stop at the isolated Tour de Broue** [10] (11 km, 10 min south of Saint-Just-Luzac through Saint-Sornin, follow the D118 towards La-Gripperie and turn left past Le Valérick camp site). The tower itself offers a 360° view over the marshes and **a trail from the car park leads inside the marshes**. In addition to the numerous waterbirds and reedbed species, hoopoe, jackdaw and little owl breed near and on the ruins.

123 Moëze-Oléron reserve

0.5 DAY; MAP 160   The Moëze-Oléron reserve is **an internationally important wetland for wintering and migrating** waterbirds, 12 km south of Rochefort (drive to Soubise, then Moëze). It covers a 50 km² tidal flat between Oléron island and the mainland north of Bourcefranc and a mosaic of mudflats, brackish lagoons and ponds, dry meadows and tamarisk hedges. Take the D3 south from Moëze and follow signposts to 'Réserve naturelle' and 'Ferme des Tannes' [1], where a **platform overlooks the meadows** where wintering greylag geese, golden plover, lapwing and curlew forage. **It is worth staying a while here looking at bird movements. Then, walk south on the road and check the ponds** looking for rarities such as marsh sandpiper (among large numbers of spotted redshank, redshank and greenshank) or Temmick's stint (within dunlin, curlew sandpiper and little stint). Even rarer species have been found on the reserve, including long-billed dowitcher, Wilson's phalarope and other Nearctic waders. You will likely encounter flocks of godwits, plovers, ruff, teal, wigeon, shoveler, pintail, shelduck and spoonbill at all seasons. Bluethroats breed on the canal at the end of the road [2] from March to May.

A second viewpoint is located at the Plaisance reserve headquarters [3] (2 km, 5 min north). From the car park, **paths lead to a hide bordering the main lagoon** [4], which is at its best at high tide when waders roost inside the reserve. Expect flocks of grey and ringed plovers, sandpipers, whimbrel, curlew, knot, avocet and oystercatcher. You can **also spot** waders **and gulls on a rising tide from the beach** at Les Sables de Plaisance [5] or Monportail lock [6].

124 The Seudre estuary

2 HOURS; MAP 159   **The Seudre estuary covers a vast expanse of salt and brackish-water marshes that attract** waterbirds **at all seasons. It is less well known than Moëze or other coastal sites, but it forms a perfect follow-up to a morning visit to the Brouage marshes** (site 122). Take the road that leads to the Moulin des Loges [2] from the D728 between Marennes and Saint-Just-Luzac. **Stop-and-go regularly along this road and stay in your car to ensure close views of most species**. Breeding birds include spoonbill, night-heron, little and cattle egret, shoveler, yellow-legged gull, black-winged stilt, common shelduck, lapwing, yellow wagtail, turtle dove and fan-tailed warbler. Herons (with the odd glossy ibis) forage mainly around the 'Moulin des Loges' [1] and the 'Prise des Gabbarets' while waders (mostly redshank, avocet, black-tailed godwit, greenshank, sandpipers and stints including the odd Temminck's) are most numerous at the 'Prise de Bercion'. Wildfowl (teal, pintail, wigeon, gadwall, garganey) are most abundant at the latter and around the 'Prise de Chevrêches'.

125 Ronce-les-Bains

< 0.5 DAY; NO DETAILED MAP   Cross the bridge over the Seudre river in Marennes and drive south towards Ronce-les-Bains, then follow the coastal road (D25) until you find the Pointe du Galon d'Or car park [1] (15 km from Marennes). Walk to the beach and cove, **where you should find wintering** brent geese, shelduck **and** waders (mostly grey and ringed plovers, dunlin, sanderling, oystercatcher, knot, redshank, bar-tailed godwit, curlew, a few greenshank and breeding

Kentish plover). Western Bonelli's warbler is easy to find in the surrounding pines, and crested lark and tawny pipit breed among the dunes.

126 Bonne Anse bay

0.5 DAY; MAP 161 **Bonne Anse forms the northern extremity of the Gironde estuary** (20 km, 30 min northwest of Royan through La Palmyre or 25 km, 30 min south of Marennes on the D25), facing the Pointe de Grave (Region 7, site 141). The bay is bordered by La Coubre pine forest, La Palmyre harbour and separated from the sea by a dune complex. From la Coubre car park [1], you can **either walk the path that passes the lighthouse to look at the sea or walk inside the bay**. In winter huge numbers of common scoter (with the odd long-tailed duck) may come within close range at high tide and there are large gulls and sanderlings along the beach. **The lighthouse and the dunes are the entrance point for migrating** passerines **that cross the Gironde from the Pointe de Grave in spring**. Loads of finches, warblers, wheatears, linnets, grey and yellow wagtails, skylarks and some raptors may fly over on clear days with light wind.

Several other car parks give access to the bay [2,3,4], which hosts hundreds of ducks, gulls and brent geese as well as thousands of waders in winter (dunlin, knot, redshank, grey and ringed plovers, both godwits, both curlews, oystercatcher, little stint and curlew sandpiper). At rising and high tide, the largest roosts form in the salt meadows facing Bonne-Anse camp site [3] and at the northwestern corner of the bay (Pointe de la Coubre). Gulls also regularly roost on sand pits in front of the harbour [5]. **In spring, visit the dunes south of the Pointe de la**

Coubre [6] for breeding Kentish plover, crested lark, tawny pipit and fan-tailed warbler and the pine woods which are full of western Bonelli's warblers (from April) and firecrests.

127 Oléron island

> 1 DAY; MAPS 162–165 Oléron is the southernmost island of the French Atlantic coast. **It is worth a day trip or more during both spring and autumn migration as** wildfowl, waders **and** passerines **of the east Atlantic flyway pass through on their way along the Bay of Biscay.** The island is connected to the mainland by a bridge in Bourcefranc. From April, bluethroats are abundant along the coastal stretch between the bridge and Le Château-d'Oléron. You can for instance park along the D734 before the Chenal d'Oulme [1] (turn right immediately past the bridge and drive 1 km north) and check the birds **singing in the vegetation on the shore**.

Several viewpoints are spread along the east coast for thousands of shorebirds, dabbling and sea ducks that gather here on a rising tide. Drive along the D734 to Le Château-d'Oléron (2 km past the bridge). Make your way along the coast and park at the 'Chenal de l'Etier Neuf' to walk to the Pointe des Doux [2]. 11 km, 15 min further north, Fort-Royer [3] and the Pointe de Boyardville [4] in Boyardville provide other good views over wader roosts and are regular spots for migrating osprey and peregrine. **Check groups of** common scoter **at sea and look for wintering** pale-bellied brent **and the odd** black brant **(casual) among large flocks of** brent geese. Bypass Saint-Georges d'Oléron and follow **'Plage de la Gautrelle' [5] (7 km further north), another viewpoint for sea ducks** where velvet scoter and common eider are sometimes seen among hundreds of common scoters and a few grebes and red-breasted mergansers.

Drive north to Saint-Denis d'Oléron and reach the northernmost extremity of the island, known as 'Pointe de Chassiron'. On your way, make stops at Le Sabia camp site [6] and La Morelière [7] for waders, gulls and sea ducks, grebes and divers. **The Pointe de Chassiron itself** [8] **is renowned for seawatching (position yourself behind the lighthouse).** With southwest and northwest autumn winds, expect Balearic shearwater, gannet, great and Arctic skuas, auks and other pelagic seabirds.

The west coast is just as good for seabirds. You should, for instance, stop at La Petite Nègrerie [9] (2 km south of the Pointe de Chassiron), where Slavonian grebes winter every year. The most reliable places for waders and gulls on the west coast are the Pointe des Trois Pierres near les Huttes [10] (Caspian gull should be searched for among the large gulls), Saint-Denis d'Oléron water treatment plant [11] (passage black tern, Ross's and Bonaparte's gulls have been found here in the past) and Conche Madame [12], 8 km south (hard to find: drive to L'Illeau from Chéray on the D734 and turn on Rue de Ponthezière to park on a track in front of the electricity transformer – worth it for purple sandpiper and Caspian gull).

The Charente valley (sites 128–129)

HIGHLIGHTS
» The last breeding sites for corncracke in the region
» A small population of breeding Savi's warbler

KEY SPECIES
BREEDING hobby, **corncrake**, **Savi's**, reed, Cetti's and fan-tailed warblers, reed and corn buntings

VISIT DURATION 1–2 days

PLAN YOUR VISIT
Your target here is corncrake, from April to June. They are easier to hear at dusk but, given the conservation status of the species, any kind of disturbance would be highly inappropriate. Even if you do not connect with the species, a visit is worth it for other marsh and meadow species.

128 Saintes
2–3 HOURS; NO MAP   Saintes alluvial meadows host the **last** corncrakes **of the Charente valley**. Park in La Palue [1], on the eastern bank of the Charente in Saintes. A walking trail runs around the Prairie de la Palue and Prairie du Maine meadows where corncrackes sing at dusk. More common breeding species include reed, Cetti's and fan-tailed warblers, reed and corn buntings and hobby.

129 L'Anglade marshes
1–2 HOURS; NO MAP   From Saintes, drive south for 10 km on the D137 towards Pons, bypass the airfield and turn left to reach L'Anglade reedbed and marsh [1]. Your target species here are Savi's warbler (a few pairs), reed warbler and water rail.

From Poitiers to Parthenay (sites 130–131, map 166)

HIGHLIGHTS
» The northwesternmost breeding population of ortolan bunting in the country
» One of the best sites of western France for breeding little bustard and other farmland and open-field species
» A large colony of breeding herons

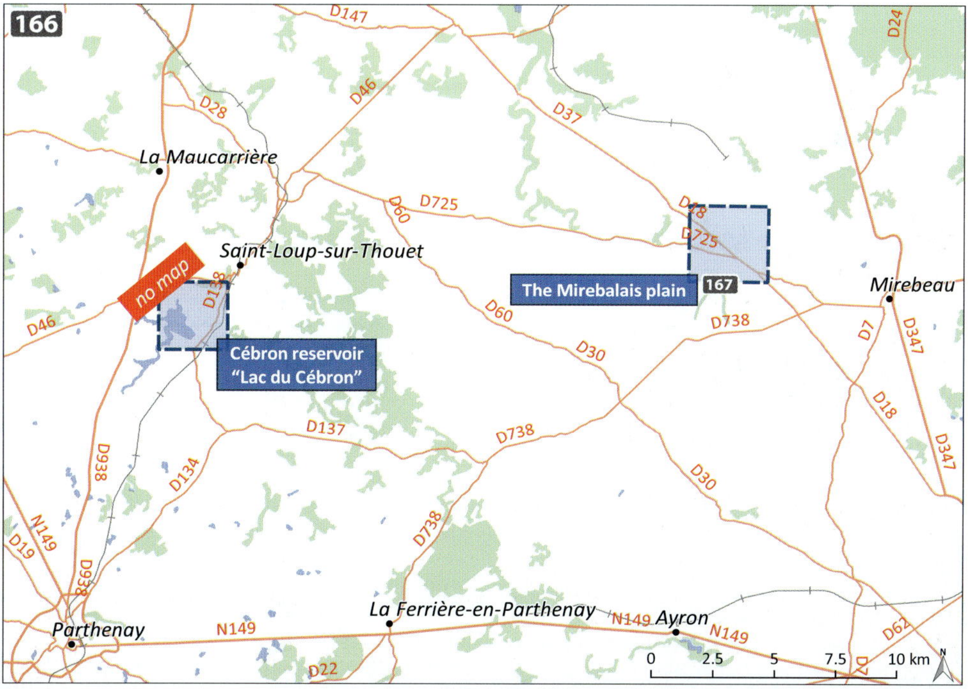

| J | F | M | A | M | J | J | A | S | O | N | D |

KEY SPECIES

YEAR-ROUND little grebe, red-legged partridge, hen harrier, crested lark

BREEDING night-heron, cattle egret, Montagu's harrier, black kite, **little bustard**, quail, **stone-curlew**, turtle dove, nightingale, tree pipit, red-backed shrike, **ortolan bunting**

VISIT DURATION   1 day

PLAN YOUR VISIT

The key species is little bustard, which has been the subject of a population reinforcement programme. The accompanying suite of farmland species is quite rich, and the area deserves a visit in spite of the rather unappealing flat landscape. Farmland bird activity peaks at dawn and dusk; avoid the heat of the day.

130 The Mirebalais plain

2–4 HOURS; MAP 167   The Mirebalais agricultural plain hosts most of the **Atlantic population of** little bustard (migratory) and the **last population of** ortolan bunting **in western France**. From Mirebeau (40 km, 40 min north of Poitiers on the D347 to Loudun and Saumur), drive west towards Mazeuil on the D725 and look for singing ortolan bunting (from May) on isolated trees and in orchards between Croix Lambert and Arpentin [1]. Little bustard and stone-curlew are easier around La Cour de

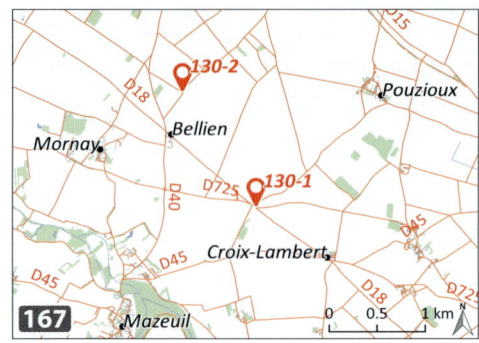

Lièvre [2] (turn right on the D40 after the Bellien junction). Other breeding birds in the area include hen and Montagu's harriers, red-legged partridge, crested lark and quail.

131 Cébron reservoir (Lac du Cébron)

1–2 HOURS; NO DETAILED MAP   The Cébron is a **reservoir** 20 km north of Parthenay [1] (take the D134 and the D138 to Saint-Loup-sur-Thouet until you find signposts). Walking paths lead to **two hides near the banks of the reservoir** [2]. The site hosts wintering and migrating dabbling ducks and some waders. Breeding species include a colony of grey heron, night-heron and cattle egret, common tern and little grebe. Surrounding hedges and fields host turtle dove, nightingale, tree pipit, red-backed shrike and black kite (all from April or May).

Poitiers area (sites 132–133, map 168)

HIGHLIGHTS

» The best place to see woodland species in the region
» A protected area devoted to waterbirds close to Poitiers and Châtellerault with breeding little bittern.

| J | F | M | A | M | J | J | A | S | O | N | D |

KEY SPECIES

YEAR-ROUND hen harrier, goshawk, middle spotted and black woodpeckers, Dartford warbler, marsh and crested tits, firecrest, hawfinch

BREEDING Montagu's harrier, short-toed eagle, honey-buzzard, hobby, **little bittern**, little ringed plover, nightjar, turtle dove, redstart, grasshopper, wood and western Bonelli's warblers, spotted flycatcher, golden oriole

VISIT DURATION   1 day

PLAN YOUR VISIT

A spring visit to the Pinail nature reserve can be combined with a stop at Saint-Cyr lake for a glimpse of breeding little bittern. Playback can be useful for woodland species (e.g. woodpeckers and passerines)

132 Saint-Cyr

1–3 HOURS; NO DETAILED MAP   Saint-Cyr lake, with its watersports centre, lies halfway (20 km) between Poitiers and Châtellerault along the D910 (signposted from La Tricherie). The western part of the lake is protected (guided tours by the LPO Vienne, +33 5 49 88 55 22). A **walking trail** starts at the car park [1] and **leads to three hides** [2]. Wintering wildfowl (teal, shoveler, pochard and tufted duck) and waders (snipe, common and green sandpipers) account for most of the birds, but there is also a pair of little bittern and breeding little ringed plover and common tern.

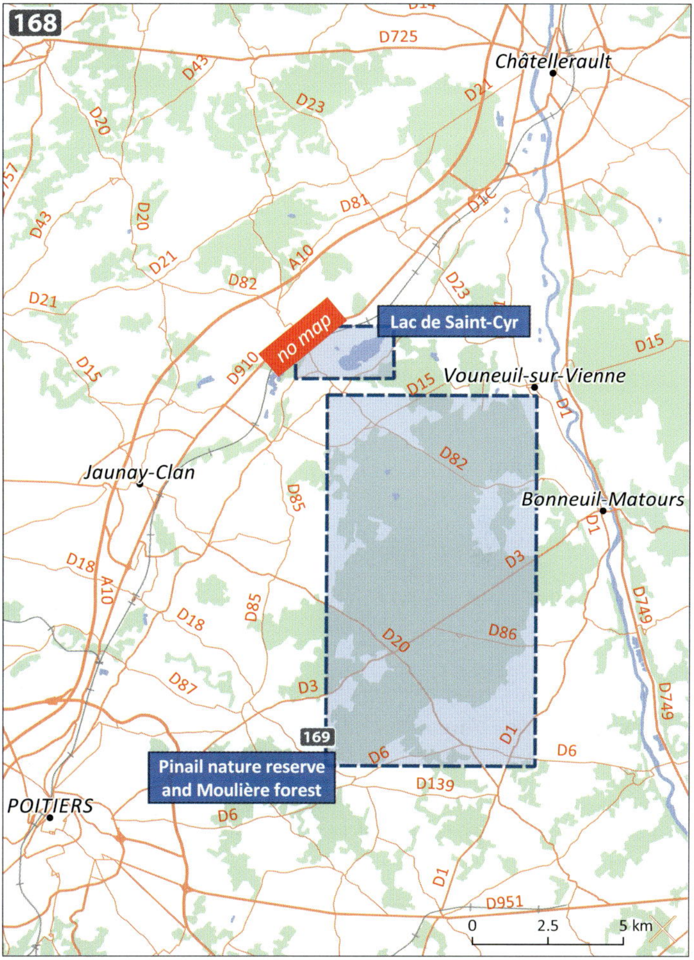

133 Pinail nature reserve and Moulière forest

2 HOURS – 0.5 DAY; MAP 169 The Pinail nature reserve lies 15 km south of Châtellerault and 30 km northeast of Poitiers (follow signposts to Vouneuil-sur-Vienne; once there drive west on the D15, turning left at signs for 'Réserve naturelle du Pinail' [1]). The reserve protects a **mosaic of woodland and moorland with thousands of waterholes, crossed by a trail best visited in the early morning.** Depending on the season, expect Dartford and grasshopper warblers, meadow pipit, stonechat, linnet, hen and Montagu's harriers, short-toed eagle and nightjar.

Moulière forest, 6 km south, is a mixed **deciduous–coniferous forest where** middle spotted woodpecker **breeds in old oak stands.** Follow the D3 for 5 km southwest from Bonneuil-Matours and turn left at the first major junction inside the forest. **Park at the gate and walk on the forest track to locate the** woodpecker [2]. Other breeding birds include western Bonelli's warbler (mixed coniferous–deciduous), crested tit (coniferous), firecrest, spotted flycatcher (mature deciduous stands), turtle dove, honey-buzzard, goshawk, hobby, short-toed eagle, nightjar (young stands and moorland). **For the same species, you can also walk around the Maison de la Forêt** [3] (700 m southeast of the junction between the D20 and the D3, towards Lavoux) or on the Piste forestière du Bignoux [4] near Bignoux at the southern edge of the forest. Middle spotted, great spotted and black woodpeckers, golden oriole, hawfinch, wood warbler, firecrest, marsh tit and redstart are all easy to find in this area.

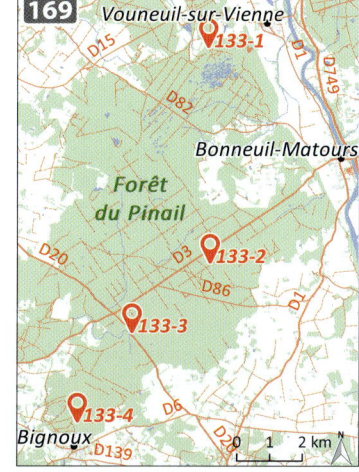

Inner Poitou-Charentes (sites 134–138, map 170)

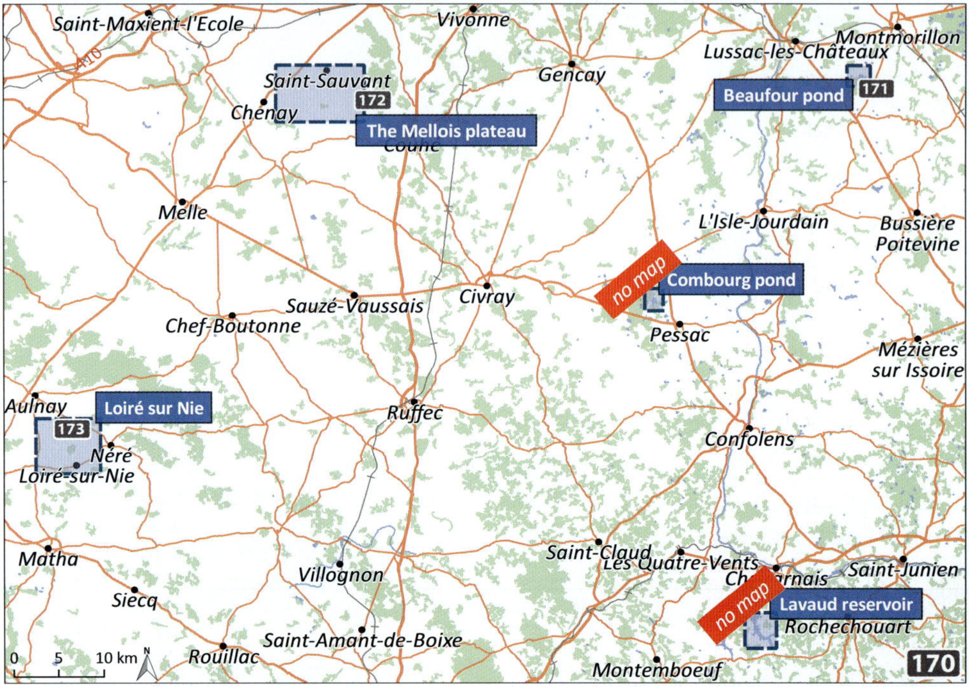

HIGHLIGHTS
» Farmland communities starring little bustard, stone-curlew and two shrike species
» Breeding herons and wildfowl on easily accessible sites

| J | F | M | A | M | J | J | A | S | O | N | D |

KEY SPECIES

YEAR-ROUND little grebe, hen harrier, red-legged partridge, cattle and great white egrets, crested lark

BREEDING red-crested pochard, **purple heron, night-heron**, black kite, honey-buzzard, hobby, Montagu's harrier, **little bustard**, quail, stone-curlew, turtle dove, hoopoe, reed and sedge warblers, red-backed and **woodchat shrikes**, golden oriole

WINTER common crane

VISIT DURATION 1–2 days

PLAN YOUR VISIT
Visit farmland areas at dawn and keep the ponds for later in the day when it is warmer. These sites are at their best in April and May when all the breeding birds have settled.

134 Beaufour pond

1–2 HOURS; MAP 171 Beaufour pond lies between Lussac-les-Châteaux and Saulgé, 50 km southeast of Poitiers (take the N147 towards Limoges, turn left to Sillars past Lussac, then right to Saulgé on the D116 until you find a signpost indicating 'Les Forêts', which will lead you to a hide at the northern end of the pond [1]). **Beaufour hosts a colony of** herons (purple and grey herons and cattle egret) and breeding or stopping-over gadwall, tufted duck, pochard, sometimes

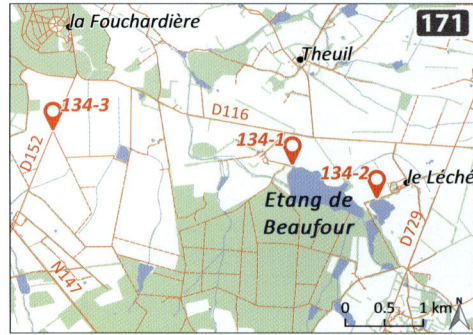

red-crested pochard and waders. Hobby, curlew, reed and sedge warblers breed at the southeastern end of the pond around Le Léché [2]. Red-backed and woodchat shrikes, buntings and warblers also breed in the surrounding hedges. Common crane can be found during migration in the nearby La Fouchardière fields [3].

135 Combourg pond
1 HOUR; NO DETAILED MAP The area between Pressac, Mauprévoir and Pleuville **is a mosaic of forest and farmland** where good densities of red-backed and woodchat shrikes, hoopoe and raptors (black kite, harriers, hobby, honey-buzzard) breed (drive south for 60 km, 1 h from Poitiers on the D741 through Gençay). The Combourg pond [1] (2 km north of Pressac on the D741, turn at the 'Combourg' signpost) hosts wintering and migrating waterbirds, ducks, waders and great white egret. Night-heron, grey and purple herons, pochard and tufted duck breed on the pond.

136 Lavaud reservoir
2 HOURS; NO DETAILED MAP Lavaud reservoir lies 50 km east of Angoulême and 60 km west of Limoges (reach Saint-Quentin-sur-Charente via the N141, then follow signposts to 'Lac de Lavaud'). **Follow 'Observatoire ornithologique' to find a hide on the western bank of the reservoir** [1]. The site hosts wintering and migrating ducks (mostly teal, wigeon, shoveler), little and great crested grebes, great white, little and cattle egrets, purple heron and crane (February and October–November). Black kite, turtle dove, hoopoe, golden oriole and red-backed shrike breed in the surrounding fields.

137 The Mellois plateau
2–5 HOURS; MAP 172 The Mellois is an open-field plain dominated by cereal crops and locally polycropping with meadows and scattered hedges, 40 km south of Poitiers towards Niort (bypass Lusignan on the D150; once in Saint-Sauvant turn on the D96 to Courgé and

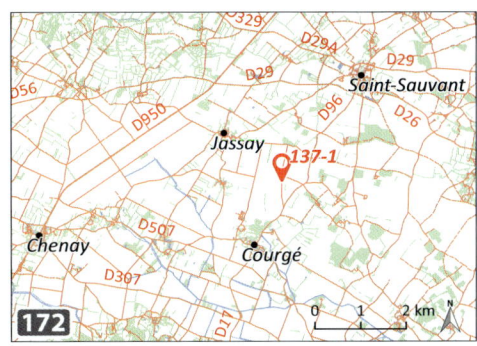

take a track to the right 1 km before entering the village [1]. Search for little bustard, hen and Montagu's harriers, quail, red-legged partridge, stone-curlew, curlew, red-backed shrike and turtle dove.

138 Loiré-sur-Nie
2–5 HOURS; MAP 173 The plains 60 km east of Rochefort host **a locally high diversity of farmland birds**. Drive to Saint-Jean d'Angely on the D739, then take the D130 to Loiré-sur-Nie. **Park along the D222 near the small wood west of Galanchat** [1]. Walk among fields through Chaudusson farm. Most farmland breeders are present around here, including little bustard, hen and Montagu's harriers, quail, red-legged partridge, stone-curlew, turtle dove, crested lark, stonechat, yellowhammer, corn bunting and red-backed shrike.

REGION 7
AQUITAINE

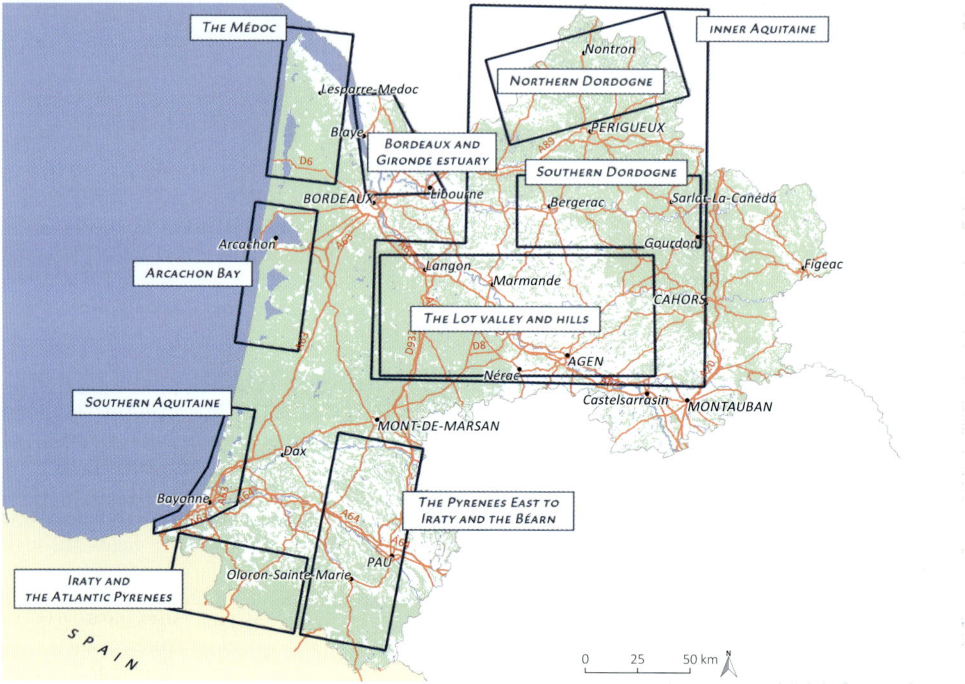

HIGHLIGHTS
» A long coastline on the east Atlantic flyway
» Several internationally renowned migration viewing points
» Extensive stopover sites for waders and wildfowl in well-preserved marshes
» Breeding area of white-backed woodpecker and Iberian chiffchaff

REVIEWERS Luc Barbaro, Franck Jouandoudet, David Simpson

LEFT: The sandwich tern colony at Banc d'Arguin is best seen from the Dune du Pilat (site 147). Elegant terns and other rarities are recorded in the colony from time to time.
RIGHT: The black-winged kite is no longer a newcomer in southwestern France, but it remains as sought-after as ever. You'll likely come across a hunting bird as you wander in the riparian lowlands of Aquitaine (sites 149–155).

AQUITAINE | BORDEAUX AND THE GIRONDE ESTUARY

USEFUL WEBSITES

www.faune-aquitaine.org
http://lpoaquitaine.org has detailed site accounts for the whole region (in French)
www.wilddordogne.co.uk for guided birding tours in Dordogne
Birding Dordogne by David Simpson available on iTunes

Bordeaux and the Gironde estuary (sites 139–140, map 174)

HIGHLIGHTS

» Extensive marshes and reedbeds within one hour's drive from the centre of Bordeaux
» Sites are easily accessible by public transport

KEY SPECIES

YEAR-ROUND marsh harrier, great white, cattle and little egrets, **Eurasian bittern**, **spoonbill**
BREEDING white stork, **purple heron, night-heron**, short-toed eagle, Montagu's harrier, black-winged stilt, **bee-eater**, bluethroat, **Savi's**, grasshopper, sedge and great reed warblers, red-backed shrike
MIGRATION garganey, **black stork**, **glossy ibis**, osprey, crakes, waders, **wryneck**, passerines
WINTER wildfowl, jack snipe, **penduline tit** 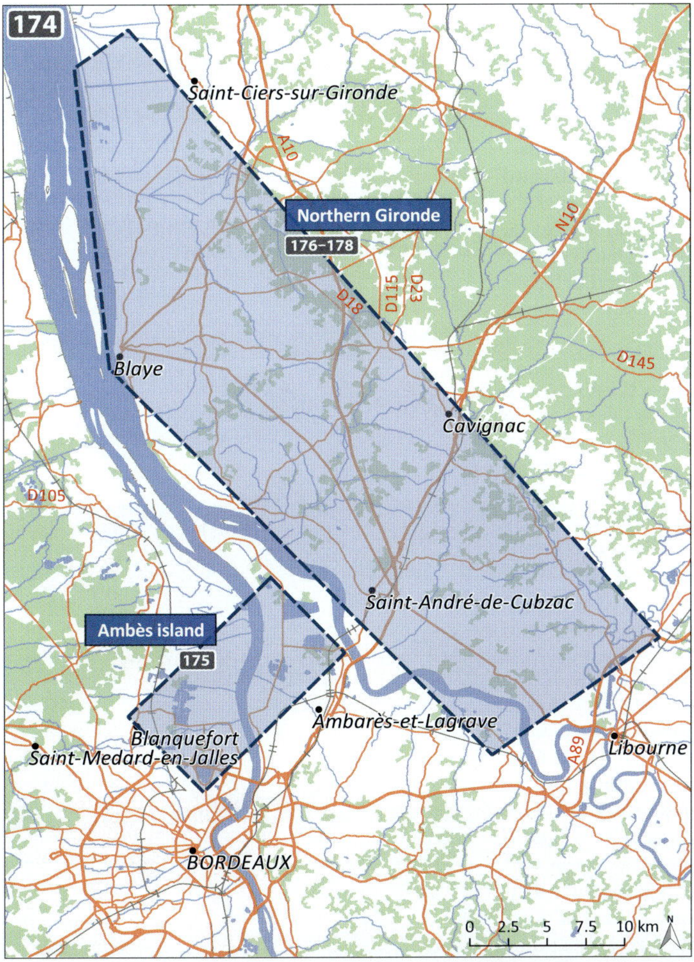
VISIT DURATION 1–2 days

PLAN YOUR VISIT

These sites are perfect local patches for those tied by a business or family trip inside Bordeaux, several of them being easily accessible by public transport. The site is mostly reedbeds and wet meadows, at their best from March to June. The local star is bluethroat, but there are several other breeding and wintering species worthy of interest such as Savi's warbler, not to mention loads of migrating waders and passerines. Start early from Bordeaux or Libourne: Ambès island is great at dawn. Then, drive from site to site on the northern banks of the Dordogne and Gironde, timing your itinerary to spend midday hours in the Isle valley or in the Moron, and dawn in the Braud-et-Saint-Louis marshes.

139 Île d'Ambès

0.5 DAY; MAP 175 Start your day in Saint-Vincent-de-Paul, especially from late March to June. This is a well-preserved expanse of marshes and pastures on the Île d'Ambès, a wide peninsula separating the Garonne and Dordogne rivers to the north of Bordeaux. Although small and embedded in an industrial area, **the Petit Marais is a hotspot for wetland** passerines, wildfowl **and** waders **just 30 minutes from Bordeaux** (exit the A10 in Ambès, take the D257 at the roundabout north of Saint-Vincent-de-Paul towards Ambès again and turn right after 1 km: a first car park is located at the end of a small track on the left 80 m after the junction [1], and a second 400 m further on near a bus stop). This site is also **great for photographing singing** bluethroats, warblers **and** purple herons. Aquatic warbler has been recorded in late April, and more regular breeding or migrating highlights include black-winged stilt, grasshopper, sedge and great reed warblers and wryneck. The bushes can be full of nightingales, redstarts and willow warblers in **spring, especially after early-morning rain**.

The extensive reedbed of the nearby Grand Marais [2] is excellent for all marsh species, including garganey, jack snipe and crakes at dawn and dusk (6 km, 5 min, go back on the D257, turn right at the roundabout and drive for 2.8 km on the D113, park at the end of a small track right after a bridge).

If you are tied to the immediate vicinity of Bordeaux, visit Bruges marshes [3] (bus: lines 6 and 57, stop at 'Les 4 Ponts'; by car take exit 6 on the A630 in Bruges and drive to Blanquefort on the D210, follow signposts 'Réserve Naturelle'; hides, open 10.00–18.00 except on Thursdays and Fridays, guided tours, contact: +33 5 56 57 09 89). Although not as good as the two previous spots, it has waders, wildfowl, spoonbill, bluethroat, red-backed shrike, short-toed eagle and osprey. White or even black stork can occur.

early spring), crakes (calls at dusk in April and May) and red-backed shrike (turn right behind Les Billaux city hall ('Mairie'), cross the motorway after 1.2 km and take the D18E3 for 1 km more, park just after crossing the D18 [1]). **From here, proceed to the Dordogne sites (sites 154 and 155) or the northern bank of the Gironde.**

Near Prignac-et-Marcamps, **the Site Naturel du Moron** [2] **covers 92 ha of wooded wetlands** where wryneck (from March) and hobby (from April) breed (30 km, 40 min from Les Billaux, take the D133 in Saint-Laurent-d'Arce, car park signposted 600 m after the sign 'Marcamps 0.8 km'). **It is suitable for a midday stop as tree cover maintains bird activity.**

140 Northern Gironde

0.5 DAY; MAP 176–178 The Isle valley has a mixture of meadows, hedgerows and marshes between Les Billaux and Montpon-Ménestérol (5 km, 11 min from Libourne on the D910, or 45 km, 40 min from Bordeaux on the A89). Several heron colonies with breeding white stork, great, cattle and little egrets, purple heron and occurrences of glossy ibis spread along the valley. **The Brizards marshes are good for most of these species**, plus jack snipe (wet grassy meadows,

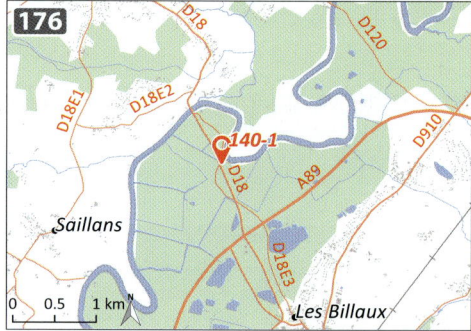

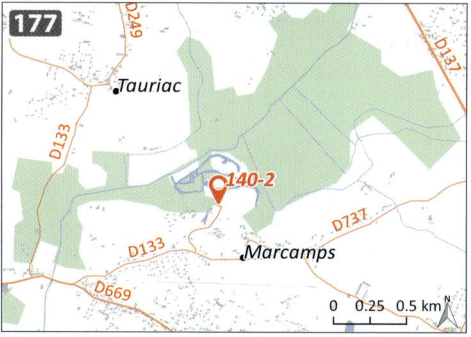

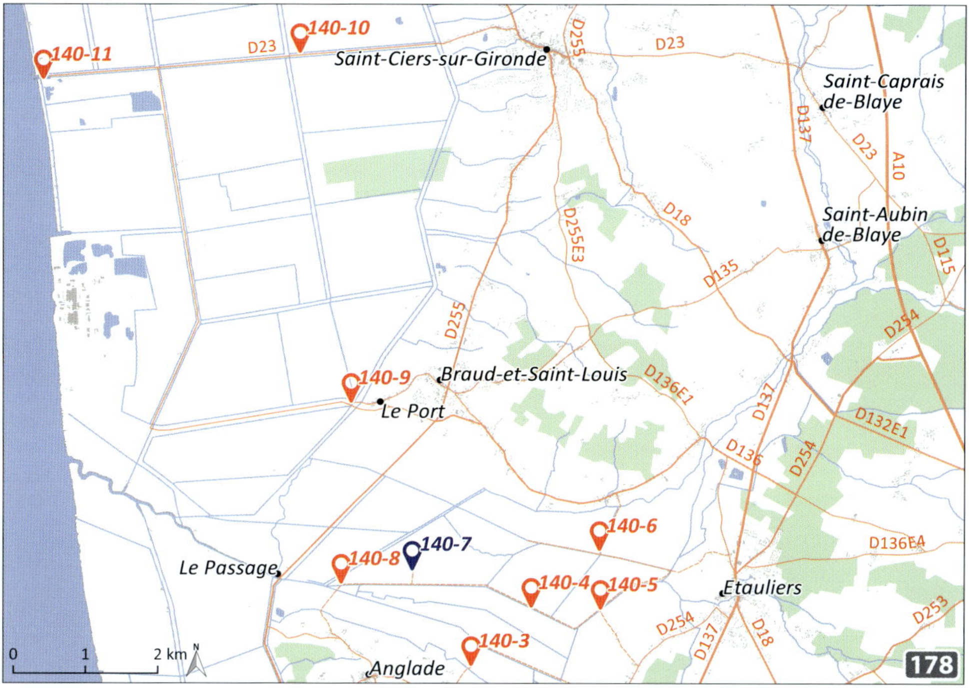

AQUITAINE | THE MÉDOC | 137

Further north, past Blaye, **La Vergne marshes** (take the D255 towards Braud-et-Saint-Louis; bird on dykes just north of Anglade, e.g. on a circuit [3,4,5,6,7,8]) **and Saint Louis marshes are extensive and well-preserved** (from Anglade, turn on the D136 towards the 'centrale EDF' just before Braud-et-Saint-Louis, the marshes start after 1.2 km near Le Port [9]). They host breeding spoonbill, night-heron, shoveler, gadwall and other ducks in spring, plus marsh harrier, hobby, bittern (up to 4 booming together), glossy ibis, osprey, Savi's warbler (among the best densities in the region) and bluethroat. **These sites are heavily hunted in winter, but there are large numbers of** wildfowl **and a few** penduline tits **in reed-beds** [10], especially in cold weather.

The **Terre d'Oiseaux domain** [11] **has several footpaths with platforms and hides suitable for photography** (signposted from the D18 in Saint-Ciers-sur-Gironde, fee, guided tours, open year-round from 10.00 to 18.00 or 20.00, http://terresdoiseaux.fr). **As a hunting-free reserve, this is the only place really worth a visit in winter** for wildfowl and waders (large numbers of lapwing and snipe), penduline tit and spoonbill. Both migrations are good for osprey (March–April and September) and waders (casual occurrences of Temminck's stint, pectoral sandpiper and other rarities). Bittern (early spring) and spotted crake (April and May) have been recorded. Breeding highlights include hobby, Savi's warbler, red-backed shrike and bluethroat. Garganey, Montagu's harrier and bee-eaters are regular, although they do not breed on site.

The Médoc (sites 141–143, map 179)

HIGHLIGHTS

» The Pointe de Grave ranks among the top French sites for spring passerine migration
» Scenic atmosphere at dawn and dusk on the banks of forest lakes and marshes

J F M A M J J A S O N D

KEY SPECIES

YEAR-ROUND marsh harrier, great white egret, **spoonbill**, avocet, crested lark, crested tit

BREEDING white stork, **night-heron, purple heron, little bittern**, black and **black-winged kites**, short-toed eagle, honey-buzzard,

141 THE POINTE DE GRAVE AND SURROUNDINGS MARSHES | AQUITAINE

black-winged stilt, Kentish plover, whiskered tern, hoopoe, nightjar, **wryneck**, tawny pipit, Dartford, grasshopper, **Savi's**, sedge and western Bonelli's warblers, red-backed and **woodchat shrikes**

MIGRATION red kite, osprey, Montagu's and hen harriers, **Eleonora's falcon**, **black stork**, all the crakes, both godwits, shanks, sandpipers, great, pomarine and Arctic skuas, gulls, terns (including black), auks, divers, **black stork**, swifts, **passerines** including larks, swallows and finches

WINTER wildfowl, black-throated and great northern divers, black-necked grebe, seabirds, **Eurasian bittern**, common crane, short-eared owl, **penduline tit**

VISIT DURATION > 1 day

PLAN YOUR VISIT

From March to mid-May, head to the Pointe de Grave if the weather is good for migrants. Otherwise, visit marshes and lakes, ending your day at Cousseau pond and surrounding lakes. Summer birding is quiet but sometimes brings scarce migrants. Southern sites can be combined with the northern sites of Arcachon bay. Most of the region is covered with pine forests interspersed with ponds that are hard to access, so stick to the described locations.

141 The Pointe de Grave and surroundings marshes

0.5–1 DAY; MAP 180   **Birds on their northward journey concentrate along the coast before crossing the Gironde estuary from a sandy headland**

AQUITAINE | **142 INNER MÉDOC** | 139

known as the Pointe de Grave. The result is a constant flow of huge numbers of migrants, especially in light northeasterly winds from March to mid-May, **making this place one of France's very best spring migration hotspots**. It is not as good in autumn – then, prefer Cap Ferret (site 144). Check the weather the day before and be on site at dawn. Park at the very tip of the headland [1] (5 km, 5 min from Le Verdon sur Mer, follow signposts) and climb onto the surrounding dunes; **a team of birdwatchers is present all spring.**

Flocks of finches, larks, swifts and swallows can pass through continuously. Among raptors, expect red and black kites (from late March), honey-buzzard (from mid-May), kestrel and hobby, Montagu's, hen and marsh harriers (late March), turtle dove (May), golden oriole (May). Osprey, white and black (rarer) storks, spoonbill and hoopoe are seen weekly during the season. Rarer raptors sighted every year include booted eagle, pallid harrier and red-footed falcon; greater and lesser spotted eagles have also been recorded. Also check for seabirds, especially skuas, gulls, terns, auks and divers. Local breeders include tawny pipit, western Bonelli's warbler, crested tit and short-toed eagle.

Passage often decreases by 10.00–11.00; then check the surrounding marshes. Start with the Marais du Logit [2] (3 km, 5 min from the Pointe de Grave, cross the railway and turn right after 960 m, then drive for 900 m more to find the car park). It can be crowded with migrating birds after rain or if winds turn adverse. Shelduck, purple heron and black-winged stilt breed here. Next, visit the Marais du Conseiller [3], 7 km, 7 min south on the D1215 (turn left towards 'Zone Portuaire' and bird along the access road to the harbour). Penduline tits (rare) and wildfowl (pintail, gadwall, shoveler, wigeon, greylag goose, scoters

at sea) winter in reedbeds until early April, and sedge warbler breeds. Park on a good sandy track before crossing a canal on the access road [4] to **scan wader flocks at high tide year-round**; they include godwits and avocets with shanks and sandpipers. Kentish plover and oystercatcher are local breeders. **Coastal dunes, hunting ponds and marshes may host local rarities** such as crakes, icterine and aquatic warblers and other scarce passerines.

142 Inner Médoc

2–5 HOURS; MAPS 181–183 **Back from the Pointe de Grave, several reedbeds and ponds deserve at least a quick stop to look for migrants and local breeders.** The La Perge marshes [1] are among the few accessible sites for breeding bittern, great white egret, avocet, hoopoe and wryneck (25 km, 30 min from the Pointe de Grave on the D1215 and D101, park just south of a small bridge 3 km from L'Hôpital and 1.1 km from Mayan). Spotted crake has been heard in the reedbeds; listen at dawn and dusk. The Marais de la Maréchale [2], 26 km, 30 min further southeast near the Gironde, is dominated by **meadows and hedgerows which host the southernmost** meadow pipit **population in Atlantic France** (turn towards Ipauillac in Lesparre-Médoc, drive

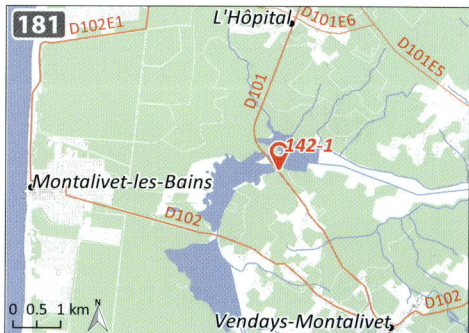

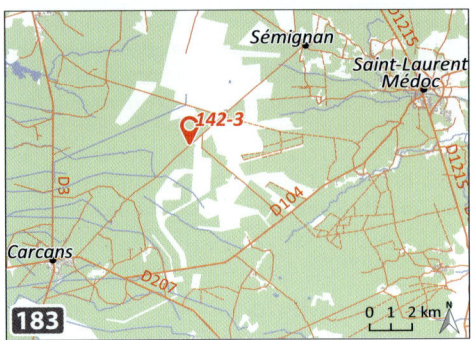

to Ordonnac, cross the village and turn right, then left after 300 m). There are good chances for hobby, white stork, barn owl, sedge warbler and red-backed shrike. South again, the Plaine du Jonc has black-winged kite, short-toed eagle, curlew and woodchat shrike (in Saint-Laurent-Médoc, turn right just past the church on 'rue Francis Fournié' and follow signs to Hourtin; turn left to Carcans and Le Jonc just past Sémignan; the best area lies 2.4 km after a large farm [3]).

143 Lakes of Southern Médoc

4 HOURS – 0.5 DAY; MAP 184   **Several lakes spread along the Médoc's southwestern coast and are worth an early-morning or evening visit, in particular in May when wetland breeding birds have settled. North of Hourtin lake, several paths cross dunes, pine forest and swamps**, for instance along the D107E1 between Cartignac and Contaut [1]. **This is an ideal stopover area for migrating birds, and breeding diversity is high from April to June**. The star species is aquatic warbler, but most records are from ringing stations (passage peaks in April and late August–September). More likely targets are little bittern (from late May) and Eurasian bittern (winter), black and whiskered terns (late April), short-eared owl (October and early spring), penduline tit (all winter to March), grasshopper and Savi's warblers (from late March, the former in low and moist pine forest stands, the latter in open reedbeds and sedges), Dartford warbler (check low pine stands and heathlands, easier in early spring). Great white egret, night-heron and purple heron (from April), nightjar, crested lark, tawny pipit (from mid-March) and western Bonelli's warbler (from March) are near-certainties.

The same species are likely on the trail leading to the Étang de Cousseau reserve (30 km, 30 min south from Hourtin, the reserve is signposted from Carcans, park under pines at [2]). **A longish, yet rewarding one-hour walk through pine forest and heathlands leads to a platform and hide [3] above a wide expanse of marshes (**a telescope is essential**). In addition to previously mentioned species, the reserve is a wintering site for crane (recent breeding attempts), waders, gulls, terns and wildfowl (especially teal). Spoonbills may forage in late winter. The platform is also a good place for honey-buzzard (from May), marsh harrier, short-toed eagle (from March) and hobby. Osprey occurs on migration in March and September, and Eleonora's falcon has been recorded in summer (search for it among hunting flocks of hobbies).

Make a final stop at Lacanau lake (Étang de Lacanau) (drive along the western bank of the lake and turn left on a track 3.6 km after La Grande Escoure to reach a car park [4]). Wildfowl and black-necked grebe winter here, sometimes joined by the odd red-necked or Slavonian grebe. The lake is also good for black-throated and great northern divers, great white egret and goosander (rare). **Walk towards the southern arm of the lake and beyond** [5], as there is a small purple heron colony, sometimes hosting a pair of little bitterns, close to the canal heading south to the Arcachon Bay. Water rails (abundant) and singing spotted crakes also occur in spring. From Lacanau, it is a 60 km, 1 h 20 min drive to Bordeaux or 25 km, 30 min to Lège-Cap Ferret on the northern end of Arcachon Bay.

Arcachon Bay (sites 144–148, map 185)

HIGHLIGHTS
» The autumn counterpart to the Pointe de Grave at Cap Ferret
» Several large marshes and lakes full of waders and wildfowl in winter and during both migration periods

| J | F | **M** | **A** | **M** | J | J | **A** | **S** | **O** | N | D |

KEY SPECIES
YEAR-ROUND great white egret, **spoonbill**, yellow-legged and Mediterranean gulls, Dartford, Cetti's and fan-tailed warblers, crested tit

BREEDING white stork, **night-heron, little bittern**, black-winged stilt, little ringed and Kentish plovers, **wryneck**, hoopoe, **nightjar**, tawny pipit, bluethroat, grasshopper, **Savi's**, melodious and western Bonelli's warblers, spotted flycatcher

MIGRATION Cory's and Balearic shearwaters, **Leach's** and storm petrels, **black stork**, red kite, osprey, merlin, all the crakes, waders, **grey phalarope**, great, pomarine and Arctic skuas, **Sabine's gull, Caspian tern, passerines** including ring ouzel, swallows, pipits, larks, finches, ortolan bunting (rare)

WINTER brent and greylag geese, other wildfowl, velvet and common scoters, eider, red-breasted merganser, black-throated and great northern divers, black-necked grebe, **Eurasian bittern**, woodcock, **Caspian and ring-billed gulls**, short-eared owl, **penduline tit**

VISIT DURATION 0.5–2 days

PLAN YOUR VISIT
Arcachon bay (le Bassin d'Arcachon) has long been one of the most renowned birdwatching sites of the Atlantic coast, owing to **the high numbers of** waders **that roost year-round on its banks, joined by thousands of** wildfowl **in winter – among which hundreds of northward-bound** pintail **in March are a truly amazing sight.** Choose either the northern or the southern part, as the area is too large to be covered in a single day. Unless you wish to try autumn migration in Cap Ferret, **be at Le Teich or Certes at high tide and get closer to the sea when the tide drops.** Seawatching fans should consider spending a day on the Wharf de la Salie, one of the region's very best pelagic spots, if winds are in the northwest.

144 Northern Arcachon Bay
0.5 DAY; MAP 186   **Cap Ferret marks the northern mouth of Arcachon Bay**, separated from the Pyla dune, on the southern bank, by less than 5 km of water that birds need to cross on their autumn migration. **Simply speaking, it is the autumn counterpart of the Pointe de Grave (site 141), to which it is equivalent in terms of diversity and counts, and rightly renowned as among the main fall migration hotspots.** It is somewhat isolated within a largely urbanised area, so head there to spot migration, but do not expect long strolls in natural landscapes. High numbers of swallows, swifts, pipits, larks, pigeons, wagtails, thrushes, warblers, starlings, buntings and finches pass through from late August to November. Osprey, hobby, merlin, red kite, short-eared owl, ring ouzel and ortolan bunting are all regular, together with rarer species.

Park in the main car park at Cap Ferret [1] and walk 200 m south to the dunes [2] (parts

145 LE TEICH NATURE RESERVE | AQUITAINE

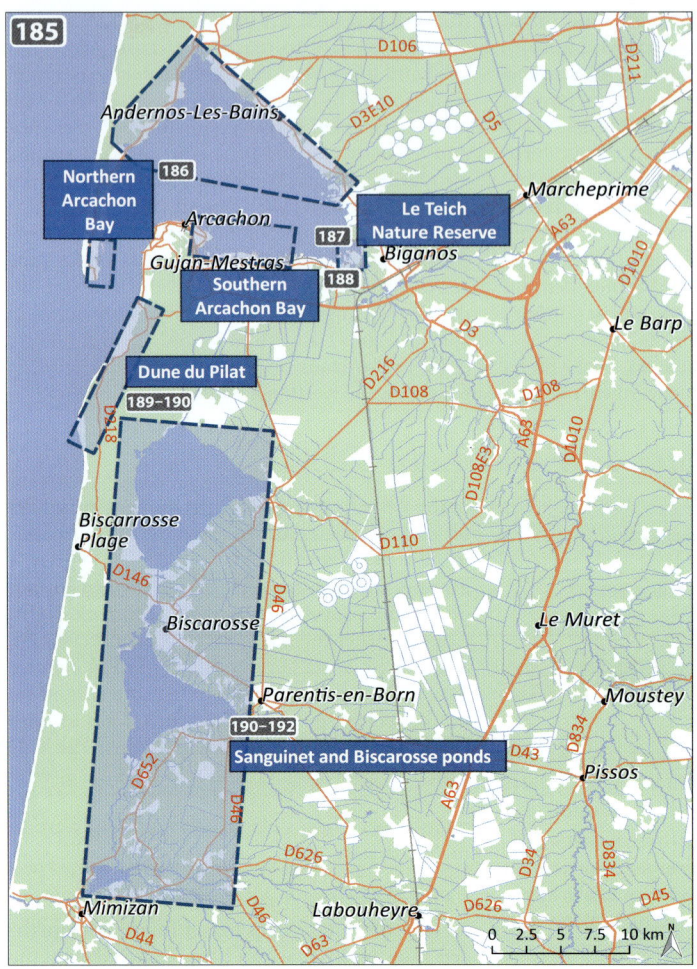

inland, **the saltmarshes in Prés Salés d'Arès reserve** are good for breeding bluethroat, hoopoe, wryneck and turtle dove. In winter, it is a roosting site for spoonbill, brent and greylag geese, wigeon, teal and pintail, plus snipe and woodcock. The surrounding mudflats host high numbers of waders at all seasons. **The reserve is accessed by footpaths from several car parks** (along the D106 2 km west of Arès [5], on the roundabout at the turning for Lacanau [6] or at the north end of Arès harbour, follow 'port ostréicole' from the church [7]). At high tide, try the Pointe des Quinconces [8] in Andernos for waders and the nearby pier for grebes, brent geese and gulls (including ring-billed in winter).

Another major wader site is the three-hour walk at the Domaine de Certes (car parking on the right of the road at the western end of Audenge along the D3 [9]; viewpoints on marshes around [10] and on the bay around [11]).

were recently fenced), preferably when the wind is southwesterly. Seawatching can be good in westerly winds for geese, ducks, shearwaters, petrels, phalaropes, skuas, gulls, terns and auks including rarer species such as long-tailed skua, Sabine's gull or little auk. Migration is not as good in spring but tawny pipit, Kentish plover and nightjar breed in sandy dunes. On your way inland from Cap Ferret, check the sand spit at La Conche [3] (turn right at the first roundabout after the Cap, then left and park near the shore): Kentish plover breeds on the sand and the site is good for migrating waders and passerines.

Piraillan fish ponds (10 km, 15 min further east on the D106, turn left at the signpost 'Les Piscicultures de Piraillan' [4]) can be visited on a footpath which allows close views of waders, herons, little grebe and kingfisher. Further

145 Le Teich nature reserve

3 HOURS; MAP 187   **Le Teich nature reserve** [1] (20 km, 30 min from Arcachon; open 10.00–18.00, fee; from Le Teich follow 'Parc Ornithologique') **is the highlight of any visit to Arcachon Bay.** For birders, it offers great wader diversity and numbers in almost any month. For photographers, convenient hides provide exceptionally close views. And for families, it is a nice two-hour stroll with plenty of easy-to-see birds and options for guided visits. This **120 ha network of ponds, saltmarshes and mudflats** is a hunting-free sanctuary for wildfowl and waders all year. Timing is critical, as **most birds are present from 2 h before to 2 h after high tide; a visit outside this slot will probably be disappointing.**

AQUITAINE | **146 SOUTHERN ARCHACHON BAY** | 143

The site can only be accessed on footpaths along which 17 hides are spread (the best are hides 2,3,7,10,11–13); **allocate a minimum of three hours for the visit**. Besides flocks of migrating birds, white stork, spoonbill, black-winged stilt, little ringed plover, bluethroat, melodious warbler, spotted flycatcher and crested tit breed in the marshes and woods inside the reserve. Rarities such as Caspian tern, Temminck's stint and marsh sandpiper are near-annual, and Le Teich has a long list of vagrants, including terek, buff-breasted and pectoral sandpipers, plus exceptional rarities such as pink-footed goose, Pacific golden plover, stilt sandpiper and Bonaparte's gull.

After a morning spring visit to Le Teich, it may be worth spending some time in the Leyre delta, a wooded river mouth with reedbeds (from Le Teich reserve, turn left after the roundabout, then left again on the 'rue du Pont Neuf' [2]). The area is hard to bird, but grasshopper and Savi's warblers breed from mid-April, and osprey and black stork are regular migrants or stopovers. Other breeders of interest include grey wagtail, Cetti's and fan-tailed warblers, bluethroat and golden oriole. **In winter, penduline tit is still possible**, although it has become rare in recent years. Small flocks of spoonbill can also be seen at the river mouth (exit Biganos west on the D3 and turn left on 'route de Vigneau' and straight on to end along the river bank [3,4]).

146 Southern Archachon Bay

3 HOURS – 0.5 DAY; MAP 188 Although the southern side of Arcachon Bay is highly urbanised, several harbours, saltmarshes and mudflat areas concentrate birds especially at high tide. **It is even better than the northern bank in winter, as birds come closer and the shore is more easily accessible.** The local flagship is ring-billed gull (October to March), but it is a surprise-prone area where any bird can turn up – something rare or more.

Visit the piers in Gujan-Mestras harbour [1] (17 km, 30 min from Arcachon, follow signposts to Port de Larros and drive among oyster farms), Port de la Hume [2] and Port du Rocher [3] (both signposted from the D650 to Arcachon); several ring-billed gulls winter in this area. An ivory gull stayed here for several days in January 2009, and there are records of glaucous and Iceland gulls. Following the D650 through La Teste de Buch, stop at Les Prés Salés Est [4] (turn right at the roundabout after the 'Comptoir Nautique Arcachonnais' shop, drive to the end of the road and walk eastwards) and Les Prés Salés Ouest [5] (park at the next roundabout and walk towards the bay). **These two brackish-water marshes are partly isolated from the bay by several embankments which reduce the influence of tides.** They are used as roosts by wintering black-tailed godwits, sanderling, dunlin and shanks. Ring-billed gull is regular from November to March; other wintering birds of interest include penduline tit (rare) and siskin.

End in Arcachon harbour [6]. Ring-billed gulls are present all winter within mixed flocks of herring, yellow-legged, Mediterranean, black-headed and common gulls. **The harbour is also a good viewpoint to look for** grebes **and** wildfowl. **Boat trips** inside Arcachon Bay (phone: +33 8 25 27 00 27) can be excellent in winter for closer views of brent geese, divers, black-necked and Slavonian grebes, red-breasted merganser, eider, scoters and auks.

147 Dune du Pilat and the coastline

2 HOURS – 0.5 DAY; MAPS 189 & 190 The Pilat dune is the highest sand dune in Europe, and a prime viewpoint for the mouth of Arcachon Bay (from Arcachon, 10 km, 15 min on the D259, signposted). It is not particularly birdy in itself but the scenery justifies the diversion at the end of a day trip around southern Arcachon bay. The northern side of the dune is accessible from a well-marked path [1] (fee), but a better option is to **park near the Camping de la Forêt** [2] **and climb on the southern side.** Crested tit and western Bonelli's warbler breed in the surrounding pine forest. The dune itself is rather birdless. More interesting at the top of the dune is to **scan the Banc d'Arguin** [3], a protected sand bank separated from the mainland by 800 m of water; it hosts a colony of Sandwich terns within which several elegant terns have been recorded in past years.

10 km, 10 min south on the D218, **the Wharf de la Salie** (turn right to 'Plage de la Salie' [4]) **is one of the region's very best sites for autumn seawatching in northwesterly winds** (expect Cory's, great (scarce) and Balearic shearwaters, storm petrels, divers, grey phalarope, Sabine's and glaucous gull and auks). In migration periods and winter, explore surrounding dunes for chances of dotterel (early September), short-eared owl (from October) and snow buntings (from November).

148 Sanguinet and Biscarosse ponds

2 HOURS; MAPS 190–192 Just behind the coastline, two large freshwater ponds are worth a visit from November to March for divers, grebes, sea ducks and bittern. Stop in Sanguinet harbour [1], Navarosse [2] and Les Hautes Rives [3] on the Étang de Sanguinet, and Biscarosse [4] and Gastes [5] on the Étang de Biscarosse for black-throated and great northern divers, black-necked and Slavonian grebes, red-breasted merganser, velvet scoter, scaup, great white egret and bittern. **At the southern end of the Étang de Biscarosse, the Courant de Sainte-Eulalie** hosts night-heron and little bittern (from late April) (turn to 'Le lac' 1.8 km south of Sainte-Eulalie on the D652 and park near the camp site [6]; rent kayaks to explore the marshes).

AQUITAINE | **SOUTHERN AQUITAINE AND THE PYRENEES** | 145

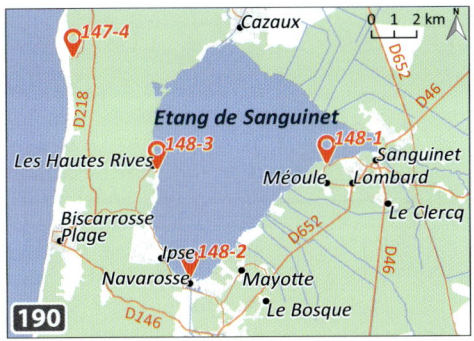

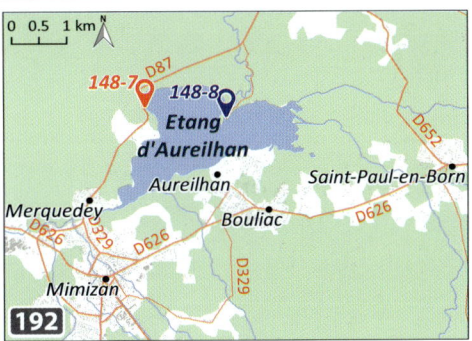

The same species can be found at the Étang d'Aureilhan (also called Étang de Mimizan; car park 1.7 km north of Mimizan on the D87 [7]; several tracks further along this road lead to the southern mouth of the Courant de Sainte-Eulalie [8]; access may be hard). Surrounding reedbeds could yield singing crakes at dusk and dawn in April and May. Caspian gulls have occurred increasingly in **roosting flocks of large** gulls **in winter and spring**. On your way back to Bordeaux, **young plantations in the Landes pine forest are worth dusk stops** for grasshopper and Dartford warblers and nightjar (from late April).

Southern Aquitaine and the Pyrenees (sites 149–152, maps 193 and 194)

HIGHLIGHTS
» An altitudinal gradient from sea level to high mountains
» Iberian endemics and regional rarities
» Large gull roosts and excellent seawatching sites on the coast

J F **M A M J** J **A S O N D**

KEY SPECIES

YEAR-ROUND great white egret, spoonbill, griffon vulture, **lammergeier**, golden eagle, **black-winged kite**, peregrine, marsh harrier, **Pyrenean grey partridge**, **Tengmalm's owl**, **white-backed woodpecker** *lilfordi*, Iberian green woodpecker, dipper, Alpine accentor, common rock thrush, red-billed leiothrix, **Sardinian**, Dartford and Cetti's warblers, Eurasian treecreeper, wallcreeper, red-backed shrike, raven, red-billed and Alpine choughs, **snowfinch**, cirl and rock buntings

BREEDING squacco, little, night and purple herons, little bittern, Egyptian vulture, black kite, short-toed eagle, stilt, avocet, nightjar, Alpine swift, **wryneck**, **bee-eater**, whinchat, **ring ouzel**, **blue rock thrush**, common whitethroat, **Savi's warbler**, **Iberian chiffchaff** and western Bonelli's warbler, spotted flycatcher, red-backed shrike, hawfinch

MIGRATION Cory's and Balearic shearwaters, **Leach's petrel**, **raptors**, booted eagle, honey buzzard, red kite, osprey, all harriers, common crane, **black stork**, **Sabine's gull**, whiskered and black terns, stock dove, woodpigeon, wheatear, great reed, sedge and aquatic warblers

WINTER wildfowl, **greater spotted eagle**, **purple sandpiper**, **Caspian**, glaucous and Iceland gulls, **penduline tit**

VISIT DURATION 2 days

PLAN YOUR VISIT
The best itinerary depends on the season. Lowland marshes and lakes are good year-round, while the coast is at its best in autumn and winter, especially when storms bring pelagics from the Bay of Biscay. The mountains are a must, as

149 Lakes and marshes of southern Aquitaine

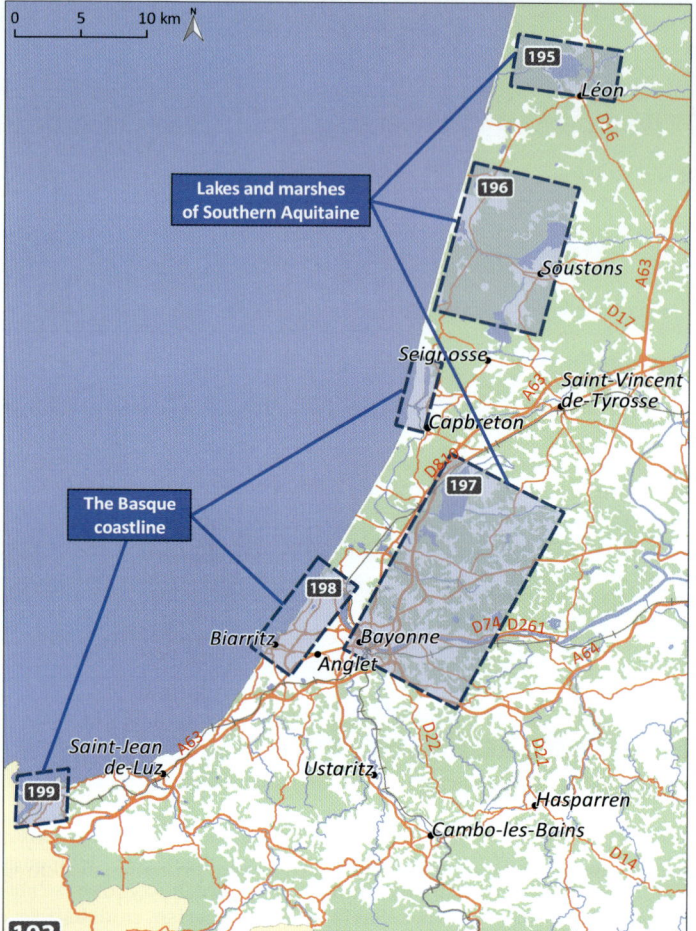

0.5 DAY; MAPS 195–197
The Courant d'Huchet reserve [1], **Lac de Moïsan** [2], **Lac de Soustons** [3] **and the Étang Blanc** [4] **are surrounded by bogs and reedbeds that host breeding** little bittern, night-heron **and** purple heron, Cetti's **and** Savi's warblers **(exit 12 on the A63 and drive to Léon).** The surrounding Landes forest has good densities of Dartford and western Bonelli's warblers, spotted flycatcher and nightjar (from late April).

Take the A63 and leave it at exit 8 (Labenne); **this will lead you to the Marais d'Orx nature reserve** [5], another protected freshwater pond surrounded by marshes and bogs (www.reserve-naturelle-marais-orx.fr, exit 8 on the A63 and follow signposts from Labenne). The site is only accessible along a 6.2 km footpath with hides. **In winter, Orx hosts up to 2000** greylag geese, **together with hundreds of** pochard, shoveler, gadwall **and** teal, **joined by** pintail **in February and March.** Although marsh harrier is the only resident raptor, greater spotted eagle and white-tailed eagle have wintered several times. During the breeding season, Orx has breeding spoonbill, great white egret, purple heron, little egret and night-heron. Squacco heron has bred a few times in the past but is now irregular at the site. Black-winged stilt and avocet do not breed but regularly roost at the lake, along with common snipe and curlew. Due to access restrictions, the site is not as good for passerines, although Savi's warbler breeds and great reed, sedge and aquatic (scarce) warblers occur on migration.

The nearby Bergusté reserve offers similar conditions and birds (17 km, 30 min from

they are the breeding grounds of the endemic white-backed woodpecker lilfordi (sometimes treated as a separate species from eastern Europe's white-backed woodpeckers), which can be either a real pain or ridiculously easy depending on luck and season (the birds become really secretive in June, when nestlings are being fed). If you fail, you'll still have the compensation of an Iberian chiffchaff or a lammergeier. The woodpecker inhabits one of Europe's most famous raptor migration sites, the Organbidexka pass, a definitive must on a sunny late-summer or autumn day. But remember that woodpeckers and raptors are not the only birds in these mountains. We describe two Pyrenean itineraries from the Organbidexka area, one westwards and one eastwards connecting with Pau, which will give you an access to all the highland specialities.

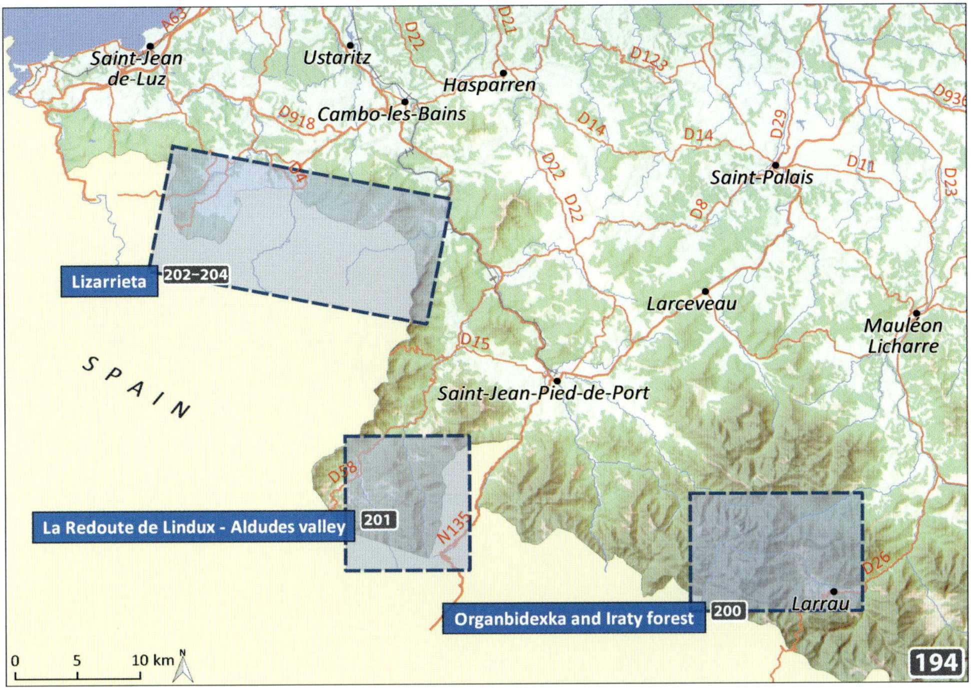

Orx, turn on the D54 in front of Saint-Martin-de-Seignanx church to Bayonne; bypass Villenave and turn left at the last road before the Adour bank, signposted 'Point de Vue – Réserve de Faune'; drive 3 km to find the car park and hide [6]). A greater spotted eagle has been wintering here for several years (late November to late March).

If you wish to head to the coastline, stop in the Ansot plain ('Plaine d'Ansot') [7], near Bayonne (09.00–17.30, hide and paths; head to Bayonne centre, then follow directions to 'Saint Léon hôpital'; the site is signposted from the roundabouts around the hospital). Penduline tit regularly winters in the area. Spring targets include migrating black stork, Savi's warbler, wryneck, night-heron, purple heron, stilt, hobby and red-backed shrike.

150 The Basque coastline

0.5–1 DAY; MAPS 198 & 199 **Hossegor pond** [1 – see GPS file] **has become a prime** gull-**watching place with** Caspian gulls **every winter in the last few years** (11 km, 20 min from the Marais d'Orx and 30 min from Bayonne, head to Hossegor and follow 'Les Plages', the lake lies 800 m inland from the coast).

Another spot for gulls **is the Adour river mouth.** The northern bank is a harbour with several viewpoints on the river up to its mouth [2] (8 km, 15 min from Bayonne, once in Boucau follow 'Tarnos Plage' or 'Zone portuaire'). Audouin's gull and lesser crested tern have been recorded since 2012, and glaucous and Iceland gulls are possibilities after northwestern winter storms. **Beyond the far end of the northern bank, a military training area** [3] dominated by low grass and bushes can be exceptionally active with wheatear, flycatchers and warblers during both migrations. **There are alternative viewpoints on the southern bank** [4,5] (follow signs to Anglet along the river), which also offers easier access to the seafront. **Izadia park ('Parc**

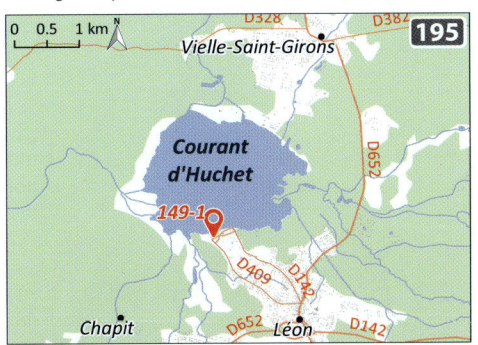

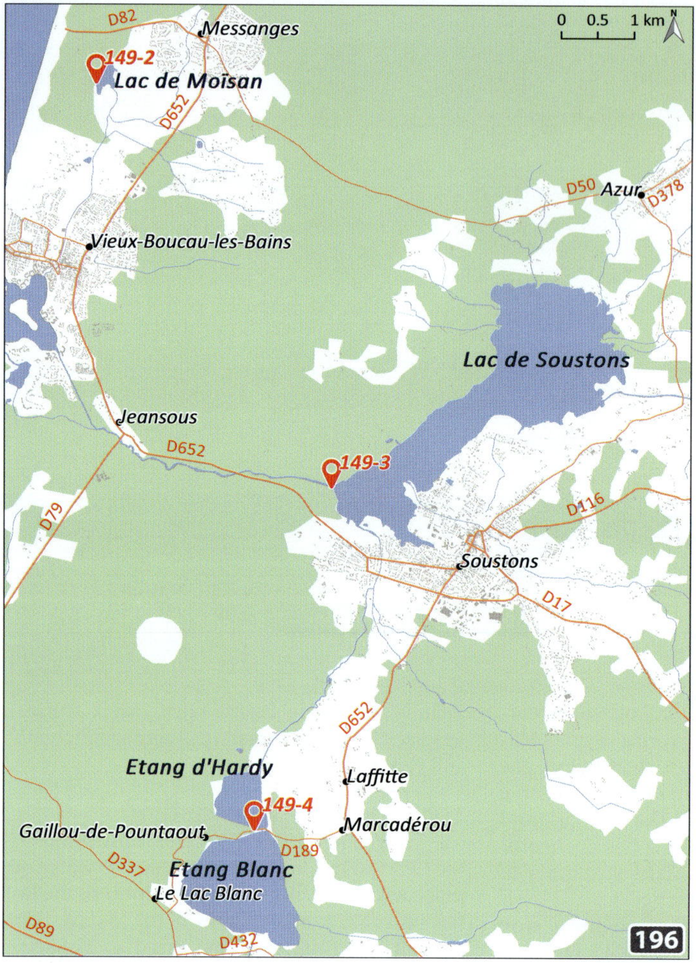

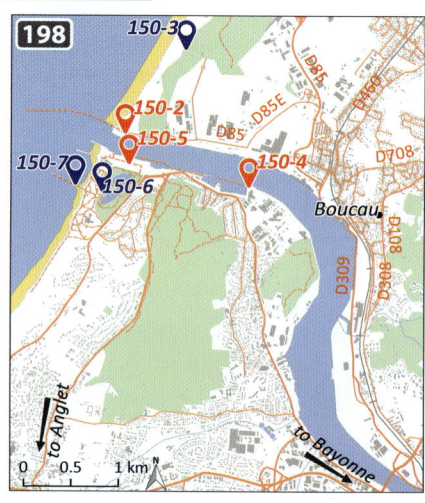

écologique d'Izadia') [6], at the river mouth, can be full of migrating passerines, and seawatching can be excellent from the nearby breakwaters [7] **in strong northwesterly winds** (expect Cory's and Balearic shearwaters and rarer pelagic species such as Leach's petrel and Sabine's gull. Purple sandpiper may occur along the breakwaters in winter.

Further south, Biarritz lighthouse [8 – see GPS file] and the Rocher de la Vierge [9 – see GPS file] in Biarritz are good alternatives for seawatching. **Pelagic trips using bait are organised in late August and early September from Saint-Jean-de-Luz harbour** (check with the LPO, http://lpoaquitaine.org).

South again, the Domaine d'Abbadia [10], 3 km north of Hendaye, is a **65 ha protected area known as the best French Atlantic**

AQUITAINE | **151 IRATY AND THE ATLANTIC PYRENEES** | 149

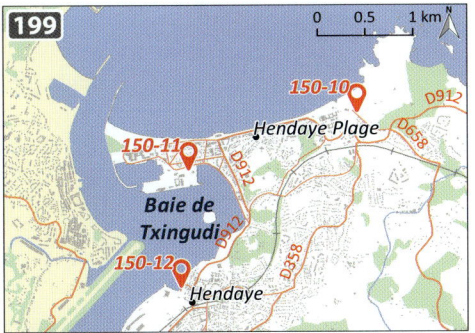

spotted flycatcher. In winter, look for hawfinch and check the rocky coast for shag, purple sandpiper, auks and divers. The nearby Hendaye bay and harbour (sometimes called 'Baie de Txingudi') [11,12] host several gull and wader roosts.

151 Iraty and the Atlantic Pyrenees

1–2 DAYS; MAPS 200–204 The Basque Pyrenees are well known for **three** raptor **migration hotspots: Organbidexka, La Redoute de Lindux and Lizarrieta**. The three passes are monitored by birdwatching teams from mid-July to mid-November and yield some of the most impressive autumn migration records in France. Black (July–August) and red (September–November) kites, short-toed eagle (August–September), honey-buzzard (August–September), osprey (August–October),

location for Sardinian warbler (signposted from Hendaye centre; 10.00–18.00 March to August, closed September to January). Other breeding species include peregrine, raven, Dartford warbler, common whitethroat and

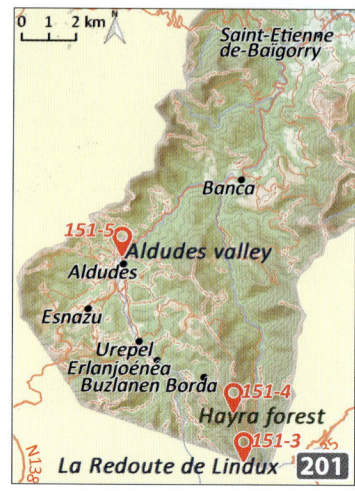

four harrier species (August–November) and black stork (September) may pass through in high numbers on sunny days with low wind. Large flocks of cranes and geese plus stock doves and woodpigeons replace raptors from October. Golden eagle and lammergeier are seen year-round in these three areas.

Organbidexka pass [1] is the easternmost and most famous of the three sites; it also has the most impressive records (news on www.migraction. net, follow signposts to Iraty from Larrau, 11 km, 20 min; or Saint-Jean-Pied-de-Port, 32 km, 1 h; the spot is located 400 m southeast of Iraty chalets on the D19). **On its northern slope, Iraty forest** [2] **is the most reliable place in the Pyrenees for** white-backed woodpecker *lilfordi*. The species is present in all mountain old-growth beech forests from the Atlantic coast to the Gavarnie area and maybe further east. Other species of interest in the Iraty area include Pyrenean grey partridge (high altitude, scarce), Eurasian treecreeper and ring ouzel (March–August). Singing Iberian chiffchaffs are sometimes recorded in younger stands and clearings. Tengmalm's owl could occur in the area in some years and dipper is present on all streams in the area.

The second migration pass, **La Redoute de Lindux** [3], **lies on the Spanish border, 60 km west of Organbidexka** (60 km, 2 h south of Saint-Jean-Pied-de-Port, head west to Saint-Étienne-de-Baïgorry on the D15, then D948 towards Pamplona through Aludes; once in Urepel turn left just before the church and navigate on narrow local roads – GPS strongly recommended; an easier and more direct 30 km route from Saint-Jean-Pied-de-Port is to enter Spain via the D933 and N135 to Pamplona and turn right on a climbing dirt road at the 'Ibañeta' signpost 1260 m before Roncevaux; stop 4 km higher up at the Spanish border). **The nearby Hayra forest** [4] **, on your way back to Aludes and Saint-Étienne-de-Baïgorry,** also has white-backed woodpecker. Iberian chiffchaff sometimes breeds 15 km north in the Aludes valley [5], and can be found sporadically in all the western Pyrenean foothills below 1000 m (currently not reliably identifiable without a song, beware of individuals singing intermediate songs with common chiffchaff, which are also common in the area).

Further west again, the trail from the **Veaux pass** [6] **to Méhatché pass** [7] **leads across high-altitude meadows** (37 km, 1 h from Biarritz, drive southeast to Cambo-les-bains on the D932 and follow signs to the 'Pas de Roland', bypass a restaurant in Laxia and climb along the stream). There is a griffon vulture colony here, and Egyptian vulture is regular. Red-billed and Alpine choughs, raven, common and blue

rock thrushes, red-backed shrike (from May), water pipit, whinchat (from mid-April) and cirl bunting all breed locally.

Slightly to the south, **Iparla ridge [8] has the same species**, plus wallcreeper and rock bunting (4 h 30 hike from Bidarray, 15 km, 20 min from Cambo-les-Bains on the D918). **Lizarrieta, the westernmost migration viewpoint [9], forms a reasonable alternative to Organbidexka** for birders with limited time or who do not wish to drive a long way from Hendaye or Bayonne (25 km, 45 min from Saint-Jean-de-Luz, drive to Sare and follow signposts on the D306 for 11 km, 22 min). Although less famous, it has impressive autumn raptor and passerine records. **Back north to Sare and up to the Pyrenean foothills again, Lezeko Gaina [10] is a reliable place for** Iberian green woodpecker, Dartford warbler and Iberian chiffchaff (April–July) (park at the 'Grottes de Sare', 6 km, 10 min from the village).

152 The Pyrenees east of Iraty and the Béarn

0.5–1 DAY; MAPS 205 & 206 **The Lapiaz de la Pierre Saint-Martin [1] has virtually all the regional mountain specialities**, including lammergeier, Egyptian vulture, Pyrenean grey partridge and Iberian green and white-backed woodpeckers (45 km, 1 h well signposted east of Iraty through Larrau or south on the D132 from Oloron-Sainte-Marie). Alpine accentor, common rock thrush, citril finch and snowfinch, rock bunting, both choughs and wallcreeper also breed on cliffs and rocky slopes.

Golden eagle, lammergeier **and** Egyptian vulture **are easier in the vicinity of the Col du Porteigt [2]** (on the D294 between Escot and

Bielle, 30 km, 40 min south of Oloron-Sainte-Marie), as well as both choughs and Alpine swift. Ring ouzel and Eurasian treecreeper are common and a few whinchats still occur. Pyrenean grey partridge can be found at higher altitudes in rocky heathlands, but the species is scarce and sensitive to disturbance: keep on tracks and refrain from using playback. **If you've missed Egyptian vulture here, you can try Arguibelle cliff** [3] (20 km, 25 min from Oloron on the D918 towards Féas, turn left just past Lanne-en-Barétous on the D632 towards the Pierre Saint-Martin), also good for peregrine, crag martin and wallcreeper.

If you've spent your morning in the mountains, drive the 54 km, 1 h to Pau around midday (see Map 206). The local exotic speciality, red-billed leiothrix, can be found in the dense woods around Artix lake [4 – see GPS file] (26 km, 30 min west of Pau: once in Artix drive towards Mourenx until you reach the lake after 1.7 km). Night-heron and squacco heron frequent the riparian woods around the lake and all along this stretch of the Gave de Pau river. Migrating ospreys, bee-eaters, whiskered and black terns (spring) and wintering wildfowl may also justify a one-hour stop along the river and lake. **Northwest of Pau,** black-winged kites are easily found around the airport [5 – see GPS file] **year-round.** Search them as they perch on top of high isolated trees or hedgerows, or when hovering over fields. Small numbers of red-footed falcons show up from time to time on the fences of the airport in May. Also visit Lac de l'Ayguelongue [6 – see GPS file] near Momas (20 km, 26 min from Pau on the D945 towards Orthez, turn right on the D201). This 60 ha pond is a wintering site for wildfowl (goosander, ferruginous duck and scaup are regular) and a stopover site for waders on both migrations. You will quickly see whether the water levels deserve a stop, but mind that small coves may have mudflats invisible from the main parking area. Temminck's stint is annual, and vagrants like buff-breasted, Baird's and White-rumped sandpipers have been recorded. The surrounding woods and hedgerows host lesser spotted woodpecker and hawfinch (winter). If you have a couple of hours left before dusk, **drive to the plain around Aire sur l'Adour** (70 km, 60 min on the A65 or the D834 from Pau) (see Map 206). Several pairs of black-winged kite breed along the D352 [7 – see GPS file], as well as quail and red-backed shrikes (from May). Booted eagle, both kites and bee-eater are easy to find in this area.

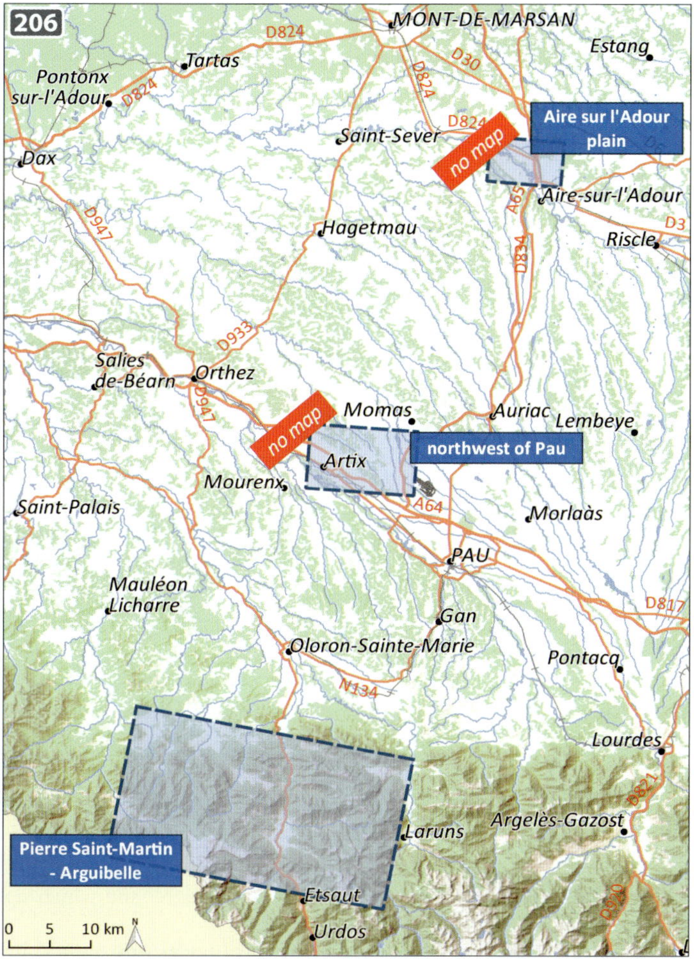

AQUITAINE | INNER AQUITAINE | 153

Inner Aquitaine (sites 153–155, map 207)

HIGHLIGHTS
» A transition region, with birds from the Atlantic lowlands and Mediterranean species
» Remote areas with very low birding pressure – make your own discoveries!

| J | F | M | A | M | J | J | A | S | O | N | D |

KEY SPECIES
YEAR-ROUND hen harrier, red and **black-winged kites**, goshawk, peregrine, great white and cattle egrets, eagle owl, long-eared owl, little owl, black, middle spotted and lesser spotted woodpeckers, woodlark, crested lark, crag martin, **dipper**, Cetti's warbler, crested tit, firecrest, raven, hawfinch, **rock sparrow**, cirl bunting

BREEDING short-toed eagle, **purple heron**, quail, stone-curlew, little ringed plover, whiskered tern, **Alpine swift**, **scops owl**, hoopoe, nightjar, tawny and tree pipits, whinchat, common whitethroat, **subalpine**, western Bonelli's and melodious warblers, spotted flycatcher, **woodchat** and red-backed shrikes, **ortolan bunting**

WINTER wildfowl, common crane, **Alpine accentor**, **wallcreeper**

MIGRATION black stork, osprey, waders

VISIT DURATION > 1 day

PLAN YOUR VISIT
This is probably one of France's least-known areas birdwise, but its location is likely to yield surprises in the future, especially if several

Mediterranean species continue their northward expansion. There is much to see, from wintering cranes and wallcreepers to breeding shrikes and subalpine warbler, **and the landscapes consists of nicely preserved mosaics of farmland and forest** in which wandering for birds is a real pleasure. The area is divided into three west–east routes that can be birded in 1–2 days each. Roughly the same species are found on the three itineraries. Concentrate on reservoirs and cliffs in winter. Mid-April to mid-June is the best time to visit warm dry habitats (including arable farmland areas) that host species with Mediterranean affinities.

153 The Lot valley and hills
1–2 DAYS; MAPS 207 & 208 – NO DETAILED MAP **Captieux army camp is an ideal start for this itinerary in winter** (leave the A65 in Captieux and follow the D932 south for 20 km, then turn right to Lencouaq). The camp is closed to the public but there is a hide on the southern boundary [1]. Large flocks of cranes winter in the camp and its surroundings and a white-tailed eagle is a regular winter visitor (presumably a returning individual). **In spring, skip this area and head directly to the Lac de Salabert** [2] (exit 6 on the A62 towards Villeneuve-sur-Lot, drive for 22 km, 25 min to Lacépède through Lafitte-sur-Lot, take the first track on the right after the village; footpath and a hide). Hobby, whiskered tern, purple heron, cattle egret, little ringed plover and golden oriole breed around the lake from April to July; osprey and red kite are also regularly seen on migration. In winter, the site has great white egret and wintering wildfowl.

In spring, **your next logical stop should be the hills around Pinel** [3] (from Salabert: 23 km, 25 min north, in Castelmoron turn on the D262 and turn 4 km after Saint-Étienne-de-Fougères to 'manoir de Pinel'; car parking and access information near the church). **Open areas with scattered juniper and oak woods and bushes host a bird community typical of dry warm inland habitats.** Woodlark, tree pipit, western Bonelli's warbler, whitethroat and cirl bunting are common breeding species. Also search for hoopoe, spotted flycatcher and golden oriole. A visit here at dusk should easily yield little owl and nightjar (from April). **The surroundings of Monflanquin** [4] (25 km, 30 min northeast) consist of a mosaic of agriculture, woodland and hedgerow excellent for barn, little and long-eared owls, quail (from April), short-toed eagle (from March) and rock sparrow.

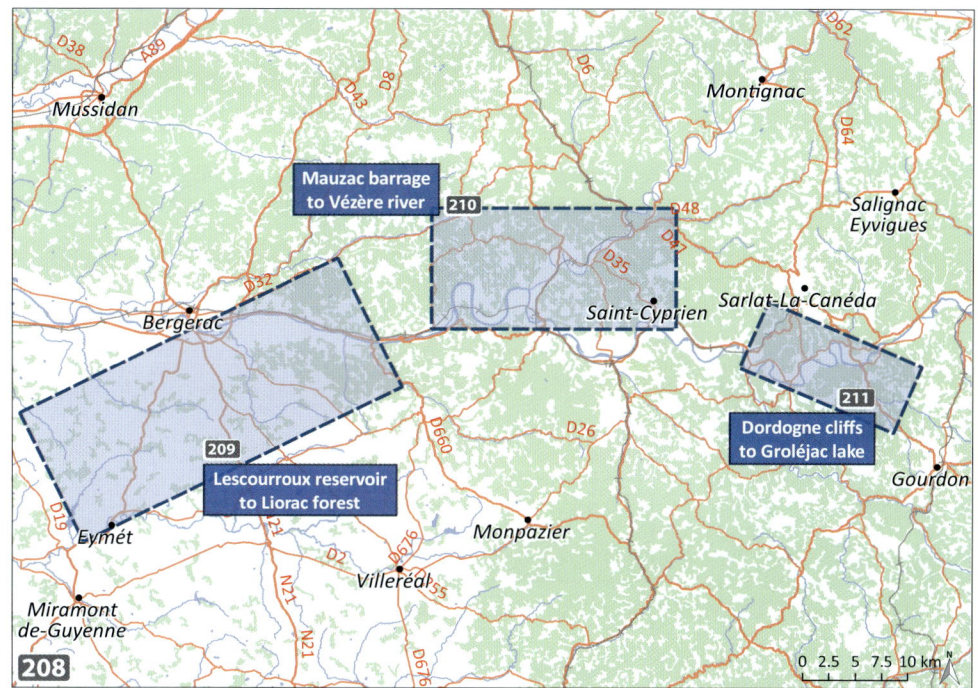

Drive 25 km, 30 min further east to reach the **dry hills around Fumel airfield** [5] (signposted 'Aérodrome', turn on the D139 between Condezaygues and Fumel). **This is one of very few sites for breeding** subalpine warbler **(from April) in Aquitaine.** Short-toed eagle, nightjar, golden oriole, western Bonelli's warbler **and** rock sparrow **also occur here in a rich bird community including other species with Mediterranean affinities**. The easternmost location on this itinerary is the agricultural mosaic around Masquières [6] (10 km, 15 min southeast of Fumel airfield; in Mauroux turn on the D4 to Tournon, then left on local roads after 4 km). Apart from other species present elsewhere, the area is good for raven, goshawk, peregrine, middle spotted and lesser spotted woodpecker.

154 Southern Dordogne

1–2 DAYS; MAPS 209–211 Start at the Lescourroux reservoir [1] (leave the A62 at exit 5, bypass Marmande and drive 33 km, 40 min northeast to Miramont-de-Guyenne; follow signposts from La-Sauvetat-du-Dropt). **In late summer and autumn (and sometimes at other times), mudflats appear in its southern part** and are ideal for waders, with chances of black stork, Temminck's stint and marsh sandpiper among more regular sandpipers and shanks. Check the open water for osprey, whiskered and black terns at migration times. In winter, wildfowl might be joined by rarities such as goosander, smew, red-throated diver and Slavonian grebe.

The agricultural plain around Faux and Issigeac [2] (30 km, 30 min northeast of Lescourroux and 25 km, 25 min southeast of Bergerac, bird along small roads and paths between these two villages) has a **typical warm-dry farmland bird community**, with black-winged kite (year-round) **as the highlight. Other species rarely found elsewhere in the area can reasonably be expected here:** scops owl, stone-curlew, woodchat shrike, rock sparrow, hen harrier, crested lark, tawny pipit, **together with commoner species such as** quail, little **and** long-eared owls, melodious warbler, nightingale, woodlark, tree pipit, corn **and** cirl buntings, nightjar **and** red-backed shrike. Yellow wagtails, whinchat and wheatear can be found during spring and autumn migrations.

Liorac forest [3] can be a good option as the temperature rises in the morning (25 km, 25 min north of Issigeac on the D21 and 15 km, 15 min east of Bergerac on the D32). The forest has short-toed eagle, honey-buzzard (from May), black, lesser spotted and middle

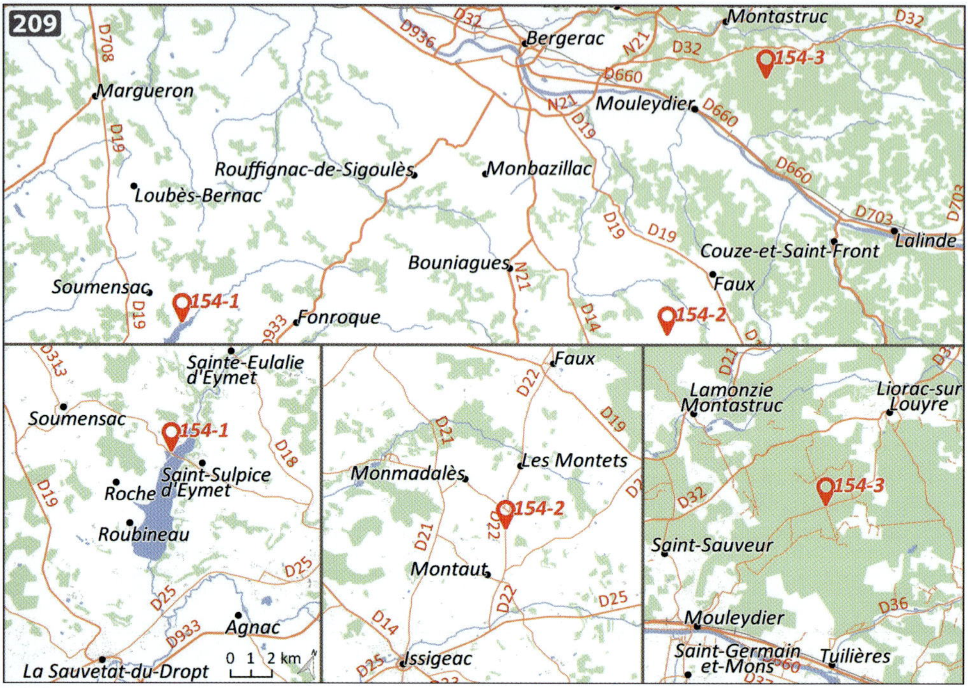
209
Bergerac
Montastruc
154-3
Margueron
Mouleydier
Rouffignac-de-Sigoulès
Monbazillac
Loubès-Bernac
Couze-et-Saint-Front
Lalinde
Bouniagues
Faux
Soumensac
154-1
Fonroque
154-2
Sainte-Eulalie d'Eymet
Soumensac
154-1
Saint-Sulpice d'Eymet
Roche
Roubineau
Agnac
La Sauvetat-du-Dropt
0 1 2 km
Faux
Les Montets
Monmadalès
154-2
Montaut
Issigeac
Lamonzie Montastruc
Liorac-sur Louyre
154-3
Saint-Sauveur
Mouleydier
Saint-Germain et-Mons
Tuilières

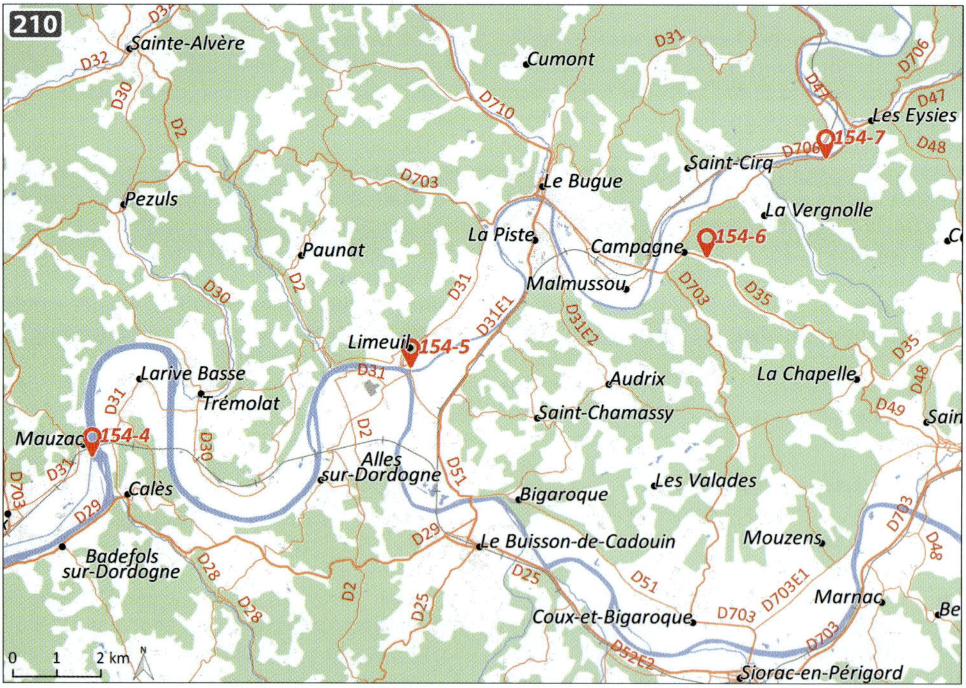
210
Sainte-Alvère
Cumont
Les Eysies
154-7
Saint-Cirq
Le Bugue
Pezuls
La Vergnolle
La Piste
154-6
Campagne
Paunat
Malmussou
Limeuil
154-5
La Chapelle
Larive Basse
Audrix
Trémolat
Saint-Chamassy
Mauzac
154-4
Alles sur-Dordogne
Les Valades
Calès
Bigaroque
Le Buisson-de-Cadouin
Mouzens
Badefols sur-Dordogne
Marnac
Coux-et-Bigaroque
Siorac-en-Périgord
0 1 2 km

spotted woodpeckers, western Bonelli's warbler (from April) and hawfinch (in winter, among hornbeams).

Alternatively, visit the Dordogne valley for osprey (spring and autumn migration), peregrine, raven, crag martin and wallcreeper (winter). To find these species, **you can start at Mauzac barrage** [4] where great white egret, osprey, common sandpiper, crag martin, grey wagtail, kingfisher and little ringed plover are regular. **Take the secondary road above the river upstream, stopping at Limeuil where the Dordogne and Vézère rivers meet** [5] – check for tree sparrow. Follow the River Vézère via Le Bugue to the Campagne chateau, forest and cliff [6], a good place for peregrine, raven, black and middle spotted woodpeckers, short-toed treecreeper, crested tit, firecrest, redstart and western Bonelli's warbler, as well as wallcreeper in winter.

Similar species can be found further along the Vézère on the cliffs at Les Eyzies [7] or on the cliffs above the Dordogne between Beynac-et-Cazenac (including the chateau) and La-Roque-Gageac [8] (which also has breeding eagle owl, peregrine, Alpine swift from April, crag martin, stock dove, raven, and wintering Alpine accentor). **Ideally, you should be here in mid-afternoon, so that you can end your day at the Groléjac lake** [9] (15 km, 18 min from La-Roque-Gageac southeast on the D50, turn right at the signpost 'Marais de Groléjac' just before joining the D704). A walk on the boardwalk in the riparian woods and wetlands south of the lake should yield Cetti's warbler, golden oriole, water rail and siskin in winter.

155 Northern Dordogne

1–2 DAYS; MAPS 212–215 In spring and summer, **start in the Auvézère valley** [1 – see GPS file] (take the A89 east from Brive-La-Gaillarde to exit 17, then on the D704 for 18 km, 20 min; turn right towards Génis; there follow signposts to the 'Circuit de l'Auvézère'; there is a small car park and walking trail along the river). This small river runs within a steep-sided wooded valley with rocks and cliffs that are home to goshawk, black and middle spotted woodpeckers and dipper. Visiting the area can take up to a full day. **In winter, start at the Miallet reservoir** [2 – see GPS file] (50 km, 50 min north of Périgueux on the N21; in La Coquille turn left and follow signposts to the 'plan d'eau'). Common wildfowl are sometimes joined by rarer species such as black-necked grebe, goosander and smew; great white egret also winters and osprey can be seen in March and September. 25 km, 30 min east, the Saint-Estèphe lake [3 – see GPS file] (follow signposts from Augignac) can be even better in cold winter conditions, especially for casual divers.

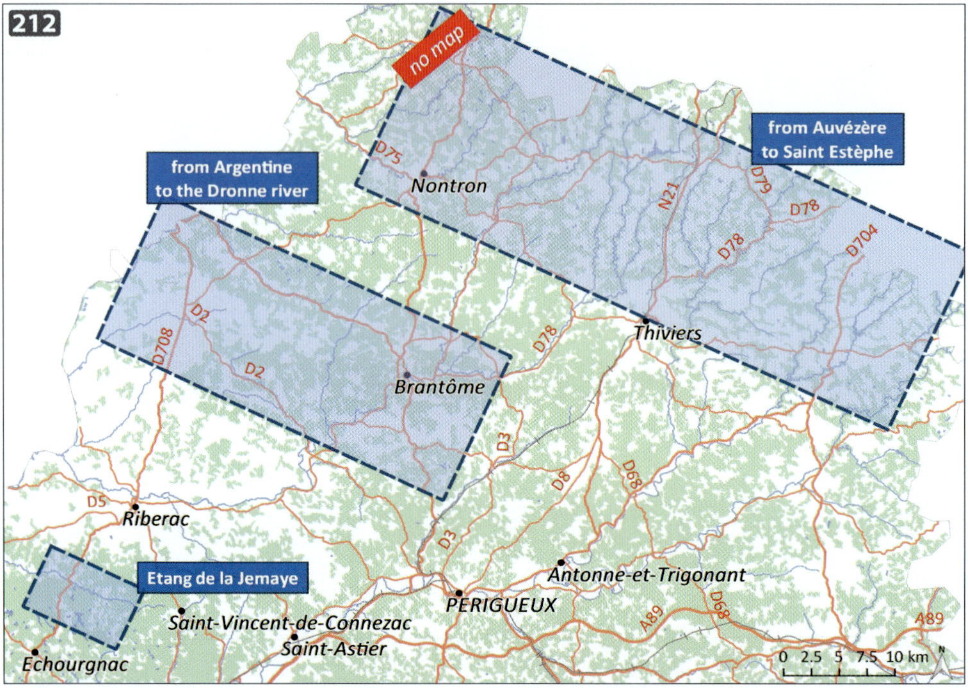

As both reservoirs are less productive in spring, replace them by a morning or evening visit to the warm-dry Argentine plateau [4] which has woodlark, tree pipit, nightjar and red-backed shrike (30 km, 40 min southeast of Angoulème, follow the D939 towards Périgueux and leave it at a roundabout towards Verteillac on the D12, the plateau is signposted on the left after 800 m; bird on small roads and footpaths around Argentine). The plain around Verteillac [5], 10 km south, hosts species such as hen harrier, stone-curlew, red-backed shrike, rock sparrow and ortolan bunting (rare).

Also spend some time in the dry hills around Paussac-et-Saint-Vivien [6] (20 km, 20 min east of Verteillac, bypass Lisle and turn left to Saint-Vivien where Paussac is signposted; park in the village and walk or drive along the small roads on both sides of the D93) for

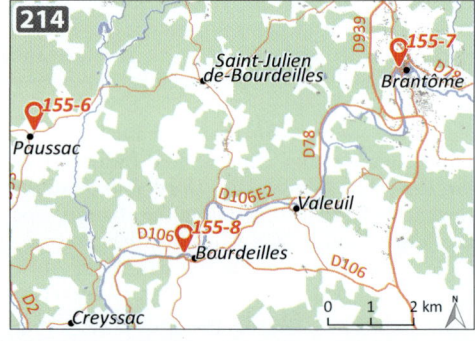

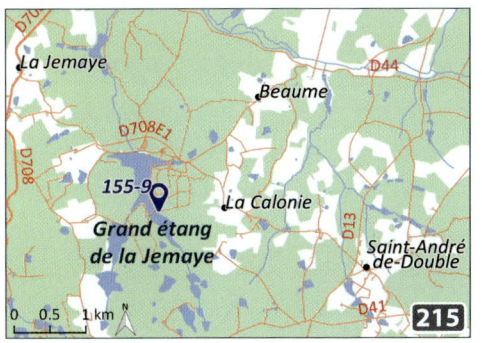

hoopoe, raven, peregrine, rock sparrow (in the village) and forest birds such as middle spotted woodpecker, tree pipit and western Bonelli's warbler. Wallcreeper **can occur here and at the nearby Brantôme abbey** [7] **in winter, and the Dronne river in Bourdeilles** [8] **gives a good chance of** dipper.

End your day at the Jemaye lake, which has black woodpecker, Dartford warbler, crested tit, goshawk and reed warbler (from April) (43 km, 50 min from Bourdeilles, drive west to Ribérac and take the D13 south to Siorac; once in Saint-André-de-Double after about 12 km turn right to the 'Grand Étang de la Jemaye', then left after 3 km at the 'Espace Naturel Sensible' signpost to reach **a hide in the marshy southern part of the lake** [9]).

REGION 8
THE PYRENEES

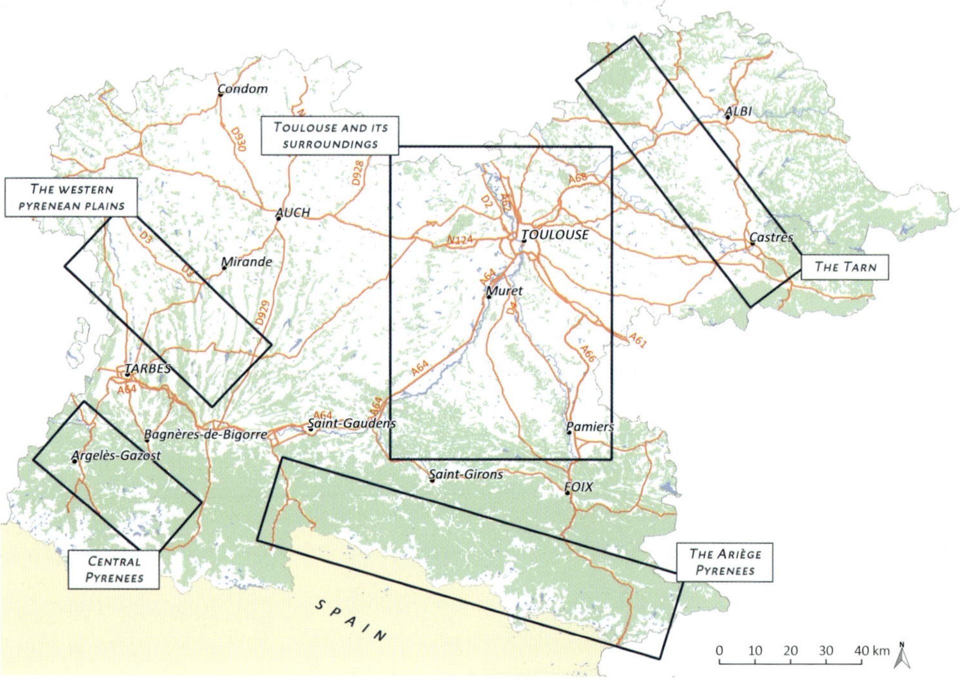

HIGHLIGHTS
» Wild mountain landscapes hosting the rare Pyrenean subspecies of capercaillie, ptarmigan and grey partridge. The most determined birders will search for the easternmost white-backed woodpeckers (subspecies *lilfordi*) in old beech forests
» A definitive raptor region with your best chance of lammergeier and Egyptian vulture in southwestern France
» A transition from continental to Mediterranean species

REVIEWERS   Jérémy Dupuis, Tristan Guillosson
USEFUL WEBSITES
www.baznat.net
www.faune-tarn-aveyron.org

Egyptian vultures can be found in all the vulture colonies of Ariège, but be patient – they can be elusive.

THE PYRENEES | TOULOUSE AND ITS SURROUNDINGS | 161

The steep slopes of the Ariège Pyrenees (sites 165–168) are hard work, but the reward is high. Cross the entire altitudinal vegetation gradient in a single walking day, from low riverine valleys and their dippers to summits frequented by lammergeier, Egyptian vulture and Alpine accentor.

Toulouse and its surroundings (sites 156–161, map 216)

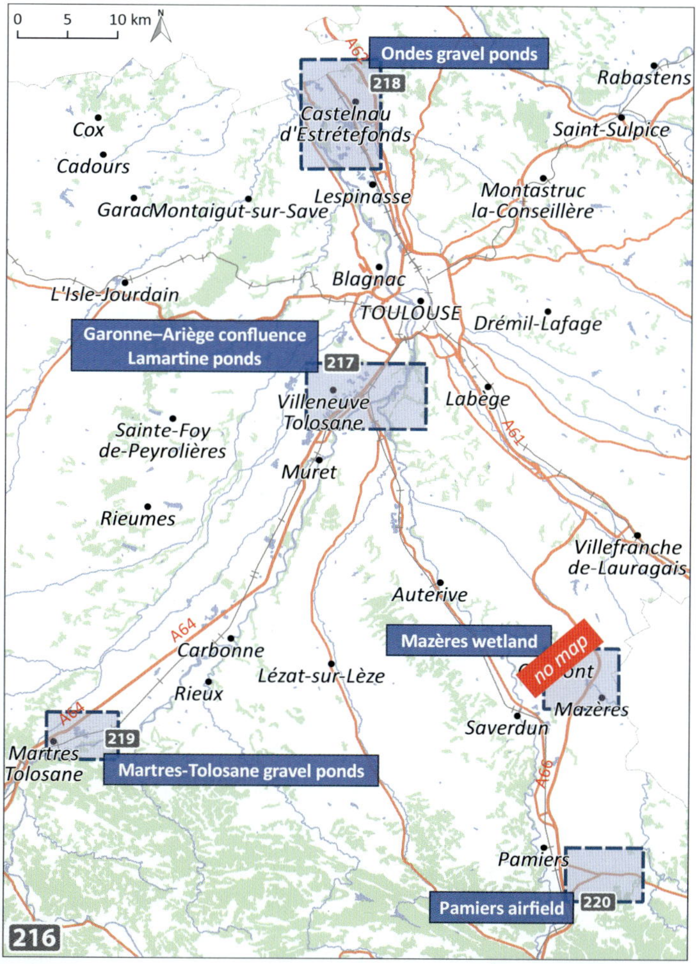

156 THE ARIÈGE–GARONNE CONFLUENCE | THE PYRENEES

HIGHLIGHTS
» Several easily accessible small wetlands for short birding sessions around Toulouse
» Viewing points for breeding and migrating waterbirds
» A photographer's heaven in Mazères

KEY SPECIES
YEAR-ROUND cattle egret, red and **black-winged kite**, peregrine, crested lark, serin, **rock sparrow**, cirl bunting

BREEDING red-crested pochard, **squacco, night- and purple herons**, white stork, **booted eagle**, black kite, hobby, **stone-curlew**, black-winged stilt, little ringed plover, green sandpiper, Mediterranean gull, black tern, **bee eater**, sand martin, great reed warbler

MIGRATION garganey, waders

WINTER all wildfowl, ferruginous duck, merlin, common crane, short-eared owl, **Richard's pipit**

VISIT DURATION 2 days

PLAN YOUR VISIT
The cultivated plains around Toulouse are not a birder's paradise. Nevertheless, a couple of sites are worth a visit at any time of the year. They are best in April when waterbirds stop over on their way north while most resident and migrating breeders are already territorial. Most of the described places can be visited in a couple of hours or less from the centre of Toulouse; step from one to another to build your itinerary. Keep half a day for Mazères, the birdiest site of the area and a photographer's favourite, worth the long drive south from Toulouse. If tied to Toulouse centre, remember that pallid swifts breed on buildings along the banks of the Garonne.

156 The Ariège–Garonne confluence
1–2 HOURS; MAP 217   In the southern suburbs of Toulouse, where the Ariège flows into the Garonne (the Confluence Garonne–Ariège), is an enclave of a **natural reserve within a highly urbanised area**. Follow the eastern bank of the Garonne southwards from Toulouse centre towards Lacroix Falgarde (D4, 11 km, 20 min). **Turn right at the 'Bac de Portet' sign to reach a car park** [1] **where several trails run into the reserve towards the river bank.** One or two pairs of booted eagle breed in nearby woods (from April) with hobby, black kite and night-heron. In winter, look for black-headed and

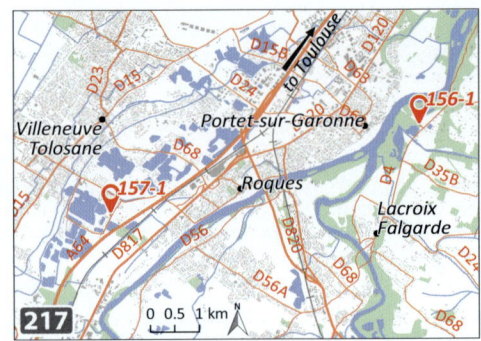

yellow-legged gulls and black or common terns, purple heron, common and green sandpiper among other migrants.

157 Lamartine ponds (Lac Lamartine)
1 HOUR; MAP 217   **The small Lamartine nature reserve includes five artificial ponds** which can be visited as a follow-up to the confluence of the rivers (site 156, 11 km, 20 min) or as a short birding break away from the crowds of Toulouse. From Toulouse, exit the A64 to follow 'Tarbes/Lourdes' (last exit before the toll) and keep this heading until you find signs to the D42 and 'Lac Lamartine' (first roundabout past IKEA), which will lead you to the reserve's car park [1]. The site hosts a **mixed heron colony** (with night-heron, grey and purple herons, cattle and little egret) and breeding little and great crested grebes. It is also a stopover site for waterbirds; expect pochard, tufted duck and dabbling ducks in winter and common tern in spring and early autumn.

158 Ondes gravel ponds
2–4 HOURS; MAP 218   **Several former and operational gravel pits are spread along the River Garonne 30 km north of Toulouse centre. The ponds themselves are fenced but they are easy

to scan from surrounding local roads. From Toulouse, follow the D820 or the A62 north to start in Saint-Jory. Once in the village, cross the canal and head 1.8 km northwest towards Saint-Caprais on the D20 to visit the first pond [1]. Drive 5 km more and turn right towards Ondes on the D20, then take the local road to the right signposted 'Parcours cyclable du canal de Garonne' [2]. The two other sites lie 3 km north of Ondes [3,4] (head north towards Grisolles and turn right at the speed control point). Although somewhat odd-looking, these ponds are a local hotspot for migrating wildfowl (pochard, tufted duck and dabbling ducks), purple heron, night-heron, egrets and black tern. They also host breeding great crested and little grebes, common tern, red-crested pochard, black-headed and Mediterranean gulls, crested lark and sand martin.

159 Martres-Tolosane gravel ponds

1–2 HOURS; MAP 219 The Martres-Tolosane gravel pits are the southern counterpart of the Ondes ponds (site 158), closer to the Pyrenean foothills. **These three ponds are located less than 1 km from exit 22 on the A64** [1,2,3] (60 km, 50 min south of Toulouse and 40 km, 30 min northeast of Saint-Gaudens, once off the highway follow signs towards 'Z.A. Carnaval'). **They are fenced but they can easily be scanned from the road.** Breeding species include little and great crested grebes, black-winged stilt, little ringed plover, bee-eater, sand martin and great reed warbler. Cranes roost in surrounding fields in March and November and sandpipers, gulls, terns and swallows stop over on both passages. Among wintering wildfowl, expect teal, shoveler, pochard, tufted duck and the odd ferruginous duck. In winter, red kite, black-winged kite, peregrine and merlin hunt or roost in the immediate vicinity.

160 Mazères wetland

0.5 DAY; NO DETAILED MAP The Domaine des Oiseaux in Mazères is the largest protected wetland in the Toulouse area and a prime stopover site for migrating wildfowl and waders in spring and autumn. Head south on the A66 from Toulouse and exit in Mazères (60 km, 45 min, exit 2), then follow signposts to the 'Domaine des Oiseaux'. **There are several parking places around the reserve, the main one lying 630 m past the highway on the way to Calmont** [1]. Hides are spread along a trail at strategic viewpoints where birds approach within photographic range. Among the most common breeding species, expect little ringed plover, black-winged stilt, white stork, common tern and bee-eater (from late April). Serin, corn and cirl bunting, rock and tree sparrow should be conspicuous on surrounding hedgerows. Teal, shoveler, garganey and wigeon form the bulk of the migrating ducks (October to April), and waders (sandpipers, stints, shanks, snipe) show up as soon as mudflats are exposed. Rarities including buff-breasted and pectoral sandpipers and regional rarities such as squacco heron have been found here in the past.

161 Pamiers airfield

1–3 HOURS; MAP 220 **The grassy meadows around Pamiers airfield are worth a few stops for wintering and migrating raptors and passerines**, particularly in autumn or when the weather is cold. Drive east for 8 km on the D119 towards Mirepoix and turn right after the D119–D12 junction. Richard's pipit winters every year in the meadows around Millet farm [1]; you can walk in the meadows if they are not fenced. Further on, on the minor road that leads south of the airfield [2], look for roosting red kite, lapwing, stone-curlew, curlew and short-eared owl.

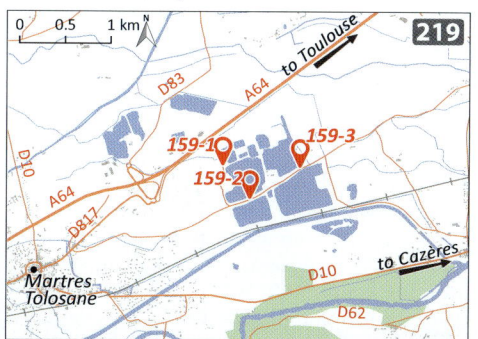

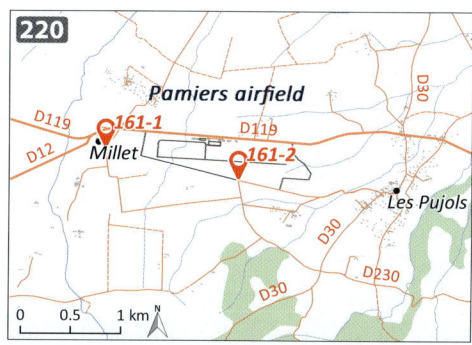

The Tarn (sites 162–164, map 221)

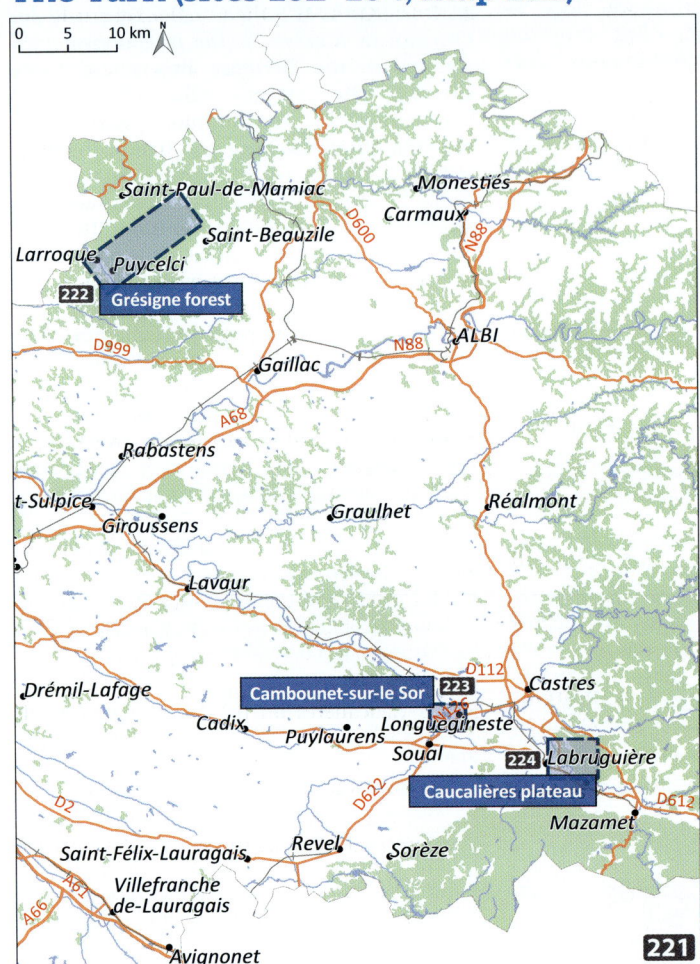

HIGHLIGHTS
» A transition from continental woodland species to the Mediterranean avifauna
» A variety of landscapes, from mature forest to dry scrub and wetland

| J | F | M | A | M | J | J | A | S | O | N | D |

KEY SPECIES
YEAR-ROUND great white egret, hen harrier, peregrine, little owl, middle spotted woodpecker, woodlark, **Sardinian** and Dartford warblers, raven
BREEDING squacco, **purple and night-herons**, cattle egret, **booted** and short-toed eagles, black kite, honey buzzard, **stone-curlew**, little ringed plover, **hoopoe**, **wryneck**, nightjar, tawny pipit, **subalpine**, melodious, wood and western Bonelli's warblers, golden oriole, red-backed shrike
WINTER Alpine accentor, **wallcreeper**

VISIT DURATION 2 days

PLAN YOUR VISIT
This is a spring area, at its best between late April and mid-May. It is remote and little watched, so you'll mostly have to find your own birds. The transition from continental to Mediterranean avifaunas is well-demarcated and results in a breeding community that is a rich mixture of species with multiple affinities. Grésigne forest

stands alone in the north; you can cover the two other sites in a single day.

162 Grésigne forest

0.5–1 DAY; MAP 222 **Grésigne is a 36 km² deciduous forest located 70 km northeast of Toulouse** (exit 9 on the A68 in Gaillac, then north on the D964, 7 km past Castelnau de Montmirail) or 50 km east of Montauban (D115 through Montricoux and Bruniquel, then head to Saint-Paul-de-Mamiac). **A trail runs across the forest from a car park along the D170 near the Maison Forestière de la Grande Baraque** [1] (turn left at a wide clearing, 13 km past Bruniquel). Middle spotted woodpecker breeds in old oak stands with golden oriole, wood and western Bonelli's warblers; wryneck is relatively common inside young stands. **In spring, spend late morning and early afternoon at Cabanes** [2] to scan the sky for soaring booted eagle, hen harrier, honey-buzzard from May, short-toed eagle from March and sparrowhawk (from the D87, turn on the D28 to Vaour, then right on a narrow road towards 'Ferme équestre' just before reaching the D15). **In winter,** Alpine accentor, wallcreeper, peregrine **and** raven **frequent the limestone cliff overlooking Larroque** [3].

163 Cambounet-sur-le-Sor

2 HOURS; MAP 223 The Cambounet-sur-le-Sor reserve lies just 10 km west of Castres (N126 towards Soual). Cross the bridge just before reaching Cambounet-sur-le-Sor and turn right following 'Réserve naturelle' until you reach one of two car parks [1,2]. This wetland consists of **several former gravel pits covered with aquatic vegetation, willows and poplars**. A path runs around the site and leads to **four hides**. The main interest of the reserve lies in its heron **colony** (the largest of the region) where night-heron, cattle egret and sometimes squacco and purple herons breed. **The typical** waterbird **community is also present** (great crested and little grebes, migrating ducks, snipe, sandpipers, little ringed plover, cormorant and great white egret). Surrounding fields and hedges are also worth some time in spring for breeding black kite, little owl, hoopoe, melodious warbler and nightingale.

164 Caucalières plateau (Causse de Caucalières)

0.5 DAY; MAP 224 **The Caucalières plateau is subject to Mediterranean influences that attract regionally rare open-land and scrub species,** halfway between Castres and Mazamet. Reach Caucalières on the N112 and turn left at the 'L' Auriol Vieux' signpost, then park after 2 km [1]. Walk the trail to the 'Vallée d'en Crabière' or take the track that crosses the plateau towards Castres–Mazamet airport. You can also access the plateau by parking at the entrance of Labruguière along the N112 [2] or on the airport car park [3]. **The plateau is at its best from mid-April to mid-June when all breeders have settled.** Expect short-toed eagle, tawny pipit, red-backed shrike, western Bonelli's, subalpine, Sardinian and Dartford warblers, hoopoe, woodlark, nightjar, little owl and stone-curlew.

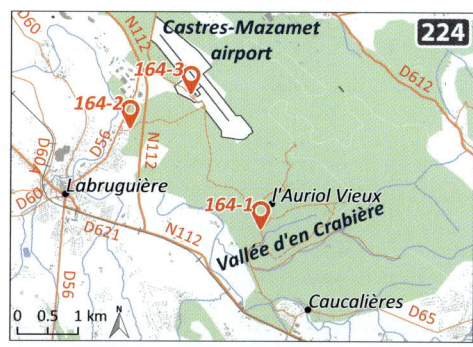

The Ariège Pyrenees (sites 165–168, map 225)

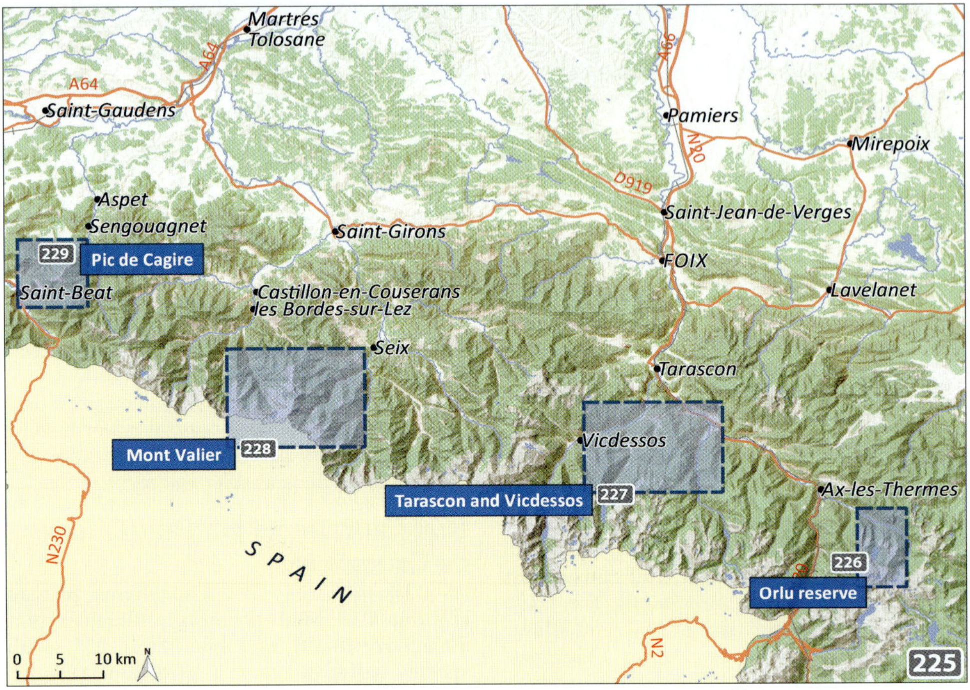

HIGHLIGHTS
» A Pyrenean hotspot for high-altitude breeders
» The main stronghold for Pyrenean capercaillie in Orlu forest
» Encounters with lammergeier, golden eagle and Egyptian vulture are all but certain, if you have enough time and the weather is sunny

KEY SPECIES
YEAR-ROUND **capercaillie** (*aquitanicus* subspecies), **ptarmigan**, **Pyrenean grey partridge**, golden eagle, griffon vulture, **lammergeier**, red kite, peregrine, eagle owl, **Tengmalm's owl**, black woodpecker, crag martin, grey wagtail, dipper, marsh and crested tits, red-billed and Alpine choughs, raven, **snowfinch**, **citril finch**

BREEDING **Egyptian vulture**, booted and short-toed eagles, black kite, Alpine swift, water pipit, wheatear, **Alpine accentor**, ring ouzel, **common rock thrush**, **wallcreeper**, rock bunting

VISIT DURATION ≥ 2 days

PLAN YOUR VISIT
The Ariège Pyrenees concentrate all the Pyrenean species along a wide altitudinal gradient culminating at the Pique D'Estas (3143 m). You are in high mountains here, so behave accordingly and expect long hikes except at a few sites. The flagships are the endemic subspecies of capercaillie and grey partridge, which are tricky to find, the former in dense beech forests and the latter on high-altitude rocky slopes. Most target species are resident year-round, but their high-altitude habitats are barely accessible in winter due to snow cover. Try to plan a visit in May or June; in summer, most birds become shy and hide on inaccessible slopes far from the crowds of hikers. It can be a good idea to plan a night in a mountain refuge, so that you can start birding at first light.

165 Orlu reserve
1–2 DAYS; MAP 226 The Orlu reserve covers a wide altitudinal gradient from mid-altitude meadows and forests to high-altitude rocky ridges, 135 km, 2 h south of Toulouse on the way to Andorra. Follow signs to Orlu from

Ax-les-Thermes and follow signposts towards 'Réserve Nationale' to park at Les Forges d'Orlu [1] or at the very end of the D22, 3 km further on [2]. Short-toed eagle, black and red kites breed at lower altitudes and can be seen from the access road. From the car parks, **the trail to Naguilles pond [3] crosses a beech wood** known as Bois des Seys. **This is one of the** aquitanicus **capercaillie's last strongholds, but encountering it will require preparation and patience.** Be on the trail before first light to maximise your chances, and be aware that the species' sensitivity means that disturbance of any kind is unacceptable. Tengmalm's owl (rare) and black woodpecker (common) also breed in the forest. **Another trail runs into a deep valley and eventually climbs to En Beys refuge (1970 m)** [4]. High-altitude breeding species in this area include ptarmigan, Pyrenean grey partridge (rocky slopes), lammergeier, golden eagle, griffon vulture, peregrine, red-billed chough, water pipit, wheatear and ring ouzel at the forest limit.

166 Tarascon and Vicdessos

1 DAY; MAP 227 Start in Sinsat, 10 km south of Tarascon-sur-Ariège along the N20 (36 km, 30 min south of Pamiers), and park behind the bridge [1]. Dipper and grey wagtail breed on the banks of the Ariège. Walk along the cliff that overlooks the village; lammergeier **and** Egyptian vulture **breed within a** griffon vulture **colony.** Other cliff species breed in the vicinity; look for peregrine, eagle owl, crag martin, Alpine swift and wallcreeper. Further along the D220, shrubby slopes host rock bunting and red kite, and booted eagles sometimes soar over the valley. **Another viewpoint over the** vulture **colony and soaring** raptors **lies 5 km south above Les Cabannes** [2].

Go back to Tarascon-sur-Ariège and head to Vicdessos (15 km on the D8). Park in the Stade de Neige in Goulier, at the very end of the D208 [3]. **Hiking trails climb above 2000 m towards Sarrasi** [4] **and Endron** [5] **peaks** (full-day hikes), offering access to **high-altitude species** (including rock ptarmigan, Pyrenean grey partridge, water pipit, Alpine accentor and Alpine chough). Check the sky regularly for a soaring red kite, griffon vulture, golden eagle or raven.

167 Mont Valier

1–3 DAYS; MAPS 228 **Mont Valier (2838 m) overlooks a wide expanse of protected wilderness, worth several days of birding hikes along a vast network of trails.** Leave the A64 at exit 20 and follow the D117 to its end in Saint-Girons

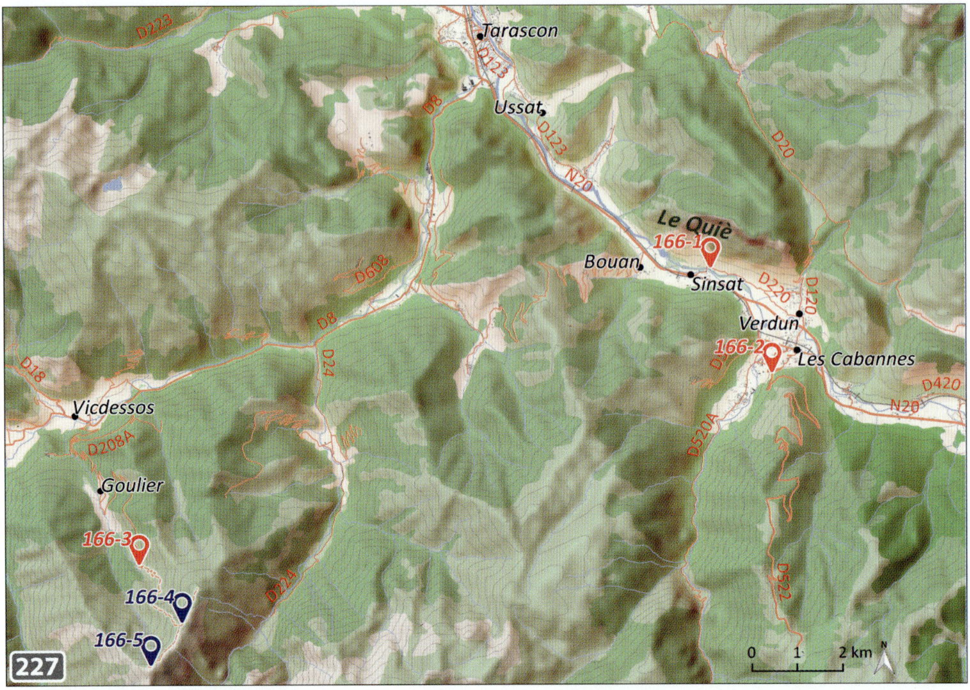
Tarascon
Ussat
D223
D123
D20
N20
D8
Le Quiè
166-1
Bouan
Sinsat
D608
D220
D120
Verdun
166-2
Les Cabannes
D420
N20
D8
D24
D18
Vicdessos
D208A
D520A
Goulier
166-3
166-4
D224
D522
166-5
227
0 1 2 km

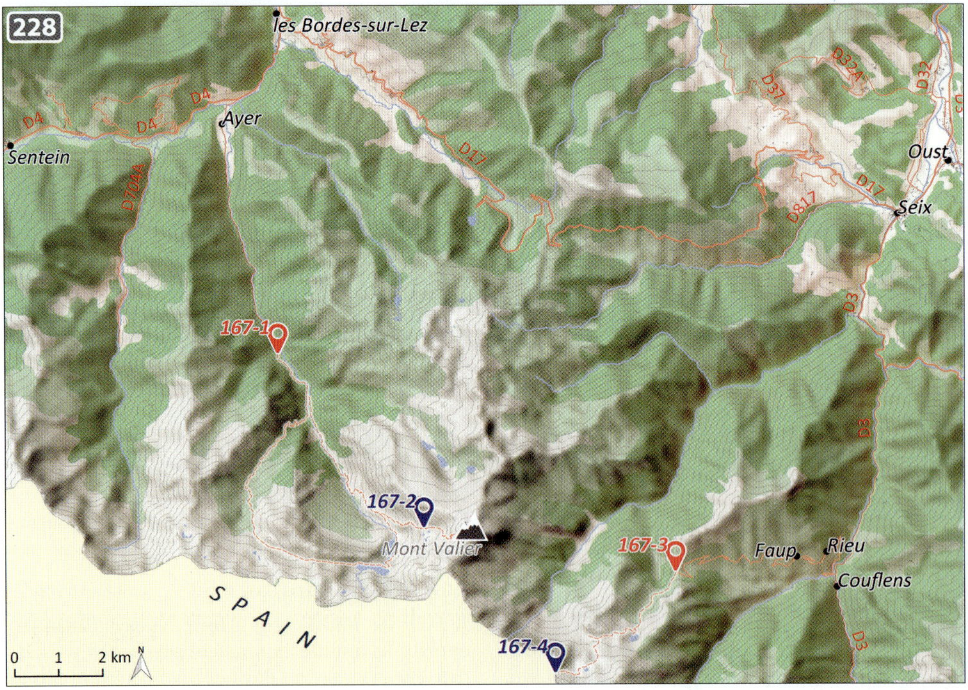
228
les Bordes-sur-Lez
D4
D4
D4
Ayer
Sentein
D704A
D17
D37
D32A
D32
Oust
D17
D817
Seix
D3
167-1
D3
167-2
Mont Valier
167-3
Faup
Rieu
Couflens
SPAIN
167-4
D3
0 1 2 km

(30 km, 30 min), then turn on the D618 towards Les Bordes-sur-Lez, where you will find signposts to the 'Réserve Domaniale du Mont Valier' [1]. **Ideally, you should get there by mid-afternoon and hike for three hours to spend a night in the Refuge des Estagnous** [2] (book ahead of time, https://refuge-estagnous.com, +33 5 61 96 76 22). **Be out at first light** for a chance of ptarmigan and Pyrenean grey partridge, both to be found on rocky slopes and screes. All other Alpine species breed around, the most common being wheatear and water pipit (grassy meadows) and ring ouzel (tree limit). Smaller numbers of common rock thrush and harder targets such as wallcreeper, Alpine accentor and snowfinch (snow limit) also breed around the refuge. As the temperature rises, golden eagle, griffon vulture, lammergeier, Alpine and red-billed choughs and crag martin should start soaring.

The eastern access to the Mont Valier reserve requires shorter hikes and can be a wise alternative if you are short of time or energy. **Park at the Col de Pause** [3] (climb to Seix on the D3 from Saint Girons). Bird along the GR10 (then GRT) trail that leads to Port d'Aula [4] at the Spanish border. **All the species mentioned earlier occur around here**, although ptarmigan and grey partridge are much harder to find.

168 Pic de Cagire

1 DAY; MAP 229 The Pyrenean foothills south of Saint-Gaudens mark a sharp transition from warm-climate species to mid-altitude forest communities. A 35 km, 45 min drive from Saint-Gaudens will lead you to the Pic de Cagire

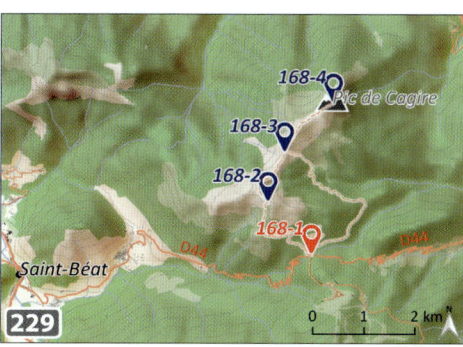

(1900 m). Park at the Col de Menté [1] (follow signposts to 'Station du Mourtis' past Aspet and Sengouagnet on the D5) and hike to the Escalette peak [2] (3 h). **The first part of the trail runs through a mixed forest** where black woodpecker, goldcrest, coal, marsh and crested tits occur year-round. You should find ring ouzel and citril finch at the tree limit and common rock thrush, crag martin, water pipit, and rock bunting higher up. Once out of the forest, expect griffon vulture and short-toed eagle to soar as soon as the sky is clear. Lammergeier, Egyptian vulture, peregrine, golden **and** booted eagles **are often seen along this trail.** An extension along the ridges towards the Pas de L'Âne [3] and the Pic de Cagire [4] should yield Alpine and red-billed chough and further chances of raptors. **Migration can be spectacular on sunny, windless days in August and September** as flocks of soaring black kites and honey buzzards with single ospreys, black storks and bee-eaters cross the mountain towards Spain.

Central Pyrenees (sites 169–171, map 230)

HIGHLIGHTS
» A full basket of raptors starring Egyptian vulture and lammergeier
» All the altitudinal gradient from lowland forests to high-altitude rocky ridges in vast expanses of wilderness
» Access by car to snowfinch, common rock thrush and other high-altitude breeders
» Look for Pyrenean ptarmigan and Lilford's white-backed woodpecker

| J | F | M | A | **M** | **J** | **J** | A | S | O | N | D |

KEY SPECIES
YEAR-ROUND ptarmigan, griffon vulture, **lammergeier**, golden eagle, peregrine, **Lilford's** white-backed and black woodpeckers, red-billed and Alpine choughs, **snowfinch**, citril finch

BREEDING Egyptian vulture, short-toed and booted eagles, water pipit, wheatear, **Alpine accentor**, **common rock thrush**, ring ouzel, rock bunting

VISIT DURATION ≥ 2 days

PLAN YOUR VISIT
This is the core area for high-altitude Pyrenean breeding species. Be prepared for day-long hikes across the wilderness, especially if you are looking for ptarmigan, which is only reliable at Néouvielle. **Choose one area and stick to it for a couple of days.** Most high-altitude breeders can

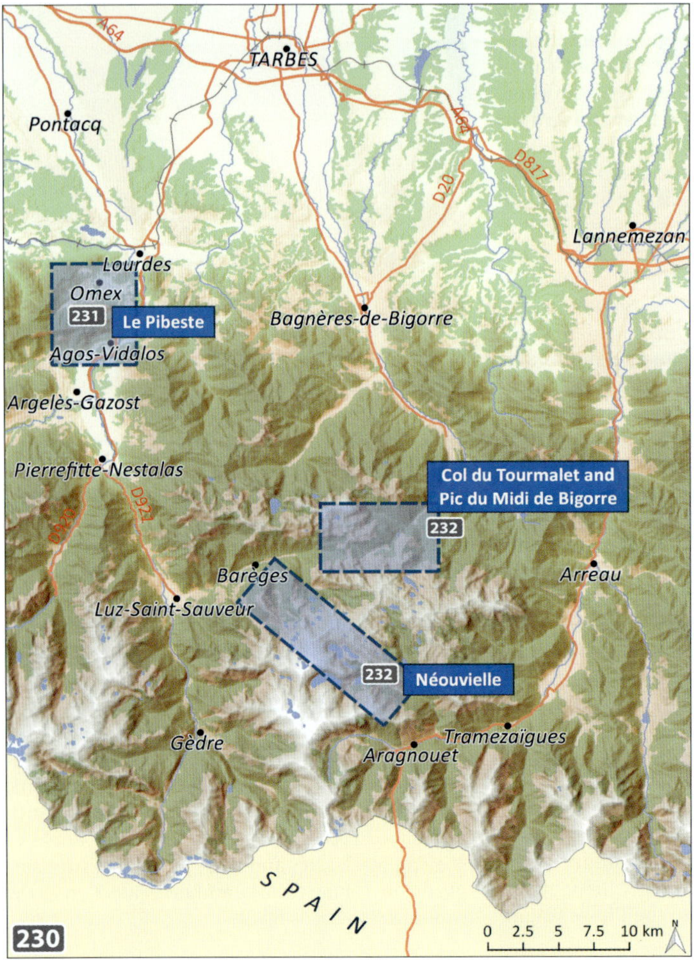

be found close to car parks or by using the Pic du Midi de Bigorre cable car, but bear in mind that snow usually renders access difficult before April. If you have enough time, spend a night in one of the many high-altitude refuges spread along the trails (some of them require prior booking), as this will maximise your chances of ptarmigan and will help you to avoid crowds of hikers. Even if you do not wish to climb high, the area has much to offer – it is full of raptors (starring lammergeier) and there are a few pairs of white-backed woodpeckers (subspecies *lilfordi*) in Le Pibeste.

landscapes typical of the Pyrenean foothills. **A few pairs of white-backed woodpeckers breed in extensive beech woods but they are not precisely located; start your day early in the**

169 Le Pibeste

1 DAY; MAP 231 If you are staying near Lourdes, the Pibeste–Aoulhet nature reserve gives you an opportunity to visit mid-altitude

morning and keep a patient ear open for any woodpecker call. Red-billed chough and rock bunting are likely on rocky slopes. The reserve headquarters are located along the D921B in Agos-Vidalos [1], just past the junction with the D821 from Lourdes (information centre and maps). Trails start from here or from Ouzous, 5 km further west [2] and climb up towards the Pic du Pibeste (1350 m) [3] through woods and meadows. There is an obvious griffon vulture colony with a few Egyptian vultures on a cliff above the valley which is easy to spot from the trail. The peak itself offers a 360° view over the reserve and is the ideal place to spot soaring raptors on a clear day: expect golden, short-toed and booted eagle, peregrine and both kites.

The northern access to the reserve is via a valley west of Ségus (15 km from Lourdes on the D13). Park at Cap de la Serre past Omex [4]. **Climb to the Col du Prat du Rey through Batsurguère valley** [5], a good viewpoint for lammergeier, which is more likely here than around the Ouzous colony. Return through the Bois de Ségus to experience the whole range of habitats that occur in the reserve.

170 Col du Tourmalet and Pic du Midi de Bigorre

0.5–1 DAY; MAP 232 For easy access to high-altitude species, drive to the Col du Tourmalet [1] and climb to the 2900 m high Pic du Midi de Bigorre (50 km, 1 h south of Tarbes through Bagnères-de-Bigorre or 45 km, 1 h east of Ouzous, site 169, through Luz-Saint-Sauveur). Snowfinch flocks might be seen around the pass itself as long as some snow remains early in season. **The whole suite of high-altitude meadow breeding birds occurs here**, dominated by water pipit and wheatear with smaller numbers of common rock thrush and rock bunting. Alpine and red-billed choughs, golden eagle, griffon vulture and sometimes lammergeier or Egyptian vulture may soar above the pass if the sky is clear. The nearby Cabane de Toue car park [2] is the starting point for a trail that climbs to **the Pic du Midi de Bigorre (3 h), where** Alpine accentor **and** snowfinch **can be expected.** The peak can also be accessed by a cable car from La Mongie [3] (June to November, operates 09.00–17.30).

171 Néouvielle

1–2 DAYS; MAP 232 Néouvielle nature reserve, 20 km south of the Col du Tourmalet (site 170), is

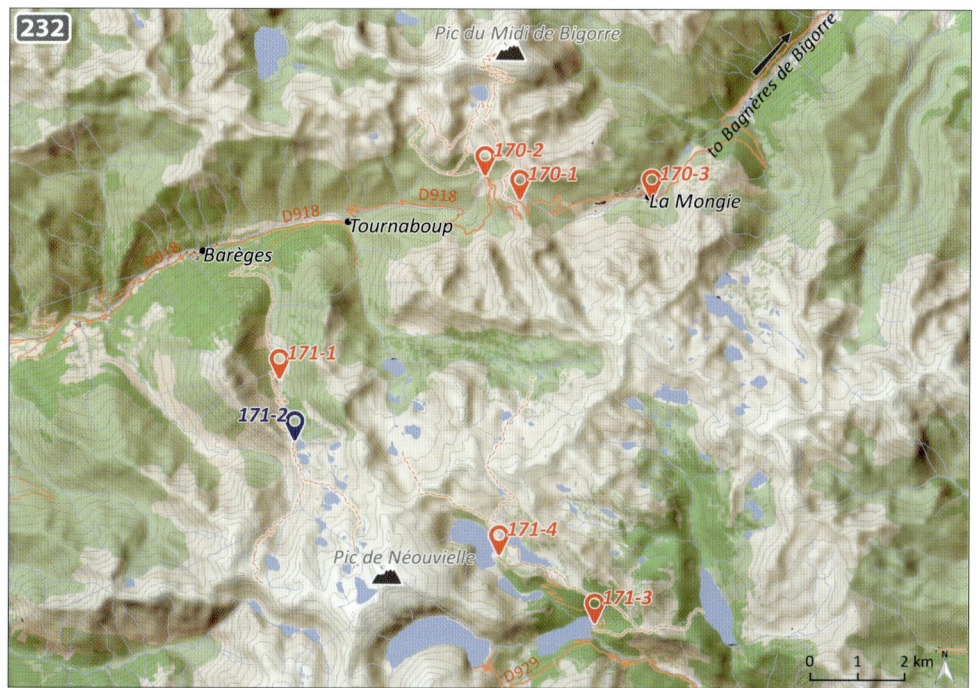

covered by **mountain pine woods, lakes, rocky cliffs and several peaks. The species list is similar to that of the Pic de Bigorre plus** ptarmigan **(rare).** Being away from the most beaten tracks, it is wilder and less disturbed, although you should plan early-morning starts to avoid hikers. Exit Barèges towards the Col du Tourmalet and turn at the 'Forêt Domaniale de l'Ayré et du Lisey' signpost to park at Les deux tonnes [1]. **From there, walk to La Glère refuge** [2] (consider spending a night up there to maximise your chances of ptarmigan, http://refugedelaglere.ffcam.fr). Trails lead to Rabiet pass or Turon de Néouvielle past several small lakes among barren slopes.

Alternatively, you can visit the southern parts of the Néouvielle reserve from Lannemezan (50 km, D929 through Arreau and Tramezaïgues). Turn right in Fabian and park at Orédon lake [3] or Aubert lake [4]. **A vast network of trails runs into the forest towards rocky ridges.** For forest species, try the Pinède des Passades d'Aumar. Expect ring ouzel, black woodpecker, citril finch, coal tit and crossbill. For higher altitudes, climb to the Hourquette d'Aubert peak, the Madaméte pass or the Cabane d'Aygues Cluses refuge (2 h 30), where golden eagle and lammergeier should be easy to find.

The western Pyrenean plains (sites 172–173, map 233)

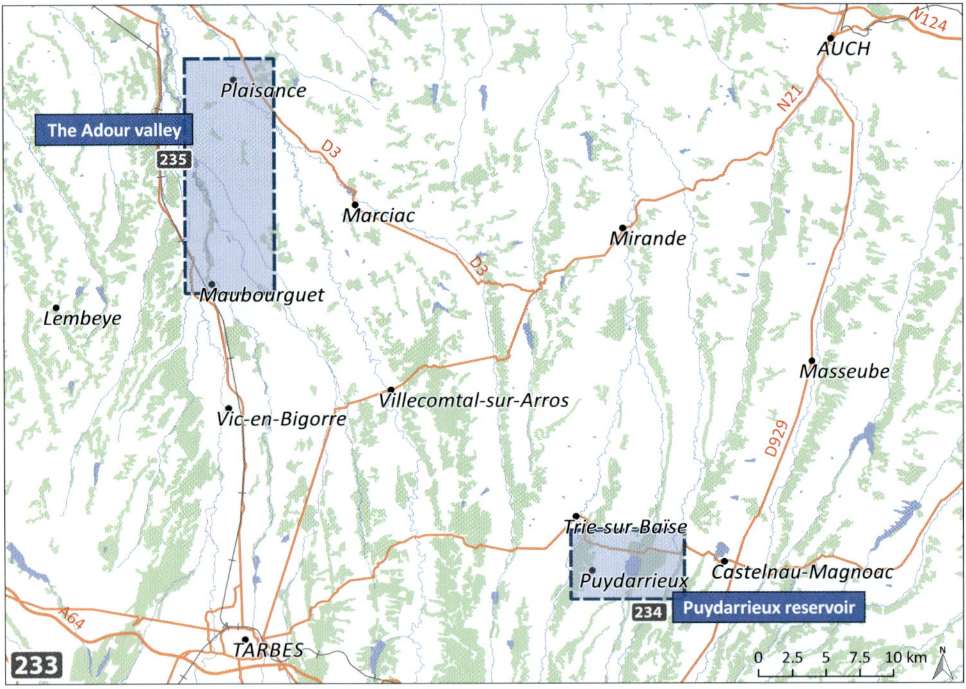

[map of the region showing The Adour valley, Plaisance, Maubourguet, Lembeye, Marciac, Mirande, Auch, Masseube, Vic-en-Bigorre, Villecomtal-sur-Arros, Trie-sur-Baïse, Puydarrieux, Castelnau-Magnoac, Puydarrieux reservoir, Tarbes]

HIGHLIGHTS
» France's historical stronghold of black-winged kite
» One of the region's largest wildfowl and crane wintering sites

KEY SPECIES
YEAR-ROUND black-winged and red kites, great white and cattle egrets

WINTER wildfowl, peregrine, merlin, common crane

VISIT DURATION 1 day

PLAN YOUR VISIT
Your main target should be black-winged kite, which has become easy to find at any season. Search for it on isolated trees or hedgerows among cultivated fields. In winter, the Puydarrieux reservoir is a major crane and wildfowl roost.

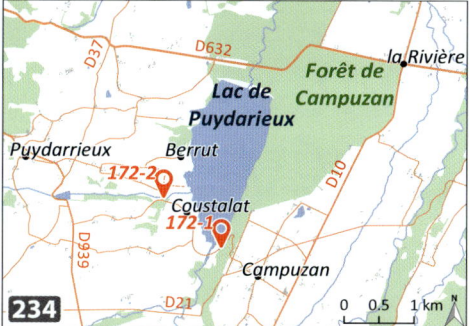

172 Puydarrieux reservoir

1 HOUR – 0.5 DAY; MAP 234 **Several thousand cranes winter around the Puydarrieux reservoir**, 36 km northeast of Tarbes (follow Trie-sur-Baïse then Puydarrieux). The southern end of the reservoir holds most birds (turn left in Campuzan to park at [1] or [2] and walk on the banks). The diversity of migrating or wintering waterbirds can be high when there is frost further north, or at both passages. Expect greylag goose, wigeon, shoveler, teal, great white and cattle egrets, lapwing, curlew, snipe, ruff and yellow-legged gull. Black-winged and red kites winter in the vicinity, and a peregrine or merlin may be seen hunting waders from autumn to spring.

173 The Adour valley

0.5 DAY; MAP 235 The Adour valley is historically the French stronghold of black-winged kite. The species is expanding towards the southwest (Region 7) and has become a common breeding species of hedgerows and isolated trees among the crop-dominated agricultural plains surrounding Tarbes. **A drive on local roads between Maubourguet** [1] **and Plaisance** [2], 30 km, 30 min north of Tarbes (D935 past Vic-en-Bigorre) should be rewarded with sightings at any season without too much effort.

REGION 9
WESTERN MEDITERRANEAN COAST AND THE CÉVENNES

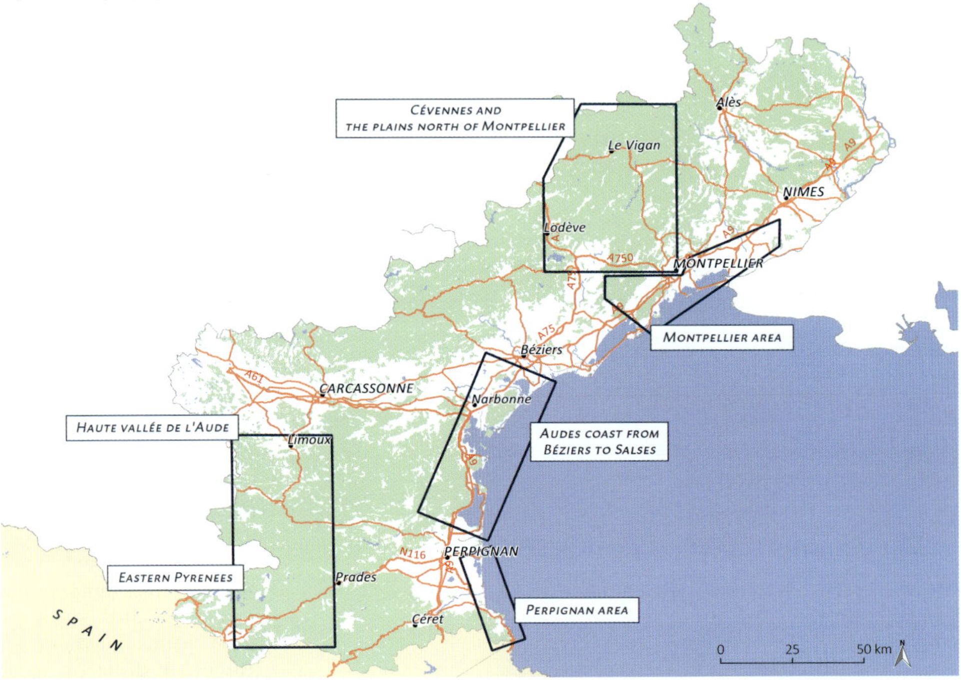

HIGHLIGHTS
» Day lists of over 100 species without much effort from April to June
» Spectacular migration in Gruissan and Eyne from late March to mid-May and from mid-August to November
» Most Mediterranean specialities easy to find
» Mountain species within an hour's drive from the Mediterranean shore

REVIEWERS Boris Delahaie, Laurence Guillosson, Tristan Guillosson, Karsten Schmale

USEFUL WEBSITES
http://faune-lr.org
http://birdinglanguedoc.com for guided tours

The black-eared wheatear is one of the most eagerly awaited migratory breeding species in late April. Although declining, it is still commonly found in most barren plateaus and vineyards of the region, for instance in the Corbières (site 183).

WESTERN MEDITERRANEAN COAST AND THE CÉVENNES | MONTPELLIER AREA

Here in the Sigean saltmarshes (site 184), spring days with light northwesterly winds bring thousands of migrants rushing to their breeding grounds. The surrounding habitat mosaic hosts France's richest breeding community.

Montpellier area (sites 174–176, map 236)

HIGHLIGHTS
» Some of the best migration stopover areas for waders and passerines in the region, half an hour from Montpellier town centre
» Most Mediterranean specialities within an hour's drive, in a hilly mosaic of brackish ponds, marshes, vineyards and oak forest

J F **M A M** J **J A S** O N D

KEY SPECIES
YEAR-ROUND western swamphen, glossy ibis, **Bonelli's** and golden eagles, **little bustard,** eagle and little owls, **moustached warbler**, rock sparrow, cirl bunting

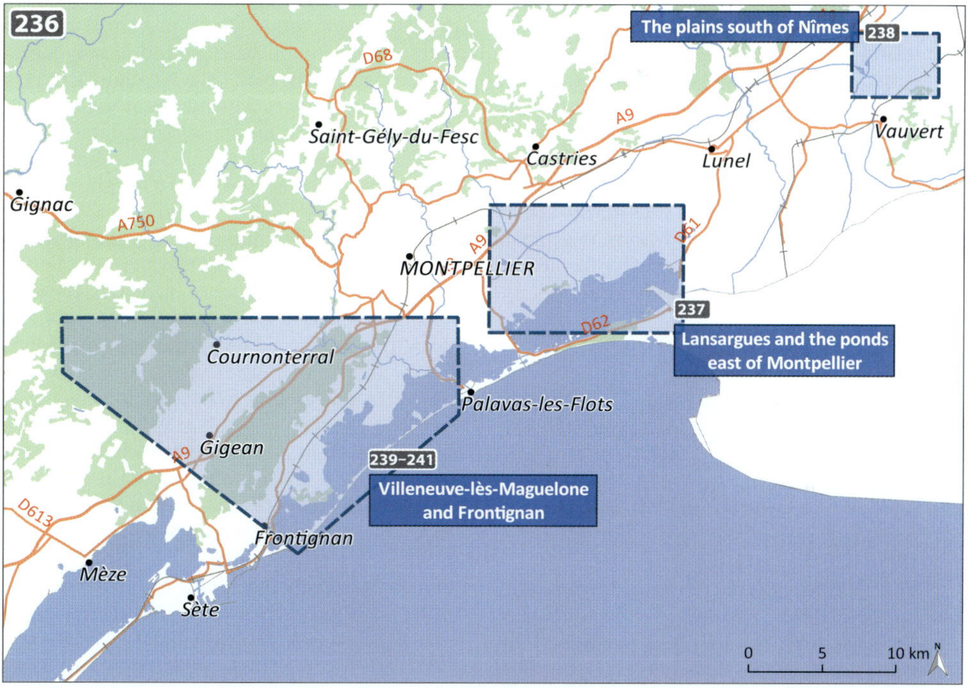

BREEDING purple heron, night-heron, **little bittern**, short-toed eagle, **lesser kestrel**, stone-curlew, **collared pratincole, slender-billed** and Mediterranean gulls, gull-billed tern, scops owl, bee-eater, roller, great spotted cuckoo, **black-eared wheatear**, great reed and orphean warblers, woodchat shrike, ortolan bunting

MIGRATION waders including **Temminck's stint** and **marsh sandpiper, Caspian tern**

WINTER wildfowl, **wallcreeper, rock bunting**

VISIT DURATION   1–2 days

PLAN YOUR VISIT

Most birders do not stay in Montpellier but head directly to the Camargue or the Aude coast. This is a mistake, especially during migration, when the marshes are full of waders that can provide much closer views than in the larger sites. **Most species found elsewhere in the region or in the Camargue can be seen in Montpellier marshes and garrigues with some effort,** including annual marsh sandpiper, breeding little bustard, collared pratincole, black-eared wheatear and Bonelli's eagle, not to mention one of France's largest slender-billed gull colonies and what is virtually the country's only breeding site for gull-billed tern. **The Montpellier coast has exceptional migrant records with several high-profile rarities in recent years,** among them white-crowned wheatear and Tristram's warbler. **If you just have a morning or evening left,** visit Lansargues (for waders), Villeneuve-lès-Maguelone (for reedbed breeding species) or Frontignan seashore (for migrating passerines). **If you have a whole day,** add raptors inland around Aumelas and Gignac or look for steppe species in the plains south of Nîmes. **If you have two days,** spend a few hours further inland for raptors and breeding passerines such as orphean warbler and ortolan bunting. **During migration rushes, you will probably need a whole day to cover just a few sites on the coast: forget inland sites in such conditions.** In general, spend time on two or three sites rather than rushing to complete the whole circuit in a day.

174 Lansargues and the ponds east of Montpellier

2 HOURS – 0.5 DAY; MAP 237   **Start or end a day in the wetlands south of Lansargues** (20 km, 30 min east of Montpellier, drive to Mauguio, then towards Lunel on the D24). **This is a key site for** wader **migration and the easiest place for breeding** slender-billed gull **and** gull-billed tern (March to August) in southern France.

Drive to Tartuguières [1], checking wires and buildings on which little owls, bee-eaters, rollers and rock sparrows commonly perch from May onwards (ask onwards ('domaine de Tartuguières' at the first

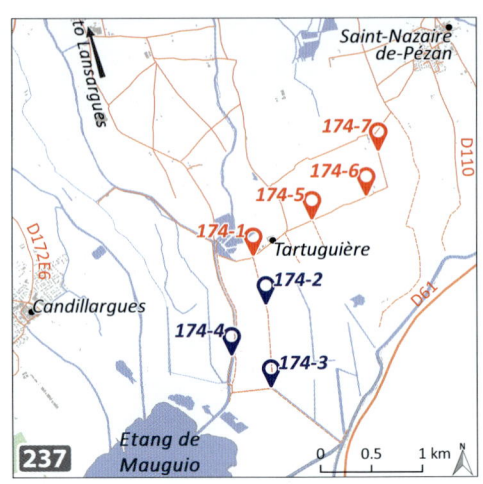

roundabout in Lansargues when coming from Montpellier, then straight on to the end of the road and park near a small bridge; walk across the bridge and turn right just before the farm). **Walk the paved track southwards inside the marshes;** waders **and** squacco herons **forage among tamarisks or in the surrounding wet grasslands** [2]. Stilt, wood sandpiper (sometimes in huge numbers), spotted and common redshanks, common, black, little and whiskered terns, Cetti's and fan-tailed warblers are common to abundant from spring to late summer. There is a small collared pratincole colony on bare ground among tamarisks just west of the path (May to August); its precise location changes depending on water levels. Singles or small groups of marsh sandpiper, Temminck's stint and white-winged tern show up at both migrations every year, and a few short-eared owls occasionally roost in meadows near the farm in winter. **The track ends at a lagoon** [3] which hosts a large mixed colony of Mediterranean gulls, gull-billed terns and slender-billed gulls, roosting flamingos, black-necked grebes and dabbling ducks. This pond can be crowded with thousands of gulls and terns in mid-spring. A few Caspian terns often stay around from April to September: search for them among gull-billed and Sandwich terns. **Turn right towards a few chalets where you can join a track along the main canal** [4], **leading back across the marshes to the car park.** On your way, check the meadows and bare soil on the west side of the canal: Temminck's stint is often recorded here together with common snipe, Kentish plover, stone-curlew (great snipe has been recorded several times), raptors, herons and glossy ibis. Red-throated pipit (April and May) can sometimes turn up during yellow wagtail migrations (beware of commoner tree pipit).

Drive back east to enter a marshy area with reedbeds, ponds and mudflats known as the 'marais du Grès' (drive east on dirt tracks from Tartuguières farm). Stop here and there (for instance at [5] or [6]), expecting red-crested pochard (abundant), western swamphen (common), purple heron, little bittern (from late May, uncommon), waders, moustached, reed and great reed warblers among the suite of common wetland birds. The area is exceptionally rich, so take your time. At the end of the last track, rock sparrows often perch on wires and stone-curlews breed in the surrounding fields [7].

On your way back to Montpellier, check the northern end of the airport runways [8 – see GPS file]: little bustards **live in grassy areas year-round (largest numbers in winter)**. From there, you can either go back to the city or drive past the 'Étang de Pérols', 'Étang du Méjean' and 'Étang de l'Arnel' to scan roosting gulls.

175 The plains south of Nîmes

2–5 HOURS; MAP 238   The cultivated plains and groves between the Camargue and Nîmes host all of the Mediterranean steppe breeding species. A morning or evening from early May to late June should be rewarded with displaying little bustard, pictures of roller and bee-eaters, plus sightings of great spotted cuckoo, little owl and ortolan bunting. In winter, the plains are worth a visit for buzzards, harriers and merlin. Virtually the entire area from Nîmes to Vauvert is suitable for these species; we only describe an itinerary in the western part which can easily be connected with some birding around Montpellier or the Camargue Gardoise (Region 10, site 190).

Pylons around the roundabout north of Vauvert [1] are a good place to look for rock sparrow (park near a small shed after the exit towards

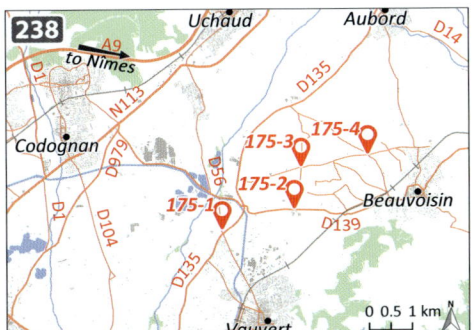

Uchaud). Then drive along the D139 towards Beauvoisin and **turn left on good driveable tracks 2 km further on** [2]. Park at [3] or [4] and walk around. Little bustards occur year-round and can be seen displaying from the tracks in May and June (there is no justification for approaching them, and remember that the fields are private land). Little owl, hoopoe, roller and great spotted cuckoo breed in the immediate surroundings, and numerous bee-eaters may well be seen hunting in the vicinity. Ortolan bunting, woodchat shrike and more common species are also present.

176 Villeneuve-lès-Maguelone and Frontignan

3 HOURS – 1 DAY; MAPS 239–241   This area can be particularly productive from March to June and during the autumn migration; it is the main alternative to Lansargues (site 174) for wetland species. It is not as good for gulls and waders, although you are still likely to find most species, but it is far better for reedbed species and migrating passerines.

A good option is to spend the first couple of hours after sunrise in the Boulas reedbed and marshes (14 km, 25 min from Montpellier, once in Villeneuve-lès-Maguelone follow signposts 'Sète par le littoral', bypass the dump and turn left on a good track to 'Mas des Quinze'; after 100 m take the right branch which leads to a car park [1]). Walk along the dyke [2] for waders, gulls and terns (including Caspian). Western swamphen can be seen from here, although it was easier from inside the reedbed [3] along a now permanently closed dyke that we still mention in the hope that it will open again in the future. Little bittern (from mid-May), moustached warbler (year-round, easiest in March and April when singing), bearded tit (year-round) and wintering penduline tit are common or abundant in the reedbed. A few shy bluethroats winter deep inside the reedbed but can occasionally be seen foraging along reed edges. Spotted and little crakes are seen every year inside the reedbed or on a small pond at its eastern edge, accessed from the main road [4] (also good for garganey, snipe and purple heron). Great spotted cuckoo, hoopoe, rock sparrow breed in surrounding garrigues; they can be found with some effort in season. Red-rumped swallow does not breed but is often seen hunting at dusk, mainly in late April when it has just returned from migration.

Halfway between the marsh and the motorway, the 'creux de Miège' cliff [5] **is known for its cooperative** eagle owl **pair**. Take the small

road that starts in front of the track to the previous spots, cross the railway and continue for 1 km until you reach a track to the right, which leads near the cliffs. Look from a distance, as the owls won't show up if you are too close. Scops and little owl also breed around here.

The tamarisks along the seashore can be crowded with passerines in sunny spring days with light southerly winds, most usually in early May and September (follow signs to the cathedral from Villeneuve-lès-Maguelone and park just before a mobile bridge, open 10.00–18.00, parking fee). Search for migrant passerines in gardens [6], in the cathedral woods [7] and in tamarisks along the beach westwards [8]: chiffchaff, willow warbler and pied flycatcher can be abundant, and melodious, icterine (rare), wood, subalpine, Moltoni's and orphean (easily mistaken for lesser whitethroat, which does not occur yearly in the region) warblers are regular.

Once done with the reedbeds or the tamarisks, visit the **dry fields and saltmarshes** around the Conservatoire des Espaces Naturels [9] (follow 'Les Salines' as you bypass Villeneuve-lès-Maguelone on the D116, then park in the main car park [10] or near 'Les Moures' [11]). Red-legged partridge, little owl and rock sparrow breed, and osprey, Caspian and gull-billed terns or slender-billed gull are likely.

If you need to go back to Montpellier, keep an hour and a half available to visit the wetlands inside the 'Maison de la Nature' in Lattes [12 – see GPS file]. In spite of their position inside a heavily urbanised area, they can be surprisingly good during migration. Records of little bittern and red-footed falcon are regular here, and great reed, moustached and melodious warblers breed (follow 'Lattes-centre' until you find signposts).

Further west, the 'Plage des Aresquiers' in Frontignan is one of the best migration spots for falls of passerines in the region (mid-April to late May, mid-August to mid-November; drive to Vic-La-Gardiole, then 'Frontignan par les plages' on the D114; park before the bridge over the canal [13] and walk along the beach; spend time around the watersports centre [14] and further east); an eastern Bonelli's warbler was here in 2017. Red-throated pipit (late April), Moltoni's (early May) and icterine (late May) warblers are found annually in tamarisks along the beach among commoner warblers and flycatchers; other surprises will certainly arise in the incoming years. A few hundred metres to the west, visit 'Mas d'Ingril' just before Frontignan's easternmost camp site [15]. **This small pasture is a real magnet for migrants** and will take 30 minutes to 2 hours depending on whether a passerine fall has occurred. Tristram's and yellow-browed warblers, red-breasted flycatcher and other rarities have been recorded on this inconspicuous spot in the few past years.

For garrigue birds, try the D114E2 which crosses **a dry, scrubby plateau known as the Causse d'Aumelas**, between Cournonterral and Plaissan (20 km, 35 min from Montpellier or 15 km, 20 min from Villeneuve-lès-Maguelone; in Cournonterral follow 'maison de retraite' and 'bergerie' until you find signs for 'Gignac vers D609'; climb for 7 km, 10 min and turn on a road left towards Mas Sainton). **The best sections are the maquis and open garrigues between the Mas Sainton [16] and Mas Barral** [17] (private land, stay on the road). Golden and Bonelli's eagles, lesser kestrel (from the Vendémian colony [18 – see GPS file]), nightjar, tawny pipit, Dartford and spectacled warblers (the latter rare and irregular), black-eared wheatear (at least one pair on ridges above Mas Barral), ortolan bunting and southern grey shrike breed on this plateau. Red-footed (May) and Eleonora's (summer) falcons might be seen hunting along the ridges, mainly near Mas Barral. The plateau gets windy in winter, but all resident species remain.

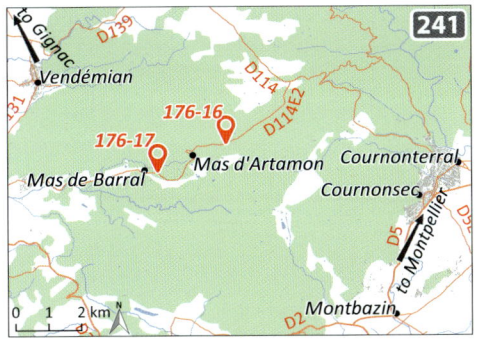

short-toed eagles, black storks, booted eagles, bee-eaters and hoopoes.

From November to early March, **you can end your circuit in Villeveyrac dump** to search for a vagrant Caspian or herring gull within loads of yellow-legged gulls, cattle egrets and white storks. In May, you would do better to end on the plain north of Poussan [20 – see GPS file] (bird along the D2/E5 heading north of the village) before heading back to the coast. **The surrounding fields host all the Mediterranean breeding species of the region**, including golden oriole, bee-eater, hoopoe and roller. You might have a chance to catch one of France's very last lesser grey shrikes, perched on the lower branches of a plane tree or on the top of a shrub. From there, you can head back to spend the last hours of daylight in the Villeneuve-lès-Maguelone marshes (20 km, 30 min east).

12 km, 20 min further south, **there is a lesser kestrel colony under the roofs of Saint-Pont-de-Mauchiens** [19 – see GPS file] (where rock sparrows also breed). The birds are often seen hunting from the top of the village, which is also worth a visit in early May for migrating

Cévennes and the plains north of Montpellier (sites 177–179, map 242)

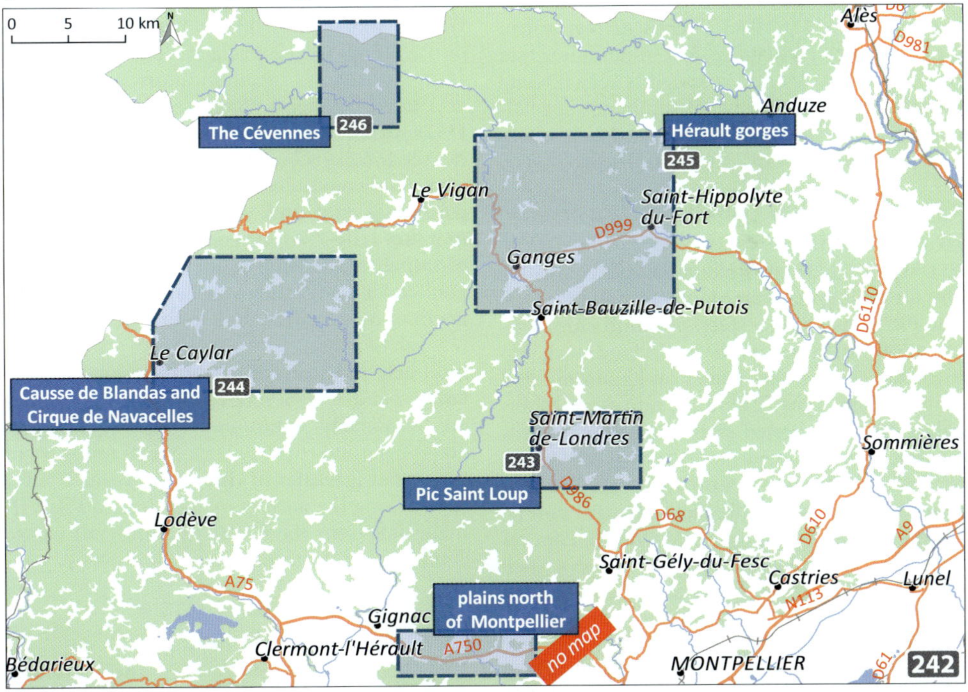

HIGHLIGHTS
» High diversity of mountain specialities along a smooth altitudinal gradient
» High raptor diversity, starring three vulture species and Bonelli's eagle

| J | F | M | A | M | J | J | A | S | O | N | D |

KEY SPECIES

YEAR-ROUND griffon vulture, golden and **Bonelli's eagles**, peregrine, eagle and **Tengmalm's owls**, black and middle spotted woodpeckers, Alpine swift, woodlark, grey wagtail, crag martin, **dipper**, blue rock thrush, crested tit, **southern grey shrike**, **red-billed chough**, rock sparrow, **rock** and cirl buntings

BREEDING Egyptian vulture, short-toed eagle, lesser kestrel, **scops owl**, nightjar, hoopoe, **wryneck**, bee-eater, roller, **red-rumped swallow**, tawny pipit, **common rock thrush**, subalpine and orphean warblers, woodchat and red-backed shrikes, **ortolan bunting**

MIGRATION Eleonora's falcon

WINTER Alpine accentor, **wallcreeper**

VISIT DURATION Up to 1 week

PLAN YOUR VISIT

The plains north of Montpellier offer opportunities for Mediterranean species that are harder to find elsewhere, including rock bunting (winter) and red-rumped swallow. Slightly further north, the Cévennes consists of low mountains that reach 1565 m at Mont Aigoual. It is not a birding hotspot, but it hosts several mountain specialities accessible on a day trip from Montpellier, with a good density of Tengmalm's owls.

177 The plains north of Montpellier

0.5 DAY; MAPS 243 & 245   Numerous places are suitable for birding between Montpellier and Ganges. The Pic Saint Loup, a steep pyramid culminating at 660 m, serves as the main landmark. Up to 10 Alpine accentors winter on its summit and are incredibly tame (25 km, 30 min from Montpellier; well signposted from the D986 towards Ganges, park in Cazevieille [1] and walk 45 min). Wallcreeper **also winters on the peak, but it is more easily found on the facing Hortus cliff** [2] (9 km, 10 min from Saint-Martin-de-Londres on the D122 to Saint-Martin-de-Tréviers; park on forest tracks under the pines at the foot of the cliff). Peregrine, eagle owl, Alpine swift and crag martin also breed on the Hortus.

The plains between Saint Martin de Londres, Mas de Londres and Notre-Dame-de-Londres are home to a wide range of Mediterranean species including hoopoe, bee-eater, woodlark, tawny pipit, woodchat shrike, roller and ortolan bunting (for instance around [3] or [4]). From

mid-April, several pairs of scops owls breed in plane trees in Saint Martin de Londres as well as in virtually all the villages on the plain (also in the north of Montpellier itself). **In winter,** larks, thrushes **and** buntings **sometimes come down from the mountains and settle in open fields. Small** rock bunting **flocks winter**, with larger numbers of cirl buntings and finches in vineyards around Pompignan and Claret (try for instance the vineyards around [5]).

Alternatively, in spring, head west to the Gignac area (30 km, 30 min from Saint-Martin-de-Londres). Check bridges, below which red-rumped swallows breed; locations change from year to year. **A good place for the species is the D619 which runs just under the A75 motorway** [6 – see GPS file]: check every small bridge or pipe for nests. **From there, either climb to the Causse d'Aumelas and end your day in the coastal marshes (site 176), or continue south, back to Montpellier**, and stop on the D27 between Saint-Paul-et-Valmalle and Murviel (20 km, 20 min north of Montpellier; park under the power line [7 – see GPS file]). Eleonora's falcons sometimes hunt here in the evening from late June to September, and several interesting breeding species are reasonably common, including southern grey shrike and rock sparrow. **Small numbers of** rock bunting **winter in vineyards** [8 – see GPS file] (drive along the narrow road to 'hameau de Valmalle' left from the centre of Saint-Paul village coming from the motorway).

178 Causse de Blandas and Cirque de Navacelles

3 HOURS – 0.5 DAY; MAP 244   **The Cévennes foothills form a transition zone between Mediterranean and mountain species**. Drive to the 'Causse de Blandas' and walk from Blandas [1] to the Cirque de Navacelles [2] on hiking trails, or drive on the D713 to visit the same sector by car

(70 km, 1 h 30 min from Montpellier through Ganges, Saint-Laurent-Le-Minier and Montdardier). Common breeding birds on this **mid-altitude garrigue** include wryneck, hoopoe, nightjar, tawny pipit, blue and common rock thrushes, subalpine and orphean warblers, southern grey, woodchat and red-backed shrikes. Bee-eater and roller are regularly seen, although they do not breed in the vicinity. On the raptor side, expect short-toed and golden eagles. Bonelli's eagle does not breed close by but sometimes soars up to the plateau when hunting. Look for Eleonora's falcon from June to September; sightings are casual but the species could be more common than it seems. The Cirque de Navacelles itself is a viewpoint for griffon vulture, eagle owl, red-billed chough, rock bunting and wallcreeper (in winter). **The same species can be found on the D158 between Blandas and Alzon [3] or along a 33 km drive on the D152 through Saint-Maurice-Navacelles [4], Saint-Michel and Le Caylar [5 – see GPS file]**, from where you can take the A75 back to Montpellier or Millau.

179 The Hérault gorges and the Cévennes

1 DAY; MAPS 245 & 246 **Half an hour north of Montpellier, the River Hérault runs into a canyon in Sainte-Bauzille-de-Putois.** First try the 'Grotte des Demoiselles' [1] (3 km north of Sainte-Bauzille, signposted). Alpine accentor and wallcreeper can be found on surrounding cliffs in winter and eagle owl breeds all around. **Walk from the caves car park to the northwest along the cliffs** [2]; Egyptian vulture breeds close by and is often seen flying over. The local Bonelli's eagle can be spotted from the supermarket car park in Saint-Hyppolyte-du-Fort [3], 15 km, 15 min from the caves. The banks of the river are hard to reach, **but the canyon widens 6 km further on, between Ganges and Valleraugues**, making it easier to find grey wagtail and dipper (for instance at [4]).

Mont Aigoual and the surrounding spruce–beech forests are the closest spot to Montpellier for several forest species (55 km, 1 h north of Ganges, follow signposts). **Expect** water pipit in spring, crested **and** coal tits **and occasionally** snowfinch **in winter at the top** [5]. The area around the Col du Minier [6] should yield black woodpecker, coal tit and crossbill (7 km, 11 min from L'Espérou, at the base of the mountain; take the road to Dourbies and turn left after 1 km). Middle spotted woodpecker **has been found and could breed, but the species is poorly known.** Tengmalm's owl is a major highlight of the area; it is rare but widespread in beech stands, for instance near the Col du Minier, L'Espérou and Camprieu [7]. It is easiest to catch while singing from February to April, either at dusk or late in the night (singing activity peaks around midnight).

Further north, dry plateaus with grassland and maquis replace forest. This area is

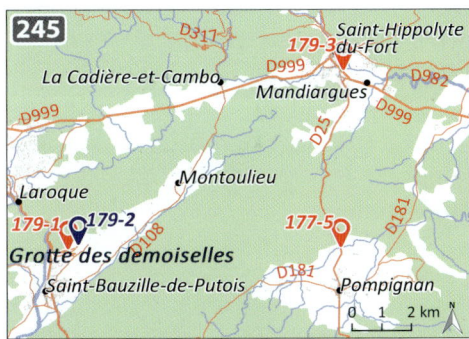

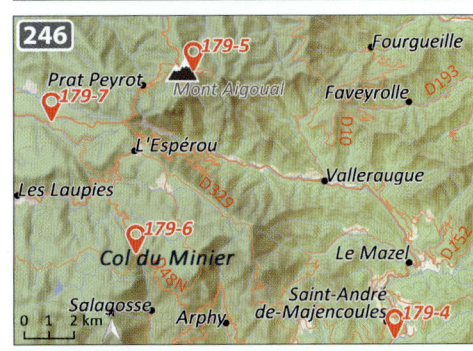

snowy and generally quiet in winter although frequented by thrushes, finches and buntings. In summer, look for short-toed eagle, orphean warbler, shrikes and ortolan bunting. Flocks of lesser kestrel from the coastal plain sometimes roost here in late summer, and red-footed falcon or dotterel could stop over in May and September. As a continuation, you can head north for a trip to the Gorges de la Jonte (Region 12, site 246) or the Lozère (Region 12, site 248).

The Aude coast from Béziers to Salses (sites 180–184, map 247)

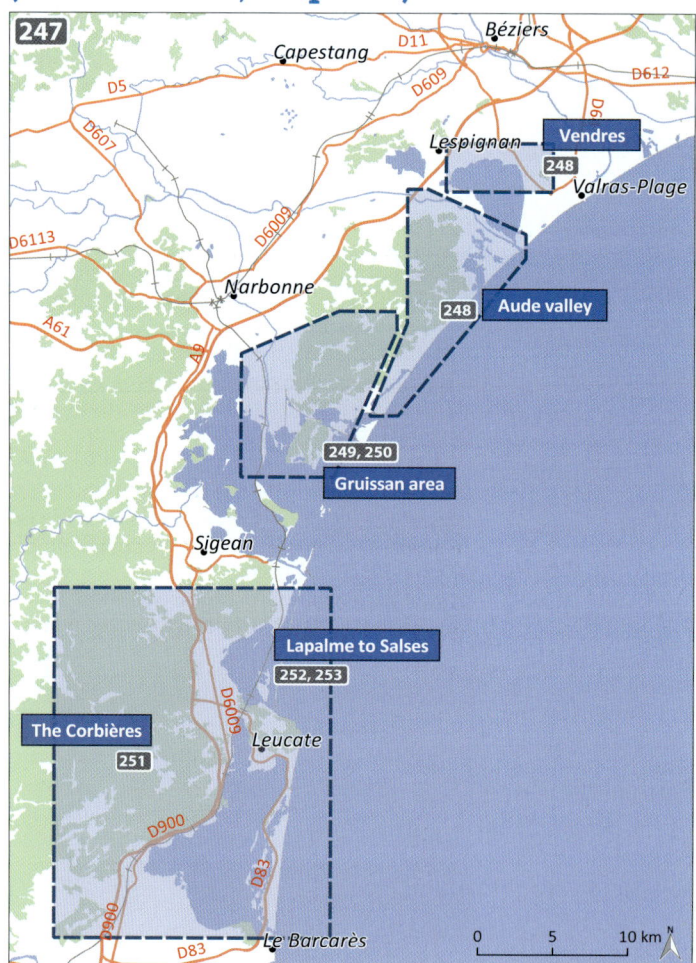

HIGHLIGHTS
» Spectacular migration in Gruissan on clear days with northwesterly winds
» All Mediterranean specialities in the garrigues and marshes
» The highest bird diversity one can expect in a birding day in France

| J | F | M | A | M | J | J | A | S | O | N | D |

KEY SPECIES
YEAR-ROUND red-crested pochard, yelkouan shearwater, Eurasian bittern, western swamphen, **little bustard**, little owl, **Thekla** and crested

larks, **moustached warbler**, bearded tit, **spotless starling**, rock sparrow

BREEDING purple and squacco herons, **little bittern**, short-toed eagle, honey buzzard, Montagu's harrier, **lesser kestrel**, black-winged stilt, Kentish plover, Mediterranean and **slender-billed gulls**, scops owl, nightjar, hoopoe, great spotted cuckoo, bee-eater, roller, **pallid swift**, **greater short-toed lark**, tawny pipit, **black-eared wheatear**, spectacled (rare), melodious, orphean and great reed warblers, woodchat and **lesser grey (rare) shrikes**, **ortolan bunting**

MIGRATION garganey, pintail, Scopoli's shearwater, black stork, osprey, **pallid harrier, Eleonora's and red-footed falcons**, Baillon's (scarce), **spotted and little crakes**, waders including wood sandpiper and **Temminck's stint**, Arctic and pomarine skuas, **Audouin's gull**, black, **white-winged and Caspian terns**, Alpine swift, **red-rumped swallow**, **red-throated pipit**, whinchat, **blue rock thrush**, sedge warbler

WINTER ferruginous duck, **penduline tit**

VISIT DURATION   2 days – 1 week

PLAN YOUR VISIT

Ask any French birder for his favourite spring birding area, and he'll probably put the Aude coast as high as the Camargue on his list. In early May, you're unlikely to do less than 100 species in a reasonable birding day, and the area is a very good candidate for breaking the 200 barrier on a big-day contest. All the Mediterranean specialities breed here, including Bonelli's eagle, moustached warbler, still a few lesser grey shrikes, and so on. **You need a minimum of two days to cover all the sites**. They can be done year-round, but **the highest diversity is found between mid-April and the second week of May**. Check the weather to decide whether to try Gruissan, the most spectacular migration site in the Mediterranean region. Birds pass through when the Pyrenees are visible and in northwesterly winds, mainly from March to mid-May and from August to mid-November. Keep breeding species for the midday hours. **If you have only one day, your best bet is to start in Leucate and head northwards to end your day in the Aude valley, unless conditions are suitable for migration.** If the weather is rainy or too windy, try garrigues inland. The whole area can easily be birded by bike from the many train stations between Montpellier and Perpignan, for instance in La Franqui.

180   Vendres and Portiragnes

2 HOURS; MAP 248   The Vendres marshes and pond are a key regional site for waterbirds and **your best bet in winter** (75 km, 1 h south of Montpellier, leave the A9 at exit 36). From Vendres, drive towards Valras-Plage and turn left on a dirt road after the 'Sainte-Germaine' farm [1], the best viewpoint over the open waters of the pond. Wildfowl numbers peak from December through February. Large groups of red-crested pochard, shoveler, gadwall and other wildfowl usually gather in the southern parts of the pond, and single ferruginous duck or a scaup often mingle with pochard. Cold weather can bring rarer species, such as a goldeneye or other northern ducks. Expect migrating garganey and pintail from March to May, as well as all possible terns (including the rare white-winged tern in May). Stop at [2], [3] or at the ruins of an old temple [4], all good places for great reed and moustached warblers (from April), Eurasian bittern (mainly winter), little bittern (May to August), western swamphen and bearded tit (year-round).

At dusk, scan the distant meadows north of the lake [5] (exit Vendres to the west and turn left just before the bridge on the motorway) for little bustard, lesser kestrel, golden oriole, waders and terns. Woodchat shrike, whinchat and ortolan bunting breed in the vicinity. **If you intend to head back north, visit Portiragnes ponds** [6 – see GPS file] (14 km, 15 min east of Vendres, bypass Sérignan and drive east on local roads; once parked walk 300 m north to fenced ponds). The ponds themselves should be checked for waders, gulls and terns, and there are western swamphen, moustached and great reed warblers in the surrounding reedbeds. The access road is surrounded by pastures full of yellow wagtails and tawny pipits in which bee-eaters hunt, sometimes at very close range. **On the way back to Béziers, pay a quick visit to the eastern end of the airport runway** [7 – see GPS file] **for displaying** little bustards.

181   The Aude valley

3 HOURS – 0.5 DAY; MAP 248   The last section of the River Aude runs across a typical Mediterranean landscape composed of vineyards, pastures and small woods surrounded by scrubby hills. **A couple of hours birding along the river down to its mouth can yield 80 species in May, but you'll need at least half a day to make the most of this exceptional place** (90 km, 1 h 10 min from Montpellier and 5 km, 7 min from Vendres).

above the road and rollers occupy conspicuous nestboxes from May.

Drive down to the mouth of the Aude, to the 'cabanes de Fleury' [3], **for seawatching**; yelkouan and Scopoli's shearwater, Mediterranean gulls, pomarine skuas and divers may pass by – you will quickly see if it's worth staying or not (wind conditions are no clue). **Next, head to Pissevaches lagoon** (Étang de Pissevaches) (follow signs to 'camping La Grande Cosse', 8 km, 10 min south of the 'cabanes de Fleury'). Slender-billed gulls, Caspian tern and a few waders often occur on the lagoon itself, but **the prime spot here is the fenced sewage farm just before the camping resort** [4]. Walk carefully around the pools: black-winged stilt, wood sandpiper, snipe, Temminck's stint and other waders stop over here, also with white-winged tern and red-rumped swallow. **Behind the sewage farm, a clearing inside the reedbed with mudflats** [5] is often frequented by purple and squacco herons (common), moustached warblers (numerous, easiest when singing in March and early April before sedge warblers show up), great reed warblers, western swamphens, bearded (year-round) and penduline (winter) tits and waders. **Scan the mountains in the background**, as Eleonora's falcons can be seen hunting above the ridges (June to September). Short-toed eagles, honey buzzards (not before early May) and hobby are usual sightings. **Also check the tamarisks near the road and the vineyards and cypress below the ridge** for migrant warblers and flycatchers as well as most of the local breeding species.

Start at dawn at the bridge over the River Aude [1] (from Vendres or from exit 36 on the A9, cross Lespignan on the D14 and pass below the motorway): bee-eaters, great spotted cuckoos and orioles are unmissable from mid-May. The last few French pairs of lesser grey shrikes breed a few kilometres away at undisclosed and changing sites: look for them on fences and along hedgerows from mid-May (report precisely any sighting to local birders). Montagu's harrier, quail, little bustard, little owl, whinchat and woodchat shrikes, melodious and orphean warblers, serin, cirl and ortolan buntings are common to abundant all around from May to August. **Cross the bridge southwards and take the first small road left** [2] (ignore the 'no trespassing' signs). There is a lesser kestrel colony on a **small white building on the hillside**

Once you're done with the area, make your way to Gruissan (20 km, 30 min from Pissevaches), stopping near the **bridge between**

Saint-Pierre-la-Mer and Narbonne-Plage [6]; a colony of pallid swift lives on the buildings. Look for greater short-toed larks that may occur on the sandy plain behind Les Mateilles beach [7 – see GPS file] (signposted; seawatching can be good as well from this place).

182 Gruissan area

UP TO 1 DAY; MAPS 249 & 250   Gruissan is the key spring migration hotspot of the entire French Mediterranean region and a definitive must between March and May or from August to October. Head for the 'Cabanes de l'Ayrolle' [1] (4 km, 8 min from Gruissan, follow signposts to 'L'Ayrolle' and drive to the end of the road; park near a ruin on the shore of the laguna) **or settle on the 'Roc de Cornilhac'** [2] **in autumn** (turn right at the bridge south of the village centre and drive 4 km, 4 min to a conspicuous hill in the middle of the plain). A team of spotters is present as soon as the weather is suitable. **Loads of raptors and passerines pass through, flying along the beach or the hillsides**; your highlights will be black stork, osprey, pallid harrier, red-footed falcon, flocks of bee-eaters, red-rumped swallow or red-throated pipit. **Optimum conditions are sunny weather with no cloud cover on the coastal Pyrenees and light to moderate northwesterly winds, from March to May and from mid-August to mid-November.**

If migration is slow, walk along dykes inside the Gruissan saltpans to look for waders [3,4,5]. Kentish plover, dunlin and little stint can be abundant, with the odd Temminck's stint or even rarer species (lesser yellowlegs and

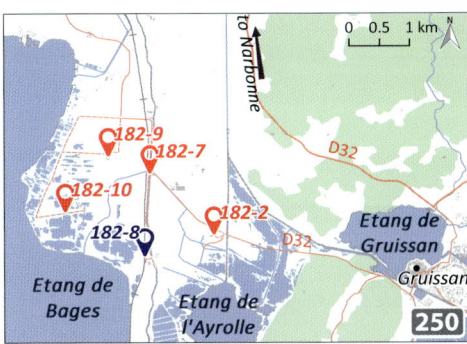

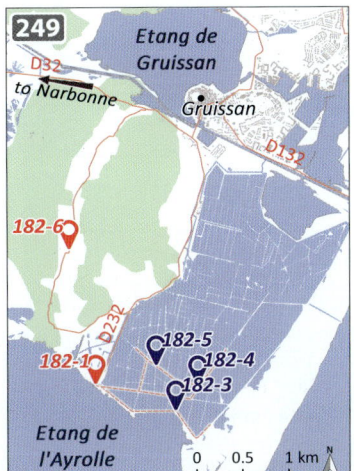

white-rumped sandpiper have occurred). A few pairs of greater short-toed larks breed among the saltpans. **Keep a couple of hours to wander around the Île Saint Martin** [6] (a small road runs to the west from the saltpans through the plateau to end at the southern end of Gruissan village) in search of woodchat shrike, ortolan bunting, black-eared wheatear, tawny pipit and nightjar, all easy to find. **Migrating raptors often follow the ridges, and the bushes can be full of migrants if the weather turns adverse or too windy**.

For squacco and purple herons, western swamphen, moustached warbler and other reed-related species, **follow the D32 and explore the area west of the canal lock** [7]. There are several access roads with several viewpoints beside ponds and marshes [8,9,10]. Raptors sometimes fly by 'La Clape' mountain; park above the radio mast [11 – see GPS file] for good chances of Eleonora's falcon (summer) and the locally breeding Bonelli's eagle (access via the D32 north from Gruissan towards Narbonne and the A9; turn right at 'Ricardelle' towards 'Chemin de la Couleuvre'). End your day with eagle owl, **easily seen from the scenic ridge along the D168 just under the radar station** (park at [12 – see GPS file]). Climb uphill and scan the top of the distant cliff at dusk. Nightjar is often heard from here in May and June, and several pairs of blue rock thrush breed on the cliffs.

183 The Corbières

0.5 DAY; MAP 251   The Corbières are low-altitude limestone hills covered by vineyards, garrigues and maquis, extending inland and aligned north–south roughly from Narbonne to Perpignan. The best period ranges from late April to mid-June when breeders are displaying well even at midday. Start in Caves [1] (4 km from exit 40 on the A9). **Stop-and-go between**

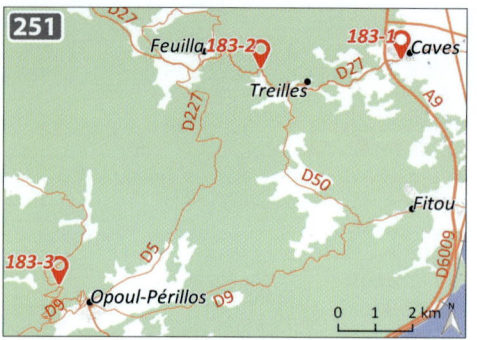

Treilles and Feuilla on the D27 [2], then to Opoul-Périllos on the D227 and D5. Stop wherever you like the look of the habitat. Black-eared wheatear and Thekla lark are conspicuous along these roads especially when singing from a perch (which crested lark, also in the area, never does). Little owl, ortolan bunting, orphean warbler, blue rock thrush and woodchat shrike are the most typical species of the Corbières; all them should be found with a bit of care, **especially around Opoul castle** [3]. Spotless starlings breed in all the villages (beware of dark common starlings and hybrids; unmarked undertail coverts are a good clue).

184 From Lapalme to Salses

3 HOURS – 0.5 DAY; MAPS 252 & 253 **The coastal section from Port-La-Nouvelle to Le Barcarès consists of bushy sand beaches and rocky plateaus where passerines and waders stop over during their migration from mid-April to**

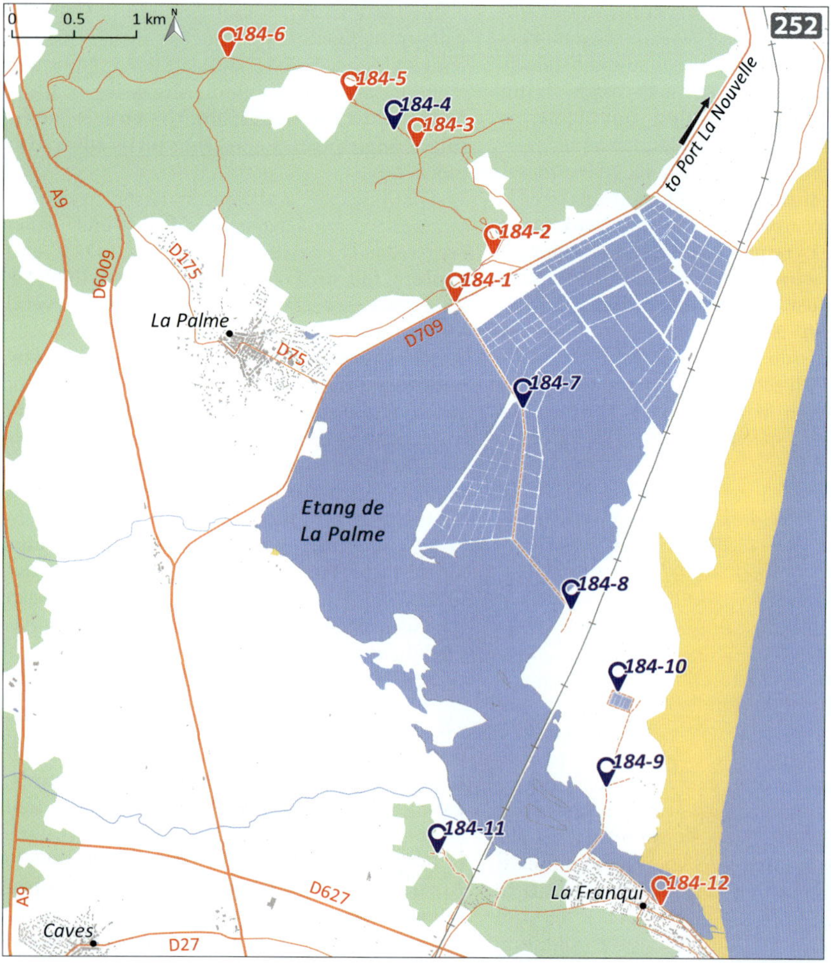

mid-May and from mid-September to November. Leave the A9 at exit 39 (111 km, 1 h 20 min from Montpellier). **The Lapalme plateau is a migration hotspot which also concentrates many breeding specialities.** The best areas are located along a bad track, driveable with caution with a standard car, from the southern end of the plateau (head to Lapalme from Port-La-Nouvelle or Sigean and turn off the D709 on a small road heading upwards among gardens in front of the access track to the saltmarshes). Blue rock thrush breeds in the disused quarry along the access road [1] and in the current quarry [2]; there are also several pairs of rock sparrow around, and an eagle owl sometimes comes hunting at dusk. Thekla lark becomes more common than crested lark from this point.

Go past the quarry and take the left branch until the track goes down into a hollow [3]. This is the first migration spot. **Check carefully the surroundings of the water tank** [4] **and the small wood to the north** [5]: golden oriole, flycatchers and warblers can fall here in high numbers. Tawny pipit, black-eared wheatear (from April), orphean warbler (from May), ortolan bunting (from May) and Thekla lark (year-round) breed in the immediate vicinity and are usually conspicuous on sunny, windless days. Territorial spectacled warblers have also been recorded in recent years but are far harder to sort out. Eleonora's and red-footed falcons sometimes hunt over the area in May.

Bypass the windfarm and **stop as the track turns inside a funnel** [6]**: this is the local viewpoint for passerine migration from mid-March to mid-May if the Pyrenees are clear and the wind is northwesterly;** be there at dawn and stay as long as birds pass through. Conditions can be even better than in Gruissan (site 182) for passerines, as they fly closer along the ridges.

South of the plateau, waders roost in irregular numbers in the **Lapalme saltpans** [7] (access track on the D709, access on foot only). Scan open water for slender-billed gulls, which are sometimes joined by the odd Audouin's gull (rare). A small wood at the south end of the saltpans [8] is a magnet for migrating warblers, flycatchers and buntings. However, **the best area to search for stopover passerines in late April and May or in late August to October is the large expanse of tamarisks and flooded meadows behind La Franqui camp site and beach** [9] (10 km, 15 min south of Lapalme or 7 km, 7 min from exit 40 on the A9): on a good migration day the bushes can be crowded with warblers and flycatchers, plus the usual hoopoes, great spotted cuckoos, shrikes and orioles. Check passing flocks of swifts for the odd Alpine or pallid, and swallows for red-rumped. **At dawn, be at the four ponds** [10], which have spotted and little crakes every year (Baillon's has also been recorded). Check the beach for Kentish plover, Audouin's gull, tawny and red-throated pipits. **Pastures on the western side of the railway** [11] **can be just as productive** (first exit from the roundabout north of La Franqui, then immediately right; turn on tracks just after a ruin).

If the weather is too windy for birding in open areas, **try the gardens in La Franqui village** [12]. **On quieter days, climb up on the Leucate plateau** [13] (4 km, 10 min from La Franqui, follow Leucate-Plage and turn to 'Phare' and 'Cap Leucate'). The plateau has known better days, but a pair or two of black-eared wheatears and spectacled warblers still breed, and rarities such as desert wheatear and trumpeter finch have recently occurred. **Migration can be active along the cliffs bordering the plateau, even at sea** where pomarine skua, shearwaters, terns and gulls sometimes pass through at close range.

If migration is poor, try inland sites in the Corbières (site 183) or drive south to Le Barcarès (15 km, 20 min from Leucate), stopping here and there; any patch of tamarisks can be full of passerines on a good day, and greater short-toed larks breed on bare sand. The ponds near L'Estaque [14] are worth a check for waders and gulls. **Spend your evening in Salses reedbeds** (15 km, 20 min from Le Barcarès, turn left while exiting Salses north on the D87, park at a gate in front of several ponds [15]). Several paths lead into the reeds; look for little bittern, moustached warbler and bearded tit. Spotless starlings are often seen in the village and on the railway wires, and little and scops owl are common in the area.

The Perpignan area (sites 185–186, map 254)

HIGHLIGHTS
» Falls of migrants can be exceptional in spring
» Seawatch from Cap Béar when fishing boats come back to Port-Vendres for Scopoli's shearwater, Audouin's gull and Mediterranean shag (subspecies *desmarestii*)
» Most Mediterranean breeding birds are easier to find in this area than anywhere else

KEY SPECIES
YEAR-ROUND yelkouan and Scopoli's shearwaters, glossy ibis, western swamphen, Eurasian bittern, little owl, **Thekla lark**, blue rock thrush, **moustached warbler**, **spotless starling**, rock sparrow

BREEDING Mediterranean shag, squacco and purple herons, **little bittern**, short-toed eagle, honey buzzard, stone-curlew, Mediterranean gull, **collared pratincole**, scops owl, **pallid swift**, **red-rumped swallow**, **greater short-toed lark**, **black-eared wheatear**, grasshopper, orphean and subalpine warblers, woodchat shrike, **ortolan bunting**

MIGRATION storm petrel, **black stork**, osprey, booted eagle, **pallid harrier**, **red-footed and Eleonora's falcons**, little and spotted crakes, pomarine and parasitic skuas, **Audouin's gull**, **Caspian tern**, **red-throated pipit**, **Moltoni's warbler**, pied, spotted and collared (scarce) flycatchers, *badius* woodchat shrike

WINTER ferruginous duck, common and velvet scoters, red-throated and black-throated divers, razorbill, puffin

VISIT DURATION
1 day to 1 week

PLAN YOUR VISIT
The Perpignan coastal plain is the main alternative to the Aude coast during spring migration (mid-April to mid-May). Although the two areas are adjacent, they differ markedly in terms of habitats and bird migration dynamics. While the Aude coast will yield day-long fluxes of birds which stop over, scattered all along the seashore, the marshes around Perpignan can yield brief but massive passerine falls after a clear night or a thunderstorm. You simply need to

be there at the right time. On heavy migration days (clear days with no to light northwesterly winds or rainy days following clear nights from mid-April to mid-May), concentrate on a single productive location on the plain. If the weather is too rainy or windy, one option is to drive one hour south to Spain and spend a day in the Aiguamolls de l'Emporda (not covered in this book). If you wish to do some seawatching, try it in the evening when fishing boats are returning to harbour. Day-long pelagic trips are organised in spring from Canet-en-Roussillon (booking required: www.decouverteduvivant.fr).

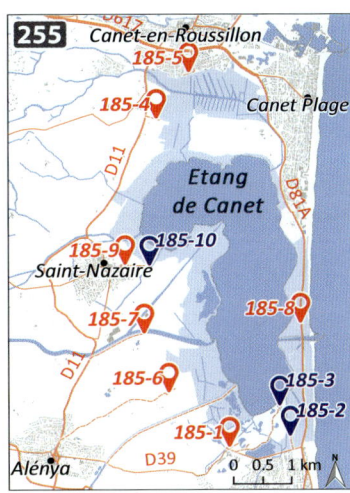

185 Canet-en-Roussillon

0.5 DAY; MAP 255   An extensive brackish-water lagoon known as the 'Étang de Canet' separates Perpignan from the sea. **Surrounding habitats are composed of a combination of pastures with small ponds, marshes with tamarisks and reedbed. Start your day at the very first light at one location and wait for the sun to warm up the bushes**: they can be crowded with birds falling from the sky after having crossed the Pyrenees. **Reedbeds host all the usual Mediterranean breeding species, including** little and Eurasian bitterns, glossy ibis, squacco and purple herons, western swamphen and moustached warbler. Little and spotted crakes stop over from March to May, even on small ponds as long as there is vegetation cover and mudflats. **In wet meadows**, look for red-rumped swallow and collared pratincole. Carefully check all bushy patches and groves: during falls, they can be crowded with all possible species of warbler, including grasshopper, orphean, subalpine (sometimes Moltoni's, especially from May), pied and spotted flycatchers (plus the odd collared, scarce and irruptive), woodchat shrike (badius subspecies is annual) and ortolan bunting. Rarities like pygmy cormorant, great snipe and Baillon's crake have occurred recently, and more surprises regularly turn up. **Bare soil and fields** are the breeding grounds of stone-curlew and greater short-toed lark; calandra lark is unusual but sometimes shows up on migration. Check buildings for pallid swift, rock sparrow and little owl, and at night for scops owl; all these species are common everywhere.

The surroundings of the golf resort at Saint-Cyprien can be a good start (10 km, 15 min from Canet-en-Roussillon, follow the coastline and turn right at the roundabout before entering Saint-Cyprien). Stop at the golf car park and look at the course itself [1] or walk along the seashore [2], focusing on tamarisks and on a small pond [3] which often hosts crakes (access by a trail which starts from the roundabout and leads to the golf resort). **Others spots to check include the Prés de la Ville** [4] (turn right before 'Camping Les Fontaines' on the D11 between Canet and Saint-Nazaire), **the Hauts-de-Canet** [5] (turn south 300 m west of Canet swimming-pool and park in front of the meadows) **and the Alénya meadows** [6], the latter especially for red-throated pipit in April and early May (park at the end of the road running northeast from Alenya cemetery).

For waders, check the Réart river east of the bridge on the D11 [7] and **the lagoon mouth** [8] (for Audouin's gull). **La Passe hill, between the lagoon and Saint-Nazaire, is a skywatching spot** (park at [9] and walk to [10]). All harriers (including pallid), red-footed and Eleonora's falcons, black stork, honey-buzzard, osprey and short-toed or booted eagles can pass through in good numbers if the eastern end of the Pyrenees is clear. **In winter,** ferruginous duck and common/velvet scoters can occur on the lagoon within large numbers of both pochard species and tufted duck.

186 The coastline from Perpignan to the Spanish border

0.5 DAY; MAP 256   South of Argelès, the coast becomes rocky as the Pyrenees fall abruptly into the sea. Birding will be less productive than on the plains, but Cap Béar [1] can be rewarding for **evening seawatching as fishing boats head** back to **Port-Vendres** (35 km, 40 min from Perpignan, follow signposts to Port-Vendres and Banyuls). Yelkouan and Scopoli's shearwaters and high numbers of Mediterranean gulls are likely in spring and autumn; in winter red-throated and black-throated divers can be seen fishing around the headland. This is also France's most reliable site for Mediterranean shag and Audouin's gull. Other uncommon but regular species include storm petrel (mainly autumn), pomarine and parasitic skuas, Caspian tern, razorbill and puffin (casual). Blue rock thrush, Thekla lark and black-eared wheatear breed on the cliffs and bushes of the headland, where Eleonora's falcon is often seen hunting in late summer.

At the extreme south of the region, **Cap Cerbère** [2] will yield similar seawatching records and should produce Thekla lark. Drive 5 km south of Port-Vendres along the coast: **Paullilles** [3] and the **Baillory valley from Banyuls** [4] are **the easiest places to see** red-rumped swallow **in the region** (turn towards 'col de Banyuls'

once in the village). Port-Vendres, Cerbères and most other villages host colonies of pallid swift and spotless starling. Audouin's gulls sometimes perch on roofs in the evening.

Eastern Pyrenees (sites 187–189, map 257)

HIGHLIGHTS
» Combine hiking and birding in the mountains around the Madrès, less visited than its famous neighbour the Canigou
» The Pyrenees foothills host a number of continental species that are hard to find elsewhere
» Spend a day watching for spring or autumn migration in Eyne

KEY SPECIES
YEAR-ROUND capercaillie (*aquitanicus*), **Pyrenean grey partridge** (*hispaniensis*), **rock ptarmigan** (*pyrenaicus*), **lammergeier** and griffon vultures, golden eagle, woodcock, **Tengmalm's owl**, blue rock thrush, Alpine accentor, Eurasian treecreeper, red-billed and Alpine choughs, **citril finch**
BREEDING Egyptian vulture, common rock thrush, ring ouzel, red-backed shrike, rock bunting
MIGRATION black stork, booted and short-toed eagles, honey buzzard, harriers, lesser kestrel, Eleonora's falcon, bee-eater, all swifts, red-rumped swallow
VISIT DURATION 1 day to 1 week

PLAN YOUR VISIT
When spring migration comes to an end in mid-May, it is time to head upwards, as mountain breeding species will likely be present and accessible. The rare *aquitanicus* capercaillie and Pyrenean grey partridge breed at the sites described, but they will require patience. While searching for them, enjoy common rock thrush, citril finch, rock bunting and breeding raptors. Visit the Pyrenees on clear, windless days, preferably not before May to avoid snow – autumn is less productive except in Eynes. **Most high-altitude sites cannot be accessed by car, so plan some hiking to reach the tree limit.** Stick to one of the described areas unless you want to spend all day driving on small, winding roads. Start your day before first light and be at high altitude for sunrise, then descend slowly – the foothills have higher species diversity and will remain active for longer. If you have some time left, you can use the late afternoon to head back to Perpignan and end with coastal sites.

187 The Col de Mantet and Pic Carlit

1 DAY; MAP 258   **The Col de Mantet** [1] **provides the easiest driveable access to mountain**

species (75 km, 1 h 20 min from Perpignan through Prades). Rock bunting occurs on rocky slopes or walls along the access road from Villefranche-de-Conflent [2]. On a clear day, you should see the local Egyptian vulture fly over without too much difficulty. Lammergeier, griffon vulture and golden eagle are often seen from the pass itself, as well as citril finch and siskin, plus more common species such as tree pipit, coal and crested tits. Capercaillie **and** Tengmalm's owl **breed in the surrounding forest**, but seeing one of them is another matter as they inhabit the deepest parts of old beech stands on steep slopes, usually well away from any trail or path.

You will have a slight chance of capercaillie on the hike between Les Bouillouses (park at [3]) and Pic Carlit. However, bear in mind that your chances of seeing any good birds decrease from late June as the trail gets crowded. Ptarmigan and grey partridge live along the highest rocky slopes of the peak. Again they will challenge your best birding skills: they are rare, secretive and sensitive to the slightest disturbance. Alpine accentor, both choughs and raptors are more likely encounters at these altitudes. Lammergeier, citril finch, Eurasian treecreeper, ring ouzel, common rock thrush and Alpine accentor (early in the season) can be found around Les Bouillouses lake.

188 Pic de Madrès and the Sault plateau

0.5–1 DAY; MAPS 259 & 260 **Pic de Madrès (2470 m) is a quiet alternative to the nearby**

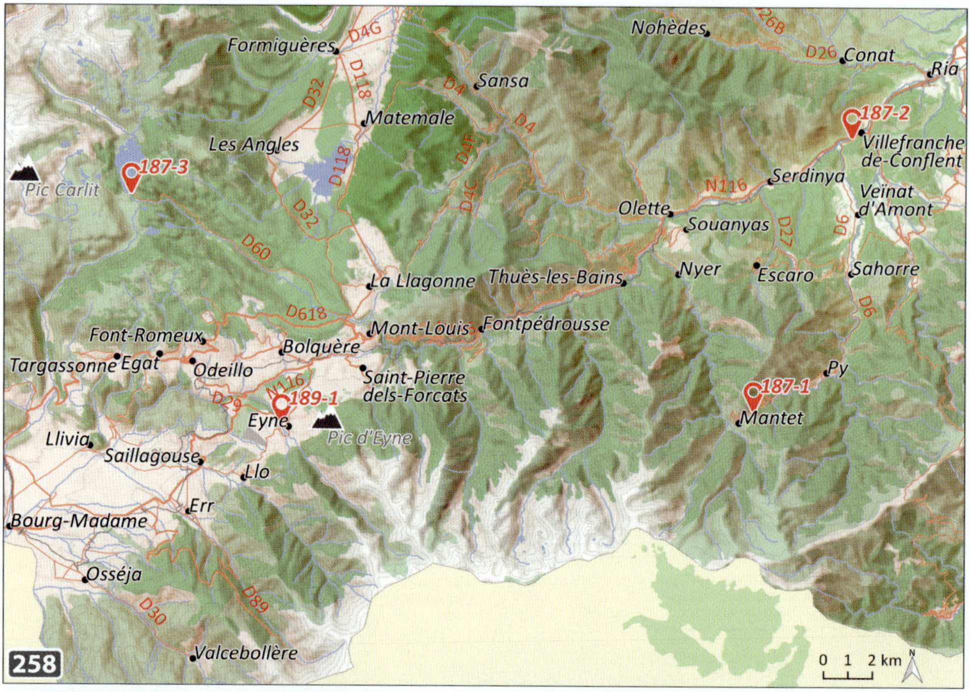

188 PIC DE MADRÈS... | WESTERN MEDITERRANEAN COAST AND THE CÉVENNES

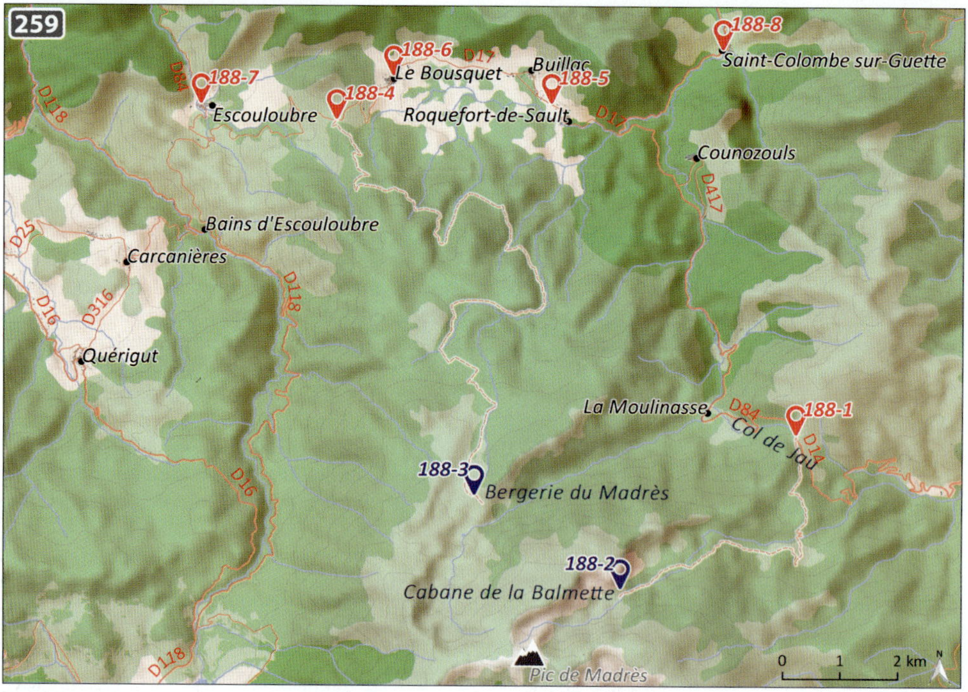

188 PIC DE MADRÈS...

famous Canigou. All the species found in the latter (not covered here due to high numbers of visitors) occur on the Madrès, except ptarmigan. **Start at the Col de Jau** [1] (30 km, 1 h from Prades on the D14 through Mosset). Citril finch, crossbill and common forest species occur on the pass, but **for Alpine species you will need to hike upwards to the Alpine meadows around the Cabane de la Balmette** [2] **or the Bergerie du Madrès** [3] (2000 m, both about 2 h from Col de Jau, the latter also accessible on 4WD tracks from the Col du Garabeil [4], above Le Bousquet). **For both itineraries, a hiking map or a GPS device is necessary; most paths are sparsely signposted.**

Tengmalm's owl breeds and can be heard while climbing inside the mixed spruce–beech forest. Its song peak is around midnight, but you might still have a chance to catch one singing at dawn or dusk. Pygmy owl has recently been reported in the area, but there is no established population (report and document any record). **Be on Alpine meadows above the tree line at first light for a chance of your chief (and difficult) target,** Pyrenean grey partridge, which occurs along stony slopes with abundant water pipit and tree pipit. Lammergeier or golden eagle breed locally and will likely soar above the ridges as soon as the sun rises. **You will need to walk 1 h 45 min more to reach the barren top of the Madrès, but in terms of birding it is unlikely to add much** – there are Alpine chough (which you can easily catch from the shepherd's house) and Alpine accentor.

Descend slowly, spending time along the tree line. The first bushes and trees host tree pipit, ring ouzel, coal, marsh and crested tit. As you enter older forest, you will likely find black woodpecker, nuthatch (absent from the lowlands in the whole Mediterranean region), Eurasian treecreeper and crossbill. Clearings give a chance for woodcock (at dusk or dawn) and capercaillie (scarce): again, be discreet and patient for these species. Siskin and citril finch, which are common as high as the tree line, can be seen down to the Col de Jau.

When you get back to your car, head down to Roquefort-de-Sault [5] (20 km, 40 min west of the Col de Jau) **and drive towards Le Bousquet** [6] **and Escouloubre** [7], **on the Sault plateau**. Stop here and there while passing through old beech forest, where nuthatch and hawfinch breed. **If you are trying to cover the whole altitudinal gradient in a single day, you should ideally be on the plateau by mid-morning. If you have more time, allocate all morning to**

this area and spend another full day at higher altitudes. The landscape here is a mid-altitude mosaic of pastures, hedgerows and fields which oscillate between temperate weather from April to October and cold, snowy conditions in winter. Unsurprisingly, red-backed shrike, yellowhammer, a few red kites and several other regional rarities typical of temperate climates breed here at relatively high densities.

Drive to the small village of Sainte-Colombe-sur-Guette [8], 4 km, 10 min east of Roquefort. Lammergeier breeds on surrounding cliffs and crag martins and Alpine swifts can be found in the village. Search for rock buntings while making your way on the D17 towards Axat, and watch the ridges for golden eagle, eagle owl, blue rock thrush, raven and choughs.

Down at Axat, the roundabout between the D118 and D117 (with a bear statue) [9] is overlooked by a cliff where peregrine and blue rock thrush breed. From here, **head to Ginoles** (15 km, 20 min from Axat through Quillan on the D117), **climb uphill and stop at a car park on a sharp bend** [10]. From here, you have an excellent view over the surrounding slopes and cliffs, where there is a griffon vulture colony. Egyptian vulture, short-toed, booted and golden eagle are often seen soaring from the viewpoint. **Be here as the temperature rises by midday.**

On your way back to the lowlands on the D118, stop along the river in Limoux [11 – see GPS file] (30 km, 30 min from Quillan) for dipper and grey wagtail.

189 Eyne

0.5 DAY; MAP 258   **Eyne village is arguably an eastern counterpart to Organbidexka** (Region 7, site 151) **in autumn**, although it does not reach the same record counts (90 km, 1 h 30 east of Perpignan through Prades, follow signs to Andorra). Park at the information office in the village centre; from there, walk a few minutes to the viewing point [1]. **If the valley is clear with light wind, a day spent here in spring can be rewarded with high numbers of all possible soaring raptors**, plus black stork, large roosts of lesser kestrel, possible red-footed and Eleonora's falcons, bee-eater, Alpine, common and pallid swifts, all the hirundines including red-rumped swallow. Lammergeier, griffon and Egyptian vulture as well as golden eagle can be seen from the spot, and red-backed shrike breeds in suitable areas in the vicinity. Other local species include ptarmigan and grey partridge: plan some hiking to the highest altitudes if you aim to see them. Check for migration sighting updates at www.migraction.net/index.php?m_id=112 (browse to 'La Cerdagne – Eynes').

REGION 10
EASTERN MEDITERRANEAN COAST, SOUTHERN ALPS AND CORSICA

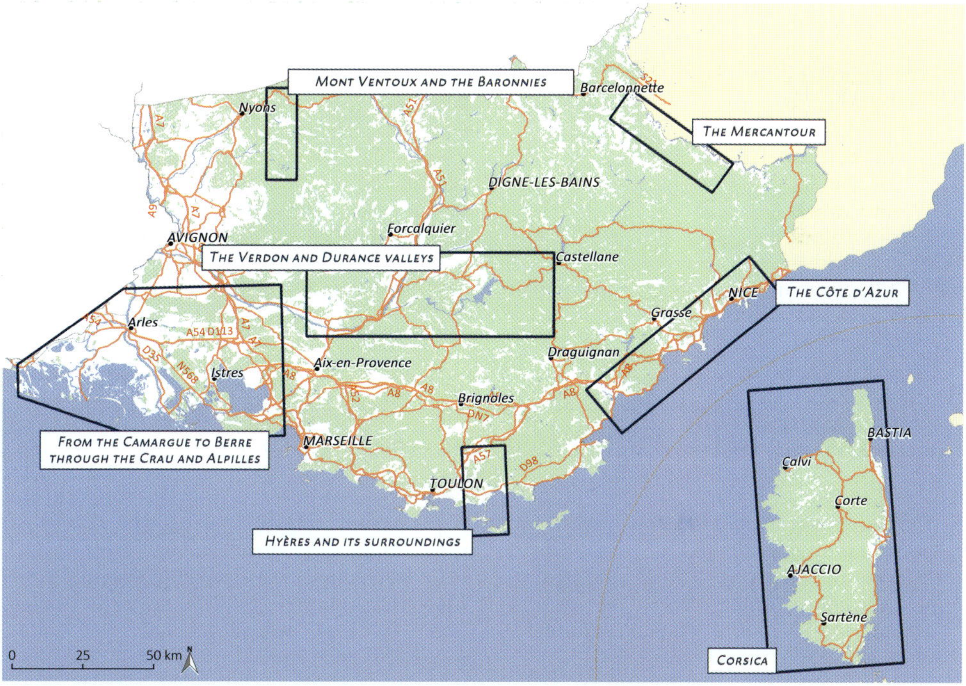

The Alpilles (site 196), an iconic Mediterranean arid landscape with blue rock thrushes, Bonelli's eagles and black-eared wheatears, make a great end to a day in the Camargue.

CAMARGUE TO BERRE | EASTERN MEDITERRANEAN COAST, SOUTHERN ALPS AND CORSICA

HIGHLIGHTS
» The whole suite of Mediterranean breeding birds, including many rare and endangered species
» An altitudinal gradient from the coastline to the snow limit in the southern Alps
» Some of France's birding hotspots, including the Camargue and Corsica
» Spectacular migration falls, a rich breeding community and major wintering sites

REVIEWERS Aurélien Audevard, Amine Flitti

USEFUL WEBSITES
www.faune-paca.org
https://paca.lpo.fr
www.cen-corse.org
www.wnat.fr/index_wnat_open.php

The greater flamingo is unmistakable in brackish water ponds of the eastern Mediterranean region, although its only French breeding colony is in the Camargue (site 191).

From the Camargue to Berre through the Crau and the Alpilles (sites 190–196, map 261)

HIGHLIGHTS
» The Camargue, France's flagship birding hotspot throughout the year
» Huge number and diversity of waterbirds and passerines in all seasons
» Migration falls in both spring and autumn, with a good selection of rarities
» Rare and endangered species such as Bonelli's eagle and pin-tailed sandgrouse

| J | F | M | A | M | J | J | A | S | O | N | D |

KEY SPECIES

YEAR-ROUND red-crested pochard, gannet, yelkouan and Scopoli's shearwaters, little, great crested and black-necked grebes, **greater and lesser flamingos**, glossy ibis, Eurasian bittern, spoonbill, **little bustard**, booted and **Bonelli's eagles**, marsh harrier, peregrine, western swamphen, **slender-billed** and Mediterranean gulls, Sandwich tern, little and eagle owls, **pin-tailed sandgrouse**, **calandra lark**, crag martin, blue rock thrush, Sardinian, Dartford, **moustached** and Cetti's warblers, bearded and crested tits, southern grey shrike

BREEDING white stork, **purple, squacco and night-herons**, **Egyptian vulture**, short-toed eagle, Montagu's harrier, hobby, **lesser kestrel**, **little bittern**, black-winged stilt, **collared pratincole**, stone-curlew, Kentish plover, gull-billed, little, common, whiskered and black terns, scops owl, great spotted cuckoo, nightjar, roller, bee-eater, hoopoe, tawny pipit, greater short-toed lark, **red-rumped swallow**, spectacled, subalpine, reed and great reed warblers

MIGRATION Eleonora's and red-footed falcons, pallid harrier, black stork, spotted and little crakes, **dotterel**, all **waders** including shanks, sandpipers and stints, scarcities including **marsh, Terek and broad-billed sandpipers**, Temminck's stint, **red-necked** and **grey phalaropes**, Audouin's gull, **Caspian** and **white-winged terns**, skuas, wryneck, Alpine swift, migrating pipits, warblers and flycatchers, red-throated pipit

WINTER Bewick's swan, common and velvet scoters, red-breasted merganser, eider, black-throated and red-throated divers, Slavonian grebe, **greater spotted eagle**, merlin, common

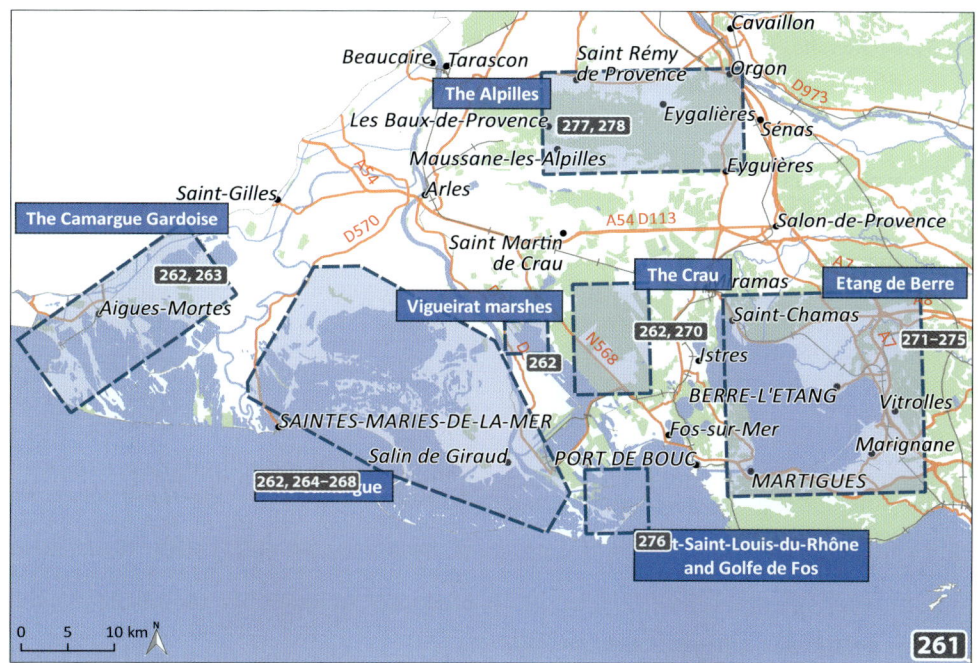

crane, **Richard's pipit**, Alpine accentor, **penduline tit, wallcreeper**

VISIT DURATION Up to 7 days

PLAN YOUR VISIT

The Camargue is France's flagship birding location, worth an extensive visit in any season. A day trip here easily yields over 100 species, especially from late April to mid-May. Allocate at least least one day per site, but a week would bring new birds every day. **Try to avoid windy days** when birds become inconspicuous and water levels rise. All the sites are at their best at dawn and dusk, but if you are short of time, **the Camargue is easier for day-round birding while the Camargue Gardoise is better for early morning and the Crau for the evening.**

190 The Camargue Gardoise

1–2 DAYS; MAPS 262 & 263 The Camargue Gardoise concentrates several of France's largest reedbeds and freshwater wetlands. Although often lumped in with the Camargue (site 191), it differs in its habitats and its bird community, dominated by huge numbers of herons and reedbed passerines. Check the weather and try to avoid winds above 30 km/h; otherwise, shift to the Camargue itself.

The central point is Scamandre reserve [1], which protects a large circular pond surrounded by reedbeds (50 km, 1 h east of Montpellier through La Grande Motte and Aigues-Mortes, follow Arles, then 'Centre du Scamandre'; or 35 km, 40 min from Nîmes through Saint-Gilles). The reserve is open Tuesday to Saturday, 09.00–18.00. Inside, a hide and platform provide great photographic opportunities at all seasons. **Scamandre pond is well known for its** heron **colony where** glossy ibis, squacco heron **and** night-heron **breed in large numbers.** A pygmy cormorant (sometimes even two) has frequented the colony in recent years, mainly in winter and early spring. Among commoner species, expect red-crested pochard, breeding stilt, stopping-over shanks and sandpipers (including the odd marsh sandpiper). Scan the reedbeds for western swamphen (abundant year-round except after cold weather), purple heron (from April), Eurasian bittern (year-round) and little bittern (from May), penduline (easier in winter) and bearded tits. In winter, good numbers of cranes, red kite, booted eagle, peregrine, merlin and up to several greater spotted eagle winter in the surrounding woods and fields (all private, stick strictly to roadsides).

All these birds can also be spotted from outside the reserve along the D779 between the Étang du Charnier and Scamandre [2,3], **a good**

190 THE CAMARGUE... | EASTERN MEDITERRANEAN COAST, SOUTHERN ALPS AND CORSICA

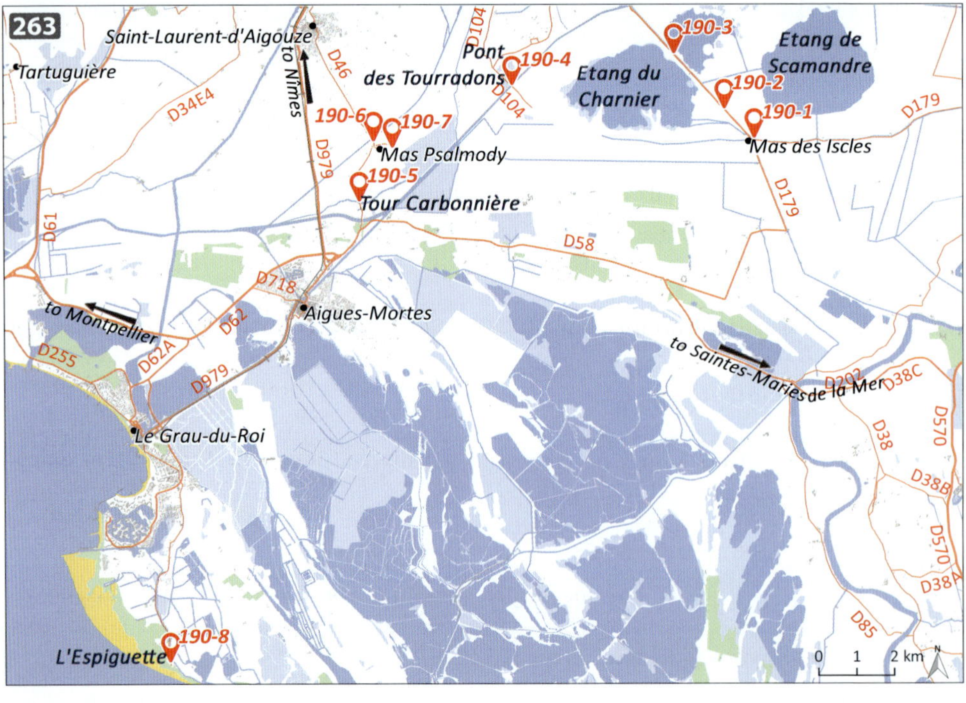

place to spend early morning hours before Scamandre opens. The canal bordering the road ('Canal des Capettes') is especially recommended for little bittern, crakes (April) and bearded tit.

The next viewpoint is the Pont des Tourradons [4], arguably France's most reliable site for western swamphen at all seasons and another place to spend your time if Scamandre is closed (turn left as you reach Saint-Gilles from the D779 and drive for 6 km on the D381; from Montpellier or Nîmes follow signposts from the first roundabout on the D6572 from Vauvert to Aimargues). **Scan the landscape with a telescope from the bridge over the canal for an overview of surrounding reedbeds, meadows and ponds. You can get closer to birds by walking a few hundred metres southeast (check the small pond on the left of the road).** Apart from western swamphen, which is abundant and generally conspicuous along reed edges, there is a heron colony to the west (all species plus glossy ibis). Expect all dabbling ducks dominated by loads of red-crested pochard and little, great crested and black-necked grebes. Little bittern breeds in reedbeds from May (not as easy as in Scamandre due to viewing distance). Other breeding species of interest include moustached, Cetti's, reed and great reed warblers, stonechat and reed bunting. During migration, good numbers of waders can show up provided some mudflats are available. In winter, penduline and bearded tits are common but tricky in windy conditions. The bridge is also frequented by white-winged (uncommon), whiskered and black terns (all in May) and collared pratincole (May to August).

End your day with waders **and** terns **at the Tour Carbonnière [5], an isolated tower overlooking flooded meadows and reedbeds** between Aigues-Mortes and Saint-Laurent-d'Aigouze on the D46 (25 km by car from the Pont des Tourradons or 7 km by bike along the canal westwards, signposted from the roundabout at the southern exit of Saint-Laurent-d'Aigouze and from the second roundabout east of Aigues-Mortes on the way to Arles). **The marshes surrounding the tower can be crowded with all possible freshwater** waders **(peaking 15 April – 10 May and August–September)** with up to several tens to hundreds of wood sandpipers, shanks (look for single marsh sandpipers in April and early September), ruffs, little stints (including Temminck's stints in May and August) and breeding stilts. Gull-billed, whiskered **and** common **terns** are sometimes joined by a white-winged or Caspian tern.

Drive 2 km north and turn right on a bad track [6] **(forget it in winter or after heavy rain – drive carefully) which leads inside the marshes behind Mas Psalmody [7].** Pass through the gate (access is tolerated, but close it carefully behind you) and walk among reedbeds for close views of wildfowl, waders (reliable for marsh sandpiper in early May and August, and for collared pratincole from May through September), terns, western swamphen and herons.

In early May and September, or if the weather is windy, an alternative to ponds and reedbeds could be to search for passerines and seawatch on L'Espiguette beach [8] (signposted from Le Grau du Roi, 40 km east of Montpellier on the road to Aigues-Mortes, parking fee except in winter). **Pine woods around the lighthouse can be crowded with stopping-over** warblers, flycatchers, redstarts, pipits **and** shrikes **especially on cloudy mornings after a clear night or in the hours following a thunder storm.** In recent years, surprises have included black-winged kite, yellow-browed and icterine warblers and collared flycatcher. **In autumn or winter, scan the sea for** common and velvet scoters, yelkouan or Scopoli's shearwater, gannet and skuas.

191 The Camargue

1–3 DAYS; MAPS 262 & 264–268 The Camargue ranks among Europe's largest wetlands and top birding sites. It is the favourite choice of many French or foreign birders for its easy access to suitable habitats, close views of birds and huge counts and diversity. **Bird activity peaks in mid-April to mid-May, but a visit is worthwhile at any time of the year.** Consider staying up to three days to fully explore the area, mostly driving or cycling from spot to spot. A day trip typically starts or ends on the Baisses de Cinq Cent Francs (waypoints 19–25) or on the Route de Cacharel (waypoints 29–33). You can also cross the Rhône in Salin-de-Giraud on the Bac de Barcarin to end your day in the Crau (site 193). **From Arles, follow signposts to Salin-de-Giraud, Le Sambuc, Les Saintes Maries de la Mer and 'Étang du Vaccarès'.**

To visit northern Camargue, start early in the morning in the Remoules paddyfields [1] and along the road that crosses the Mas d'Agon marsh [2] (14 km south of Arles, head to Les Saintes-Maries-de-la-Mer on the D570 and turn left towards 'Mas d'Agon'at the Auberge des Plaines, 5 km past the last roundabout). **These extensive reedbeds and flooded meadows attract** glossy ibis, night-heron, squacco **and**

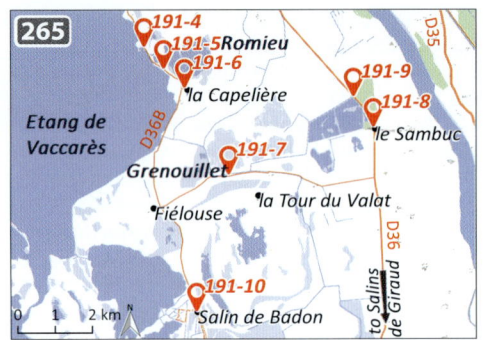

purple herons, both bitterns, spoonbills, black-winged stilt, gull-billed, whiskered, black and a few white-winged terns (early May and early September), migrating waders (including the sought-after marsh sandpiper), collared pratincole, red-rumped swallow, great reed warbler and bearded tit.

Drive south to the D37 and turn right to scan flying herons, cranes (winter), storks and ibises from the **Mas Neuf platform** [3]. **Resume your drive eastwards along the Étang de Vaccarès on the D37 and D368, stopping at platforms along the road** [4,5]. Greater flamingos are abundant throughout. In winter, expect loads of black-necked grebe, coot, teal, common and red-crested pochard, tufted duck and a few red-breasted mergansers. Flocks of shoveler and gadwall usually rest on protected freshwater ponds a few hundred metres inland (a fence prevents any access but some of these ponds can be scanned from the main road). In spring, tawny pipits breed on bare ground. **For guided visits, stop at the Camargue reserve headquarters in La Capelière** [6]. A few hides (fee) allow photographic-range views of wildfowl, herons, spoonbills and waders (marsh sandpiper is annual here, mostly in early May and late August).

Further east, **turn left towards Le Sambuc and stop after 6 km at the Grenouillet platform** [7]. This freshwater pond and surrounding flooded meadows can be crowded with wildfowl (mostly shoveler, teal, wigeon, pintail, gadwall, greylag goose), shanks, sandpipers, curlews, godwits and other waders, all terns (including good numbers of gull-billed and sometimes a white-winged) and large flocks of glossy ibis. All the herons (including squacco) occur here. Single black storks are not unusual, especially in spring and late summer, as well as collared pratincole. **In winter, the Grenouillet marsh is hunted but a stop is still worthwhile** for marsh harrier and, with luck, one of the few wintering greater spotted eagles.

Drive 5 km east to Le Sambuc and visit the Verdier marsh. Park in the village near the football ground [8] and walk north or park in front of the main gate [9] along the D36 (limited space). Eurasian and little bitterns, gull-billed tern, glossy ibis, moustached warbler (year-round) and penduline tit (winter) frequent these flooded meadows and reedbeds. Rollers and bee-eaters breed in the area and perch conspicuously on wires. **There is a collared pratincole colony close by in a protected area; its precise location is classified but the birds are often seen flying around.** Head back to the D368. **You can stay overnight in Salin de Badon** [10], a former salt farm with several hides that allow close views of waders (marsh sandpiper is regularly seen), wildfowl and herons (access permits and room booking in La Capelière).

To reach southern Camargue, leave the D36b to the right towards 'La Gacholle' and 'Digue à la mer', 3 km south of Salin de Badon. Stop-and-go on the track at the end of the road to scan vast expanses of bare ground and mudflats, for instance **at** [11] **between the Enfores de la Vignolle and the Fangassier.** This is the closest driveable point to the flamingo colony (look south), huge numbers of which forage around (remember that one or two lesser flamingos have occurred in the Camargue for several years and could be mixed up in these flocks). Park at [12] at the end of the track and walk towards the Galabert [13] (bring drinking water). **It concentrates** waders, gulls **and** terns, **of which numbers vary with water levels.** Expect abundant breeding Kentish plover and yellow wagtail. Look for groups of dunlin, curlew sandpiper and little stint, which sometimes roost close to the dykes. If they are close enough, you might find

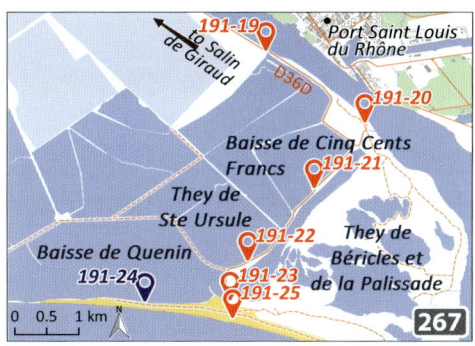

a broad-billed sandpiper or a Temminck's stint amongst them (May and August). Red-necked phalaropes are sometimes seen on remnant open waters. Spectacled warbler **and** tawny pipit **breed in salt scrub between La Comtesse car park** [14] **and La Gacholle lighthouse** [15]. Light playback may help if warblers do not show up by themselves. **The lighthouse attracts migrating passerines** (warblers, flycatchers, redstarts, larks, pipits and shrikes) among which red-throated pipit, collared flycatcher, icterine and eastern subalpine warblers have been found in the past.

On the other side of the Fangassier, a bad track leads to Beauduc beach (drive towards Salin de Giraud on the D36c and turn right at [16] to Mas de la Belugue, bypass it and turn right on the track at the end of the road). **It runs between the Fangassier and Grand Rascaillan ponds** [17] **on which large numbers of** flamingos **and** waders **roost if the water is sufficiently low** (same species as in previous locations). Water levels and wader numbers are not necessarily consistent across all parts of the ponds – if you failed at La Gacholle, try here. Slender-billed gulls and Caspian terns sometimes forage within groups of yellow-legged gulls and Sandwich terns. The track ends at Beauduc beach [18]. **Try seawatching in autumn and winter for** black-throated and red-throated divers, yelkouan and Scopoli's shearwaters, gannet, velvet and common scoters, black-necked and Slavonian (rare) grebes. **In early May, the surrounding tamarisks can be crowded with migrating passerines** (wryneck, **a variety of** warblers, flycatchers, redstarts, shrikes **and** buntings) **especially in light southwesterly or northwesterly winds.** Virtually anything can turn up in these bushes, including rarities (icterine warbler and collared flycatcher being the most likely). Check the sky for a passing Montagu's harrier, hobby or even Eleonora's falcon.

The southeastern corner of the Camargue is the classic place to start or end the day during wader migration. Bird records vary daily with water levels and can range from incredibly rewarding to miserably frustrating (the latter especially in strong winds). Bypass Salin-de-Giraud and follow the D350 towards Piémanson. The first viewpoints start just after the salt dump [19]; there is usually no more here than a few yellow-legged or Mediterranean gulls, but red-necked or grey phalaropes or slender-billed gulls sometimes feed on the canal.

A visit to the Domaine de la Palissade [20] (fee, open 09.00–18.00) will get you close to wildfowl, waders, herons and migrating passerines. **Head to the last stretch of the road and stop along the Baisse de Cinq Cent Francs** [21]**, the They de Sainte-Ursule** [22] **and the Baisse de Quenin** (walk 1 km along the beach for another look at the latter) [23,24]. These three ponds can be crowded with dunlin and little stint flocks, among which you should search for broad-billed sandpiper and Temminck's stint (annual) or even rarer species (stilt and least sandpipers have been seen here in the last decade). **Carefully check the canal bordering the road even if all the waders are far inside the ponds,** as a few stints often remain with little ringed or Kentish plovers. **Scan the banks of all small islets**, where single red-necked phalaropes forage in late summer. Gulls and terns roost on these islets with small numbers of slender-billed gulls and a few Caspian terns; the sight of birds flying over the ponds at dusk can be exceptionally photogenic. Eleonora's falcons sometimes hunt over the ponds from May to September (beware of hobbies and peregrines). **If the weather is windy, try seawatching on Piémanson beach** [25]. Yelkouan and Scopoli's shearwater, gannets, scoters (winter, sometimes with an eider), skuas and gulls (possibly with an Audouin's gull) sometimes fly past at close range.

The area is generally raptor-rich. Besides marsh harrier and common buzzard, booted and Bonelli's eagles, peregrine and merlin are often recorded here between October and March. The open view towards the north will also yield huge numbers of cranes and a few accompanying storks (black stork sometimes winters around here). The mudflats east of [32] **are the local viewpoint for** dunlin, grey plover **and** curlew. Terek sandpiper and broad-billed sandpiper have been recorded here (scarce).

A further 9 km north, **Basses Méjanes** [33, see map 264] **is Camargue's most reliable spot for wintering** Bewick's swan (up to 100, November to March). A stop here could also yield geese, black stork and spoonbill. From this point, either head back to Arles (20 km) or connect with northern Camargue (waypoint [3], 3 km).

Saintes-Maries-de-la-Mer marks the south-western corner of the Camargue, 70 km from Piémanson, 38 km south of Arles and 30 km east of Aigues-Mortes (site 190). **The Pont de Gau ornithological park** [26] **is renowned among photographers for its hides** (5 km north of Saintes-Maries towards Arles, open 09.00–19.00 in spring, fee). Glossy ibis and spoonbill are easier to approach here than anywhere else in the Camargue, and there are chances of marsh sandpiper at both passages. **Check the dykes west of Saintes-Maries** [27] **for** seabirds (yelkouan shearwaters sometimes come close to the shore), then walk from the entrance of the camp site (check tamarisks for migrating passerines) towards a water treatment plant [28]. High numbers of dunlin and other waders may feed on the mudflats to the north, among which a broad-billed sandpiper or other rarities might occur (a white-rumped sandpiper stayed for a while in mid-summer 2017).

In winter, drive along the Route de Cacharel [29] (5 km north of Saintes-Maries on the D85a), a good track running north for over 10 km among vast expanses of saltmarsh, reedbed and pasture. **Stop after 800 m beside a reedbed** [30] where bearded tit are particularly easy to see in windless conditions. There are western swamphens, abundant reed buntings and smaller numbers of penduline tits around. Head further north to park at **Pont des Cinq Gorges** [31]; wildfowl and waders rest here if there are mudflats (spot from the gate). **The next 4 km are an open grassy plain, and the best place in Camargue to look for** greater spotted eagle**, which winters annually in the surrounding woods and can be seen soaring on clear windless days**. Recent years have seen a rise in records of unidentified spotted eagles that are usually considered to be hybrids – any large eagle should therefore be carefully documented.

192 Vigueirat marsh (Marais du Vigueirat)

2 HOURS–0.5 DAY; MAPS 262, 269 & 270   The Vigueirat reserve gives an opportunity to see most of the Camargue's wetland breeding birds on the east bank of the Rhône [1], which will save time if you wish to focus on the Crau (site 193) or if your main interest is photography (signposted from the D35 halfway between Arles and the Bac de Barcarin, 20 km either side; free entrance, open 09.30–17.00, guided visits, hides). All the herons including squacco and both bitterns, moustached and great reed warblers, bearded tit, red-crested pochard, western swamphen, bee-eater and great spotted cuckoo breed here. Thousands of ducks stop over on both spring and autumn migration and some of them overwinter. It is also a classic spot for wintering penduline tit.

193 The Crau (Plaine de la Crau)

0.5–1 DAY; MAPS 262 & 270   The 75 km² Crau nature reserve is the remnant of a 600 km² steppe formed by the former Rhône delta east of its current course. It is now essentially a flat stony desert with sparse vegetation, crossed by a canal ('Canal de Centre-Crau') and bordered by poplar woods, groves and pastures. A visit to the Crau can easily be combined with a birding day in eastern Camargue, to which it is connected through the Bac de Barcarin. **Be on the plain before first light and stay until full darkness to fully enjoy its scenic atmosphere. The Crau's flagship attractions are the only French population of** pin-tailed sandgrouse

EASTERN MEDITERRANEAN COAST, SOUTHERN ALPS AND CORSICA | 193 THE CRAU | 203

Start at Peau-de-Meau sheepfold [1] (drive to Saint-Martin-de-Crau, 20 km, 20 min east of Arles on the N113, and follow signposts to 'Étang des Aulnes' south; the car park is located 8 km from the railway station at the end of the concrete road). A walk on tracks and paths around the sheepfold may yield pin-tailed sandgrouse, little bustard, greater short-toed lark, tawny pipit, lesser kestrel, southern grey shrike, stone-curlew and little owl. **Walk or bike 5 km east (cars are not allowed anymore), stopping here and there, for instance near the canal** [2]. If you are biking, follow the tracks up to Mas Chauvet [3]. To get there by car, bypass the Crau by the north through Saint-Martin-de-Crau and Entressen (use your GPS once there as the road is tricky to find). The surroundings are the core of the local wintering grounds of Richard's pipit. **The main population of calandra larks is to be found year-round 1 km to the south** [4]. Sandgrouse can occur anywhere in the vicinity: have their calls in mind, as you are most likely to catch them while flying in small groups of up to a dozen individuals. Little bustard, stone-curlew, little owl, roller, bee-eater and lesser kestrel form the bulk of local breeding species, together with tawny pipit and corn bunting, which are truly abundant. In May and September, look for Montagu's or pallid (rare) harriers, red-footed (uncommon) and Eleonora's falcons (rare) and dotterel (up to several dozen in good year). **It is absolutely essential to remain strictly on the tracks as the entire steppe is protected for ground-nesting species.** If you are biking, turn right after 4 km south [5] to join the N568. If your car is parked in Mas Chauvet, get back to Entressen and follow

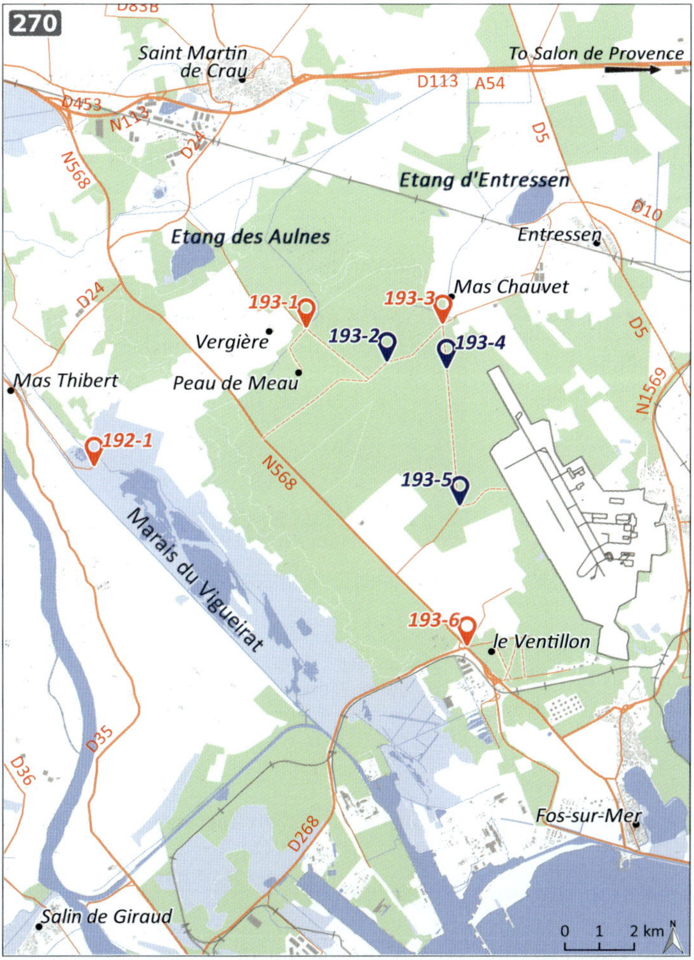

and calandra lark. They should not be your only focus, however, as most of the Mediterranean steppe breeding species are easier to find here than anywhere else.

signposts to Fos-sur-Mer, then Saint-Martin-de-Crau on the N568, to park near **Le Ventillon roundabout** [6], **one of the Crau's easiest sites for** pin-tailed sandgrouse, although it has become harder to find in recent years.

194 Étang de Berre

3 HOURS – 2 DAYS; MAPS 271–275 The Étang de Berre is a vast and deep saltwater pond lying east of the Crau and 20 km west of Marseille. It is surrounded by large expanses of wetlands in which over 250 bird species have been recorded at all seasons. **Starting from the north, park south of Saint-Chamas** at La Pointe (drive 2 km south on the D10 and turn right on the bank) and walk on the path that leads into La Petite Camargue and Le Palous [1]. Moustached (year-round) and great reed (from April) warblers breed in the reedbeds. Among the most regular non-breeding species, you should expect greater flamingo, purple heron, red-crested pochard, gadwall, and with some luck little and spotted crakes (April).

Drive 8 km further east on the D10 and turn left to Lançon. After 2 km the road reaches the **Calissanne** [2], **a scrubland plateau with all garrigue breeders**. Expect Sardinian, Dartford and subalpine warblers, southern grey shrike, nightjar, Alpine swift, bee-eater and woodlark. Roller breeds south of La-Fare-les-Olivers in the vineyards near the River Arc [3].

Berre-L'Étang saltmarshes are the richest site in the area. Follow signposts to 'Port du Passet' along the bank [4]. Common and little terns, redshank, Kentish plover, avocet, black-winged stilt and shelduck breed here. Furthermore, it is a stopover area for waders and terns, great crested and black-necked grebes, greater flamingo and yellow wagtail. In winter, red-breasted merganser, velvet scoter and black-throated diver frequent open waters around La Pointe [5] (long-tailed duck and eider are regular rarities among diving ducks).

Finish at Les Pâtis ponds and shore [6,7] **for more** waders, wildfowl **and chances of** crakes. Further east, the Arbois plateau, 2 km east of Vitrolles, had all the Mediterranean breeders until it burnt during summer 2017. Depending on vegetation recovery in the next five years, it could become a nice place again to spend a couple of hours while waiting for your flight from the nearby Marseille-Provence airport (exit 30 on the A7, head towards Aix-en-Provence on the D9 and follow 'Valbacol' up to Les Collets Rouges) [8]. Great spotted cuckoo breeds in the

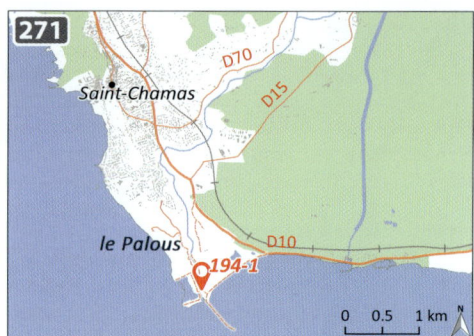

equestrian centre, and a walk in the garrigue in May should yield nightingale, Sardinian and Dartford warblers, little and scops owls, nightjar, roller, bee-eater and red-legged partridge. **At the northern end of Marseille airport, the Salins du Lion [9] are former saltmarshes where** wildfowl **and** waders **stop over in both spring and autumn** (head to the airport and follow the D20 until you find a signpost). Expect greater flamingo, little and black-necked grebes, little and cattle egrets, common pochard, shelduck, stilt, common snipe, green, wood and common sandpipers and terns. If you have more time, head south of the airport, bypass Marignane towards Châteauneuf-les-Martigues and **visit the Étang de Bolmon brackish pond.** All the aforementioned waterbirds occur here in higher numbers, plus night-heron, squacco heron and stone-curlew. Park at the dump ('déchèterie') [10], from where a trail runs into the Grande Palun marshes [11,12].

195 Port-Saint-Louis-du-Rhône and the Golfe de Fos

2–4 HOURS; MAP 276 The industrial plants and harbour surrounding the Golfe de Fos on the eastern bank of the Rhône mouth might look uninviting, but they offer a decent follow-up

to a winter day in the Camargue (site 191) or the Crau (site 192). Eider (uncommon), red-breasted merganser, velvet scoter, black-throated diver, good numbers of black-necked grebe, Sandwich tern, Mediterranean and slender-billed gull all winter around the gulf. In Port-Saint-Louis-du-Rhône, **check the gulf from the Darse Gloria** ('Telline Gloria/Plage Olga' signposts) [1] **and Carteau bay, on the opposite side of the harbour** (follow 'Plage de Carteau') [2]. Further south along the Rhône, **the brackish ponds along 'Route Napoléon' are worth a check for** waterbirds [3]. At the Rhône mouth, **Plage Napoléon beach [4] is a good winter or spring seawatching spot** (divers, red-breasted merganser, eider, scoters, Scopoli's and yelkouan shearwater, gannet, grebes, skuas).

196 The Alpilles

1–2 DAYS; MAPS 277 & 278 **The Alpilles limestone hills are an obvious follow-up to a trip to the Camargue and the Crau, offering the prospect of adding garrigue breeding species and** Bonelli's eagle. The main access is through Mouriès or Aureilles, 40 km, 35 min east of Arles (exit 12 on the A54 at the eastern end of Saint-Martin-de-Crau). Park at the junction of the D24 and D25 south of Eygalières [1] (little and scops owl breed in the village) or 3 km further southeast along the D25 [2,3]. Tracks lead north to Le Gros Calan ridges [4] and La Vallongue plateau [5]. Red-legged partridge, blue rock thrush, Dartford and Sardinian warblers, crested tit and raven are easy to find year-round. April adds short-toed eagle, tawny pipit, Alpine swift and subalpine warbler. Check the sky – there are four Bonelli's eagle pairs in the Alpilles and Egyptian vulture is not unusual. Eagle owl is quite common on cliffs and can be heard on clear windless evenings (easiest February–April).

The same species can be found 10 km west around La Caume plateau. Exit Saint-Rémy-de-Provence south towards Maussane and park before the archaeological site [6] (eagle owl breeds here) or 3 km further south on the D5 [7]. **A walk towards the plateau [8] will yield all the above-mentioned species**, and there are breeding eagle owl and wintering wallcreeper and Alpine accentor at the nearby Peirou dam [9]. **For another chance of wintering** wallcreeper **and** Alpine accentor, **visit Les Baux-de-Provence** [10], one of France's most charming villages. Crag martin (resident), Alpine swift and blue rock thrush (resident) breed on the

surrounding cliffs and on the citadel, and roller occurs throughout the surrounding plain.

End a spring day in the Alpilles with a drive on the plains around Eyguières, 12 km northwest of Salon-de-Provence. **At dusk, take the narrow road to Les Glauges from the D17 and stop before the farm** [11]. Eagle owl breeds on the cliff which overlooks the road and there are little and scops owls on surrounding farms. Other Mediterranean species breed in the valley, among them short-toed eagle, stone-curlew, roller, nightjar and southern grey shrike. If you need to head north to Avignon or Cavaillon, you can also stop in Notre-Dame-de-Beauregard [12] (signposted from Orgon, 13 km north of Eyguières or 9 km east of Eyguières). Walking up the track will probably yield blue rock thrush, Alpine swift, crag martin and southern grey shrike; this is also another reliable place for Bonelli's eagle.

Hyères and its surroundings (sites 197–199, map 279)

HIGHLIGHTS
» The spring equivalent to Ushant island is Porquerolles island, where spectacular falls of passerines can occur in early May
» Breeding shearwaters and otherwise rare Mediterranean species including slender-billed gull and red-rumped swallow
» A major stopover site for waders in Hyères

KEY SPECIES

YEAR-ROUND Mediterranean shag, **Reeves's pheasant**, little bustard, greater flamingo, glossy ibis, great white egret, marsh harrier, peregrine, avocet, raven

BREEDING Scopoli's and **yelkouan shearwaters**, **little bittern**, purple heron, black-winged stilt, Kentish plover, **slender-billed gull**, scops owl, great spotted cuckoo, hoopoe, Alpine and pallid swifts, roller, great reed warbler, **red-rumped swallow**, greater short-toed lark, tawny pipit

MIGRATION Eleonora's falcon, **little and spotted crakes**, waders including **Temminck's stint**, broad-billed sandpiper (rare) and phalaropes (rare), white-winged tern (rare), **passerines** including **red-throated pipit**, wood, western Bonelli's, garden, subalpine and **Moltoni's warblers**, Balearic woodchat shrike, collared flycatcher (rare)

VISIT DURATION 2–3 days

PLAN YOUR VISIT

This area is at its best from mid-April to mid-May. For a one-day visit, head to Porquerolles island. If you have more time, devote two days to the island and one day to birding around Hyères and Giens. Pierrefeu airfield can be a nice break on your way east to Fréjus.

197 Hyères

0.5–1 DAY; MAP 280 A narrow peninsula stretches south of Hyères, 20 km, 30 min east of Toulon. **Although heavily urbanised, its southward orientation and its wide complex**

grassy ground around the runways is worth a scope check for collared pratincole, lesser kestrel, marsh harrier and migrating raptors or passerines.

Drive past the airport and park at the first roundabout on the D197, then walk carefully on the roadside to reach the southern end of the Marais Redon pond [3]. Expect black-winged stilt, Kentish plover, avocet, little, common and Sandwich terns and possibly shanks, stints (including Temminck's) and sandpipers.

To the south lie wide expanses of protected saltpans known as 'Les Pesquiers', which serve as a stopover and breeding site for large numbers of waders **in both spring and autumn.** The site can usually not be visited but there are several viewpoints along the 'Route du Sel' which runs on the coastal dyke bordering the saltpans to the west [4,5,6] (closed in winter). A long list of rarities has accumulated here over the years, starring royal tern, bimaculated lark, several buff-breasted and Terek sandpipers to name just a few. You will more likely see greater flamingo, slender-billed gull, common, little and Sandwich terns and plovers, shanks, sandpipers and stints on passage. Temminck's stint, marsh and broad-billed sandpipers, red-necked phalarope, collared pratincole and white-winged tern occur annually.

Resume your drive south to bypass Giens and park at La Madrague harbour [7]. **From there, walk to the Pointe de l'Ermitage, Pointe des Chevaliers** [8] **and Pointe des Salis** [9] (2 km). Peregrine, raven, blue rock thrush, Alpine and pallid swifts breed on the rocky cliffs. In spring, scan the sea, as **hundreds of** Scopoli's **and** yelkouan shearwaters **roost** from late afternoon to dusk.

of protected saltmarshes makes it an obvious landing point for spring migrants flying in from the Mediterranean. Start in the early morning along the fences of Hyères airport. Follow 'aéroport', then 'Z.A. du Palyvestre' and finally 'Centre Technique Municipal' to park at the end of the road, where **a path leads along a canal which borders the runways** [1]. This is France's most reliable site for little and spotted crakes from mid-March to late April. Later in spring, little bittern, purple heron and great reed warbler frequent the narrow reedbed along the canal. **You can also reach the canal from the D98 east of Hyères** (follow Macany and park at the end of 'Chemin Traversier de Macany') [2]. **Check the meadows north of the canal** for purple heron, great white egret, marsh harrier, roller and red-rumped swallow. The

198 Porquerolles island

≥ 1 DAY; MAP 281   If you had to pick just one site to visit on the French eastern Mediterranean coast in early May, it would probably be the Île de Porquerolles. Spectacular passerine falls in early May provide a unique chance of finding exciting rarities within crowds of common (and less-so) species. **Allocate at least one full day to the island during the main migration period.**

The main access is by boat from La Tour Fondue (at the eastern end of Giens on the Hyères peninsula, 20 min, several return journeys per day, booking only for large groups, €20 return trip). **There are no cars on the island.** Walk from the pier [1] among pine woods, scrubland, meadows, vineyards, olive groves and along rocky cliffs, wandering where you will. **Migrants are usually concentrated in the orchards and along woodland edges** around the Carrefour des Oliviers [2,3], the Bastide [4], the Courtade [5], Notre-Dame and Jonquière [6] plains. The most common migrants include Balearic woodchat shrike (subspecies *badius*), great spotted cuckoo, hoopoe, greater short-toed lark, tree, tawny and red-throated pipits, various subspecies of yellow wagtail (including the rare *feldegg*), spotted and pied flycatchers (possibly the odd collared flycatcher), wood, western Bonelli's, garden, subalpine and Moltoni's warblers, black-eared and northern wheatears and hirundines. **Check the treatment plant** [7] for red-rumped swallow, spotted and little crake and waders. **Local breeding birds** include the introduced Reeves's pheasant (mainly around Bréganҫonnet [8] and in olive groves between the Bastide and the treatment plant), peregrine, Mediterranean shag (subspecies *desmarestii*), Scopoli's and yelkouan shearwaters (spot them from the lighthouse [9] and the Grand Langoustier headland [10]), scops owl, pallid swift, Dartford and Sardinian warblers. Eleonora's falcon is regular in summer and early autumn.

199 Cuers/Pierrefeu-du-Var airfield

0.5 DAY; NO DETAILED MAP   Pierrefeu-du-Var airfield (30 km, 30 min north of Toulon or Hyères on the A57 towards Fréjus, exit 10) is worth an hour looking for breeding little bustard, little owl, roller, tawny pipit, hoopoe and red-rumped swallow (the latter breeds near the aeroclub along the chemin de la Sermette). Red-footed falcon and lesser grey shrike sometimes stay here for a few hours or days on their spring migration (mid- to late May). **Birds are easiest to find around the northeastern end of the runway** [1].

The Verdon and Durance valleys
(sites 200–203, map 282)

HIGHLIGHTS
» The scenic Verdon gorges and their vulture colonies
» Typical lavender fields with the Alps in the background to search for Mediterranean breeding birds and red-footed falcon on passage – and maybe even a black-headed bunting in late May
» Several viewpoints for waterbirds and crakes on your way between the Mediterranean coast and the Alps

| J | F | M | A | M | J | J | A | S | O | N | D |

KEY SPECIES
YEAR-ROUND Eurasian bittern, **black and griffon vultures**, golden eagle, little owl, crag martin, grey wagtail, **dipper**, moustached and Cetti's warblers, southern grey shrike, red-billed chough, **Italian sparrow**

BREEDING red-crested pochard, night-heron, purple heron, little bittern, **Egyptian vulture**, short-toed eagle, Montagu's harrier, hobby, quail, **little bustard**, stone-curlew, scops owl, nightjar, Alpine swift, roller, western Bonelli's, **spectacled**, subalpine, orphean and great reed warblers, red-backed shrike, **rock and ortolan buntings**

MIGRATION ferruginous duck, **red-footed falcon**, spotted and little crakes, **black-headed bunting (scarce)**

WINTER common pochard, tufted duck, **penduline tit**

VISIT DURATION 2–3 days

PLAN YOUR VISIT
This area marks the southern foothills of the Alps. You will be birding in wide plateaus with the Alpine ridges as a background or in scenic canyons. In May, start with an early morning on the Valensole plateau, then spend the warm hours watching vultures in the Verdon gorges before heading back to Valensole at dusk. For those on their way to or from the Alps along the Durance, even short stops in Vinon-sur-Verdon and Le-Puy-Sainte-Réparade should be rewarded with a good range of wetland and grassland species.

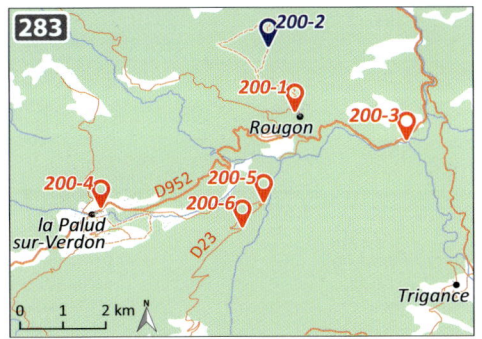

200 The Verdon gorges

0.5–1 DAY; MAP 283 The Verdon gorges host several vulture **colonies and regionally rare mountain breeders.** The main colony faces Rougon (85 km, 1 h 40 north of Fréjus or 80 km, 1 h 40 east of Manosque; signposted from La Palud-sur-Verdon). Park at [1] and cross the village to scan the cliffs. **The birds are regularly fed at the top of the enclosure that is easily visible east of the village, causing spectacular roosts of** griffon vultures joined by a few black vultures (most of them from the reintroduction programme), the odd Egyptian vulture and ravens. **Walk on the track running north towards Suech plateau** [2] looking for short-toed and golden eagles, Alpine swift, crag martin, western Bonelli's, subalpine and orphean warblers, red-billed chough and rock and ortolan buntings. Down the river (6 km along the D952), dipper and grey wagtail breed near the camp site [3]. **A walk in the surroundings of La Palud-sur-Verdon** [4] should yield Italian sparrow, woodlark, western Bonelli's warbler and scops owl at dusk. **You can choose a belvedere along the winding D23 for a midday break** [5,6]. Vultures, red-billed choughs, Alpine swifts and crag martins fly over at close range above dramatic scenery.

201 The Valensole plateau

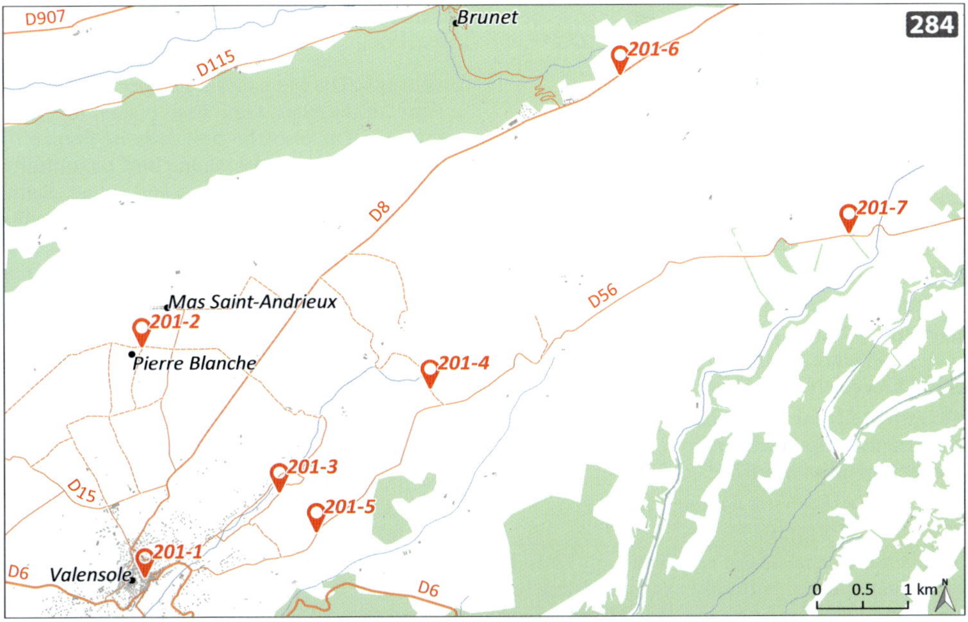

0.5 DAY; MAP 284 **The Valensole plateau has attracted attention for having produced records of singing** black-headed bunting **annually over the last decade (scarce, late May).** It has much more to offer, however. Stop-and-go on roads and tracks among endless lavender fields, cereal crops, almond orchards, scrubland and dry grassland. **A circuit east from Valensole** [1] (20 km east of Manosque) should target roller and spectacled warbler [2,3], stone-curlew and little bustard [4], Montagu's harrier, southern grey shrike and ortolan bunting [5]. Additional common breeding species include woodlark, quail, red-legged partridge, red-backed shrike, scops and little owls, nightjar and raven. **The fields 6 km northeast past Brunet road** [6] are a regular stopover site for red-footed falcon in May and host breeding hobby, little bustard,

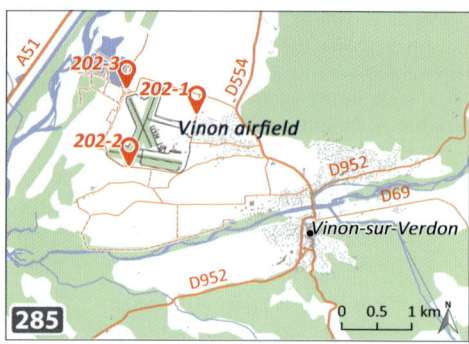

stone-curlew and little owl. Densities of ortolan buntings are impressive in this area, for instance along the track south from [7], where spectacled warbler also breeds.

202 Vinon-sur-Verdon airfield

1 HOUR; MAP 285   Vinon-sur-Verdon airfield has breeding little bustard, stone-curlew, tawny pipit, calandra lark (rare) and migrating red-footed falcon in mid-May (15 km south of Manosque on the eastern bank of the Durance). **The birds are easy to find from the roads bordering the airfield** [1,2]. **The gravel pits at the northern end of the runway** [3] host breeding little bittern and red-crested pochard. Ferruginous duck is a regular winter visitor among common pochard and tufted duck.

203 Le Puy-Sainte-Réparade

2 HOURS; MAP 286   **Several gravel pits** [1,2] **are spread along a 3 km stretch of the River Durance north of Le-Puy-Sainte-Réparade** (20 km north of Aix-en-Provence; exit 15 towards Pertuis from the A51, follow 'Les Gravières' from the roundabout after the toll and drive 2.5 km west along the river). Search for spotted and little crakes in March–April, night-heron in spring, Eurasian bittern and penduline tit in winter. Breeders include great reed, moustached (rare) and Cetti's warbler, little bittern, great crested and little grebes. Migrating great white and cattle egrets, purple heron, common and green sandpipers, common snipe and common teal are often recorded on both spring and autumn passage.

The Mercantour (site 204, map 287)

HIGHLIGHTS
» Endless wilderness with breeding rock partridge and ptarmigan
» Access to high-altitude species within reach of the Côte d'Azur

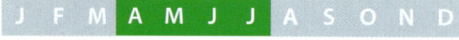

KEY SPECIES
YEAR-ROUND griffon vulture, golden eagle, **rock partridge**, **black grouse**, **ptarmigan**, pygmy owl, nutcracker, coal tit, red-billed and Alpine choughs, crossbill, citril finch, **snowfinch**

BREEDING Alpine accentor, ring ouzel, **common rock thrush**, lesser whitethroat, willow tit, **wallcreeper**, **ortolan bunting**

VISIT DURATION   2 days to 1 week

PLAN YOUR VISIT
This is the southernmost access to high-altitude mountain species, certainly not the easiest but among the most exciting for its wild and scenic landscapes. The Mercantour is especially famous as one of France's very best places for rock partridge. It is also the best area in France to look for wolves. All the Alpine bird community breeds here along wild slopes and ridges. Settle either in the south around Saint-Martin-Vésubie or in the north in Saint-Etienne-de-Tinée and be prepared for long hikes. As elsewhere in the Alps, most sites will not be accessible before the snow has melt in May. Start early morning to avoid other hikers and go straight to higher altitudes, then walk slowly down.

204 Mercantour National Park

> 2 DAYS; MAPS 288–290   The southeasternmost limit of the Alps is protected by the vast Mercantour National Park, which provides access to Alpine species only 65 km, 1 h 30 north of Nice. The main access is through Saint-Martin-Vésubie (follow 'Digne/Grenoble' and turn right 4 km past Saint-Martin-du-Var, following

signposts). Bypass Le Boréon (9 km) and turn left across the river to reach the last car park in Le Boréon (called Salèse) [1]. From there, **walk on the GR52 trail to the Col de Salèse** [2], along which you will likely encounter nutcracker, crossbill, citril finch, willow and coal tits. **Take the trail left towards the Col de la Vallette des Adus** (2356 m). Be there early in the morning for chances of black grouse. Later in the day, look for golden eagle, griffon vulture and Alpine chough. **Head down to the Adus refuge [3] and eventually back to the car park (8 km, 4 h).**

You can also hike from Millefonts (20 km, 40 min west of Saint-Martin-Vésubie) [4] to the Col de Veillos [5], Lac Gros [6], Mont Pépoiri [7] and Mont Pétoumier [8] (1 day, max. altitude 2670 m). **Your efforts will be rewarded by a wide range of Alpine breeding species, starring** rock ptarmigan **on the highest parts**, with lesser whitethroat, ring ouzel and citril finch at the tree line, common rock thrush and Alpine accentor on grassy slopes. You can reach the same areas from the car park at Adréchas above La Colmiane [9] (12 km west of Saint-Martin), but the ascent is steeper.

Rock partridge **is one of Mercantour's highlights. It breeds on rocky slopes all across the national park; one of the most reliable places to see it is the Vallée des Merveilles**, a strenuous seven-and-a-half-hour (return) hike from Gordolasque [10] (head to Belvédère, 12 km south of Saint-Martin-Vésubie, then 13 km north). Partridges breed with wallcreeper and other high-altitude species on the rocky slopes around the Refuge des Merveilles [11] (keep strictly to the trails).

The north of the Mercantour is easier to access from Gap (130 km, 2 h 30 via Barcelonnette or 60 km, 1 h 30 northwest of Saint-Martin-Vésubie). **There are plenty of areas to explore,**

for instance the trail to the Col des Fourches [12] (20 km north of Saint-Etienne-de-Tinée on the D64). Look for common rock thrush, ortolan bunting, Alpine accentor, griffon vulture, golden eagle and red-billed chough. Higher altitudes will yield snowfinch and wallcreeper. There are black grouse and quail at lower altitudes along the roadside, but be there at dawn as the road quickly becomes crowded.

8 km further west, **the Col de la Bonette and the nearby Cime de la Bonette** [13] **are a reliable spot for** snowfinch **in spring but it becomes crowded in summer.** The species then becomes more likely around the Col de la Moutière [14] (a wild 11 km uphill walk from Saint-Etienne-de-Tinée through Saint-Dalmas, or 7 km along a track starting 3 km north of the Col de la Bonette). Closer to Saint-Etienne-de-Tinée, **the GR trail leaving from Pratois restaurant** [15] crosses wide expanses of abandoned pastures where rock partridge is common within a diverse mixture of open-land and woodland species including black grouse, common rock thrush, Eurasian treecreeper, citril finch and ortolan bunting. The area serves as a hunting ground for the local golden eagle. Pygmy owls **breed in larch forests above Saint-Etienne-de-Tinée**, but there is no specific location for them.

Mont Ventoux and the Baronnies (sites 205–206, map 291)

HIGHLIGHTS
» Access to Alpine passerines in winter
» One of France's largest vulture colonies

KEY SPECIES
YEAR-ROUND griffon and black vultures, black woodpecker, woodlark, coal and crested tits, raven, **citril finch**, crossbill
BREEDING Egyptian vulture, ring ouzel, **common rock thrush**, wheatear
WINTER Alpine accentor, snowfinch
VISIT DURATION 1 day
PLAN YOUR VISIT
Mont Ventoux is the place to visit for Alpine birds in winter, when snow makes it difficult to get up into the Alps. It can easily be arranged as a day trip from the Camargue or the Côte d'Azur.

If you have more time, visit Rémuzat, one of France's most famous vulture colonies.

205 Mont Ventoux
1 DAY; MAP 292   Mont Ventoux is an isolated limestone pyramid (altitude 1900 m) between the Rhône and Durance valleys. It has long been a textbook case to illustrate the vegetation altitudinal gradient and associated bird communities. Drive 20 km north from Carpentras towards Malaucène, then 16 km to Mont Serein. **Park near the Chalet Liotard** [1] **and walk into the woodlands to the north** to look for breeding black woodpecker, raven, woodlark, citril finch, crossbill, ring ouzel, coal and crested tits. Head to the summit, 6 km further east, and **park at the Tempêtes pass** [2]. Common rock thrush and wheatear breed along the path heading to the Tête de la Grave [3], and Alpine accentor and snowfinch often

winter in this area. The same species occur 6 km southeast around Chalet Reynard [4], which gives an access to bare ridges or to woodlands where citril finch breeds.

206 Rémuzat

2 HOURS; MAP 293

Rémuzat is the place to visit for the full complement of vultures. Exit the A51 in Sisteron and take the E712 northwards to Serres, where Rémuzat is signposted along the D994 (80 km, 1 h from Sisteron). **The best viewpoint for vultures is the cliff ridge above Rémuzat.** Park at [1] (4 km from Rémuzat, follow Orange and turn right in Saint-May for a steep 4 km climb) and walk a few minutes to the top of the cliff near a large iron cross [2]. Griffon vultures often soar at a close range and can be spotted resting on the other side of the valley. Egyptian and black vultures also occur here and can reasonably be expected with a bit of patience.

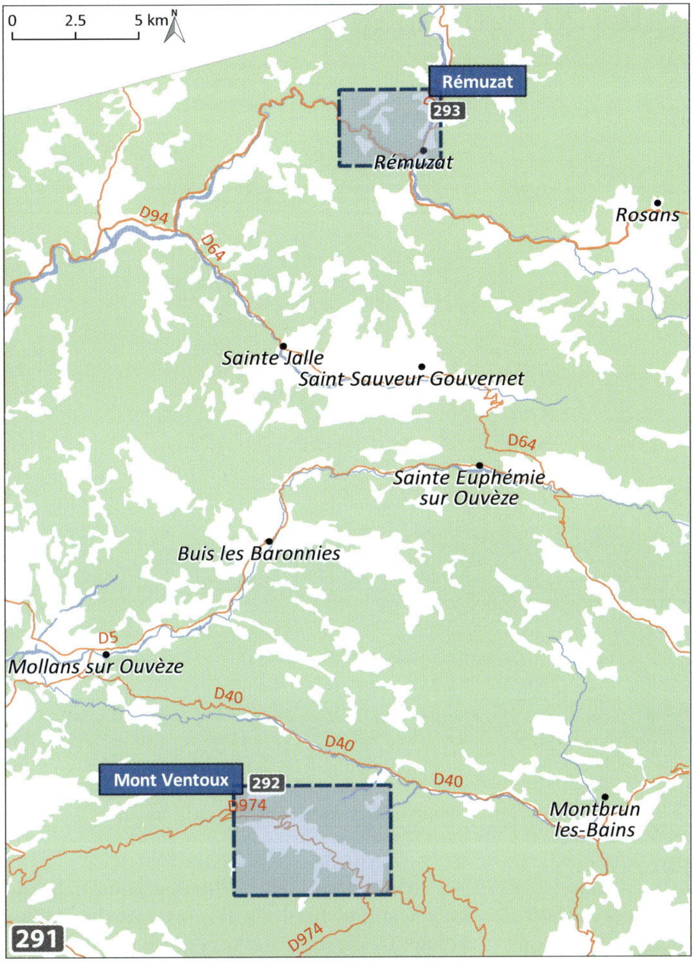

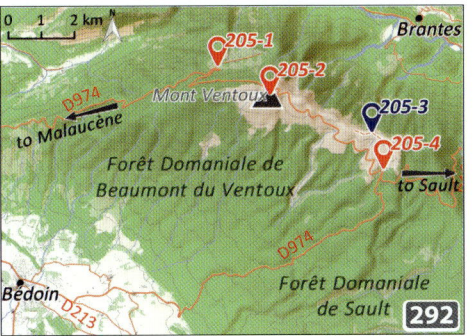

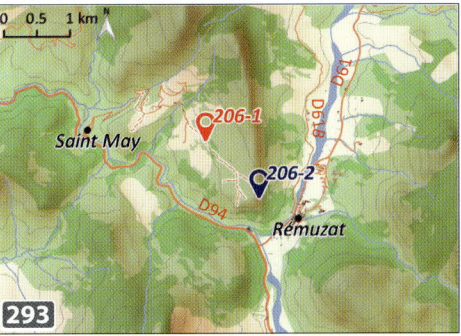

The Côte d'Azur (sites 207–209, map 294)

HIGHLIGHTS
» A few wetlands scattered along a highly urbanised coast concentrate waders and wildfowl
» A viewpoint for raptor migration at the Fort de la Revère

KEY SPECIES
YEAR-ROUND wildfowl, grebes, **Indian silverbill** (feral)

BREEDING squacco heron, short-toed eagle, hobby, bee-eater, pallid swift

MIGRATION black stork, **raptors** including osprey, honey-buzzard, booted eagle, **Eleonora's falcon**, **spotted and little crakes**, waders including marsh sandpiper (rare), Caspian tern (rare), woodpigeon, stock dove, **passerines** including hirundines and finches

VISIT DURATION 1–2 days

PLAN YOUR VISIT
Most of the seashore along the Côte d'Azur is heavily urbanised, but a few sites deserve a visit. The Fort de la Revère can be particularly rewarding during autumn migration.

207 Fréjus – Villepey lagoon
1–2 HOURS; MAP 295 **Villepey lagoon is the only wetland between Hyères and the Italian border** (5 km southwest of Fréjus towards Saint-Aygulf, stop in one of the car parks behind the beach [1,2]). **Walking trails around the ponds lead to hides and viewpoints**. Spring migrants include wildfowl, grebes, herons (including squacco), waders (including marsh sandpiper,

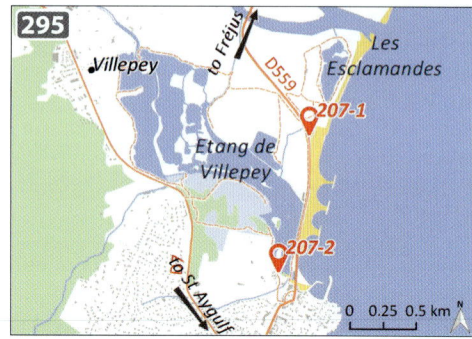

rare), spotted and little crakes (rare) in March and April, and osprey.

208 The Var estuary

1 HOUR; MAP 296   The Var estuary, just behind Nice airport, is France's main site for feral Indian silverbill. **Park at the Cap 3000 shopping mall** [1] on the west bank of the Var and walk towards the estuary. Besides silverbills, there are small numbers of waders, gulls and terns (including Caspian) at both migrations.

209 Fort de la Revère

≥ 0.5 DAY; NO DETAILED MAP   The Fort de la Revère [1] is a **major autumn migration site for raptors from mid-August to late October** (15 km, 30 min east of Nice and 13 km west of Monaco, head to Èze and 'Col d'Èze', park near the

citadel). The most abundant migrants are short-toed eagle, honey-buzzard, bee-eater, wood-pigeon, stock dove, swallows and finches, with lower numbers of booted eagle, hobby, osprey, Eleonora's falcon, black stork and pallid swift.

Corsica (sites 210–219, map 297)

HIGHLIGHTS

» Numerous endemic species and subspecies including the legendary Corsican nuthatch
» Migration hotspots, with spectacular passages and rarities to search for in wetlands
» Wide expanses of wilderness in the central mountains, with breeding lammergeier

KEY SPECIES

YEAR-ROUND waterbirds, yelkouan and Scopoli's shearwaters, Mediterranean shag, greater flamingo, **lammergeier**, golden eagle, **Corsican goshawk**, peregrine, crag martin, blue rock thrush, dipper, **Marmora's**, Dartford and Sardinian warblers, **Corsican nuthatch**, Eurasian treecreeper, wallcreeper, **spotless starling**, Alpine chough, hooded crow, **Spanish and Italian sparrows**, **Corsican finch**

BREEDING red-crested pochard, purple and squacco herons, **little bittern**, osprey, red kite, **Audouin's gull**, bee-eater, common, pallid and Alpine swifts, **red-rumped swallow**, **Moltoni's warbler**, reed and great reed warblers, **Tyrrhenian spotted flycatcher**, red-backed and **Balearic woodchat shrikes**

MIGRATION booted eagle, all harriers, lesser kestrel, **Eleonora's and red-footed falcons**, **little and spotted crakes**, waders, great snipe (rare), yellow wagtail (including feldegg, rare),

red-throated pipit, bluethroat, sedge warbler, **other warblers and flycatchers**, whinchat, other passerines

WINTER wildfowl

VISIT DURATION   > 1 week

PLAN YOUR VISIT

Corsica would deserve a book in itself, so we only cover the most famous sites, acknowledging that many other good places exist, both for breeding species and for migrants. Spring migration should lead you to either the southern or the northern extremity of the island, where birds can be passing through in huge numbers. Check wetlands carefully – you never know what can turn up, and the list of rarities recorded in the past is long. For endemic birds, visit old pine forests in the scenic central mountains.

210 Liamone estuary

2–5 HOURS; MAP 298   Liamone estuary lies 30 km, 50 min north of Ajaccio. **Park near the bridge** [1], **on the southern bank** [2] **or on the northern bank before San Petru** [3]. The flooded meadows bordering the river and the beach attract a wide range of migrating waterbirds including cattle egret, purple heron, wildfowl, waders, water rail, osprey, yellow wagtail (chances of feldegg among flocks of cinereocapilla, flava and thunbergi). The site has a good list of national and local rarities, among which the most regular are great snipe, little and spotted crakes. Surrounding fields

... and banks host breeding red-backed and Balearic woodchat shrikes and bee-eater.

211 Capu Rossu
0.5 DAY; MAP 299 Capu Rossu peninsula are **scrub-covered rocky cliffs where** Marmora's, Dartford, Sardinian **and** Moltoni's warblers **breed** (75 km, 40 min north of Ajaccio; turn towards Arone in Piana and park at the end of the road [1]). Warblers occur along the walking path to Turghiu tower [2] (3 hours return) along with woodlark, osprey, pallid and Alpine swifts.

212 Aïtone forest and Col de Vergio
0.5–1 DAY; MAP 300 **Aïtone pine forest is one of the best locations for** Corsican nuthatch. Drive from Porto to Evisa on the D84 (20 km) and park after 5 km at the start of the 'Sentier de la sittelle' [1]. You should easily connect with the nuthatch and other forest species such as Corsican goshawk, crossbill, Corsican finch, Tyrrhenian spotted flycatcher, Eurasian treecreeper and coal tit. **The same species can be found along the trail from Aïtone to the Col de Vergio** [2] (park 5 km further north).

214 Haut-Asco

0.5–1 DAY; MAP 302 The deep and narrow granite gorge of the Asco valley, north of Monte Cinto, is surrounded by high mountains (over 2000 m). The landscapes consist of wild arid scrubland, juniper and pine forests where most Corsican specialities breed. Reach Asco by driving 40 km, 1 h north from Corte or 70 km, 1 h 30 southwest from Bastia through Ponte-Leccia. **Park in Haut-Asco [1] and hike along the GR20 trail to the Carrozzu refuge [2]. You can also hike from the Maison du Mouflon [3] near Giunte towards the 'Bergerie de la Tassineta' [4].** The trails continue higher towards Cima di a Statoghia. On your way, expect red kite, lammergeier, golden eagle, peregrine, crag martin, dipper, Tyrrhenian spotted flycatcher, Corsican nuthatch, coal tit, red crossbill, Corsican finch, Alpine chough (near Haut Asco), wallcreeper (in the Cirque de la Solitude south of Haut Asco [5]) and raven.

215 Biguglia

0.5 DAY; MAP 303 The Étang de Biguglia is Corsica's largest protected lagoon and consequently a major migration stopover site on both spring and autumn migration. It consists of two large ponds separated by the San Damianu peninsula. **The reserve headquarters are located at the northern end of the northern pond** (7 km, 15 min south of Bastia centre and 6 km, 10 min north of Biguglia, follow 'Lido de la Maradana' and cross the railway at Furiani roundabout, then drive south to the last car park in Puntale [1]). Common species include

213 Corte and the Restonica valley

≥ 1 DAY; MAP 301 The Restonica valley lies in the central mountain range, 70 km from Bastia and 80 km from Ajaccio. **The area hosts most of the Corsican Alpine specialities. From Corte, take the D623 into the valley to the last car park at the Bergerie de Grottelle** [1] (16 km, 30 min southwest of Corte), where a trail leads to Melo lake [2]. Expect Alpine chough, blue rock thrush, golden eagle, Alpine swift, crag martin, raven, hooded crow, Corsican finch and Tyrrhenian spotted flycatcher. **Closer to Corte, you can walk from the Tuani camp site [3] to the Alzo plateau** [4], where Corsican nuthatch and crossbill breed. Dipper is easy to find along the Restonica river.

greater flamingo, red-crested pochard, black-necked, little and great-creasted grebes, waders and gulls. You can also spot from the eastern bank by following 'Lido de la Marana' from the Furiani roundabout. Stop at the first car park [2] and walk to the small estuary where waders and gulls forage. From the second car park [3], the **Tombulu Biancu trail runs for 1.7 km beside the lagoon. To reach the southern lagoon**, take the D107 from Borgo to the south of Bastia airport and turn left alongside the runway before the canal and La Canonica church. This is a good spot for migrating red-footed falcon in May [4].

216 Macinaggio

0.5–1 DAY; MAP 304   The location of Macinaggio wetlands, near Corsica's northernmost point, makes them a leading hotspot for spring migration. **Park at Macinaggio harbour [1] and walk north for 1 km on the 'Sentier des Douaniers' [2] to Tamarone beach** [3]. Bird numbers depend on water levels. All herons and sandpipers are likely to forage here. Breeding or migrating passerines include bluethroat, reed, sedge and great reed warbler, yellow wagtail and pipits (including migrating red-throated pipit around horse pastures). Great snipe, feldegg yellow wagtail, little and spotted crakes are seen near-annually here. **Further north, on a 2 km stretch of coast,** Finocchiarola nature reserve [4] and Santa Maria tower [5] are good places for breeding Marmora's, Dartford and Moltoni's warblers. Seawatching should yield Mediterranean shag, Audouin's gull, yelkouan and Scopoli's shearwaters.

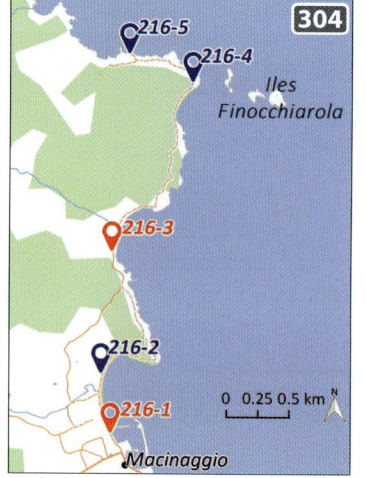

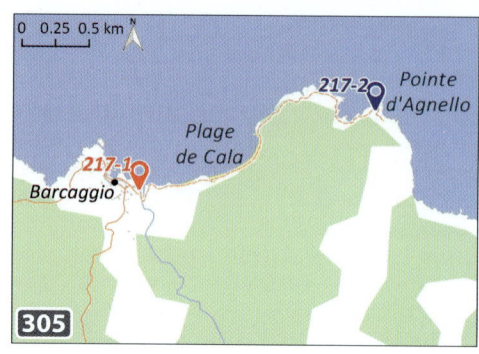

217 Barcaggio

0.5–1 DAY; MAP 305   **Barcaggio, at the island's northern tip, is a ringing station and Corsica's hottest migration spot (March to May)**. Drive for 8 km from Botticella on the D253 to Barcaggio and turn right to the car park [1]. **Check waders on the river estuary.** Besides sandpipers and shanks, great snipe (scarce), squacco heron, little bittern, little and spotted crakes can stop over on the lagoon and in the surrounding reedbeds. **Then head eastwards along the 'Sentier Douanier' looking for stopping-over passerines in gardens, scrubs and open areas.** These places can be crowded with warblers (including garden, willow, wood, melodious and icterine), flycatchers (including collared in April), whinchats, turtle doves, pipits (meadow, tree, red-throated and tawny), yellow wagtails and shrikes. **Huge numbers of migrating** raptors (honey-buzzard from early May, booted eagle, falcons, harriers, black kite, sparrowhawk) and other diurnal migrants (red-rumped swallow, Alpine, common and pallid swifts, bee-eater) **fly over the area.** A surprise can always turn up in this area. Pallid harrier, lesser kestrel, tens of red-footed falcon and Eleonora's falcon pass through yearly. **A seawatching session** should yield the same species as in Macinaggio (site 216). There are breeding Marmora's warbler and blue rock thrush along the trail to Pointe d'Agnello [2].

218 Col de Bavella

1 HOUR – 0.5 DAY; MAP 306   **For those visiting southern Corsica, the Col de Bavella [1] should yield most mountain specialities** (48 km, 1 h northwest of Porto-Vecchio, 100 km, 2 h east of Ajaccio and 60 km, 1 h 30 north of Figari, all through Zonza). **Several trails lead into the surrounding forests and to the peaks.** Golden eagle and lammergeier breed around Aiguilles

de Bavella [2] and Punta Aracale [3] (both around 1600 m). Corsican nuthatch breeds in old black pine woods with Corsican finch, crossbill, Eurasian treecreeper, coal tit and Tyrrhenian spotted flycatcher.

219 Bonifacio

0.5–1 DAY; MAP 307   Bonifacio stands on rocky cliffs at Corsica's southern tip, facing Sardinia. **Walk to the citadel from town-centre car parks** [1]. A few Spanish sparrows breed among Italian sparrows, but beware of the presence of numerous hybrids. Scan the sea for Mediterranean shag, Scopoli's and yelkouan shearwaters. **Then drive for 4 km on the D260 east of the town or walk along the coastal trail to Capo Pertusato semaphore** [2]. **Check the sea from the lighthouse** [3] for Mediterranean shag, Scopoli's shearwater, Audouin's gull (uncommon), pallid and Alpine swifts. Cliff and scrub breeders include blue rock thrush, Marmora's, Dartford, Moltoni's and Sardinian warblers, spotless starling and Corsican finch.

REGION 11
JURA AND THE ALPS

HIGHLIGHTS
» Charismatic and elusive species of mid-altitude old-growth forests: capercaillie, black grouse and three-toed woodpecker
» Snowfinch, rock partridge, Alpine accentor and wallcreeper reward strenuous, yet unforgettable hikes amid some of Europe's highest mountains
» A habitat gradient covering lowland Mediterranean habitats, spruce and fir forests, Alpine meadows and rocky slopes

» Migration viewpoints and stopover sites for waders, raptors and passerines around Lake Geneva

REVIEWERS Luc Barbaro, Marc Crouzier, Thierry Joubert, Vincent Palomares

USEFUL WEBSITES
http://haute-savoie.lpo.fr
www.faune-savoie.org
www.faune-isere.org

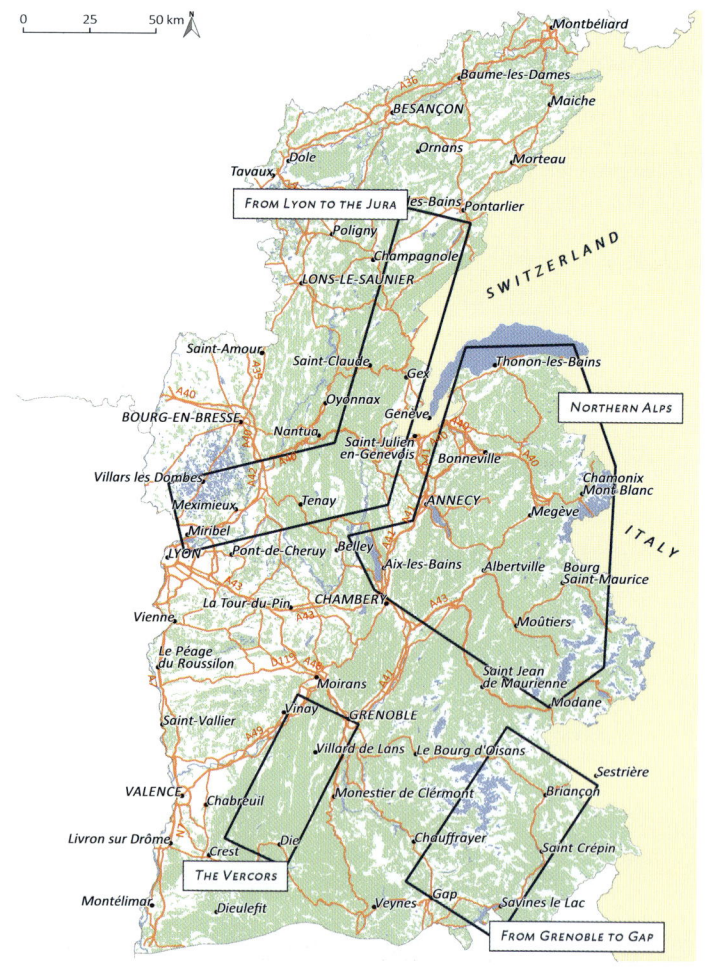

The Alps are a first-choice playground not only for their unique avifauna, spreading from lowland plains to barren summits, but also for the countless hiking opportunities in scenic landscapes.

The shy black grouse frequents high-altitude willow belts, where it is most easily found while displaying in May and June, for instance at the Col de la Colombière (site 228).

From Lyon to the Jura (sites 220–225, map 308)

HIGHLIGHTS

» Wildfowl and wader migration in the Dombes
» Dense old-growth forests host France's capercaillie and hazel grouse strongholds
» Tengmalm's and pygmy owl easy to find in the Risoux forest
» The most accessible three-toed woodpecker breeding sites in the country

| J | F | M | A | M | J | J | A | S | O | N | D |

KEY SPECIES

YEAR-ROUND capercaillie, hazel grouse, golden eagle, woodcock, **Tengmalm's, pygmy** and eagle owls, middle spotted and **three-toed woodpeckers**, dipper, nutcracker, citril finch, redpoll, cirl bunting

BREEDING night-heron, **squacco heron**, little bittern, **lesser spotted eagle**, black, whiskered and white-winged (rare) terns, Alpine swift, bee-eater, wryneck, bluethroat, ring ouzel, lesser whitethroat, **marsh and great reed warblers**, great grey and red-backed shrikes

MIGRATION little and spotted crakes, osprey, red-footed falcon, waders

WINTER ferruginous duck, scaup, Eurasian bittern, gulls, Alpine accentor, **penduline tit, wallcreeper**

VISIT DURATION > 4 days for the proposed itinerary. Minimum 1 full day from dawn to full night for the Jura.

PLAN YOUR VISIT

The sites are organised as a southwest–northeast transect from Lyon to the Jura through the Dombes wetland. **The Dombes is at its best from autumn to spring migration, but is still worth a day trip in the breeding season for** reed warblers **and** herons. Heading east in May and June will lead you to the wild mid-altitude forests of the Jura. You can spend a morning or evening in the Bugey for mountain forest species if you are short of time. **However, the core area for all the regional specialities is further north in the Risoux forest, on the Swiss border.** There, your targets will be capercaillie, **mountain** owls **and** three-toed woodpecker, **and it will be a real challenge to find them.** Apart from the owls (February and March), they are easier from mid-May to mid-July. **Be extremely cautious when birding in this area: most target species such as** capercaillie **are protected and any kind of disturbance would be highly inappropriate.** If you have only one birding day left, **try the Risoux early in the morning and shift to Frasne by midday before heading back to the forests at dusk.** In two days, you can get deeper into the forest or try to visit other areas in the region – some of them are not described here because of access restrictions. From the onset of autumn migration in late July to November, spend a day in the Défilé de l'Écluse.

220 Lyon area

0.5 DAY; MAP 309 For those staying in Lyon, the 2200 ha Miribel-Jonage resort deserves a visit at any time of the year (10 km east of Lyon centre along the Rhône, leave the A42 at exit 5 and follow signposts). **The site includes several ponds, gravel pits and islets, some of which are**

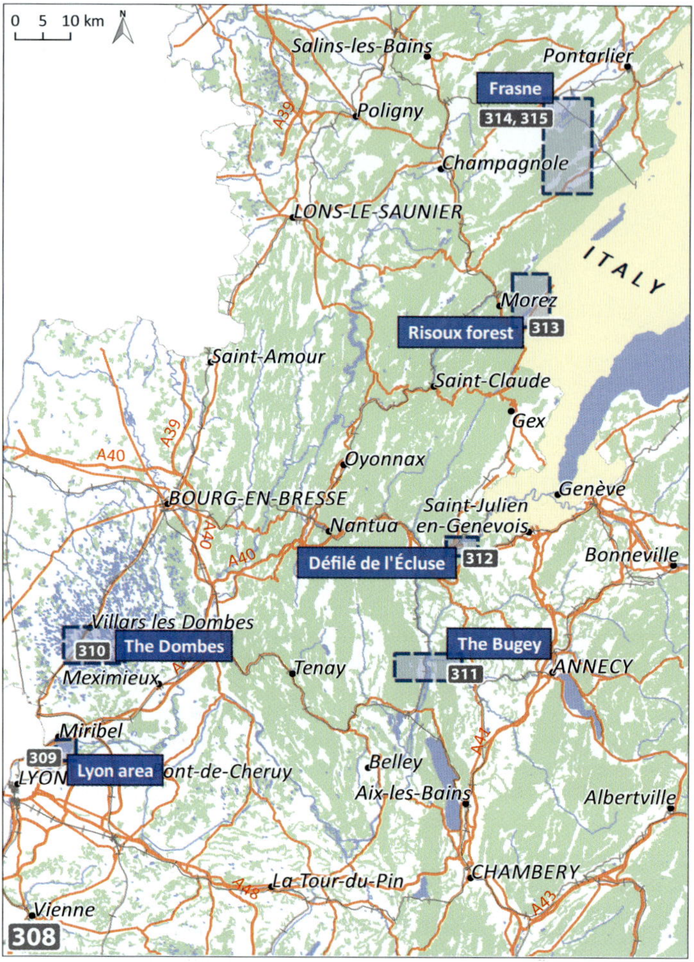

[3] and the Lac des Pêcheurs [4]. In spring/summer, you can expect close views of breeding night-heron, little bittern, gadwall, red-crested pochard, goosander, hobby, bee-eater, black kite, common tern and great reed warbler.

This wetland is also a stopover area during migration and in winter. Expect little and great crested grebes, great white egret, large numbers of dabbling and diving ducks (single ferruginous ducks and scaup are seen every year along with hundreds of common and red-crested pochards, tufted duck, gadwall, teal, shoveler, a few goldeneye). Waders show up mostly in autumn/winter, while spring is better for terns (whiskered, black and white-winged in May) and gulls occur year-round (wintering black-headed and common, migrating little and resident yellow-legged). **Roosting gulls, ducks and grebes can also be found in winter at Le Grand Large** [5] (drive to La Petite Camargue, and follow the road through the Chemin de Cornalon and the Chemin de Cheyssin to reach the banks of the Rhône).

protected. From the car park [1], **walk the trail beside the main pond** (Lac des Eaux Bleues) that leads to Les Grands Vernes [2] and the Île des Mouettes. Also take a look at the Lac du Drapeau

221 The Dombes wetlands

0.5–1 DAY; MAP 310 **The Dombes is a network of ponds embedded in a mosaic of arable farmland, meadows and woods** 40 km northeast of Lyon (exit 2.1 to Bourg-en-Bresse on the A66, then 20 km, 20 min to Villars-les-Dombes). In some respects, it is the eastern counterpart of the Brenne and Sologne wetlands (Region 5) for wader and wildfowl migration, breeding reedbed species and a taste of warm-climate birds. **Most of the ponds and marshes are privately owned, and tracks are usually access-restricted (enforced); do your birding from the roadsides.**

Ponds are drained once every three years, uncovering large mudflats suitable for crakes and waders; black-winged stilt, lapwing and little ringed plover breed, and all the shanks, stints and sandpipers are possible as stopovers, including several records of Nearctic rarities. Check hedgerows for red-backed shrike (from May, replaced by a few great grey shrike in winter), little owl, cirl (year-round) and ortolan (migrant) buntings. **Large numbers of** red-crested pochard **breed, and the diversity of** wildfowl **is great year-round. Two pairs of** whooper swans **have bred for several years** (the only ones in France, sites confidential, but the birds can be seen in flight anywhere). Raptors of interest include passing red-footed falcon (May), osprey, short-toed eagle, Montagu's harrier. The whole reedbed community is present, with night-heron, purple and squacco herons, both bitterns, great reed warbler (marsh warbler only as a spring stopover), penduline tit (winter). You'll have a slim chance of middle spotted woodpecker in the woods (easier in Seillon forest, 30 km north in the vicinity of Bourg-en-Bresse). Orioles are common (May).

Start in a small hide at the Grand Turlet [1] (turn right just after the large D1083 signpost at the western exit from Villars-les-Dombes). Ferruginous duck may occur in winter, and it is a good place for little bittern, squacco heron and all reed warblers. Back on the D1083, 100 m south, the Petit Turlet (car park at [2]) is not as good in spring, but as a hunting-free reserve it hosts large numbers of wildfowl in winter, possibly including scaup.

Head back to Villars and drive east to Versailleux (7 km, 6 min). Park 300 m before the village near the Étang Chapelier. A trail [3] leads to two hides that offer close-range views of birds year-round.

Further west, past Birieux (7 km, 10 min), two ponds [4,5] **sometimes attract** white-tailed eagle **and other large** raptors, plus smew, black stork, crane **and** penduline tit **in the reedbeds. They are just as good for all local breeders. Cross the D1083 to the west and stop at the eastern end of the Grand Glareins** [6] **for the same species** plus casual spoonbill, ferruginous duck, waders, gulls and terns. From there, either head back to Villars-les-Dombes or try random stops beside the many other ponds in the vicinity.

222 The Bugey

0.5–1 DAY; MAP 311   The Bugey allows quick access to some of the Jura's mountain specialities from Lyon, and is worthwhile for those who cannot allocate enough time to visit the Risoux forest (site 224) or early in season when access is restricted by snow. It also has the advantage of being close to the Dombes (site 221) and to a busy migration corridor (site 223).

Start at dawn on the Grand-Colombier, 60 km, 1 h east of Ambérieu-en-Bugey on the D1504 to Chambéry. Park at the pass [1] and walk 300 m in meadows towards a cross at the top of the ridge [2]. Migrating raptors will fly over as soon as the sun rises; most of them will be black kites and honey-buzzards in August, replaced by buzzards, sparrowhawks and other species later in the autumn. Short-toed eagle breeds on the wooded slopes below the pass and non-local golden eagles may be seen soaring around on sunny days. The surrounding forest hosts nutcracker, crossbill and bullfinch.

Drive down east towards Anglefort (15 km, 25 min). Hazel grouse, Tengmalm's **and** pygmy owls **breed in the upper parts of the forests; open areas at lower altitudes are wintering grounds for** rock bunting. Once down on the plain, cross the Rhône 6 km north in Seyssel and follow signposts to the 'Espace sports et nature du Fier', 5 km south on the eastern bank [3]. **The resort has a nature trail where migration can**

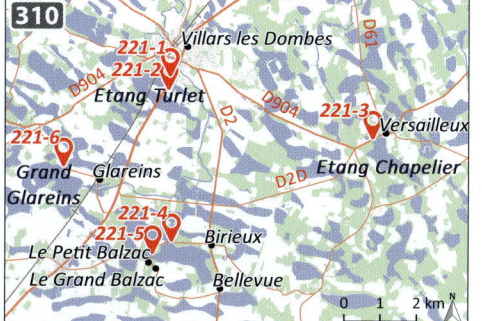

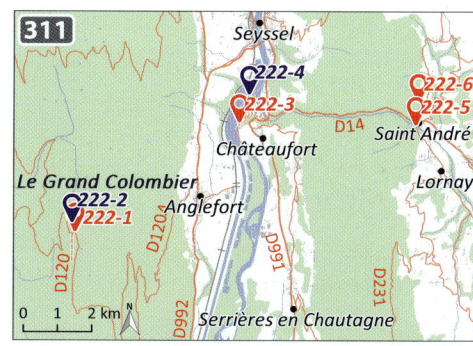

be intense during both passages. Spotted and little crakes are recorded yearly in April, May and late August. Penduline tit is uncommon, but regular, in March, April, September and October. Mudflats [4] may attract waders, including single Temminck's stints, casual gull-billed and little terns and commoner little bittern and bluethroat. The Rhône banks are a local hotspot for wildfowl in winter, with locally common goosander, regular ferruginous duck and goldeneye along with gadwall, tufted duck and common pochard.

From here, follow the River Fier east to Saint-André (7 km, 8 min). Turn right towards Lornay at [5] and walk along a small path on the right just after the bridge. It leads to a quarry where Alpine accentor and wallcreeper winter. Check the river for breeding dipper, grey wagtail, spotted flycatcher (May) and wintering goosander. **Get to the top of the cliff opposite by following the D31 towards Clermont and turning left towards Chavanne after 2 km to end on the Chemin de la Croix** [6], where you should focus on red kite, short-toed eagle, honey-buzzard, goshawk, eagle owl, Alpine swift, crag martin and western Bonelli's warbler. From here, you can try autumn migration or lowland breeding species in the Défilé de l'Écluse (site 223, 40 km, 50 min north through Bellegarde-sur-Valserine), head east to the Plateau des Glières (site 231, 60 km, 1 h 20 through Allonzier), or drive south to the Lac du Bourget (site 232, 22 km, 30 min along the Rhône).

223 Défilé de l'Écluse

2 HOURS – 1 DAY; MAP 312   **This is arguably the most impressive autumn migration viewing point of eastern France for soaring birds, raptors and passerines.** Although less well known than Organbidexka in the Pyrenees, it is arguably its equal, in terms of both diversity and numbers.

Drive along the Rhône towards Geneva on the D1206 from Bellegarde-sur-Valserine. After 9 km, park just before a tunnel at the 'Fort l'Écluse' sign [1]. Crag martin breeds just down from here. From the car park, cross the D1206 and **walk on a trail which climbs into the forest just in front of the traffic lights; it leads to a fort after a steep 500 m climb** [2]. Wallcreeper and Alpine accentor winter here, and golden eagle breeds in the vicinity.

Go back to the D1206 and follow it for 5 km. Turn right 1.5 km past the Rhône at an 'i' (information) sign in front of a line of poplar trees. After 400 m alongside a small railway, **a car park [3] serves as the migration watchpoint, where a spotter is permanently stationed from 15 July to 15 November. The site is at its best with northerly winds and cloud cover, but most weather conditions can yield good migration.** Common buzzard and red kite counts can exceed those of Falsterbo in southern Sweden, and overall raptor and stork numbers are sometimes really impressive. Migration starts with black kite and common swift in July, followed by honey-buzzard in August and red kite and common buzzard at the end of the season. Black stork numbers can reach a few dozen on good August days, possibly joined by single ospreys. Both spotted eagles are near-annual, as well as other rarities such as pallid harrier and red-footed falcon. The site is also the scene of a strong passerine passage of swallows, larks and later finches. **The surroundings are worth a check even if the weather is not right for migration.** Look for wheatear, larks, warblers on passage, and breeding species (red-backed shrike, cirl bunting, yellowhammer and middle spotted, lesser spotted and black woodpecker). A few pairs of wryneck breed along woodland edges.

Back at the Rhône [4], search for bee-eaters, reed passerines, herons and osprey (exit the D1206 on a small road 2 km north of the migration watchpoint and stop near the stud farm). Also check the farm itself for passage passerines. **From here, head 4 km south to the fields around Vulbens** [5], which can yield red-footed falcon, whimbrel, godwits and rarities on both spring and autumn migration.

224 Risoux forest

> 1 DAY AND 1 NIGHT; MAP 313   The wild Risoux forest covers a 3 × 8 km plateau which extends north into Switzerland (access from Les Rousses,

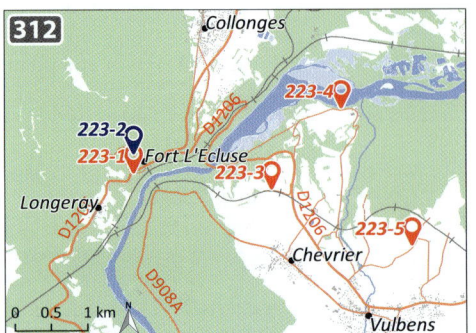

50 km, 1 h north of Geneva via the Col de la Faucille or 70 km, 1 h 30 southeast from Lons-le-Saunier, follow signs to Geneva (Genève) through Clairvaux-les-lacs on the D678). **It is dominated by old spruce stands mixed with varying proportions of beech and fir and a dense deciduous understorey, interspersed with numerous forest tracks and trails**. Most of the plateau is protected under some of France's strongest regulations against logging, hunting and disturbance. Once there you easily understand why **this mid-altitude plateau is known as the best place in France for some of the rarest mountain species**, in a taste of central Europe unique within the country. You can spend days wandering inside the forest or along large clearings at dusk or dawn looking for capercaillie and hazel grouse, the possibility of coming across the tracks of a lynx adding to the thrill.

The area is at its best from March to mid-July, but access is limited until the snow melts in April. Spend time on a few spots and tracks instead of switching rapidly from place to place, although it may be tempting when the forest gets silent in summer. **Bird anywhere along the edge of clearings and old tracks, but do not enter forest stands themselves.** Capercaillie, hazel grouse and three-toed woodpecker **should be your targets, and they will be challenging. You'll maximise your chances by being on site before first light and at dusk until full night.**

There are hostels just down from the plateau in Les Rousses [1] and Bois d'Amont [2], but it is possible to camp for the night on the plateau, for instance at an ideally located small refuge [3] (free access). Midday hours are desperately quiet in the forest and are better spent elsewhere, for instance in Frasne marshes (site 225) or in Les Rousses marshes and lake [4,5] in search of redpoll, lesser whitethroat, marsh warbler or common rosefinch (scarce).

To reach the plateau from Les Rousses, head to Bois d'Amont (9 km, 10 min) and take the first road left at the bus stop when entering the village [6] or the 'Route du Risoux' from the village centre [7] (follow the 'Forêt du Risoux' signposts). Along the latter, the wide Plan de la Citerne [8] meadow is an excellent start or end point. Be there before first light and until night has fallen. Spend time on this large clearing walking along forest edges east towards the Swiss border; capercaillie, hazel grouse, woodcock and the two owls are regularly seen here. Yellowhammer, whinchat, tree

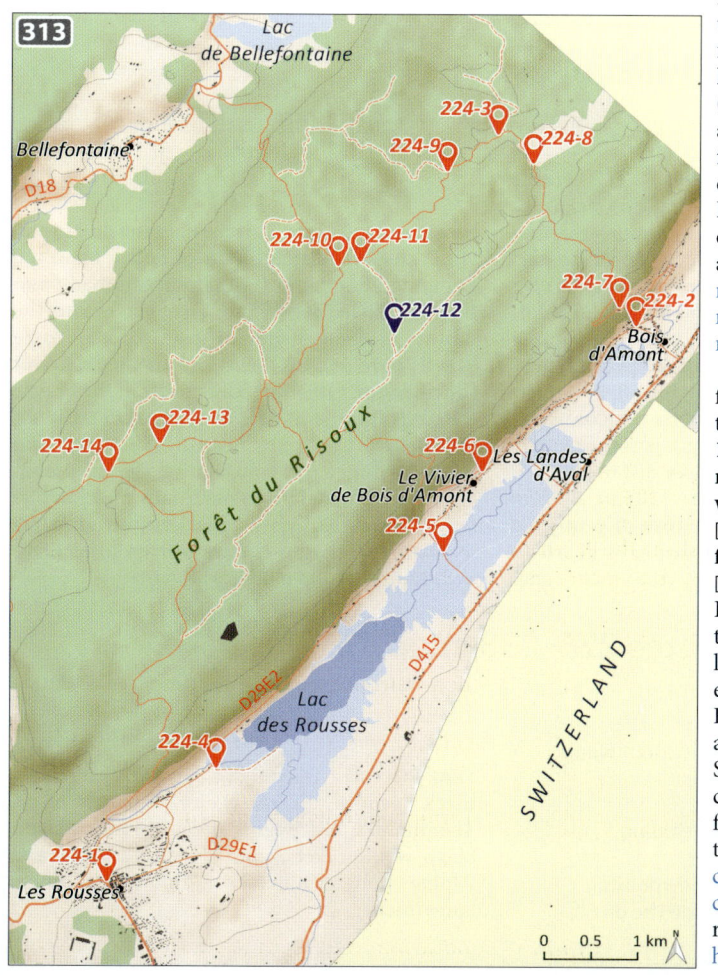

pipit, coal tit, redpoll and citril finch commonly frequent the meadows and along the forest edges. Check tree tops with a telescope for ring ouzel or nutcracker. All these species can be found virtually anywhere else in the forest.

The nearby Plan des Buchaillers [9] is well known for Tengmalm's and pygmy owls, and there are tracks around that lead inside the forest. Another good place for owls, hazel grouse and capercaillie is **the Chalet des Ministres [10] and surrounding clearings, which marks the centre of the Risoux forest. This is an excellent place to listen for** Tengmalm's **and** pygmy owls **in spring** (playback is not allowed here, or anywhere inside the protected area). A few hundred metres south, park in a small car park [11] and **walk along a good track heading to the Plan Pichon [12]: this is one of the most regular places for** three-toed woodpecker **in the Risoux.** Back to Les Rousses, several clearings and tracks are also worth some time, for instance around [13] or [14], especially for capercaillie and hazel grouse.

225 Frasne

2 HOURS – 0.5 DAY; MAPS 314 & 315 Frasne wetlands is another of these central-European-like habitats inhabited by a rich community starring a now famous breeding pair of lesser spotted eagle. (1 h north of Les Rousses via the D437/ N5 towards Dijon, turn to Foncine-le-Haut in Saint-Laurent-en-Grandvaux). On your way, make a midday stop at the shady source of the Doubs in Mouthe [1] to photograph breeding dippers (50 km, 50 min from Les Rousses, follow signposts to 'sources du Doubs', right past the church, park at the ski resort and walk for 300 m to [2]).

The area south of Frasne (20 km, 30 min from Mouthe) is a large expanse of protected marshes and bogs visible from several tracks and observation platforms, the main ones

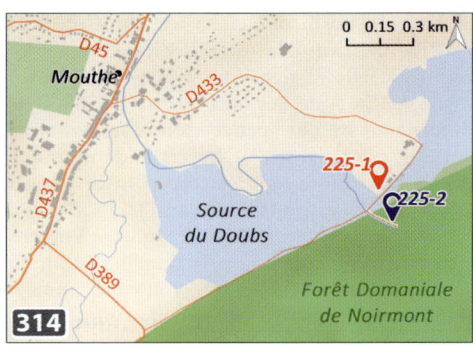

being near the Étang Lucien [3], Étang du Moulin [4] (both immediately south of Frasne) and the Étang de l'Entonnoir [5,6] (along the road from Bonnevaux to Bouverans; turn left before the railway bridge). **A pair of** lesser spotted eagle **breeds here and are best seen as they fly over the marshes.** Other breeding birds in the Frasne wetlands include curlew, dipper, lesser whitethroat, marsh warbler, willow tit, nutcracker and redpoll within a rich community of species associated with low-altitude mountain meadows. A few pairs of great grey shrike still breed, while red-backed shrike is common from May. Eagle owls hunt over the marshes at night.

Northern Alps (sites 226–233, map 316)

HIGHLIGHTS

» Black grouse, ptarmigan and rock partridge in scenic landscapes around Mont Blanc
» Lammergeier and golden eagle near-certain on sunny days
» All high-altitude species including snowfinch and wallcreeper
» Migration and wintering viewpoints around Lake Geneva and the Lac du Bourget

| J | F | M | A | M | J | J | A | S | O | N | D |

TARGET SPECIES

YEAR-ROUND ptarmigan, **rock partridge, black grouse, lammergeier,** golden eagle, **Tengmalm's and pygmy owls, three-toed** and black woodpeckers, crested tit, red-billed and Alpine choughs, nutcracker, dipper, citril finch, **snowfinch**

JURA AND THE ALPS | NORTHERN ALPS | 229

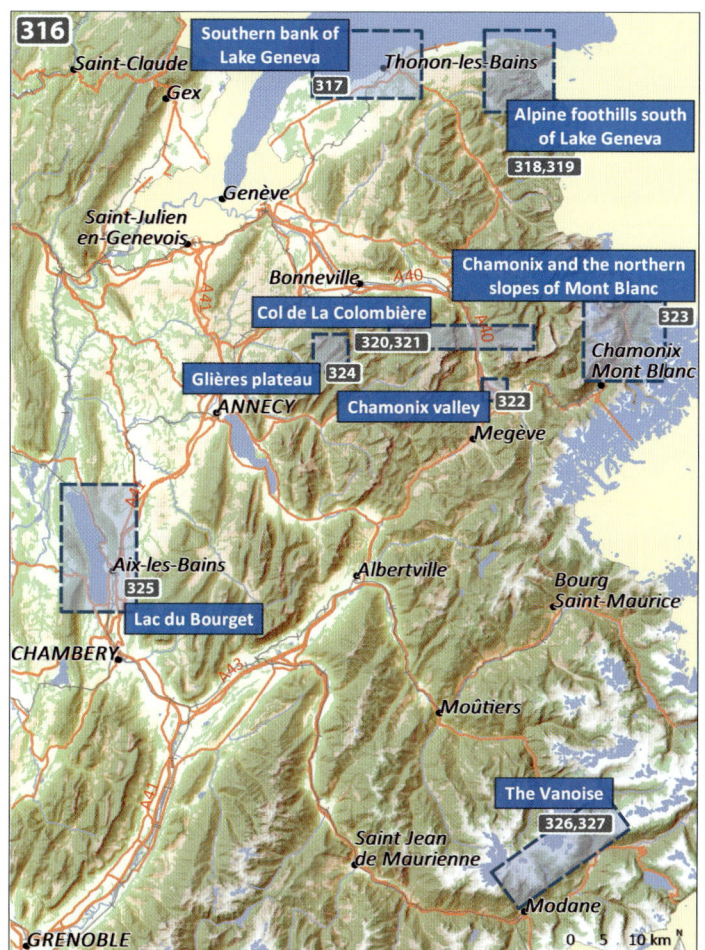

with some luck. A full week will yield a much richer experience, especially between mid-May and late June; earlier, access is impaired by snow, and later, the area is overcrowded with hikers. Get to the birding spots before the first light – do not expect black grouse, rock partridge or ptarmigan once the first hiker has shown up, and dusk won't be as good. **While in some sites most of your targets can be sighted from the car park, be prepared for mountain hikes** with a telescope and food and water for the day if you wish to maximise your chances of finding the most difficult species: rock partridge (dry rocky slopes), ptarmigan (high-altitude screes), wallcreeper (moist cliffs) and snowfinch (snow limit) will probably challenge you the most. **If you have just one day**, concentrate either on the mountains south of Lake Geneva or on the Col de la Colombière, where all your target species except ptarmigan are within walking distance of the car park. **With two days in late May or June**, spend one full day in the Risoux forest, possibly paying a visit to lesser spotted eagle in Frasne, then head at night to the Col de la Colombière to add all the Alpine species, and finish up in the Chamonix valley for migrants and wetland species. **With more time**, pay attention to the spots east of Chamonix and the Plateau des Glières. If the weather is rainy, head to Lake Geneva, the Lac du Bourget or the Chamonix plain, which can be especially productive during migration. In winter, Lake Geneva and the Lac du Bourget are thronged with wildfowl, while resident mountain species can be found around ski resorts.

BREEDING night-heron, purple and squacco herons, little bittern, black-winged stilt, bluethroat, Alpine accentor, ring ouzel, **common rock thrush**, whinchat, **marsh warbler, wallcreeper**, red-backed shrike

MIGRATION black stork, spotted and little crakes, osprey, red-footed falcon, waders, **common rosefinch** (scarce)

WINTER divers, goosander, eider, scaup, **ferruginous duck**, smew, black-necked grebe, **Eurasian bittern**, Caspian gull, **penduline tit**

VISIT DURATION > 2 days

PLAN YOUR VISIT
Here you get into Europe's highest mountains at the foot of Mont Blanc. **This is a key area for mountain birds in France; all high-altitude specialities can be found in two days**

226 The southern bank of Lake Geneva (Lac Léman)

1 DAY; MAP 317 The southern bank of Lake Geneva (Lac Léman) is flat and covered by a mosaic of fields and small woods that are worth a visit in winter or if the weather is cloudy at higher altitudes. **The most productive areas are concentrated around Thonon-les-Bains** (35 km, 50 min southeast of Geneva).

In winter, start in the Domaine de Rovorée, 5 km past Sciez on the D25 to Yvoire. Park at the entrance [1] and walk alongside the lake to a small harbour [2], a viewpoint for black-throated diver, goldeneye, red-breasted merganser, goosander, eider and good numbers of black-necked grebes. **During migration, the nearby Excenevex beach** [3] **is a better plan**, especially for waders (breeding little ringed plover, wintering turnstone and black-winged stilt, sandpipers and shanks on both passages) and gulls (including Caspian and yellow-legged).

From the D1005 towards Thonon, turn left immediately after the 'Bonnatrait' signpost to park at the start of a track leading inside the Domaine de Guidou reserve [4]. Walk 2 km to a hide [5] where spotted and little crakes (April, May and late August), bluethroat (April), penduline tit (March to April), waders (both passages) regularly occur. Marsh warbler breeds in shrubs around the hide, and the surrounding pine woods host long-eared owl, spotted flycatcher and crested tit.

The Delta de la Dranse reserve hosts a similar bird community (follow 'St Disdille' camp site from the large supermarket north of Thonon and park near 'Le Must' night club [6], then walk to a hide [7] or to the Dranse river mouth [8]). This site also has a large black-headed gull colony with a few common gulls, red-crested pochard and sometimes night-heron. In winter, scaup, smew, ferruginous duck and bittern may show up.

227 The Alpine foothills south of Lake Geneva (Lac Léman)

1 DAY; MAPS 318 & 319 From February to May, clear weather should tempt you to try migration at Le Hucel (20 km, 30 min past Thonon, head to Thollon, turn left in front of the church to park at [1]). The migration watchpoint is located below the telecom mast [2], **a high point above Lake Geneva where** raptors **may pass exceptionally close** (spotters are present some days). Expect good numbers of common

towards the Col de Bise [5] (good for Alpine accentor and wallcreeper) and the Col de Floray [6], where rock partridge breed. **Alternatively, drive to Ubine**: pygmy owl breeds along the way with commoner forest species (stop near [7]), and there are black grouse, wallcreeper, both choughs and golden eagle around the refuge at the top [8].

228 Col de la Colombière

0.5–2 DAYS; MAPS 320 & 321 This is the place to go for a chance to add all the Alpine specialities in a single morning, and arguably the northern Alps' most accessible viewpoint for rock partridge and black grouse, which can both be seen from the access road in late May and June. This high-altitude pass lies only 1 h from Geneva and 2 h from the Risoux forest, allowing you to combine a day in the Jura and one in the Alps with only a short night drive across the plain.

From Le Grand Bornant, drive for 12 km, 20 min towards Le Reposoir or Cluses on the D4 and park at La Colombière [1]. **Be there at first light, or better still pitch your tent at the pass, as birds become quieter as soon as the first hikers show up.** Evenings are never as productive. Walk along the small concrete road which heads towards the southern side of the mountain [2] to look for displaying black grouse, which you will probably hear from the pass: they usually settle on grassy patches among willows at mid-slope. Once done with grouse, head back to the pass and ascend 700 m on a wide track to the shepherds' hut [3]. Common rock thrush, water pipit, Alpine accentor and citril finch are common in this area (May), and both choughs, golden eagle and lammergeier should fly over as soon as there is a bit of sun. Wallcreeper breeds on the cliffs to the south, a 10 min walk on poorly marked trails. Then you have two options. **Very early in the morning, turn east towards 'Grottes de**

buzzard, both kites and smaller numbers of all harriers, hobby, black stork, plus the odd osprey or griffon vulture. Local golden eagle and Alpine choughs may also show up. **The surrounding woods are worth an hour walking in search of citril finch and nutcracker in early spring.** Local breeding species include black woodpecker, marsh tit, crossbill, hawfinch, bullfinch and siskin. Tengmalm's owl breeds at lower altitudes below Thollon.

Higher altitudes above Chevenoz become accessible from May (20 km, 30 min from Thonon or Le Hucel on the D32 and D22). Larch woods around the Lac de Fontaine [3] host pygmy owl, nutcracker and good numbers of citril finch and redpoll (from Vacheresse, 4 km, 5 min past Chevenoz, turn towards Le Villard and La Revenette and climb for 7 km). **Drive 3 km further up to reach a car park and chalet just above the tree line (1500 m)** [4]: black grouse is locally common around here, as well as marsh warbler and water pipit. The chalet is the starting point for an easy half-day hike leading

Montarquis' [4]. This easy 1.5 h mid-slope trail leading to several caves is the easiest place in the northern Alps for rock partridge. As you'll be walking along the tree line for a long time, you won't miss citril finch, lesser whitethroat and ring ouzel in late May or June. Detect partridges as they sing at the top of rocky ridges and look for them with a scope – sometimes they may even wander along the trail itself: be as discreet as possible. At higher elevations near the caves, you reach high cliffs with breeding red-billed and Alpine choughs, wallcreeper, and even higher, snowfinch. Another, steeper trail leads upwards to 'Lac de Peyre' [5] (1 h). You will not find partridges here, but expect close views of both choughs, Alpine accentor and snowfinch.

The same species plus ptarmigan occur in Flaine (26 km, 45 min drive through Arâches from the D1205 between Cluses and Sallanches). Park in the resort [6] and walk up past 'La Cascade' restaurant along a good track that leads to the top of the Aup de Véran cable car (2200 m) via a steep 40 min climb [7]. This is your chance to see ptarmigan, Alpine chough and Alpine accentor. Nutcracker, willow tit, citril finch and black grouse occur along the trail and should not be hard to find. From here, either wander westwards along the bottom of the cliffs in search of wallcreeper or climb up again towards the restaurant at the top of the Grande Platières cable car [8] (long and steep). Lammergeier and golden eagle are not as easy as at the Col de la Colombière, but still possible. Alpine accentor and wren breed in crevices all along the way, and you could encounter a flock of snowfinches. If you missed ptarmigan earlier, walk towards the Col de Plaré [9], where you can loop down to Flaine. For those who wish to avoid the long walk, the cable car (from early July, 09.00–17.00) is a shortcut from Flaine to the restaurant, near the main ptarmigan breeding grounds.

229 Chamonix valley

2 HOURS – 0.5 DAY; MAP 322   The plain around Sallanches and Chamonix is a well-known migration corridor, at its best in early to mid-May. Try the western end of the Lac de Passy (7 km from Sallanches town centre, follow signposts towards camp sites from the D1205 towards Chamonix and park in front of 'Les Îles' camp site [1]). The lake itself [2] is a stopover area for terns (common, little, whiskered, black and white-winged) and waders. **Cross the railway and walk between two private ponds** which can be partly viewed from the road [3]. Little bittern breeds on the left side from May. You can ask politely to access the pond to the right at the nearby café: a heron colony hosts great white egret, purple and squacco herons. Fields east of the Lac de Passy [4] (walk the 'Chemin de Fénils' eastwards) are a regular stopover area for red-footed falcon in mid-May, sometimes in large groups of over ten individuals. Among other species that may stop over in the valley, look for quail, red-throated pipit, citrine wagtail (near annual), whinchat (breeding), red-backed shrike and Mediterranean overshoots.

230 Chamonix and the northern slopes of Mont Blanc

2–4 DAYS; MAP 323   The area around Chamonix offers everything on the birder's Alpine species wish-list in exceptional scenery dominated by the Mont Blanc chain. The altitude ranges from 1000 m to above 4000 m, with a permanent snow line around 2500 m, covering a wide habitat range from low-altitude fields and meadows to rocky slopes through larch and spruce forest and Alpine meadows. Birding is a hard task: expect steep ascents to reach the breeding grounds of ptarmigan and snowfinch, **and try to be there before the first hikers**.

JURA AND THE ALPS | 230 CHAMONIX AND THE NORTHERN SLOPES OF MONT BLANC | 233

For easy access to high altitudes, use the cable car from Chamonix to Planpraz [1]. Three-toed woodpecker breeds around here. Walk up to the 'Lac Cornu' [2] (1 h 30): this is a first location for ptarmigan, lammergeier (also possibly golden eagle), Alpine chough and snowfinch. If you have several days available, stay at Argentière, 10 km northeast of Chamonix. Park at the 'Auberge La Boerne' on the D1506 towards Vallorcine [3]. Nutcracker is common, and three-toed woodpecker, Tengmalm's and pygmy owls and hazel grouse breed on the steep slope to the west. **An easier alternative for hazel grouse and pygmy owl is to climb up east in the forest towards grassy ridges signposted 'Les Frettes' [4] and 'Les Posettes' [5] (0.5 day).** Black grouse occur around the tree line, and there are common rock thrush, lesser whitethroat, garden warbler and rock bunting near the ridge.

Descend through Le Tour [6], where dipper, grey wagtail, marsh warbler, citril finch and redpoll breed in the village (a singing common rosefinch was recently recorded nearby).

From June, Le Tour is also the starting point of a **long, yet easy climb towards the Albert Premier refuge** [7] (2700 m, 2 h one way). All the high-altitude species breed on the rocky slopes around, including Alpine accentor, snowfinch and commoner species such as water pipit and even wren, which nest in rock crevices. **Do the ascent during the afternoon, stay overnight in the refuge (fee, book ahead of time) and descend via the Col de Balme** [8] **early the next morning**, looking for ptarmigan on the way. From there, either descend directly to Le Tour (expect quail, tree pipit, marsh warbler on the way) or connect with Les Posettes and Les Frettes.

231 Glières plateau

1–2 DAYS; MAP 324 The Plateau des Glières (1450 m), a valley surrounded by fir and spruce forest and covered with flower meadows, is best visited in late May and June (35 km, 40 min northeast of Annecy, signposted from the D1203 towards Chamonix or from Annecy-le-Vieux on the D5). It is a well-known World War 2 heritage site and therefore much visited, but most people remain around the memorial. **Spend some time in the meadows around the main car park** [1]. Quail, whinchat and marsh warbler breed here, as well as nutcracker, citril finch and redpoll, which you should find along the forest edges. Single corncrakes are recorded approximately once every two years: be there at the first light or at dusk for a try, but do not count on it. **Walk northwest along the path which starts from the car park and enters the forest** [2]. Hazel and black grouse, Tengmalm's and pygmy owls and three-toed woodpecker breed at low densities all along the hill and around Spée refuge [3] (stay overnight). For rock partridge, try the southern edge of the plateau [4], which can be reached from paths starting behind the memorial near the car park.

232 Lac du Bourget

0.5–1 DAY; MAP 325 The Lac du Bourget, France's largest natural waterbody, extends for 18 km northwards from Aix-les-Bains and Chambéry airport. **It is the region's most significant wetland area, for** wildfowl **in winter, for** waders **during migrations, and for reedbed** passerines **from April.**

Start from Châtillon harbour [1], 3 km south of Chindrieux (18 km, 25 min from Aix-les-Bains or 25 km, 30 min from Seyssel, site 222). The harbour itself can yield dabbling or diving ducks. **Climb towards Chambotte pass up to 'Le Belvédère' restaurant** [2] (7 km, 15 min from the harbour). Scan the cliffs for peregrine, eagle owl, raven, wintering wallcreeper, Alpine accentor and migrating raptors in April and late summer. Back at the lake, **check the harbour just to the north of Conjux** [3]. A platform is located at the end of the main pier. Common and red-crested pochards and tufted duck winter around here, sometimes with the odd goosander or ferruginous duck. Checking the lake from a distance with a scope may yield rarer species (eider, velvet scoter, goosander and divers occur annually). In April and May, the reedbed to the north is a stopover point for whiskered, black and sometimes white-winged terns, hobby and red-footed falcon (casual but annual). Check the sky for red kite, short-toed eagle and raven, which breed in the nearby hills. **Similar species can be seen from the pier just north of Hautecombe abbey** [4], a good spot for the five grebe species and white-tailed eagle (irregular) in winter.

In April and May, and to a lesser extent in winter, **the main hotspots on the lake are spread along the southern bank near Chambéry airport, so head there directly if you are short of time.** Turn left just before the 'Savoy Hotel' when entering Le-Bourget-du-Lac from the north on the D1504, then turn right to reach the 'Château de Thomas' car park [5]. Walking across the canal, **a nature trail leads to a tower-hide in front of a** grey heron **colony** which also hosts a few pairs of purple heron, cattle egret and one or two squacco herons. This is also a good spot for goshawk, which sometimes hunts over the colony. From mid-April to mid-May, check the reedbed for singing bluethroat and great reed warbler or for passage spotted and little crakes. In winter, a hide can yield close views of ferruginous duck, scaup, other wildfowl, bittern, common and jack snipe. Great grey shrike winters in the surrounding meadows.

If you have enough time, head back to the Aix-les-Bains road and stop just before the airport in front of 'Fleet Technology' [6]. Walking along the canal that borders the airport might reveal crakes, curlew, bluethroat, sedge and reed warblers in spring and penduline tit in winter. Check the grassy areas of the airport: merlin, peregrine, short-eared owl and even pallid harrier and red-footed falcon (May) sometimes hunt here, and there are sometimes large numbers of lapwing and yellow wagtail, possibly joined by a single dotterel, tawny pipit or ortolan bunting.

Past the airport on the D1201, stop at 'Les Mottets' car park [7], 200 m before the roundabout

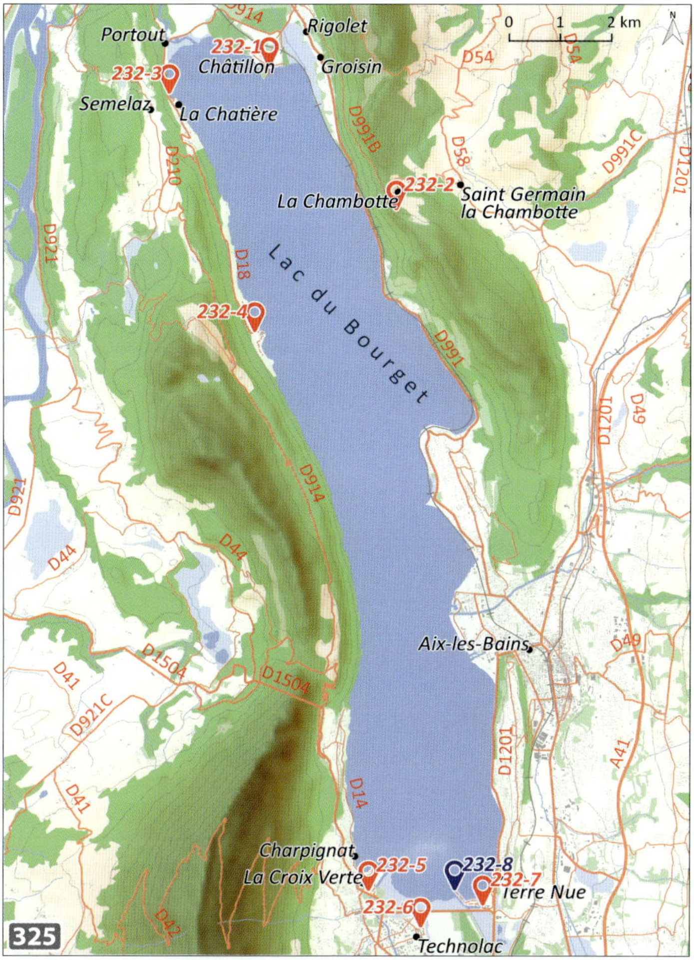

JURA AND THE ALPS | **233 THE VANOISE** | 235

towards Aix-les-Bains and walk to the lake [8]. There is an obvious yellow-legged gull colony on a small islet to the left, and the reedbed has breeding little bittern, spotted and little crakes (second half of April), bluethroat (easiest in April), great reed and Cetti's warblers (recent autumns have seen several yellow-browed warblers). In winter, bittern, all the grebes, great northern diver and waders including jack snipe should occur.

233 The Vanoise

> 2 DAYS; MAPS 326 & 327

The wild mountains of the Vanoise encompass a large expanse of protected habitats within barren landscapes hosting all Alpine species. It is not an easy birding location, however; expect to spend several days of long hikes from Modane, Val d'Isère or Bonneval-sur-Arc to find your target species. **The easiest place for birding is at the end of the D126, above Termignon** (signposted 'Bellecombe' from the D1006 towards Mont-Cenis). Park befor the end of the road above a chapel [1] and walk along the slopes westwards; many French birders have twitched rock partridge along this path. The main car park [2],

1 km above, is a viewing point for golden eagle, lammergeier, wallcreeper and Alpine chough. Another spot you can reach by car is the Orgère refuge [3] (16 km, 30 min west of Modane, signposted from the first roundabout past Fourneaux on the D1006 towards the motorway). Spending an evening or early morning on trails in the vicinity may produce black grouse, Tengmalm's owl, nutcracker as well as commoner species.

The Vercors (sites 234–236, map 328)

HIGHLIGHTS
- Easier access to mountain species than in most other areas of the Alps
- Winter birding is possible with some equipment
- The easiest area in the Alps for mountain owls

| J | F | M | A | M | J | J | A | S | O | N | D |

KEY SPECIES

YEAR-ROUND griffon vulture, **lammergeier**, golden eagle, **black grouse, ptarmigan, hazel grouse**, woodcock, eagle, **Tengmalm's and pygmy owls**, black woodpecker, crag martin, Eurasian treecreeper, raven, nutcracker, crossbill, citril finch, **snowfinch**, rock sparrow, rock bunting

BREEDING short-toed eagle, **scops owl**, nightjar, Alpine swift, bee-eater, **Alpine accentor, common rock thrush**, ring ouzel, lesser whitethroat, subalpine, **orphean**, western Bonelli's and wood warblers, willow tit, **wallcreeper**, red-backed shrike, **ortolan bunting**

MIGRATION booted eagle, **Egyptian vulture**

VISIT DURATION
> 3 days

PLAN YOUR VISIT

The Vercors stands out as the most accessible mid-altitude mountain plateau for those coming from the south. It is best known for its wild forests where Tengmalm's and pygmy owls are reasonably easy in late spring, but there is far more to find in endless Alpine meadows and on barren slopes. **Given its location and wide altitudinal range, the Vercors hosts among the richest bird communities of all the Alps, from Mediterranean species in the lowlands to** ptarmigan and rock partridge **along high-altitude ridges.** The sites are organised along a

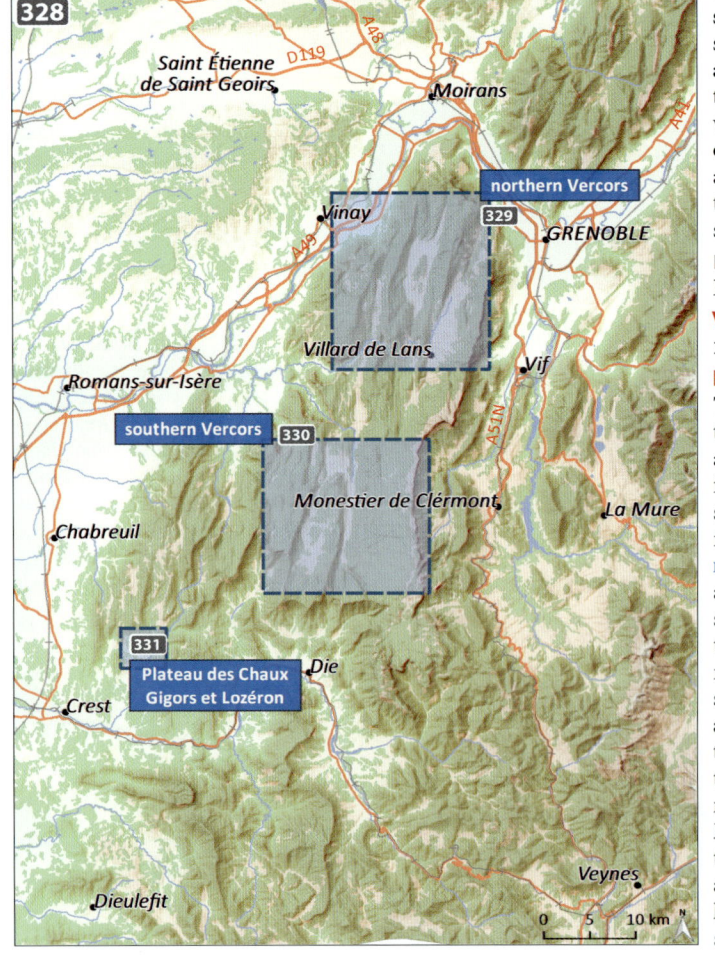

north–south transect, but target species can be found in every suitable habitat. **The best option is to park somewhere and climb up towards the tree line along hiking trails, stopping here and there when habitats look suitable for your target species.** Early in the season, access roads may not be accessible other than by walking or skiing. Bird the tree line at the very first light and climb in meadows and on rocky tops in the early morning before hikers show up. Keep forest habitats for the evening, and consider staying in refuges for owls. Hazel grouse and rock partridge are intentionally excluded from the site accounts as they are highly sensitive to disturbance, but they can be found in suitable habitats all across the plateau.

234 Northern Vercors

> 1 DAY; MAP 329 **The Ramées plateau at Lans-en-Vercors is the nearest place to Grenoble where all the mountain target species can be found in a spring birding day** (30 km, 40 min south of Grenoble on the D106 through Seyssins). Park just above the mountaineering school [1] and walk up along the numerous hiking trails towards the Cabane des Ramées [2] (a small refuge where you can stay overnight) or along ridge to the Croix des Ramées [3], then explore the plateau to the east. Common species include Eurasian treecreeper, black woodpecker, raven and willow tit. Expect black grouse (early morning), woodcock (morning and evening, mostly flying by above the trees) and nutcracker, crossbill or citril finch (both active all day) up to the tree line. Alpine meadows host water pipit, common rock thrush (rare), ring ouzel, lesser whitethroat (rare) and rock bunting, and wallcreeper breeds on surrounding cliffs. Tengmalm's and pygmy owls also breed in the surroundings.

An alternative for all these species plus hazel grouse is the Combe de Gève [4], at the extreme north of the Vercors above Autrans (10 km from Lans-en-Vercors). **This option works best if you stay overnight at the refuge to try for** owls (from Autrans, bypass the church and drive 5 km north; book ahead of time). Be out at very first light for grouse.

The main western exit from the plateau towards Romans-sur-Isère is via the deep Bourne gorges (17 km from Autrans on the D106), which give access to lower-altitude deciduous forests and cliffs in which peregrine, short-toed eagle, eagle owl, crag martin, wallcreeper and rock bunting breed. Good viewpoints include the Goule Noire [5] and the Cirque de Bournillon [6], but any location along the road can yield these species.

235 Southern Vercors

> 1 DAY AND 1 NIGHT; MAP 330   The pastures west of La Chapelle en Vercors [1] are worth a morning or evening stop for quail, woodlark, whinchat and red-backed shrike. Surrounding deciduous woods host western Bonelli's and wood warblers, while yellowhammer, cirl and rock bunting may be found along hedgerows and in rocky areas.

The Grand Veymont area, above Saint-Agnan-en-Vercors, has a wide habitat gradient from fir and spruce forests to Alpine meadows and rocky ridges with all the possible mountain targets, including black grouse, ptarmigan, both choughs, wallcreeper and snowfinch plus citril finch, Eurasian treecreeper and crossbill among commoner forest breeders. Park at the Maison Forestière de la Coche [2] (turn left 6 km south of Saint-Agnan on a narrow road) and **ascend very early in the morning towards high altitudes along hiking trails towards the Pas de la Ville [3] and the Grand Veymont [4]**. Expect a rather steep and difficult six-hour hike to get to the highest areas where ptarmigan breeds. To be on site at dusk, stay overnight at the Pré Grandu refuge [5] (booking mandatory). Back at the forest in mid-afternoon, scan the sky for griffon vulture, lammergeier and golden eagle. Note that the access road was recently closed during the breeding season in response to inappropriate behaviour by visiting birdwatchers, and that it may be closed again in the future: if visiting, be exemplary in your behaviour and refrain from any kind of disturbance. **Be aware that Alpine species are ultra-sensitive and occur in low densities.**

West of the Grand Veymont area, drive to Vassieux en Vercors and continue for 7 km to park at the Station de Fond d'Urle [6]. **This skiing resort is one of the few that remain accessible year-round, with a good chance of locating resident Alpine accentor, citril finch and snowfinch even in winter. Cross-country skiing can also yield winter-plumage ptarmigan.** In spring, the area is covered by a mosaic of woods and Alpine flower meadows where water pipit, wheatear, ring ouzel and lesser whitethroat are abundant. Walk along edges for citril finch and crossbill and try for the two owls at dusk. Rocky areas host common rock thrush (rare, from May) and rock bunting. On a sunny day, you should get good views of soaring Alpine swift and both chough species.

Heading south again, the Col de Rousset [7] is the last high-altitude location before you descend to the Alpine foothills on the D518 to Die. **It is**

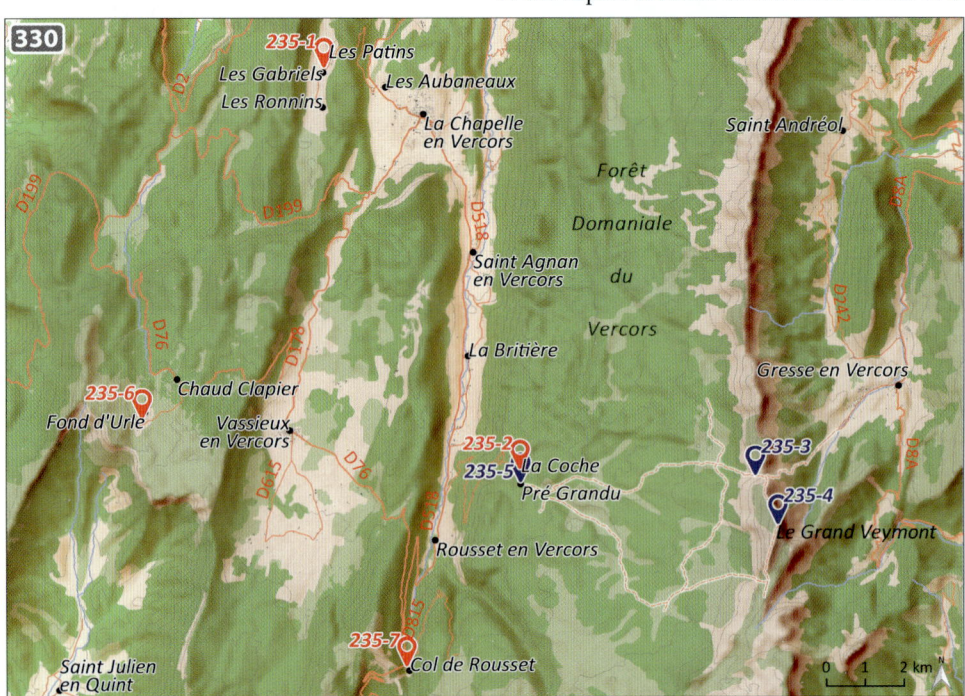

also one of the few opportunities to attempt the main mountain target species along a driveable road (black grouse, wallcreeper and snowfinch in winter). The car park at the pass offers a wide view and is a good place to scan the sky for soaring raptors around midday (griffon and black vultures, lammergeier, short-toed and golden eagles and peregrine).

236 Plateau des Chaux – Gigors et Lozéron

2 HOURS – 0.5 DAY; MAP 331 The Chaux plateau [1] lies at the extreme south of the Vercors between Gigors-et-Lozéron and Beaufort-sur-Gervanne (17 km on the D70 from Crest). It is a mid-altitude plain near the transition zone between the Alps and the Mediterranean part of the Rhône valley, dominated by extensive crops and pastures with hedgerows, at its best in May and June. **It is therefore a strategic place to add lowland, warm-climate birds after a high-altitude trip, or to spend your time if the weather is inhospitable in the mountains.** The best strategy is to stop-and-go

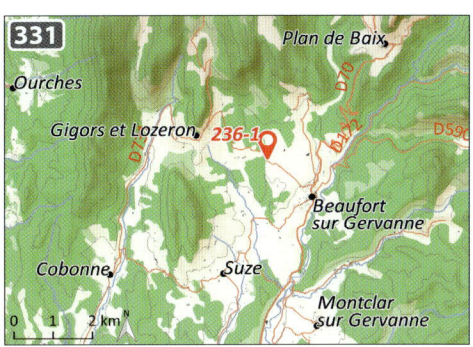

along the small roads that run across the plateau. Common breeding birds in the area are a mixture of continental and Mediterranean species, including quail, short-toed eagle, scops owl, nightjar, bee-eater, woodlark, tawny pipit, subalpine and orphean warblers and rock sparrow. Booted eagle and ortolan bunting are scarcer but quite likely, while truly Mediterranean species such as Egyptian vulture, roller and great spotted cuckoo may occasionally occur, especially during spring migration.

From Grenoble to Gap (sites 237–241, map 332)

HIGHLIGHTS
» Vast wild areas from mid-altitude hills to some of the highest summits of the Alps
» High bird diversity and densities
» Mountain target species relatively easy to find in all suitable habitats
» The best places in the Alps to mix family hiking holidays and birding

| J | F | M | A | M | J | J | A | S | O | N | D |

KEY SPECIES
YEAR-ROUND lammergeier, griffon vulture, golden eagle, **rock partridge**, **black grouse**, **ptarmigan**, **Tengmalm's and pygmy owls**, black woodpecker, crag martin, dipper, nutcracker, crested, willow and marsh tits, Alpine chough, crossbill, **citril finch**, **snowfinch**, **Italian** and rock sparrows, rock bunting

BREEDING short-toed eagle, nightjar, tawny pipit, Alpine accentor, ring ouzel, **common rock thrush**, marsh and western Bonelli's warblers, **wallcreeper**, red-backed shrike, **ortolan bunting**

MIGRATION red-footed falcon

WINTER goosander, brambling

VISIT DURATION 2 days to 1 week

PLAN YOUR VISIT
These high-altitude areas are wilder than the northern Alps but less suited to short visits. **All the mountain target species occur in good densities in suitable habitats, most of which fall within core areas of national parks that are only accessible by day-long hikes**. Good physical condition and some preparatory study of the maps is necessary, as the terrain is quite rough and precise spots for target species are mostly lacking. Do not neglect the plains, especially during migration or on rainy days; red-footed falcon and dotterel sometimes stop over in good numbers (check the news). In winter, Alpine passerines gather around ski resorts and become incredibly tame.

237 The Lautaret and Galibier passes

2–5 HOURS; MAP 333 The Col du Lautaret (2060 m) and Col du Galibier (2640 m) are two passes on the D902, the road that connects the Ecrins to the Maurienne and the southern Vanoise. To reach the Lautaret [1], drive 90 km, 1 h 40 from Grenoble towards Briançon, or

snowfinch (early spring). Commoner species (lesser whitethroat, marsh warbler, redpoll and citril finch) are also present. Black grouse breeds but may be more difficult to find. **Being significantly higher, the Galibier offers a better chance of** ptarmigan (probably the easiest site for this species in the whole of the Alps) and snowfinch.

238 Névache

> 2 DAYS; MAP 334 The Névache valley is reached by a dead-end road, 25 km north of Briançon (N94 to Montgenèvre, then D994G). Névache village [1] lies at an altitude of 1600 m and has several camp sites and hostels, making it a suitable starting point to reach **species such as** rock partridge **and** ptarmigan, **which both breed on high-altitude rocky slopes**. The village itself has rock sparrow even in winter as Italian birds come down from the mountains. **The Chemin de Ronde is a short introductory walk above the village along the tree line** (park at [2], 2 km above the village, and walk towards [3] on the 'Tour du Mont Thabor'). **Its location on the sunny side makes it the best place to encounter a mixture of low- and high-altitude species** like wryneck, ring ouzel, common rock thrush, lesser whitethroat and ortolan bunting. Rock partridge can sometimes be sighted from the trail, but do not count on it. On the opposite side from the same car park, there are black grouse in the vicinity of the Chalets de Buffère [4].

For higher altitudes, park at the starting point of the trails 10 km above the village [5] and climb upwards. **The reward for a steep half-day hike to the Lac des Béraudes [6] and the Sagnes Froides [7] will likely be good views**

30 km, 30 min in the other direction. **All the Alpine specialities can be expected from the pass,** including lammergeier, golden eagle, Alpine accentor, common rock thrush and

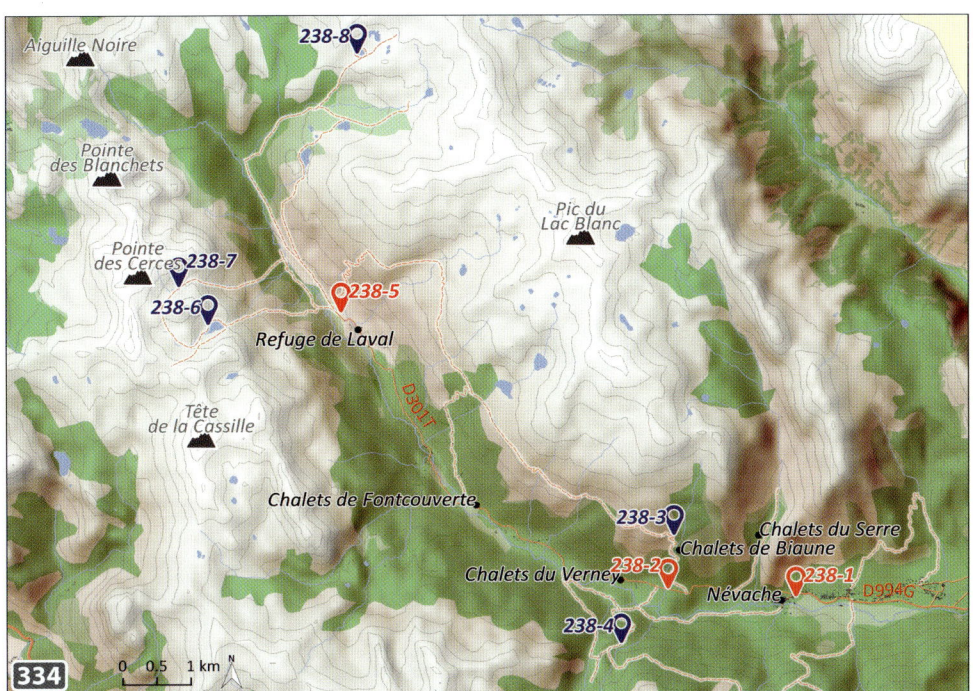

of ptarmigan, Alpine accentor, Alpine chough, wallcreeper **and** snowfinch. On the other side, the easy day-long hike to the Lacs de la Madeleine [8] is another option for the same species plus rock partridge (scarce), but the longer approach walk provides more opportunities for mid-altitude species (e.g. water pipit, whinchat and citril finch).

239 Montdauphin and the Serre-Ponçon reservoir

0.5 DAY; MAP 335 On the way to or from Névache, make a 30-minute stop in Montdauphin [1 – see GPS file], a medieval village perched on top of a small hill 30 km, 35 min south of Briançon on the N94 to Gap. Breeding rock sparrows can be very showy in the castle ruins or in the fields further downhill. Ortolan bunting breeds in the surrounding fields from early May.

A further 30 km, 30 min south towards Gap, the **Serre-Ponçon reservoir is one of the largest artificial waterbodies of the southern Alps.** Small numbers of migrating and wintering wildfowl and waders rest here, but the reservoir will probably not deserve more than a few quick stops, for instance along the eastern tail [2] (virtually the only site for waders), in the Saint-Michel bay to the north [3] or in the Ubaye [4] to the east of the dam. In winter, expect large numbers of yellow-legged gulls and great crested grebes, and small numbers of wildfowl depending on weather conditions and ice cover. A few goosander, common and Caspian gulls also winter here. **The reservoir is better for passerines, especially in winter** when Alpine breeding species such as wallcreeper and Alpine accentor come down. Also search for citril finch and brambling in large finch roosts. Grey wagtail and dipper breed on the surrounding streams and can sometimes be seen along the reservoir banks, and rock sparrow is locally common. **In spring and autumn, various species of warblers may stop over around the eastern tail**, as well as ortolan bunting. All the vultures, golden and short-toed eagles and scarcer raptors such as migrating red-footed falcons (in May) are regularly recorded flying over the reservoir or its surroundings. **From the western side of the reservoir, the upper Durance valley down to Sisteron is a migration corridor** and deserves a few random stops. For instance, the fields around Rousset [5], just above the dam, are a stopover area for red-footed falcons (peaking mid to late May).

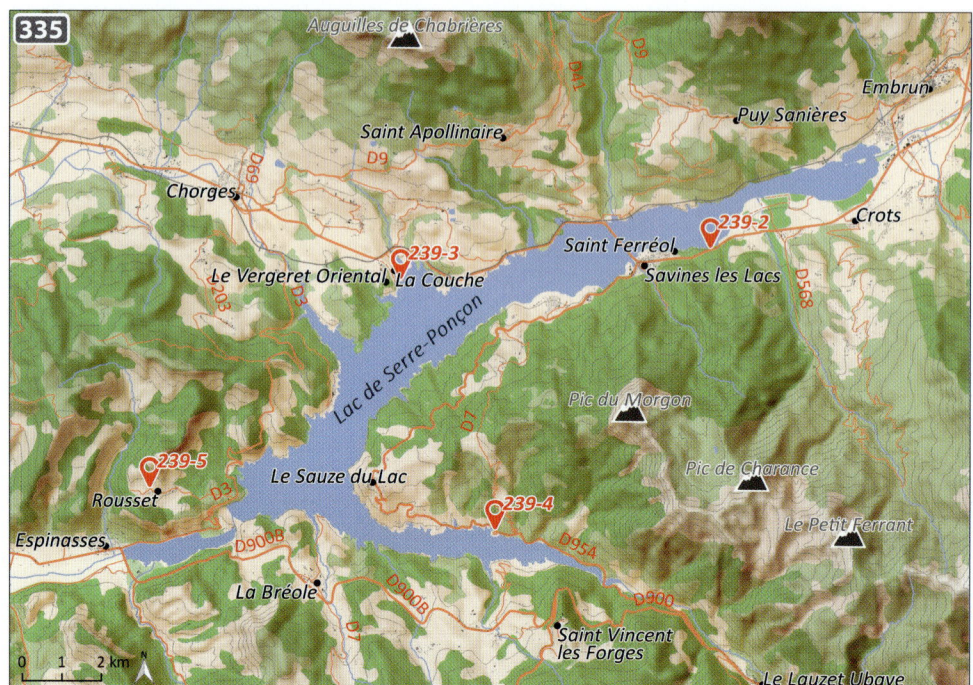

240 The Ecrins

> 2 DAYS; MAPS 336 & 337   The Ecrins are a spectacular barren mountain chain reaching an altitude of 4100 m and spreading from Grenoble to Gap east of the Champsaur valley. The area is protected by a 900 km² national park (the Parc National des Ecrins) which hosts **most of the Alpine bird community plus good numbers of warmer-climate species that are absent or harder to find further north**. The Ecrins are a top choice for those who have time and want to combine family hikes with Alpine birding. **Expect to spend several days in the national park, as most sites will require half- to full-day hikes from side valleys accessed by the N85 between Gap and Saint-Firmin (30 km). Most sites are only accessible from late May to October.**

For mid-altitude species, **try the easy two-hour trail above Prapic** (leave the N85 eastwards to Orcières, 30 km, 40 min, park below the village [1] up to the Saut du Laire [2]. There are a few Italian sparrows in the village along with hybrids, and Alpine choughs breed in the church. The trail climbs slowly along the tree line; expect crested, willow and marsh tits, whinchat, common rock thrush, ring ouzel, dipper, lesser whitethroat, western Bonelli's warbler and rock bunting. There are rock partridges on the southern slope and a few black grouse on the northern slope. **The same species plus crag martin, rock sparrow and ortolan bunting can be found on the trail from the Orcière-Merlette ski resort** [3] **to the Estaris lake** [4] **(half-day hike).** Alpine accentor and snowfinch breed around the lake itself.

Being slightly lower (1400 m), **Chaillol station** [5] **has a broader range of forest species** including pygmy owl, nightjar (June), nutcracker, western Bonelli's warbler and bullfinch. Climb to the Cabane des Parisiens [6] (1 day there and back with a climb of > 1500 m, 7 km above Saint-Julien-en-Champsaur) for black grouse, common rock thrush, citril finch and rock bunting. Higher altitudes towards Le Vieux Chaillol are home to rock partridge, ptarmigan, Alpine accentor, wallcreeper and snowfinch. Both choughs, golden eagle and griffon vulture are easy to see in sunny weather.

For another place to see rock partridge and ptarmigan, try the Vallonpierre pass from Gioberney refuge (25 km, 30 min east of Saint-Firmin through La Chapelle en Valgaudémar) [7,8]. **All Alpine passerines breeding elsewhere are present in this remote area, including** Alpine accentor**,** wallcreeper **and** snowfinch. From the same car park, the Lauzon lake [9] (1 h

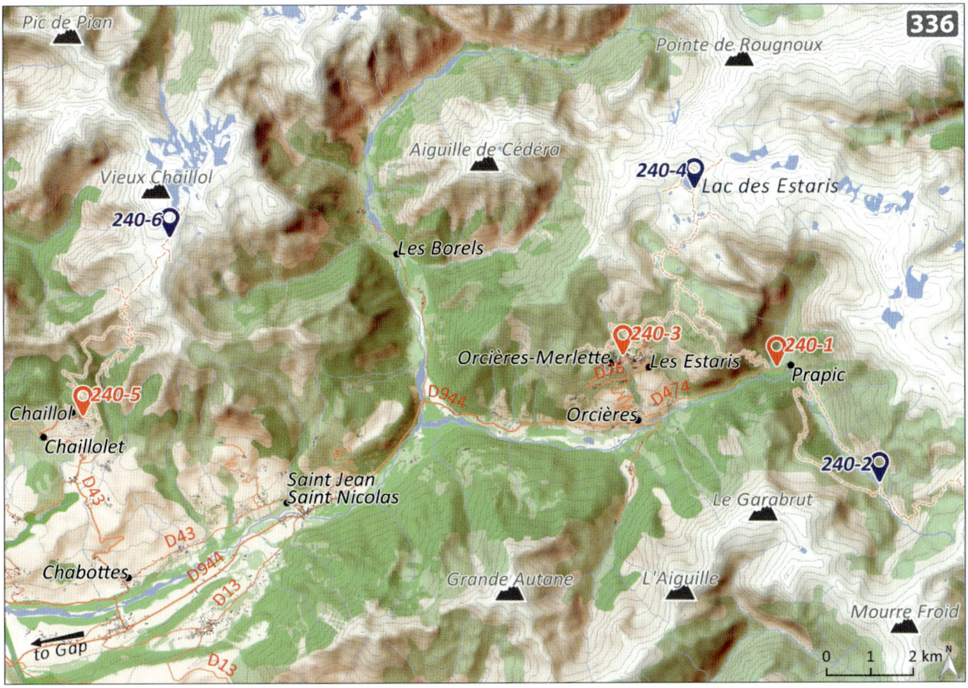

JURA AND THE ALPS | **240 THE ECRINS** | 243

30) is an easier option for dipper, lesser whitethroat, both choughs and rock bunting; black grouse and lammergeier are rare but quite likely on this trail.

For forest species, drive back to La Chapelle en Valgaudémar and park 2 km above the village in Les Portes [10]. The surrounding spruce woods have Tengmalm's owl, black woodpecker, all the tits, nutcracker and crossbill. While climbing above the forest to the Pic de Pétarel [11] (full day), expect black grouse at the tree limit and all high-altitude breeders. If the weather is cloudy in the mountains, try the meadows of the Champsaur valley around Saint-Bonnet-en-Champsaur: it has good numbers of whinchat, red-backed shrike, and it is excellent in both spring and autumn for passerines on migration.

241 The Dévoluy

> 2 DAYS; MAP 338   **The Dévoluy is the western counterpart of the Ecrins and may be a better choice if you have little time and are not looking for high-altitude species.** Start at the Col Bayard [1], 9 km, 10 min north of Gap on the N85 towards Saint-Firmin. Turn left to the

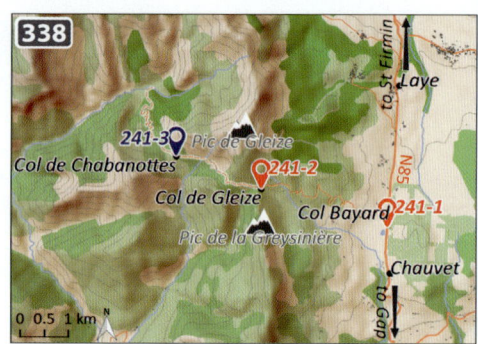

Col de Gleize [2] (5 km): the marshes at the start of the road have breeding marsh warbler, whinchat and red-backed shrike, and higher altitudes host tawny pipit (most reliable site in the Alps), western Bonelli's warbler, all the tits, ortolan bunting and commoner species. A short walk to the Col de Chabanottes [3] among larch and spruce woods should add common rock thrush, nutcracker, rock bunting and good chances for rock partridge, black grouse (early spring), golden eagle, both choughs and pygmy owl.

REGION 12

MASSIF CENTRAL

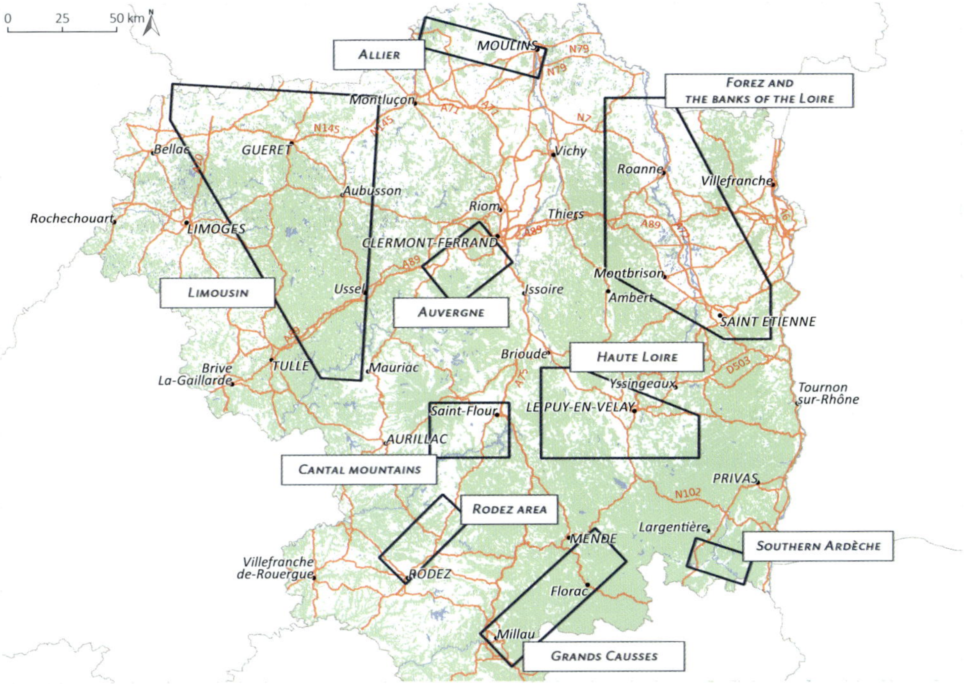

HIGHLIGHTS
» A great diversity of vast wild landscapes – mountains, plateaus, deep gorges, alluvial valleys and forests, with continental and Mediterranean climates
» A hotspot for breeding raptors, cliff species and waterbirds
» A transition region from the central European avifauna to the Mediterranean community

REVIEWER   François Legendre

USEFUL WEBSITES
www.faune-auvergne.org
http://faune-lr.org
www.faune-tarn-aveyron.org
www.faune-loire.org
www.faune-rhone.org
www.faune-ardeche.org
www.faune-limousin.eu

A quick glimpse of a Tengmalm's owl will reward long nights spent in search of this lovely bird on the slopes of the Puy de Dôme (site 261) and the other ancient volcanoes of Auvergne.

Mont Lozère (site 248) and its well-preserved pastures have something that only remote places have. Here you will find a mixture of Mediterranean and mountain species, such as citril finch and ring ouzel.

Rodez area (sites 242–244, map 339)

HIGHLIGHTS
» Small sites, easy to explore
» Steppe birds and southern species, showing Mediterranean influences

KEY SPECIES

YEAR-ROUND hen harrier, black and lesser spotted woodpeckers, woodlark, rock sparrow, cirl bunting

BREEDING short-toed eagle, Montagu's harrier, black kite, hobby, stone-curlew, little ringed plover, hoopoe, wryneck, tawny pipit, orphean and melodious warblers, whitethroat, red-backed shrike

MIGRATION griffon vulture, **dotterel**, green and common sandpipers, **Alpine accentor**

WINTER red kite, **wallcreeper**, **great grey shrike**

VISIT DURATION   1 to 2 days

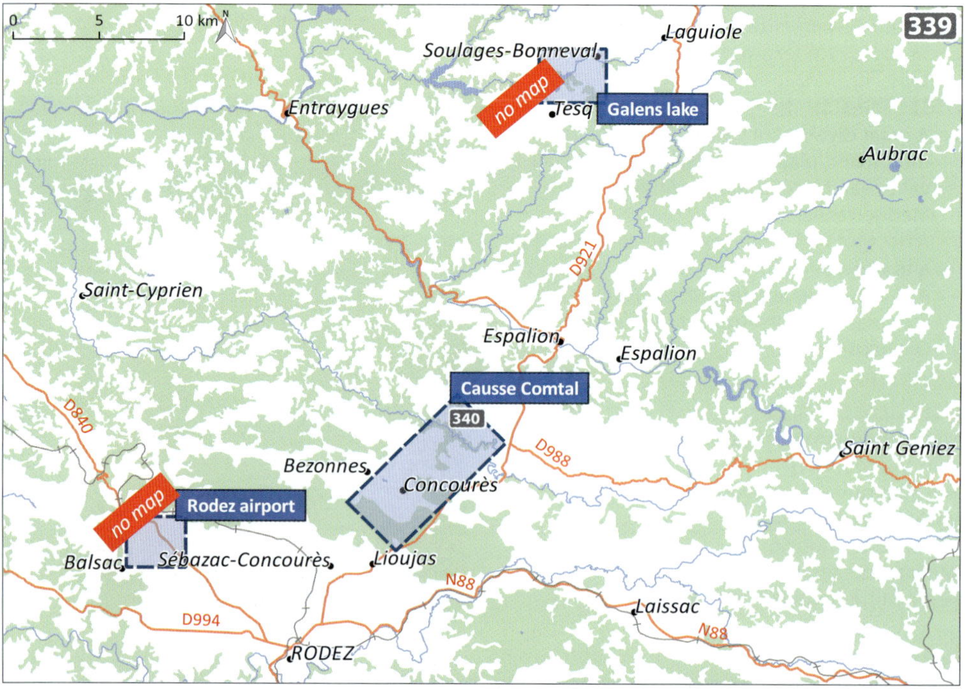

PLAN YOUR VISIT

These sites can be visited independently as day or half-day trips from Rodez. They are best between May and August when Mediterranean species are breeding.

242 Rodez airport

0.5 DAY; NO DETAILED MAP The grasslands around **Rodez airport** (6 km from city centre) are attractive for open-habitat breeding birds and migrants. Take the D840 from Rodez to the airport and turn left on the D626 towards Balsac. Once in La Combe d'Auribal, turn left at the sign for 'Carrière de Capdenaguet'. After 300 m, **take the track to the left beside the quarry and park** [1]. In spring, look for breeding hen and Montagu's harriers, stone-curlew, hoopoe, red-backed shrike, linnet, stonechat, rock sparrow, corn and cirl buntings; little ringed plover breeds in the quarry. In winter, look for lapwing, red kite and the occasional great grey shrike.

243 Causse Comtal

0.5 DAY; MAP 340 The Causses are limestone mid-altitude plateaus dominated by stunted vegetation with scattered bushes, small woods and rocks. They host a unique mixture of steppe and warm-climate species reminiscent of central European bird communities, within a generally well preserved scenic landscape. Follow the D988 from Rodez to Sébazac-Concourès for about 10 km. Once in Lioujas, turn left on the D581. **There are two car parks along the road from where trails run into the Causse**, 500 m [1] and 1 km [2] past the junction. There is another on the D68 between Sébazac-Concourès and Bezonnes [3]. Walk around looking for breeding stone-curlew, hoopoe, wryneck, tawny pipit, red-backed shrike, orphean warbler, woodlark,

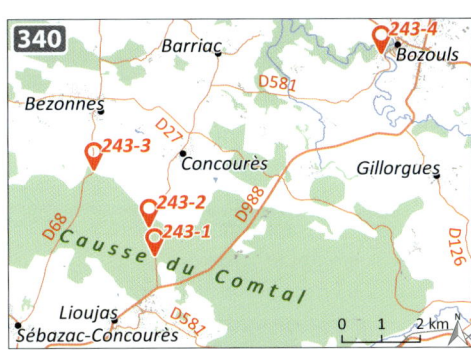

corn bunting, whitethroat, melodious warbler, linnet and wheatear. Short-toed eagle, hobby, kites and griffon vulture can be seen soaring on warm sunny days. Scarce migrants and winter visitors (southern grey and woodchat shrikes, dotterel and black-winged kite to name just a few) have occasionally been recorded. **Further north, the Trou de Bozouls** [4] in Bozouls (access via the D988) is a horseshoe-shaped gorge where Alpine accentor and wallcreeper are regular in winter.

244 Galens lake

0.5 DAY; NO DETAILED MAP Galens lake lies some 50 km north of Rodez. Take the D988 and D921 towards Laguiole then the D70 to Soulages-Bonneval. From La Vayssière, follow signposts to the 'Centre de Loisirs de Galens' on the D541 and the D213. After 1 km, a small road to the left leads to a car park on the northern bank of the lake [1]. Migrating dabbling ducks (teal, pintail, shoveler) and waders (common snipe, green and common sandpiper) stop over here. **The surrounding grasslands and woodlands** may also yield great grey shrike (winter), red (all year) and black kites, breeding hoopoe, resident crossbill, bullfinch, black and lesser spotted woodpeckers.

The Grands Causses (sites 245–248, map 341)

HIGHLIGHTS
» Large vulture colonies (four species)
» Wide and wild limestone plateaus
» A unique mixture of mountain and Mediterranean bird communities

KEY SPECIES

YEAR-ROUND griffon and black **vultures**, golden eagle, red kite, peregrine, red-legged partridge, little, **Tengmalm's and eagle owls**, crag martin, **dipper**, blue rock thrush, Dartford warbler, raven, red-billed chough, **citril finch**, rock sparrow

BREEDING Egyptian vulture, short-toed eagle, black kite, stone-curlew, **scops owl**, nightjar, Alpine swift, tawny pipit, ring ouzel, **common rock thrush**, subalpine, melodious and orphean warblers, spotted flycatcher, **wallcreeper**, red-backed shrike, **ortolan bunting**

MIGRATION red-footed falcon, lesser kestrel, **dotterel**, short-eared owl

WINTER Alpine accentor

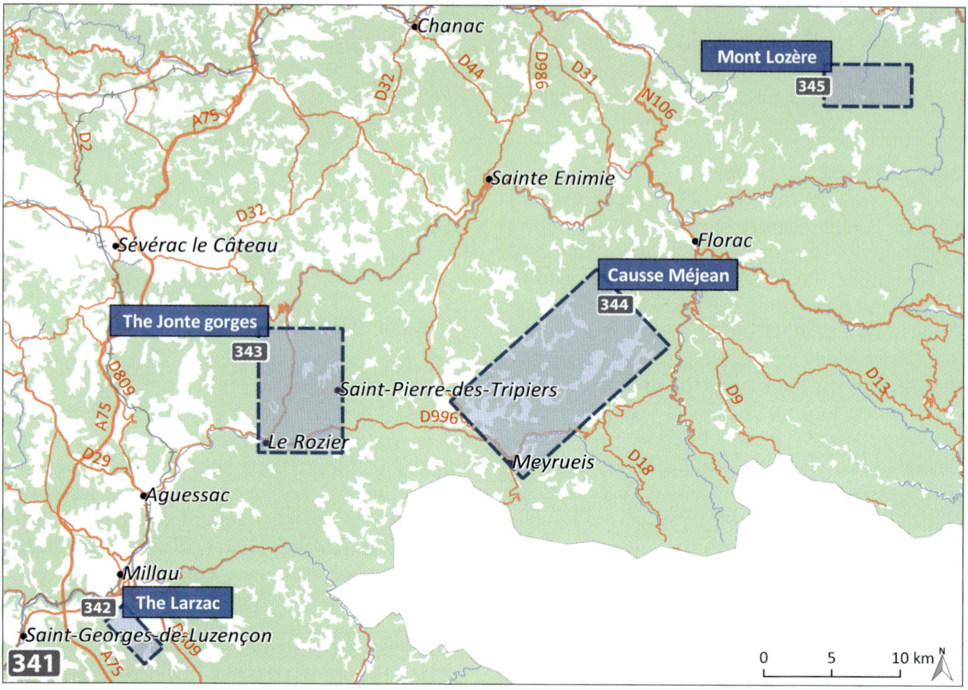

245 THE LARZAC | MASSIF CENTRAL

VISIT DURATION 2–3 days

PLAN YOUR VISIT

Although visiting the region makes sense at all seasons, given the presence of several resident species, bird richness is at its peak between May and July when trans-Saharan migrants are back on their breeding grounds. Le Rozier and Les Vignes are key sites for vultures and other cliff breeders. The plateaus largely overlap in terms of species, so chose one and cover it at length.

245 The Larzac

0.5–1 DAY; MAP 342   The karstic Larzac plateau (altitude 600–900 m) lies south of Millau (take the D809 towards La Cavalerie and turn at the 'Fermes du Larzac Nord-Ouest' signpost). You can either **stop-and-go on small roads** between La Cavalerie, Saint-Michel-du-Larzac and Devez Nouvel farms **or park at Brunas** [1] **and hike** on one of numerous trails around Bouissan and La Bouissière [2]. **Be on site early in the morning for singing passerines, and stay for the raptors that start soaring by midday.** Griffon and black vultures, golden eagle, red-legged partridge, little owl, Dartford warbler, southern grey shrike and red-billed chough are easy to find year-round. Spring and summer should yield Egyptian vulture, kites, short-toed eagle, harriers, stone-curlew, red-backed shrike, tawny pipit, rock sparrow, subalpine and orphean warblers, corn and ortolan buntings. Passage red-footed falcon (May), lesser kestrel, dotterel (May and August/September) and wintering short-eared owl are among the regional rarities to search for.

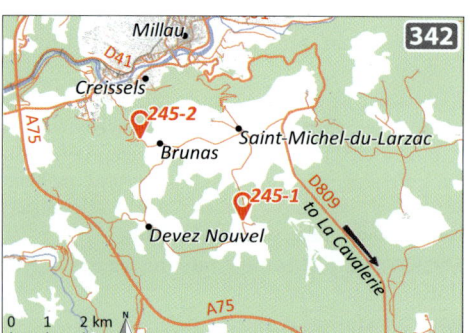

246 The Jonte gorges

0.5–1 DAY; MAP 343   Le Rozier stands at the junction of the Tarn and Jonte rivers inside Les Grands Causses national park. **The site is renowned for its large** vulture **colonies that have been helped by reintroduction programmes in the past decades.** From Millau (30 km) take the D809 to Aguessac and head to Le Rozier on the D907.

MASSIF CENTRAL | **247 CAUSSE MÉJEAN** | 249

Turn on the D996 towards Le Truel until you reach the 'Maison des Vautours' car park [1]. **The viewpoint** (a 150 m walk; fee, but the car park itself is free and still offers good views) **is excellent for** raptors **and other cliff species**. Expect dozens of griffon vultures with a few Egyptian and black vultures as well as a good diversity of other raptors including golden and short-toed eagles, kites, peregrine, eagle owl and other cliff breeders such as blue rock thrush, raven, chough, crag martin and Alpine swift.

A narrow footpath climbing upwards provides another overview of the site well away from the tourist crowd. From October to March, you can also search for wallcreeper and Alpine accentor on the 'Rocher de Capluc' and surrounding cliffs [2]. Another way to find the local species is **to hike on the plateau above the gorges**. From Le Truel, drive to Saint-Pierre-des-Tripiers and turn left towards Cassagnes. There is a car park 500 m before the first houses [3]. Walk on the road for 1 km, then take the track to the right: it leads to Le Cinclegros [4], a rocky headland that offers a nice overview of the Tarn gorges. **It is a stakeout place for** vultures **and other** raptors **which may fly past at close range along the cliffs.** Another viewpoint for vultures and other soaring raptors is Blanquefort castle [5], reached on the D16 a few kilometres from Les Vignes. A hike along the river between Le Rozier and Les Vignes can yield additional species related to riparian woodlands, such as dipper, spotted flycatcher and melodious warbler.

247 Causse Méjean

0.5 TO > 1 DAY; MAP 344 **The Causse Méjean is a vast and high limestone plateau** (800–1248 m)

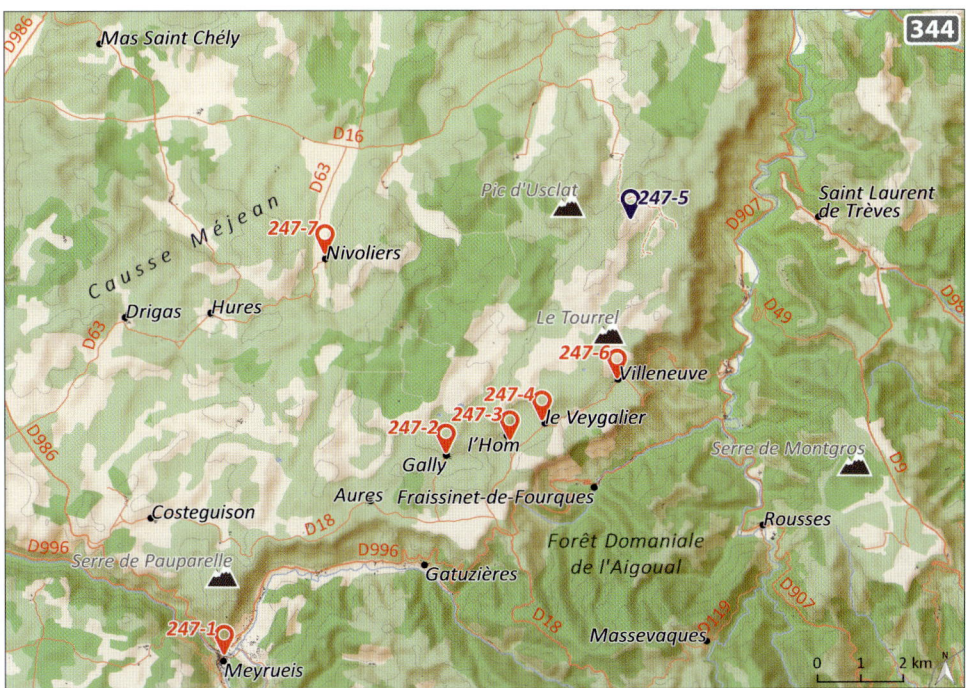

at the southern limit of the Massif Central. This large open landscape consists of stunted grasslands with scattered bushes, trees, rocks, conifer plantations and juniper and boxwood moorlands. Drive to Meyrueis (20 km, 20 min from Le Rozier on the D996). Dipper breeds in the village near the bridge [1].

From Meyrueis, take the D986 north to climb onto the plateau, and turn right after a few kilometres at signs for 'Mas de Lafont – Costeguison'. **One way to visit the area is to walk on the numerous trails** that start from Gally [2], l'Hom [3] or Le Veygalier [4] (10 km, 20 min east past Costeguison on narrow local roads). **Migrant passerines tend to concentrate in the vicinity of livestock waterholes.** Among rarer species, great spotted cuckoo is regular in the area. Scops owl breeds in all the villages, and nightjar can be heard and seen without much effort in surrounding open habitats from May. Tengmalm's owl is trickier to find, but it is a rather common resident of old pine stands (song peaks on clear, windless nights in March and April).

The area around Gargo hill [5] is a classic stopover site for dotterel in May, August and September (4–6 hours round-walk from Villeneuve [6], 3 km northeast of Le Veygalier). Even if you do not find the plovers, you will probably be rewarded by red and black kites, harriers, vultures, short-toed and golden eagles. Other rarities that can occasionally be found in the area include Eleonora's falcon (every year, late summer to autumn), Bonelli's eagle, pallid harrier and black-headed bunting (vagrants). Breeding specialities are tawny pipit, ortolan bunting, red-backed and southern grey shrikes,

northern and black-eared (rare) wheatears, stone-curlew, rock sparrow, little owl, red-legged partridge and common rock thrush, for instance on the rocky hills around Nivoliers [7]. **If you do not want to walk, the same habitats and species can be found, albeit less easily, by birding along the D63** to Drigas, Hures and Nivoliers.

248 Mont Lozère

0.5 DAY; MAP 345 **Mont Lozère is a granite hill reaching an altitude of 1700 m, above the Causses**, 26 km, 30 min northeast of Florac (D998 to Le Pont-de-Montvert and the D20 to Finiels, follow the latter and stop at the car park on the top of the 'Col de Finiels' [1]). **Hike towards the summit past Finiels hill** (30 min) [2]. Local breeding species include ring ouzel, water pipit, citril finch, crossbill and ortolan bunting. In late August to September, dotterel is regular at the summit. The forest hosts Tengmalm's owl and a handful of capercaillie, remnants from an introduction in the 1970s.

The Cantal mountains (sites 249–252, map 346)

HIGHLIGHTS
» A rich community of mid-altitude mountain species, birds of prey and waterbirds
» Varied landscapes with mountains, plateaus and gorges

| J | F | M | A | M | J | J | A | S | O | N | D |

KEY SPECIES
YEAR-ROUND red kite, golden eagle, marsh harrier, peregrine, little owl, woodlark, crag martin, **great grey shrike**, raven, rock sparrow
BREEDING booted and short-toed eagles, honey-buzzard, black kite, Montagu's harrier, hoopoe, wryneck, **common rock thrush**, ring ouzel, red-backed shrike, siskin, **rock bunting**
MIGRATION garganey, griffon vulture, osprey, goshawk, **red-footed falcon, dotterel**, short-eared owl

VISIT DURATION 2–3 days

PLAN YOUR VISIT
Explore the sites by hiking on the numerous trails that start from minor localities; expect relatively long access drives. April and May are the best months for breeding species and migrating waterbirds (at La Narse de Lascols), while a visit between August and October would be better for migrating raptors and passerines.

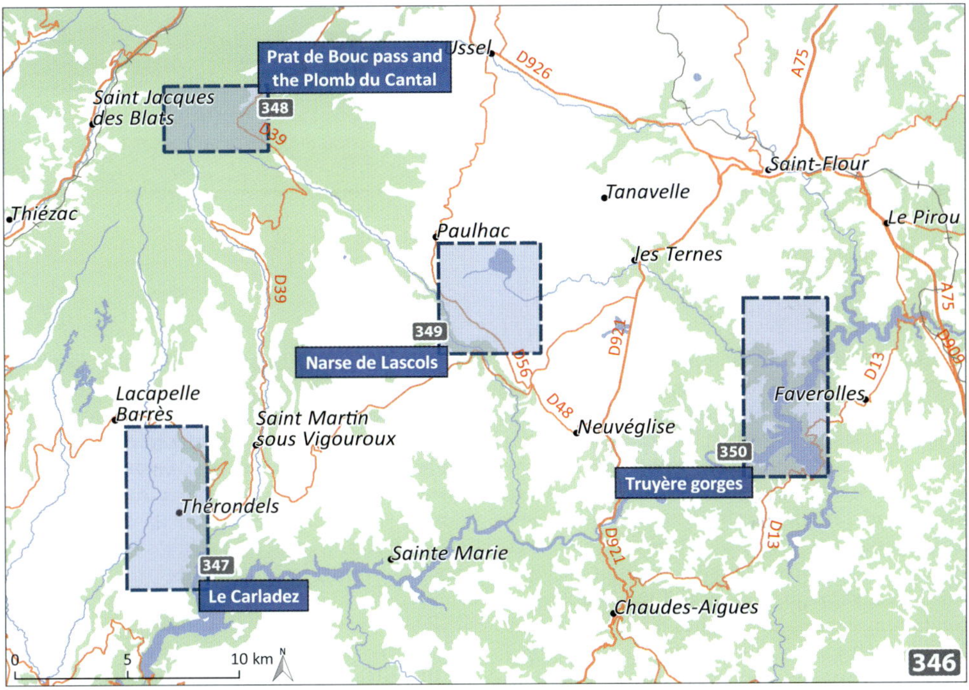

249 Le Carladez

0.5 DAY; MAP 347 Le Carladez consists of a **series of valleys** on the southern slopes of the Cantal mountains, close to the Truyère gorges and about 40 km east of Aurillac on the D990. **The mosaic of grassland with hedges and isolated trees attracts good numbers of** raptors **and open-country species.** Resident birds include woodlark, peregrine, red kite, raven and great grey shrike. Lower numbers of booted and short-toed eagles, black kite and red-backed shrike breed in the vicinity from April. **All the places shown on the map should be explored by roaming along tracks or minor roads**. Park in Thérondels [1] or drive north towards Frons [2] to walk on a track leading to Pont-la-Vieille (starts at [3], usually not driveable in a standard car). The same birds can be found on the 'Montagne de Greffeuille' [4] (from Jongues, drive 6 km, 10 min south from Thérondels via the D138 and D98).

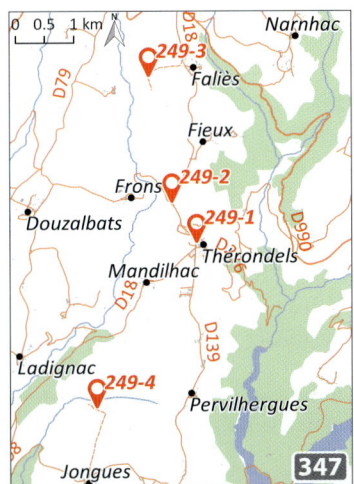

250 Prat de Bouc and the Plomb du Cantal

1 DAY; MAP 348 The 'Plomb du Cantal' is the highest volcanic summit of the Cantal mountains. Drive 50 km, 1 h east from Aurillac on the N122 or 25 km west from Saint-Flour on the D926 to Murat. Turn onto the D39 and drive through Albepierre-Bredons until you reach the Col de Prat de Bouc [1]. The site consists of a **mosaic of peat bog, moorland, stunted shrubs, subalpine meadows and rocky ridges. Its geographical position is strategic for autumn bird migration**.

Take the blue ski trail named 'Les Moutons' to the south between the restaurant 'Le Buron'

and the lodge. The track leads first to the main migration axis on the slopes of the 'Puy de Grandval' [2], where dotterel is sometimes seen in August and September. From there, it goes via the 'Col de la Tombe du Père' [3] towards the 'Plomb du Cantal' summit [4] before returning to the car park. This hike is likely to yield good view of breeding water pipit, wheatear, common rock thrush in scree areas, as well as ring ouzel and raven plus short-toed eagle and red kite. Raptor migration begins in August with black kite followed later in season (September to October) by honey-buzzard, marsh harrier, common buzzard, sparrowhawk and a few falcons (kestrel, merlin, hobby, peregrine), hen and Montagu's harriers, osprey, goshawk and booted eagle. Migrating passerines are dominated by large numbers of chaffinch, siskin, goldfinch, woodpigeon, hirundines and larks.

251 Narse de Lascols

0.5–1 DAY; MAP 349   La Narse de Lascols is a **mid-altitude marshland and peat bog** (1000 m) on the 'Planèze de Saint-Flour', a basaltic plateau at the base of the Cantal mountains, 15 km west of Saint-Flour. This wetland hosts good numbers of waterbirds and raptors in surrounding grasslands and fields, including harriers, kites, hobby, short-eared owl and sometimes red-footed falcon (May). The best season to visit the site is from March to May, when breeding and migrating species occur simultaneously. Take the D921 from Saint-Flour to Les Ternes, then the D57 towards Cussac (16 km, 20 min from Saint-Flour) to eventually turn right towards Lascols [1] (2 km, 5 min before Cussac). **From there, walk on the path that runs left to the north after the stone cross.** Look for waterfowl (teal, garganey, shoveler, gadwall), common snipe, sandpipers, curlew, lapwing, golden plover, water rail and the odd singing spotted crake. Also expect little owl, hoopoe and farmland passerines such as whinchat, warblers, shrikes, buntings and yellow wagtail. Search for the rare red-throated pipit in late April to May. **The visit can be continued by exploring surrounding hedgerows.** Another option is to drive on the minor roads between Cussac [2], la Jarrigue [3] and Farges [4] where red-backed and great grey shrikes are among the most interesting breeding birds.

252 The Truyère gorges

0.5–1 DAY; MAP 350   The Truyère is a tributary of the River Lot which runs in **deep gorges surrounded by a scenic mixture of woodlands, hedgerows and scattered trees especially attractive for** raptors. Along the river, the Grandval and Lanau dams have created **two large reservoirs** that are worth exploring for waterbirds. From Saint-Flour, drive for 10 km on the D40 to La Barge [1]. **A trail runs from the car park to the castle at Alleuze** [2], where raven and crag martin breed. From there, **walk on**

the trail towards Languiroux [3] and look for peregrine, red and black kites, rock bunting and breeding booted eagle on the southern slopes. Further raptors can be added while driving on the minor roads between Surgit, Bessols and Le Barry (stop for instance near [4]): expect short-toed, booted and golden eagles, griffon vulture, kestrel, common buzzard, sparrowhawk as well as farmland species, among which the most interesting are hoopoe, great-grey and red-backed shrikes, wryneck, warblers and several buntings. Rock sparrow breeds locally in the villages. You can also reach Fridefont via the D40 on the opposite bank, where the 'Belvédère du Mallet' [5] offers a scenic view over the gorges and soaring raptors (eagles, kites).

The Haute Loire (sites 253–256, map 351)

HIGHLIGHTS
» A variety of wild landscapes from deep gorges to mid-altitude mountains
» A good diversity of uncommon breeding species and a chance of wintering Alpine birds

J F M A **M J** J **A** S O N D

KEY SPECIES
YEAR-ROUND golden eagle, red kite, peregrine, eagle, **Tengmalm's and little owls**, crag martin, woodlark, **dipper**, crested tit, nutcracker, raven, firecrest, goldcrest, cirl and rock buntings
BREEDING short-toed and booted eagles, black kite, honey-buzzard, Montagu's harrier, Alpine swift, **common rock thrush**, ring ouzel, orphean warbler, **great grey** and red-backed shrikes, **citril finch**
MIGRATION osprey, **dotterel**, bee-eater
WINTER Alpine accentor, wallcreeper, snowfinch
VISIT DURATION 2–3 days
PLAN YOUR VISIT
Birding here requires time spent hiking in vast natural landscapes. Although spring (April to June) is the best time, there are several good wintering Alpine species to look for in mid-winter.

253 The Allier gorges
0.5–1 DAY; MAP 352 The River Allier has carved a **deep and wild canyon surrounded by wooded hills with scattered grassland and**

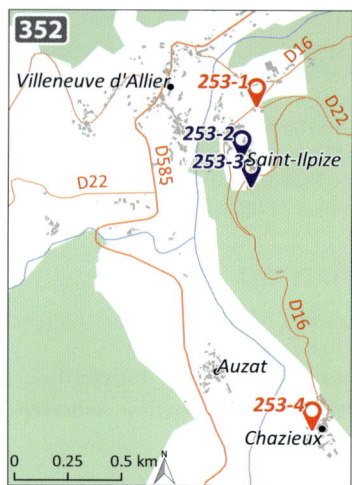
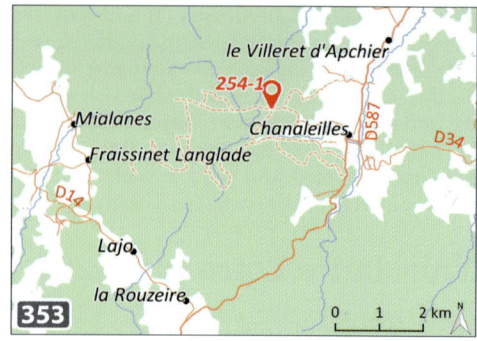

densities of crossbill, woodlark, fieldfare, mistle thrush, linnet, crested and coal tits, red-backed and great grey shrikes, meadow and tree pipits, stonechat, whinchat, citril finch, firecrest, goldcrest, both treecreepers and raven. A good selection of migrants cross the chain here; expect kites, storks, bee-eater, osprey at suitable seasons. Golden eagle has become regular in recent years.

255 Prades hill

0.5 DAY; MAP 354 The basaltic hill in Prades overlooks the River Allier and is surrounded by a nice mosaic of open, rocky and riparian habitats (30 km west of Le Puy-en-Velay: bypass Bains on the D589, then take the D481 to Vergezac and the D48 to Prades). You can either **hike from Prades** (park near the bridge [1]) **or from Saint-Julien-des-Chazes** [2], **or stop-and-go by car along the river**. The area hosts red kite, short-toed and booted eagles, peregrine and cliffs species (including crag martin, Alpine swift, rock bunting and wintering wallcreeper). A good viewpoint for the latter is **La Roche Servière**, on the northern bank of the Allier in Prades [3]. Dipper and common sandpiper are common along the river banks.

moorland. Drive to Saint-Ilpize (55 km, 1 h from Le-Puy-en-Velay on the N102, or 80 km, 1 h 30 south of Clermont-Ferrand by the A75 and N102; bypass Brioude and turn on the D16 at Vieille-Brioude along the River Allier). **Park in Saint-Ilpize** or near the sharp turn of the D22 in front of the town hall [1]. **Walk towards an old castle** [2] **where** rock bunting, crag martin **and** raven **breed**. One productive way to explore the area is to hike on the trail that starts near the cemetery [3]. Look for red-backed shrike, woodlark, stonechat, linnet, orphean warbler and yellowhammer, cirl and rock buntings. Dipper and common sandpiper occur along the River Allier between Chazieux [4] and Saint-Ilpize. By midday, short-toed eagle, kites, harriers, honey-buzzard, peregrine and hobby can be seen soaring or hunting if the weather turns sunny and warm.

254 La Margeride

0.5 DAY; MAP 353 The mid-altitude granitic La Margeride chain is a remote **mid-altitude mosaic of open country interspersed with pine–beech and spruce woods**, 60 km, 1 h 30 west of Le Puy-en-Velay. Take the D585 past Saugues and turn in Esplantas onto the D587 to Chanaleilles. From there, take the minor road to the right before leaving the village; **it leads to a wild area which can be explored by car or on foot, on good tracks** [1]. Most of the locally breeding raptors can be found here, including short-toed eagle, Montagu's harrier, honey-buzzard, black and red kites, Tengmalm's and little owls. There is a fine community of breeding passerines as well, with locally high

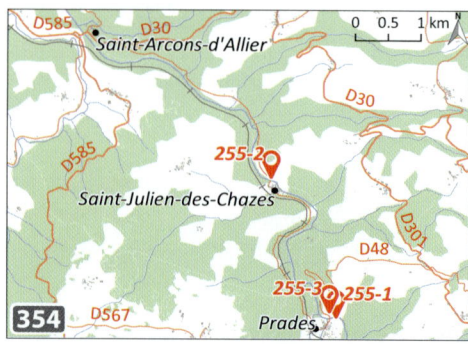

256 Mont Mézenc

0.5–1 DAY; MAP 355   Mont Mézenc is a volcanic hill (1750 m) surrounded by meadows and subalpine moorland which separates the high cold plateaus of the Velay from the Mediterranean hills of the Haut-Vivarais (30 km southeast of Le-Puy-en-Velay on the D36 through Lantriac and Laussonne; from the D36 in Les Estables, follow the D631 east and turn onto the D274 towards Chaudeyrolles). Stop at a large car park [1]: common rock thrush **breeds on the rocky hill between here and Jacassy** [2]. **Walk the trail that starts at the forest house** and runs to Croix de la Plonge (look for ring ouzel), Costebelle and Croix de Peccata (a good nutcracker spot). Citril finch, crossbill, firecrest, goldcrest, coal and crested tits are common in the conifer woods.

A trail from the Croix de Peccata car park [3] **leads to Mont Mézenc** [4], **where you should look for wintering** snowfinch, Alpine accentor **and sometimes** wallcreeper. Breeding species include water pipit and wheatear, and dotterel is annual on meadows and heather moorland. Follow the D274 for 1 km further north to les Dents du Diable [5]. Crag martin, ring ouzel, raven and eagle owl **breed on the rocky peaks** around. Follow the road to Chaudeyrolles and park near the first houses [6]. From there, a 3 km walk leads to a peat bog [7] where meadow pipit, whinchat, red-backed and great grey shrikes breed. Also expect black and red kites, hen and Montagu's harriers and short-toed eagle. Leaving the area on the D274 south, stop **near the small waterfall in front of Matagot** [8] **for** dipper.

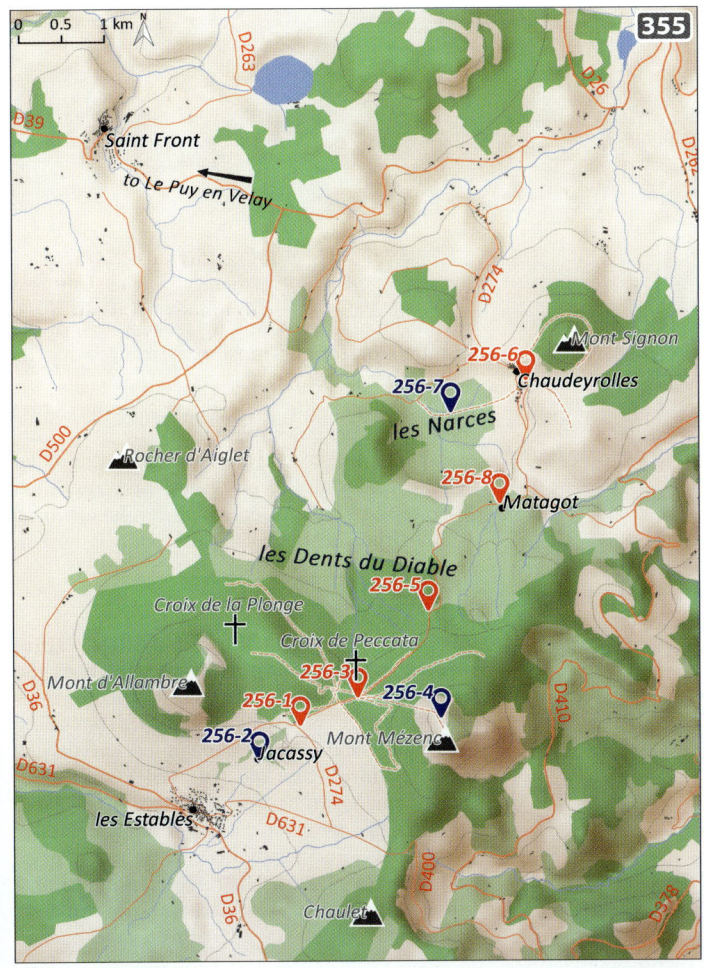

The Allier (sites 257–259, map 356)

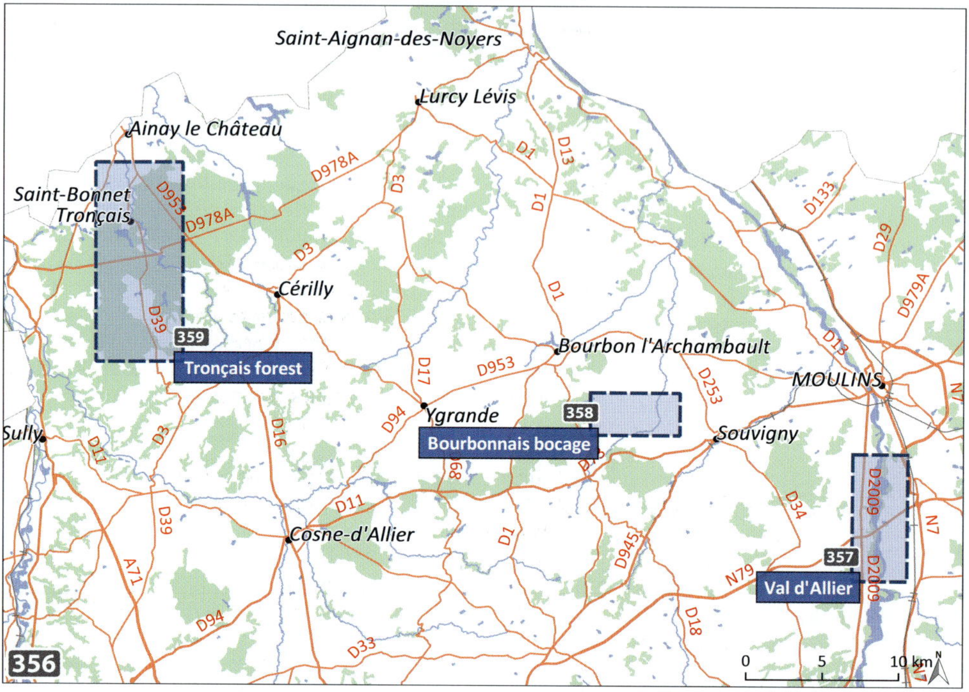

HIGHLIGHTS
» Breeding woodland species (including six woodpeckers) and raptors in Tronçay forest and surrounding open landscapes
» Breeding and migrating river birds on the banks of the Allier

KEY SPECIES

YEAR-ROUND red kite, goshawk, black, **grey-headed**, middle spotted and lesser spotted woodpeckers, marsh and crested tits, firecrest, goldcrest, hawfinch, cirl bunting

BREEDING night-heron, black kite, booted eagle, honey-buzzard, stone-curlew, little ringed plover, common and little terns, little and long-eared owls, nightjar, **bee-eater**, hoopoe, sand martin, woodlark, western Bonelli's, wood, garden and melodious warblers, common and lesser whitethroats, red-backed and **woodchat shrikes**

MIGRATION white stork, osprey, **waders**

WINTER common crane

VISIT DURATION 1–2 days

PLAN YOUR VISIT

Spend the early morning in Tronçay, one of the country's oldest and most extensive oak forests with several remarkable trees protected for their heritage value. The forest becomes quiet soon after sunrise, so visit it in the first hours of daylight and later in the morning roam along hedgerows for raptors and farmland birds. Then drive to the River Allier, which can be explored at any time, although late afternoon is best in spring.

257 Val d'Allier

0.5–1 DAY; MAP 357 **The Val d'Allier nature reserve extends along a 28 km stretch of the banks of the River Allier.** It encompasses a wide range of riparian habitats including gravel and sandy beaches, meanders and oxbow lakes resulting from old tributaries, meadows, moors and riparian woods. The most notable breeders are stone-curlew, common and little terns (gravel and sandy beaches), sand martin, bee-eater (sand banks), black kite, hobby, kingfisher, night-heron (riparian woods), common sandpiper, little ringed plover, red-backed

MASSIF CENTRAL | **259 TRONÇAIS FOREST** | 257

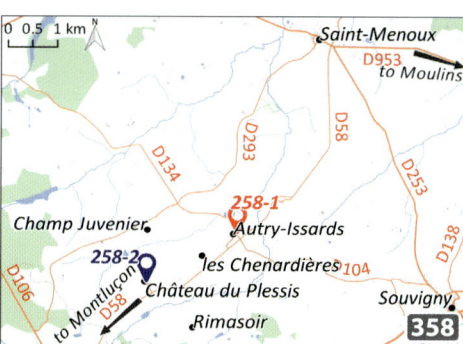

shrike and white stork. Migrating osprey, waders (sandpipers, stints…) and wintering common crane feature among the non-breeding species.

There are several possible itineraries along the river, depending on whether you prefer to walk or drive. Drive to Bressolles, 5 km south of Moulins on the D2009, and turn left 1.5 km after the village at the signes for 'Longvé/Château du Lys'. Turn left towards 'Les Girodeaux' and park at the nature reserve sign [1]. From here, you can walk on the little path that runs along the river bank. The sandy beaches can also be scanned from the opposite bank south of Toulon-sur-Allier [2]. Another option is to explore a section 8 km further south between Chemilly and Châtel-de-Neuvre where some tracks lead to the bank (e.g. in front of the Château de Prévia [3]) where grassland with hedges hosts farmland passerines (shrikes, warblers, buntings, nightingale).

258 The Bourbonnais bocage

0.5 DAY; MAP 358 The Bourbonnais bocage consists of **traditional landscapes of wooded pasturelands with hedgerows separating small agricultural plots** between Montluçon and Moulins. Drive 20 km, 20 min west from Moulins on the D945 to Souvigny then on the D104 to reach Autry-Issards [1]. You can either **explore the area west of the village** between the D58, D293 and D73 **by car or on foot** towards the 'Château du Plessis' [2]. Among the range of farmland and open-country species, look especially for woodlark, buntings (corn, cirl, yellowhammer), warblers (garden, melodious, common whitethroat, lesser whitethroat), hoopoe, red-backed and woodchat shrikes, stonechat, turtle dove and little owl. Raptors usually appear on warm days; expect booted eagle, kites and honey-buzzard.

259 Tronçais forest

0.5–1 DAY; MAP 359 **Tronçais ranks among the oldest and most extensive French oak forests** (100 km²), 40 km north of Montluçon (drive along the D2144 to Saint-Bonnet-Tronçais). The succession from regeneration to mature plots with very old and dead trees hosts one of France's richest woodland communities, featuring six woodpeckers (grey-headed, green, black, great spotted, middle spotted and lesser spotted), warblers (garden, western Bonelli's, wood), short-toed treecreeper, hawfinch, marsh and crested tits, firecrest, goldcrest, nuthatch, spotted flycatcher, cuckoo, golden oriole, nightjar (from late May), long-eared owl. Raptor diversity is at the same level; expect honey-buzzard (from May), booted eagle, goshawk, hobby, black and red kites.

The area can be explored by the many trails that cross the forest. Among the multiple options, you could drive to Tronçais on the D978a, head north at the bend on a forest track and park at the nearest junction [1]. Alternatively, take the D312 from Meaulne to Le Brethon and park at the 'rond du Pré Désert' [2]: **this area is especially good for** grey-headed woodpecker, to be searched in the oldest stands, preferably in the early morning in March and April. North of Tronçais forest, the landscape between Saint-Bonnet-Tronçais and Saint-Mamet deserves some slow driving along the minor roads either side of the D28 [3]. The surrounding grasslands and hedgerows are attractive for red-backed and woodchat shrikes, stonechat, bush warblers and buntings.

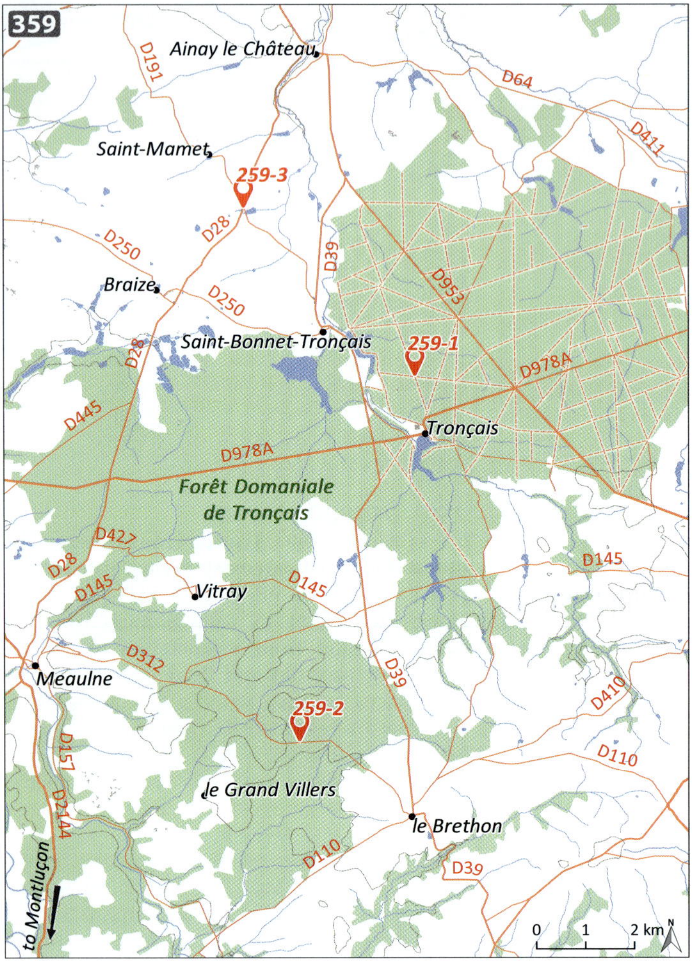

Auvergne (sites 260–262, map 360)

HIGHLIGHTS
» Several excellent sites for Tengmalm's owl
» Large diversity of mid-altitude mountain birds, from woodlands to hedgerows and cliffs
» Many uncommon breeding species

KEY SPECIES
YEAR-ROUND red kite, marsh harrier, peregrine, **Tengmalm's owl**, woodlark, crag martin, dipper, crested, marsh and willow tits, firecrest, Eurasian treecreeper, **great grey shrike,** cirl and rock buntings

BREEDING short-toed eagle, black kite, honey-buzzard, Montagu's harrier, nightjar, hoopoe, Alpine swift, ring ouzel, **common rock thrush**, common whitethroat, wood, western Bonelli's and melodious warblers, red-backed shrike

MIGRATION griffon vulture, goshawk, black stork

WINTER Alpine accentor, wallcreeper

VISIT DURATION 2–3 days

PLAN YOUR VISIT
These mid-altitude volcanoes are not well known as birding places, although they host some of France's best densities of Tengmalm's owl among other mid-altitude species. Most sites require a whole day to be fully explored and to locate all the species. Expect long hikes with occasional steep climbs to find mountain birds.

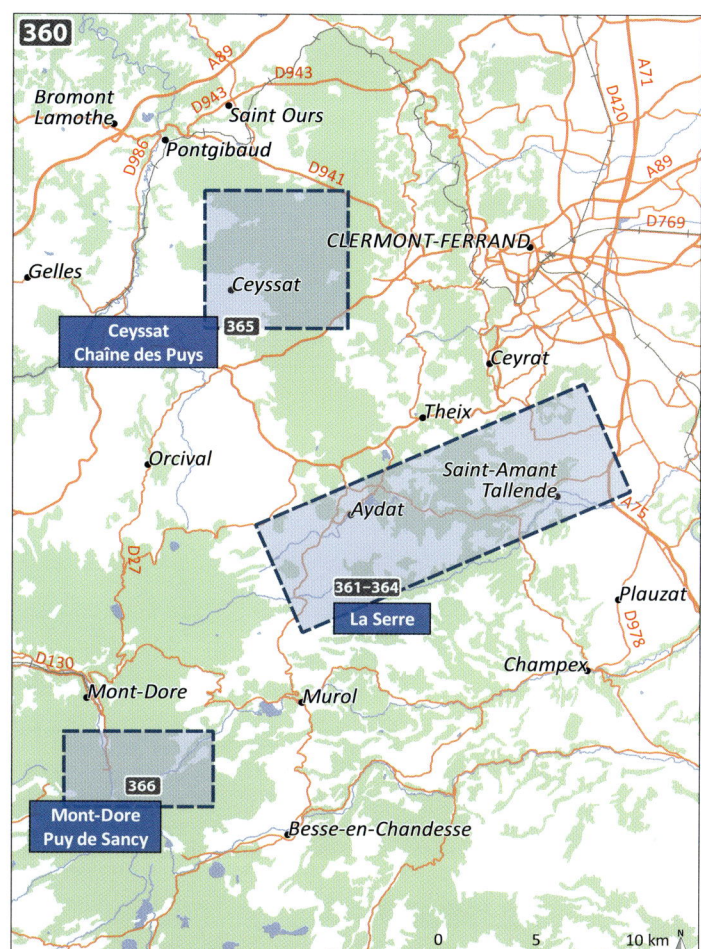

260 La Serre

0.5–1 DAY; MAPS 361-364   The Montagne de la Serre is a low-altitude (below 1000 m) 10 km long mountain chain located 20 km south of Clermont-Ferrand (exit 5 on the A75). Drive on the D213 towards Aydat for 6 km and turn right on the D96 to Chadrat. Drive for 2 km to a car park along the road [1]. From there, walk for 1 km to the Chapelle Sainte-Anne [2]. Alternatively, explore the area between Ponteix [3] and Saint-Saturnin [4]. **The mosaic of cultivated fields and open country with scattered bushes and woodlands hosts a good diversity of birds**: woodlark, honey-buzzard, buntings (corn, cirl, yellowhammer), Montagu's harrier, short-toed eagle, nightjar, kestrel, hoopoe, common whitethroat, melodious warbler, black and red kites, red-backed shrike, tree pipit, western Bonelli's warbler, nightingale, stonechat and turtle dove.

The pass northeast of Chadrat (park at [5] and walk east) **is a good spot for autumn migration** (late August to late October), dominated by black and red kites, honey-buzzard, marsh harrier, sparrowhawk, falcons, woodpigeon, stock dove, black stork, skylark, meadow pipit and finches.

To the west of the Montagne de la Serre, Aydat lake (Lac d'Aydat) is Auvergne's largest natural lake. There are few breeding birds, but **the site attracts a good diversity of migrating waterbirds** (sandpipers, common snipe, ducks, grebes) **in small numbers**. There is a car park along the D90 between Aydat and Sauteyrat [6]. From there, a path on the western bank of the lake leads to an observation platform [7] (best

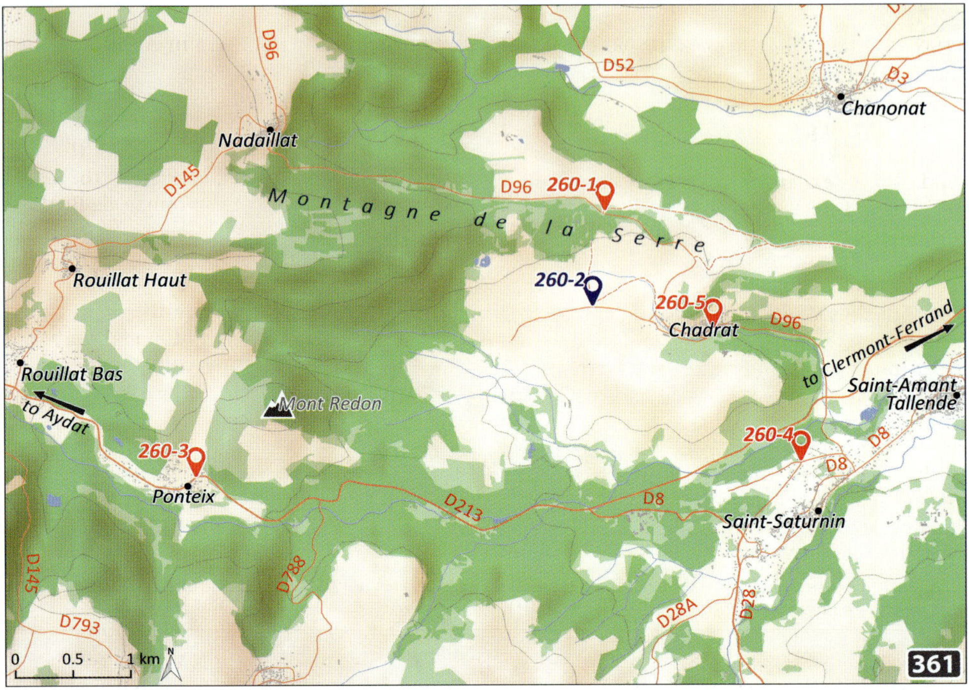

260 LA SERRE | MASSIF CENTRAL

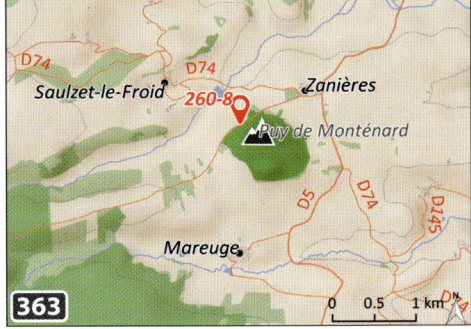

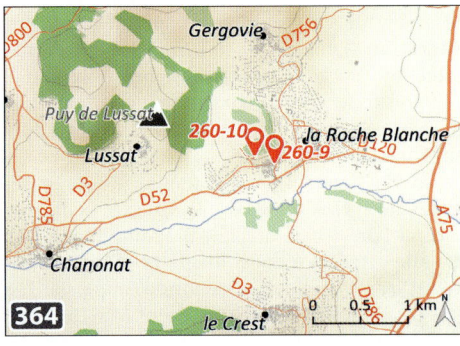

in the morning). Dipper and kingfisher which breed along the surrounding rivers are also regular visitors.

From Aydat, drive southwest for 5 km on the D5 to Zanières, then turn right on the D74 towards Saulzet-le-Froid. Near the exit from Zanières, Tengmalm's owl **breeds on a forested hill called 'Puy de Monténard' (park at the roadside near the trail that leads to a woodland** [8]**).** Search for red-backed (spring) and great grey (winter) shrikes in the surrounding meadows. From November to March, make a last stop in La Roche-Blanche [9] on your way back to Clermont-Ferrand, in the hope of catching a wallcreeper on the cliff above the village [10].

261 Ceyssat/Chaîne des Puys

1 DAY; MAP 365   The Chaîne des Puys is a narrow chain of volcanic hills running for 45 km on a north–south axis at the eastern edge of the Massif Central. The highest point is the Puy de Dôme (1450 m), 13 km, 25 min west of Clermont-Ferrand. **Wide expanses of mixed coniferous–deciduous forests on shallow slopes host a diversity of mid-altitude species in all seasons.**

To explore the surroundings of the Puy de Dôme from Clermont, follow signposts to the 'Puy de Dôme' and Ceyssat on the D68. The eastern slope is steep, highly urbanised in its lower parts, and has little to offer, while the western slope descends gently to Ceyssat through an old beech forest and pastures. Turn right 550 m before entering the village on a good track, drive 670 m and turn right. Drive as far as you can and park near a fork where the track climbs steeply into the forest [1], then walk. **Your main target here should be** Tengmalm's owl (song peak on clear, windless nights in March and April, usually most active around midnight). During the day, explore the woods and climb towards the tree limit. Expect black woodpecker, short-toed and Eurasian treecreepers, firecrest, goldcrest, seven tits (including marsh, willow at upper altitudes, abundant crested and coal), nuthatch, wood warbler (from April), crossbill, goshawk, bullfinch, hawfinch.

A good alternative for Tengmalm's owl is to take the D559 towards 'Vulcania' park. From Ceyssat, head towards Pontgibaud on the D52 and turn right after 3 km. Park along the D559, for instance at the 'Forêt sectionnale de Beuloup Les Roches' signpost [2]. A trail leads to the fir–beech forest and further towards the slopes of the Puy de Dôme, where all the species mentioned above can be found. **Yet another option for the owl is the Parc d'Allagnat, a mature beech stand southeast of Ceyssat along the D52** (park at the junction about 5 km after the village [3] and wander in the woods).

In the immediate surroundings of Ceyssat there are pastures. Great grey shrike **sometimes winters here, and red-backed shrike breeds from May; look for both on wires and fences.** Yellowhammer, stonechat, several warblers, hoopoe, red and black kites are all common in the area, for instance around [4,5].

Higher altitudes on the Puy de Dôme host mid-altitude and occasionally Alpine birds. Go back towards Clermont-Ferrand and stop at the car park on the 'Col de Ceyssat' [6]. From there, **a 45-minute trail ending with a steep 10-minute**

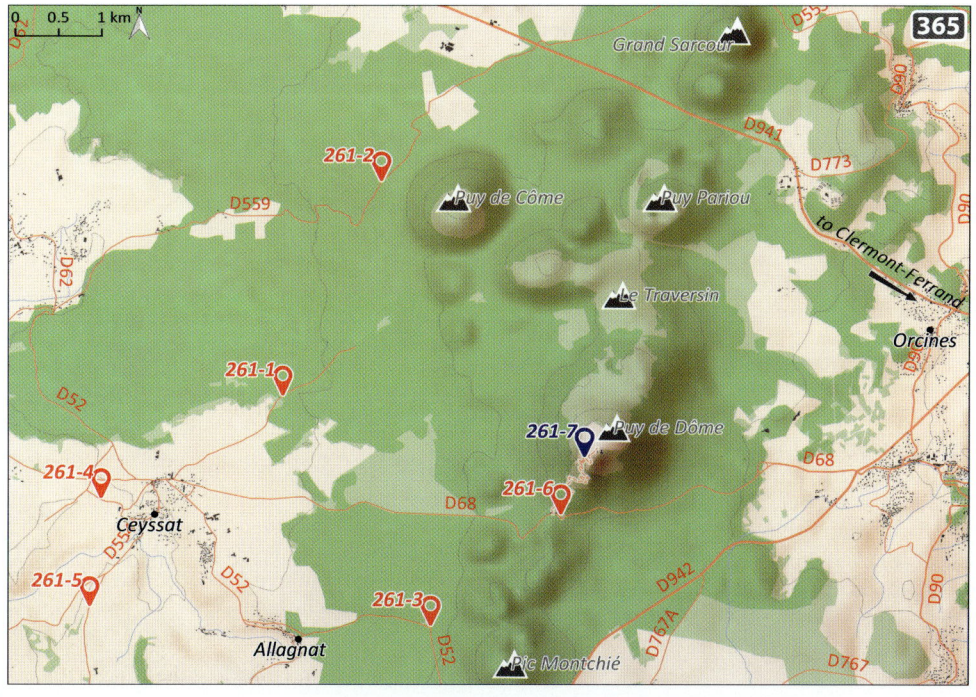

climb will take you to the summit of the **volcano** [7], where you should expect breeding rock bunting and water pipit and wintering Alpine accentor. Ring ouzel is regular from March to May and in October in the higher parts of the surrounding woods. Other rare yet regular visitors include griffon vulture, wallcreeper (winter), common rock thrush and Alpine swift.

262 Mont-Dore/Puy de Sancy

1 DAY; MAP 366   The Puy de Sancy is the highest peak (1885 m) of the Monts Dore volcanic chain, 45 km southwest of Clermont-Ferrand. Follow La Bourboule and Le Mont-Dore on the D2089 and D983, then from Le Mont-Dore follow 'Le Sancy' and park at the end of the road [1]. **In spite of severe alteration** due to mass tourism and a ski resort, **the Sancy remains the main location for mountain species** in the Massif Central. **Be on trails at first light in May and June, and avoid weekends**. Take the hiking trail to the west and climb to the Puy Redon and Tour Carrée ridges [2] through the Val de Courre. **The valley on your way to the top hosts** dipper, grey wagtail, pipits (water, meadow **and** tree) **and** crag martin. Wheatear and common rock thrush breed on the surrounding rocky screes. From here, either return to the car park or continue your hike to the Puy de Sancy, where Alpine accentor **breeds on the southern slope**.

You can also access the ridges from the Vallée de Chaudefour nature reserve (15 km,

30 min from Mont-Dore). From the southern end of Mont-Dore, follow 'Besse par col' and 'Croix Saint Robert' on the D36, then 'Besse' and 'vallée de Chaudefour' until you reach the car park at the reserve entrance [3], from where many trails lead to the ridges. **The main route towards the Sancy is long** (15 km, 5 h and a total climb of 935 m) but ranges across a nice selection of habitats from meadows and woodlands at the bottom of the cirque to subalpine grasslands and rocky slopes. Expect dipper and meadow pipit in the valley, Eurasian treecreeper, firecrest, goldcrest, tits (including marsh, willow, crested and coal) and black woodpecker in mixed deciduous–coniferous forest, ring ouzel on the northern slopes, crag martin, water pipit, wheatear, common rock thrush (rare), peregrine, raven and wallcreeper (rare) on ridges and cliffs.

Southern Ardèche (sites 263–265, map 367)

HIGHLIGHTS
» Many uncommon Mediterranean and cliff species concentrated in a few sites
» Scenic landscapes with gorges and plateaus typical of the Mediterranean biome

KEY SPECIES

YEAR-ROUND   red-legged partridge, **Bonelli's eagle**, peregrine, little and eagle owls, woodlark, crag martin, **dipper**, blue rock thrush, Dartford and Sardinian warblers, raven

BREEDING Egyptian vulture, short-toed eagle, black kite, **scops owl**, hoopoe, **roller**, bee-eater, Alpine swift, tawny pipit, subalpine, orphean and western Bonelli's warblers, woodchat and red-backed shrikes

VISIT DURATION   2 days
PLAN YOUR VISIT

The Ardèche gives you an introduction to arid landscapes inhabited by warm-climate species reminiscent of Mediterranean garrigues and maquis. Exploring the sites requires hikes on tracks and trails preferably at first light when bird activity is at its peak. Even Mediterranean species become quiet soon after sunrise.

263 Païolive/Chassezac

0.5 DAY; MAP 368   The forest of Païolive and the gorges of Chassezac, between Les Vans and Berrias (40 km, 1 h north of Alès on the D904), belong to the southern part of the Gras plateau, **typical of Ardèche landscapes**. This rocky limestone plateau covered by scrub and oak woodland was formerly cultivated and still shows the remnants of small walls and field plots. From

MASSIF CENTRAL | 264 LES GRAS PLATEAU

Les Vans, drive east on the D901 and turn left on the D252 after 5 km towards Casteljau. Stop in the roadside car park at 'Bois de Païolive' [1]. **A 5 km hike on the plateau above the Chassezac river should yield Mediterranean** warblers (orphean, Dartford, Sardinian and subalpine), western Bonelli's warbler, nightingale, tawny pipit, hoopoe, black kite and short-toed eagle. Grey wagtail, kingfisher and dipper breed on the river banks. It is also a good spot for eagle owl, Alpine swift, crag martin, blue rock thrush and raven, which all breed on the cliffs of the gorge [2]. Roller (rare) breeds around La Lauze, in the southernmost parts of the Bois de Païolive.

264 Les Gras plateau

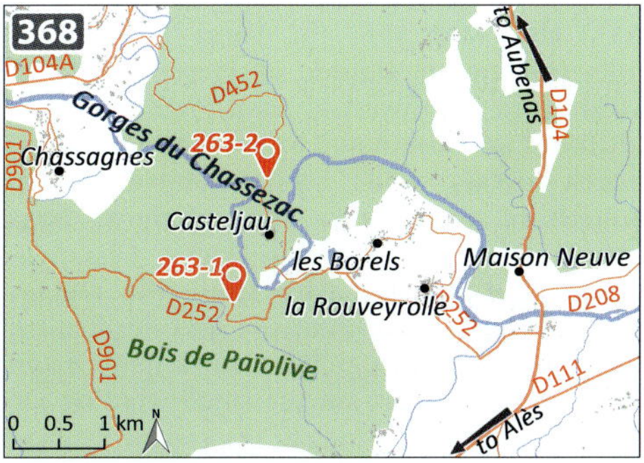

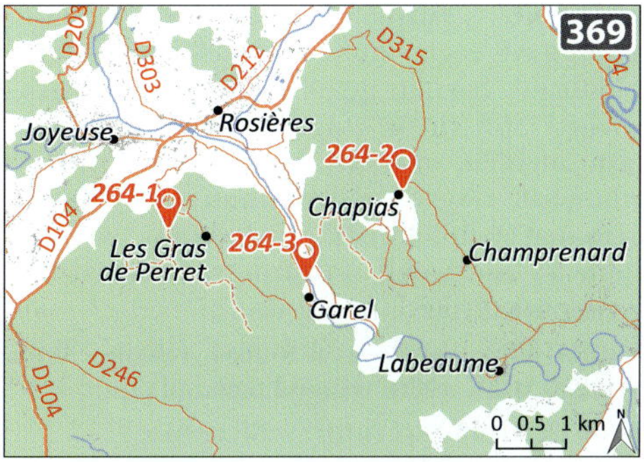

0.5 DAY; MAP 369   Les Gras plateau is located in the foothills of the Ardèche mountains, 25 km south of Aubenas on the D104. **This limestone plateau is covered with garrigues, dry meadows, scattered oaks, boxwood and lavender.** Once in Joyeuse, drive south following signs to 'Les Gras' and park after 3 km on the roadside at the start of a trail [1]. You can also walk from Chapias [2] (go back east towards Aubenas on the D104; 1 km past Rosières, turn right and drive towards Labeaume on the D315 for 4 km). Mediterranean and open-country species to be found on the plateau include short-toed eagle, black kite, little owl, red-legged partridge, woodlark, tawny pipit, hoopoe, blue rock thrush (rare), Sylvia warblers (orphean, Dartford, Sardinian and subalpine), crag martin and woodchat shrike. Scops owl breeds in the village and hamlets from April. Dipper can be found year-round on La Baume river, for instance near the 'Domaine Arleblanc' [3].

265 Ardèche gorges

1 DAY; NO DETAILED MAP   The Gorges de l'Ardèche nature reserve lies to the west of the Rhône, 77 km, 1 h 20 northwest of Avignon or 98 km, 1 h 30 southwest of Valence (from either direction, leave the A7 on exit 20 towards Mondragon, cross the Rhône in Pont-Saint-Esprit and drive 25 km to get to the nature reserve headquarters [1]). The gorges themselves run for 40 km between Vallon-Pont-d'Arc and Saint-Martin-d'Ardèche.

They are overlooked by limestone cliffs which mark the northern edge of the Mediterranean biome. Cliffs are covered with holm oaks, garrigues and maquis, while riparian willow and poplar woods border the river. The headquarters provides useful information on recent sightings, together with maps and advice on birding spots. You can either explore the reserve by walking on the many trails along the river or by driving on the D290, stopping on any of the 11 belvederes located on the eastern bank. Various Mediterranean and cliff species breed in the area, starring Bonelli's eagle (two pairs). Other breeding birds of interest include Egyptian vulture, peregrine, short-toed eagle, black kite, crag martin, Alpine swift, blue rock thrush, raven, dipper, Sylvia warblers (orphean, Dartford, Sardinian and subalpine), woodchat and red-backed shrikes, woodlark and bee-eater.

The Forez and the banks of the Loire (sites 266–270, map 370)

HIGHLIGHTS
» Good numbers of breeding and migrating waterbirds, including several dozen red-crested pochard
» Wide range of habitats, from valleys, ponds and forests to gorges and mid-altitude mountains

J	F	M	A	M	J	J	A	S	O	N	D

KEY SPECIES
YEAR-ROUND red-crested pochard, goshawk, hen harrier, red kite, woodcock, eagle and **pygmy owls**, middle spotted and black woodpeckers, woodlark, grey wagtail, crag martin, **dipper**, marsh, crested and coal tits, raven, **citril finch**, rock and cirl buntings

BREEDING   purple heron, night-heron, short-toed eagle, honey-buzzard, black kite, Montagu's harrier, quail, stone-curlew, black-winged stilt, little ringed plover, whiskered tern, nightjar, wryneck, hoopoe, sand martin, whinchat, ring ouzel, **common rock thrush**, common whitethroat, melodious, western Bonelli's and wood warblers, red-backed and woodchat shrikes

MIGRATION osprey, **red-footed falcon**, **Temminck's stint**, **white-winged tern** (all rare)

WINTER Alpine accentor, wallcreeper

VISIT DURATION   3–4 days

PLAN YOUR VISIT
If you stay around Lyon or Saint-Etienne, the Forez will keep you busy for a couple of days searching for mountain owls or looking for migrating waders that stop over along the Loire. The plain is also worth a visit in winter when wallcreepers come down from the Alps to stay on the ruins of the numerous medieval castles that spread along the hilly western side. Select your itinerary and the sites to explore according to the season and your target species. The wetland of the Ecopôle du Forez attracts wildfowl and waders at all seasons, but gorges and mountains will be at their best from March to June and may not be accessible in winter.

266 Hautes Chaumes

0.5–1 DAY; MAP 371   The Forez mid-altitude hills (maximum altitude at Pierre-sur-Haute, 1634 m) lie northwest of Saint-Etienne. **Lower altitudes are dominated by pastures and pine-beech forests**, while habitats **above 1400 m consist of vast moors and subalpine bogs** known as the Plateau des Hautes Chaumes. From Saint-Etienne, drive along the D8 for 40 km, 40 min to Montbrison and head north for 17 km more to reach Boën-sur-Lignon. From there, a 30 km drive on the small winding D6 through Chalmazel leads to the Col du Béal [1]. From there, **walk to the Col de la Chamboite**

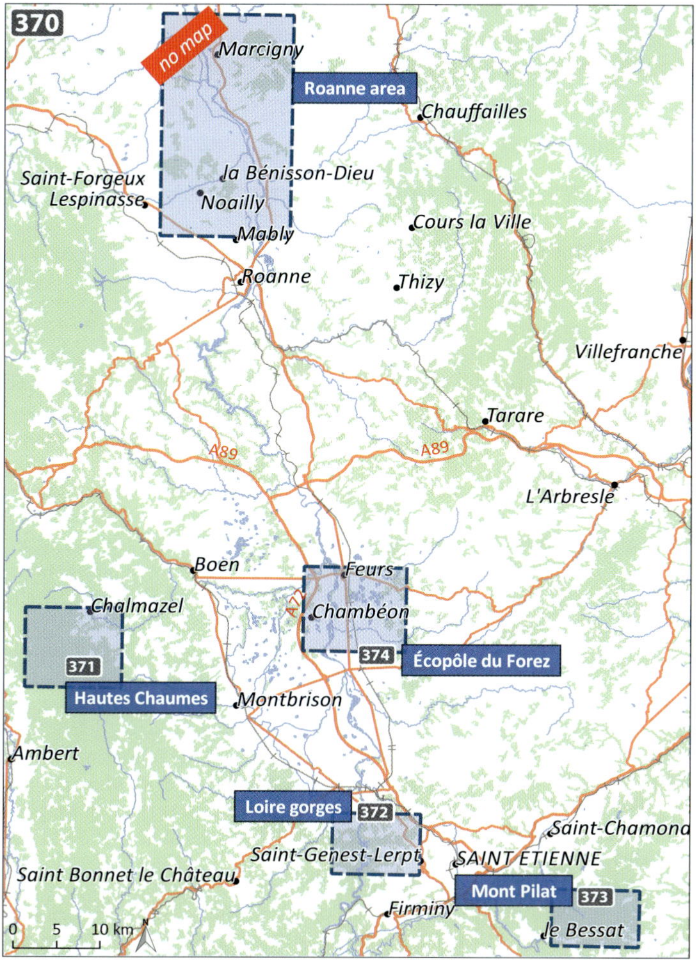

MASSIF CENTRAL | 267 LOIRE GORGES

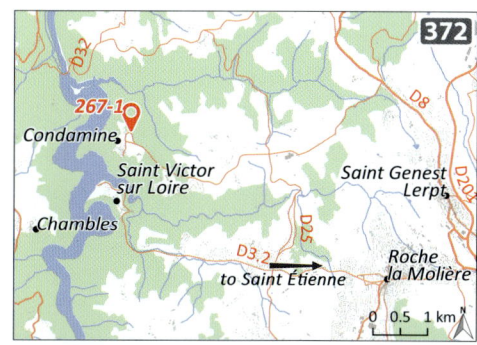

look for breeding hen and Montagu's harriers, short-toed eagle, red kite, quail, woodlark, linnet, common whitethroat, tree and meadow pipits, whinchat and stonechat, red-backed shrike and yellowhammer in open country. Forests host pygmy owl (rare, best sought at dusk). More likely encounters include honey-buzzard (from May), woodcock, black woodpecker, ring ouzel, fieldfare, both treecreepers, marsh, coal and crested tits, citril finch (uncommon), bullfinch and crossbill. Open areas at higher altitudes should yield common rock thrush (rare), water pipit, wheatear and raven (Alpine accentor winters in the area), and there are grey wagtail and dipper along mountain streams.

267 Loire gorges

0.5 DAY; MAP 372 **The Loire gorges, 15 km west of Saint-Etienne, consist of a mosaic of habitats with rocky cliffs, woods, pasture, meadow and moorland** which hosts a good range of lowland species. A birding day in May will usually yield black woodpecker, eagle owl, raven, red kite, short-toed eagle, western Bonelli's warbler, nightjar, hoopoe, wryneck,

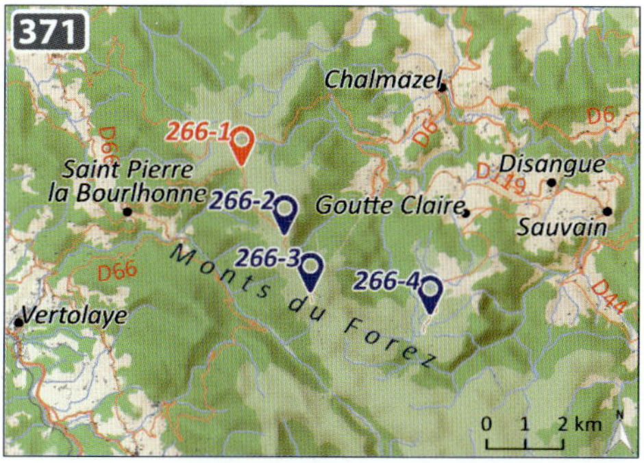

[2], **Pierre-sur-Haute** [3] **or the Jasseries de Colleigne nature reserve** [4]. The reserve can also be reached from Disangue (7 km from Chalmazel on the D101 towards Sauvain). In spring,

crag martin and rock bunting. In winter, look for wallcreeper on cliffs. A central starting point is the 'Gorges de la Loire' reserve headquarters in La Condamine [1] (signposted from Le Berland, take the D3 from Saint-Etienne through Roche-la-Molière). Several hiking trails head down to the gorges.

268 Mont Pilat

0.5–1 DAY; MAP 373 Mont Pilat is a mid-altitude massif that rises to 1400 m, 20 km southeast of Saint-Etienne. Its slopes are covered by mixed fir–beech–oak woods, moors and meadows. Follow the D8 to Le Bessat, then turn on the D8.6 to park at La Jasserie [1]. From here, **several trails lead to the ridges** via the 'Crêt de la Perdrix' [2] or the 'Crêt de Botte' [3]. The area hosts breeding honey-buzzard, woodcock, pygmy owl (rare), mistle thrush, crested and coal tits, crossbill, bullfinch and citril finch in low and mid-altitude woodlands, and raven and rock bunting on rocky slopes.

269 Ecopôle du Forez

0.5–1 DAY; MAP 374 **The Ecopôle du Forez** [1] **is a 160 ha reserve based on gravel pits bordering the River Loire** (fee; opening times and site maps at www.ecopoleduforez.fr). The surrounding **Forez plain is a wide mosaic of fields and small woods interspersed with over 300 ponds**, 30 km north of Saint-Etienne (take the A72 north and exit 6 in Chambéon, then follow signposts to the 'Ecopôle'). **Several hides** along a trail offer good views of the ponds and allow photographic opportunities for wildfowl, grebes and terns. Red-crested pochard, greylag goose, little grebe, black-winged stilt, little ringed plover, common and whiskered terns, purple heron and night-heron breed in the reserve around a colony of black-headed gulls – plus the usual wetland species. The reserve is a well-known stopover site for waders at both migrations (curlew, common snipe, ruff, shanks, stints, sandpipers), with records of regional rarities such as osprey, Temminck's stint and white-winged tern (May).

From May, spend time in the heat of the day exploring the surrounding fields and meadows between Chambéon, the Ecopôle and Feurs-Chambéon airfield [2]: you are likely to find breeding red-backed shrike, red-legged partridge, lapwing, curlew, stone-curlew and tree sparrow. The airfield itself sometimes hosts red-footed falcons on passage in May, do not hesitate to visit it several times a day in peak migration season.

270 Roanne area

0.5–1 DAY; NO DETAILED MAP **The Gravière Aux Oiseaux** [1] **is a 40 ha area of disused gravel pits beside the River Loire** (3 km from Mably hospital in the northern suburbs of Roanne). **A trail around the ponds leads to hides and viewpoints**. The site is at its best during waterbird migration (red-crested and common pochard, teal, shoveler and a few waders including common snipe, sandpipers and shanks). It is not as good as a breeding-season site, but grey heron, night-heron, little and great crested grebes, gadwall, little ringed plover, common tern, hobby, sand martin and reed bunting will justify a late-spring visit.

Further north, you will reach the **Lespinasse oak–hornbeam forest** (20 km, 20 min from the Gravière Aux Oiseaux, exit the D43 towards La Bénisson-Dieu and drive west for 10 km; 25 km, 30 min from Roanne, take the N7 to Saint-Forgeux-Lespinasse, then north on the D47 and

D18 towards Vivans). Local signposts to 'Forêt de Lespinasse' lead to a car park [2] from where several trails start. Walk along any of them in search of middle spotted and black woodpeckers, wood warbler, marsh tit, golden oriole, goshawk, honey-buzzard, short-toed treecreeper and spotted flycatcher. The surrounding hedgerows and meadows host black kite, hoopoe, melodious warbler, cirl bunting and red-backed shrike (also a few pairs of woodchat shrike).

The Limousin (sites 271–274, map 375)

HIGHLIGHTS
» Several uncommon breeding species along a gradient covering cultivated plains, peat bogs on low-altitude plateaus, bocage, ponds and gorges
» A sparsely populated region with a range of habitat mosaics

KEY SPECIES
YEAR-ROUND peregrine, woodcock, eagle and **Tengmalm's owls**, woodlark, crag martin, **dipper**, crested tit, firecrest, **great grey shrike**, raven, cirl bunting

BREEDING black-necked grebe, garganey, black and red kites, short-toed and booted eagles, honey-buzzard, night-heron, purple heron,

| J | F | M | A | M | J | J | A | S | O | N | D |

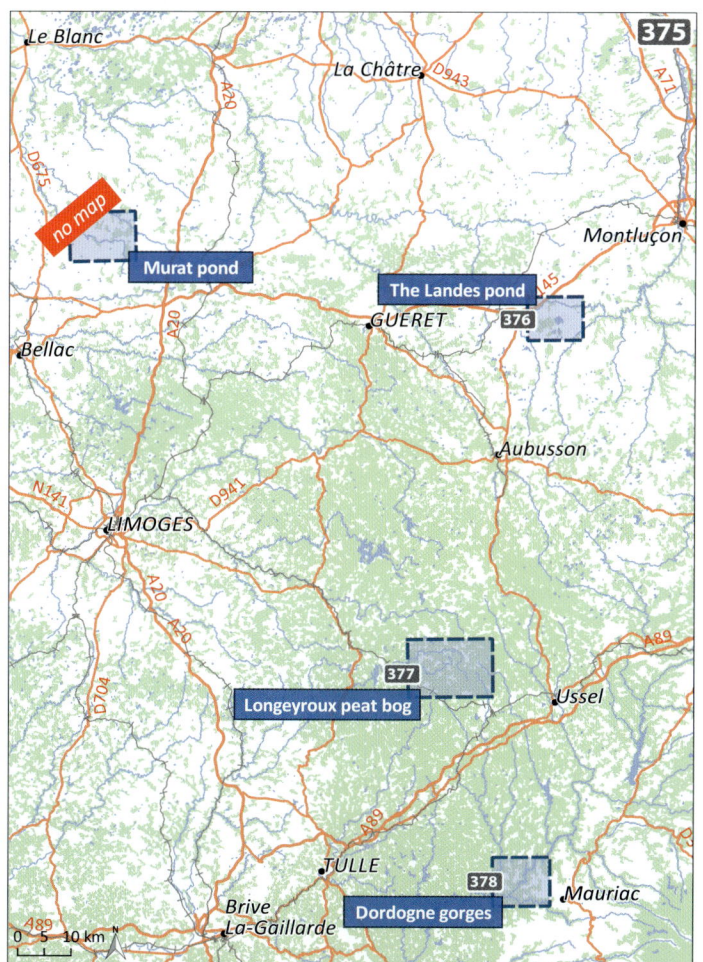

great white, cattle and little egrets, white stork, nightjar, whitethroat, grasshopper, sedge, melodious, western Bonelli's, wood and reed warblers, red-backed and woodchat shrikes

MIGRATION osprey, **black stork**, spotted crake, green sandpiper, waders

WINTER common crane

VISIT DURATION   2 days

PLAN YOUR VISIT

Wetlands are worth a visit all year round. The Longeyroux peat bog and the Dordogne gorges are at their best during the breeding season.

271   The Landes pond (Étang des Landes)

0.5 DAY; MAP 376   **The Landes nature reserve is a 100 ha natural pond bordered by reedbeds, willows and aquatic vegetation**, at the heart of a sedimentary basin covering the northwestern corner of the region (35 km southwest of Montluçon on the N145, once in Gouzon turn southeast on the D915 to Lussat and follow signposts). **The reserve is surrounded by meadows** with hedgerows and scattered trees. From the headquarters [1], **a trail (6 km/1 h 30) leads to four hides spread around the pond**. Breeding species include good numbers of night-heron, grey and purple herons, cattle and little egrets, white stork, pochard, water rail, little and great crested grebes, hobby, reed bunting, grasshopper, sedge and reed warblers. Summer visitors are replaced by wildfowl (teal, tufted duck, shoveler, gadwall, wigeon, garganey) from October to March, plus wintering or migrating great white egret, black-necked grebe, crane, peregrine, osprey, black stork (rare), spotted crake (rare), water pipit, common and green sandpipers (do not expect many waders on this site). **Take time to explore the fields around the pond** for red-backed and woodchat shrikes (May), woodlark, buntings (cirl, corn, yellowhammer, year-round), melodious warbler, whitethroat, turtle dove (from April).

272   Murat pond

1 HOUR; NO DETAILED MAP   **Murat pond is another option for waterbirds at the northwestern corner of the region close to the Limoges-Châteauroux motorway** (A20; take exit 22 and head to Saint-Sulpice-les-Feuilles, then the D912 westwards until the pond is signposted to the left, 1 km before Lussac; the road leads to a hide after 2 km [1]). There are few breeding birds (little grebe, grey heron and moorhen), but small numbers of wildfowl, herons and waders (common snipe, sandpipers) usually stop over at both passages.

273   Longeyroux peat bog

0.5 DAY; MAP 377   **The 8000-year-old peat bog at Longeyroux (tourbière du Longeyroux) is a 255 ha protected site** at an altitude of 850–900 m, lying on the Plateau de Millevaches, at the western edge of the Massif Central. Concentrate on a polygon delimited by the small villages of Meymac, Saint-Merd-les-Oussines, Chavanac and Saint-Sulpice-les-Bois (85 km, 1 h 30 east of Limoges, 70 km, 1 h northeast of Tulle, and 115 km, 1 h 30 west of Clermont-Ferrand). The easiest access is to drive south on the D109 from Saint-Merd-les-Oussines until you reach a car park beside the road after 5 km [1]. **Two trails ('sentier de la Linaigrette', 1 km, and 'sentier de la Bruyère', 9 km) allow you to explore a mosaic of bogs, moorlands, meadows and conifer woods**. Look for meadow and tree pipits, nightjar, woodlark, red-backed and great grey shrikes, stonechat, yellowhammer, woodcock, short-toed eagle, kites, short-toed and Eurasian treecreepers, firecrest, coal and crested tits. Tengmalm's owl is present but elusive.

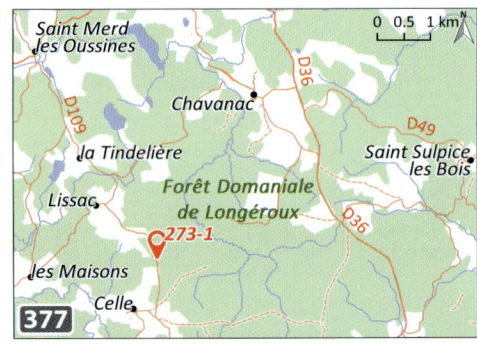

274 Dordogne gorges

0.5 DAY; MAP 378   The upper section of the River Dordogne runs from the heights of the Massif Central and is characterised by **deep gorges bordered by wooded hills which are mainly interesting for** raptors **and cliff species**. Drive to Chalvignac, 60 km east of Tulle on the D978 and 60 km north of Aurillac on the D922. **A 6 km trail (2 h 30 walk) starts at the village** [1] **and overlooks the gorges** and the impressive Aigle dam, providing several good viewpoints. Alternatively, you can reach the river bank by driving along the D105 towards Aynes [2]. Search for breeding short-toed and booted eagles, black and red kites, honey-buzzard, sparrowhawk, hobby, peregrine, raven, crag martin, eagle owl, woodland species (western Bonelli's and

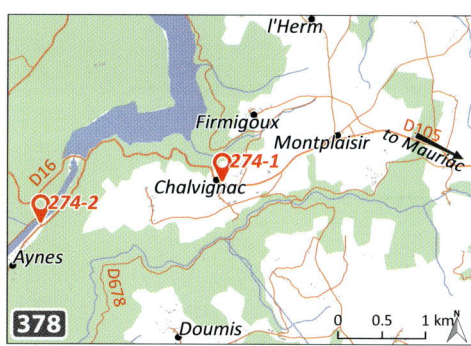

wood warblers, firecrest, woodpeckers), grey wagtail and dipper on the river and farmland birds (red-backed shrike, stonechat, warblers, buntings) in open areas.

REGION 13
BURGUNDY

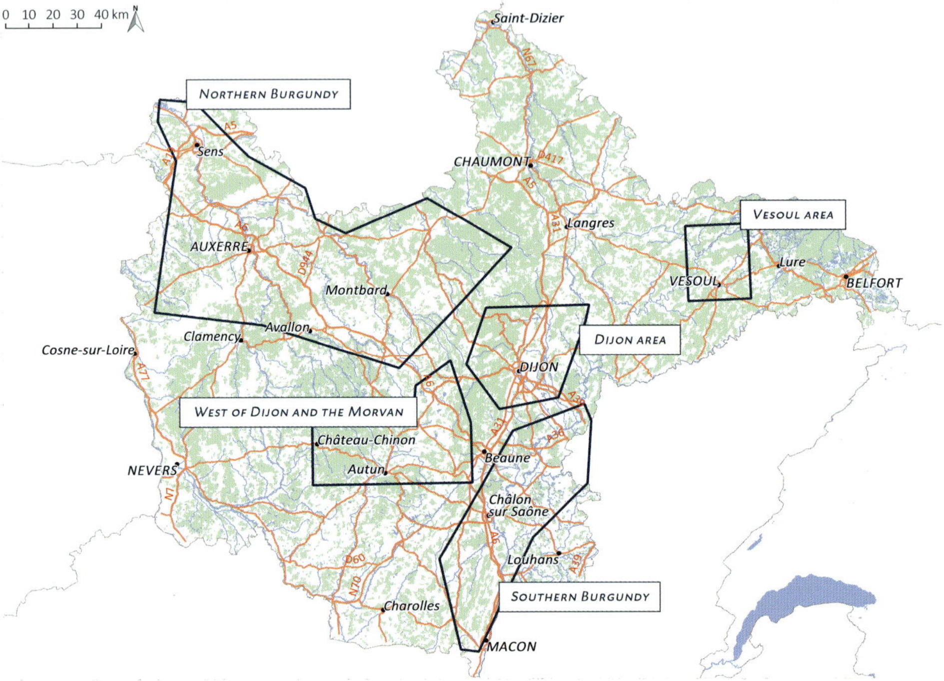

HIGHLIGHTS
» French strongholds of grey-headed woodpecker in old oak–beech forests
» A transition area from warm-climate to continental communities
» Access to several sought-after low-mountain birds

REVIEWERS François Bouzendorf, Antoine Rougeron

USEFUL WEBSITES
www.oiseaux-cote-dor.org
A detailed guide with maps and additional sites can be found at https://cote-dor.lpo.fr, and a list of guided birding tours and additional information (in French only) at www.faune-yonne.org

Cercey reservoir (site 289) is one of several ponds scattered in Burgundy's forested plains; they serve as stepping stones for migrants and host wildfowl in winter.

If you are looking for a challenge, try to locate grey-headed woodpecker in Burgundy's forests (e.g. in Is-sur-Tille forest, site 285). It will be hard, but the reward is worth the effort.

Northern Burgundy (sites 275–282, map 379)

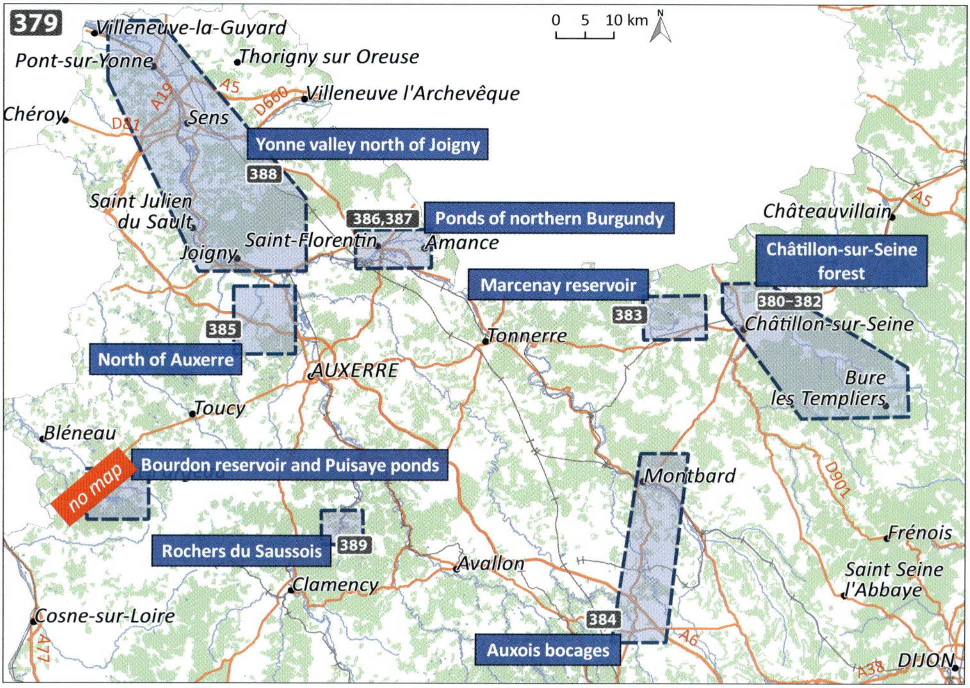

HIGHLIGHTS
- Extensive old oak and beech woodlands hosting bird communities typical of central European forests
- Passage migrants and wintering wildfowl on several large ponds and reservoirs

| J | F | M | A | M | J | J | A | S | O | N | D |

KEY SPECIES

YEAR-ROUND goshawk, hen harrier, woodcock, eagle, **pygmy and Tengmalm's owls**, black, middle spotted and **grey-headed woodpeckers**, crested lark, woodlark, **dipper**, crested tit, hawfinch

BREEDING black-necked grebe, short-toed eagle, honey-buzzard, red and black kites, Montagu's harrier, **black stork**, night-heron, **little bittern**, **corncrake**, stone-curlew, little ringed plover, wryneck, nightjar, hoopoe, **bee-eater**, sand martin, grasshopper, **Savi's**, marsh, sedge, great reed, reed, melodious, western Bonelli's and wood warblers, lesser whitethroat, spotted flycatcher, red-backed and woodchat shrikes

MIGRATION osprey, wood sandpiper and other **waders**, Mediterranean gull, black and whiskered terns, aquatic warbler (scarce)

WINTER whooper swan, **white-fronted and tundra bean geese**, **wildfowl** including goldeneye, smew, goosander and ferruginous duck (rare), **white-tailed eagle**, Eurasian bittern, **bearded and penduline tits**, wallcreeper

VISIT DURATION 1 to 2 days

PLAN YOUR VISIT

Your main targets should be breeding grey-headed woodpecker and black stork. These rare and secretive species are easier to find in this largely forested region than elsewhere in France. There is more to find along narrow rivers and isolated wetlands, which serve as stepping stones to migrating waders. **Visit forests from March to early June** or in winter; summer and early autumn are desperately quiet. **A first itinerary** (sites 275–277) starts and ends in Dijon or Troyes and can be covered in one or two days. From Dijon, start in Châtillon, visit Marcenay and drive back to Dijon through the Auxois bocages, possibly connecting with western spots. Sites 277–279 form **an alternative itinerary** from Auxerre, which easily connects with the southernmost sites of Region 1. In the vicinity of Auxerre, the Yonne valley should come up with most of the forest and open-country breeding species in a half-day spring visit.

275 Châtillon-sur-Seine forest

0.5 DAY; MAPS 380–382 Châtillon-sur-Seine forest (80 km, 1 h 15 min northwest of Dijon) ranks among the largest woodlands of Burgundy, **dominated by a mixture of old oak, beech and hornbeam stands of various ages and densities. Your target here should be** grey-headed woodpecker, **especially in clear stands with sparse old oaks.** Pygmy owl has possibly become established in the forest recently, while only a few pairs of Tengmalm's owl remain from a formerly well-established population. Woodcock, black and middle spotted woodpeckers, stock dove, wood warbler and hawfinch breed in older stands and are fairly reliable with a bit of effort. Spring birding in young stands will likely yield wryneck (rare), grasshopper warbler and nightjar (from May). Goshawk breeds in old beechstands preferably on northwest or northeast slopes; your best chance for this species will be to stay near a clearing from mid-morning, waiting for the birds to start hunting or displaying in March or April. In general, walking on tracks along edges and clearings will be the most productive approach in all seasons.

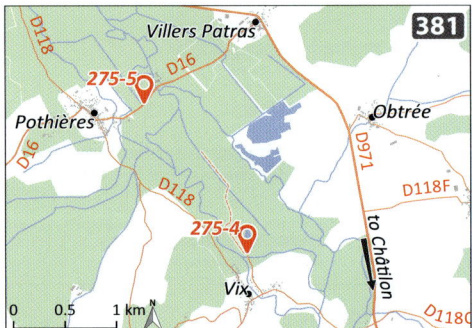

Enter the forest on the small road heading north 1.5 km past Rochefort-sur-Brévon on the D16 and stop-and-go on the 15 km stretch to Vanvey. **Park at [1] or in a similar place and walk along paths and tracks (driving is forbidden) preferably just after sunrise to catch drumming** woodpeckers. Another option is to explore the surroundings of the Val des Choues abbey [2] (turn left at Essarois church, 7 km, 10 min east **of** Rochefort-sur-Brévon). Black stork **and** honey buzzard breed nearby. The most reliable places for these two species are Vanvey forest ponds, along the River Ource [3].

There are two other options for black stork. One is to stop-and-go along the D118 from Vix to Pothières and Villers-Patras, spending time at [4] or [5], especially in late August and September. The other will be **the Brévon valley between Rochefort-sur-Brévon and Beaulieu** [6]. Besides storks, you should also expect grey wagtail, dipper and other riparian species associated with meadows and forest edges. Spend time in the marshes near Bure-les-Templiers (from Bure, take the D102J towards Chaugey for 2 km, turn left to the forest following signs for 'Marais du Côsnois', park at [7], and walk for 700 m to the main spot [8]) for bocage and forest species (goshawk, tree pipit, all tits, hawfinch, yellowhammer, woodcock).

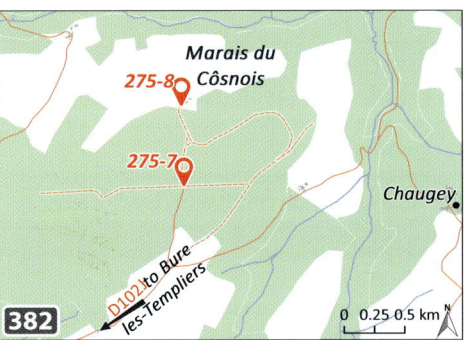

276 Marcenay reservoir

2 HOURS; MAP 383 One of the best sites for wintering waterbirds in the region lies 14 km, 15 min west of Châtillon-sur-Seine. Marcenay reservoir [1] is **a 100 ha artificial lake with deep water surrounded by reedbeds and small mudflats**, forming an ideal wintering or stopover area for migrating wildfowl **and** waders (D965 towards Laignes and turn right at the sign for 'Marcenay', then follow signs to 'Camping Les Grèbes' and spot the pond from the restaurant car park). The highest-profile regular winter visitors are Eurasian bittern, bearded (rare) **and** penduline tits, goldeneye, smew **and** goosander, which all occur irregularly as isolated birds or in

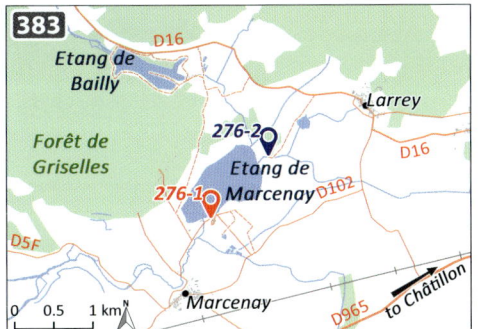

black stork (August and September, especially in marshes along the D32). A further 30 km, 30 min further south, a rich community breeds in a forested mosaic around Montigny-sur-Armançon. A drive between Roilly and Brianny [2] or in the surroundings of Précy-sous-Thil [3,4,5] might be rewarded by breeding red kite, red-backed and woodchat shrikes, hoopoe, woodlark, spotted flycatcher and melodious warbler from May. From there, you can either head back to Dijon (80 km, 1 h) or connect to the Val-Suzon (site 288) or to the Auxois reservoirs (site 289).

small groups alongside more common wildfowl. Divers, white-fronted and tundra bean geese have shown up after cold spells or during both migration periods. Osprey is a regular passage visitor from March to early May and from late July to September. A hide on the eastern bank [2] is well suited for scanning the reedbeds and the surrounding forest (walk northeast for 1.1 km on a trail beside the lake from the car park or drive to Larrey village and back to the lake along tracks): between April and June, expect sedge, great reed and reed warblers, little bittern, hobby, short-toed eagle and passing black tern (light is better in the morning). The reedbeds have recently hosted regionally rare migrating passerines such as aquatic, Savi's and marsh warblers.

278 North of Auxerre

1–3 HOURS; MAP 385   **Park along one of the tracks that run from the road bordering Auxerre airfield and explore the surrounding forest early in the morning.** Western Bonelli's warbler is common in this area from mid-April, especially around [1]. It can also be encountered on the short path which starts from the road linking Charbuy and Branches, at the north-western end of the airfield [2]. Wood warbler, middle spotted and black woodpeckers, crested tit and spotted flycatcher breed in surrounding stands in good numbers. As the sun rises, leave the forest northwards and check the cultivated plain north of the A6 on tracks [3 – see GPS file]. Montagu's and hen harriers and stone-curlew are likely in spring and summer, among more common breeding and migrating species. In winter, these fields serve as wintering grounds for lapwing, golden plover and merlin. From here, head to the Yonne valley either east (site 279) or north (site 280) of Joigny.

277 Auxois bocages

2–5 HOURS; MAP 384   Several places are worth a stop or even a couple of hours on your way back to Dijon from Marcenay (95 km, 1 h 30), especially from mid-April to June. **The ponds and woods surrounding Fontenay abbey** [1 – see GPS file] (signposted from the D905 in Marmagne, near Montbard) should yield little grebe, great white egret, goshawk, middle spotted woodpecker, dipper (year-round) and, with some luck,

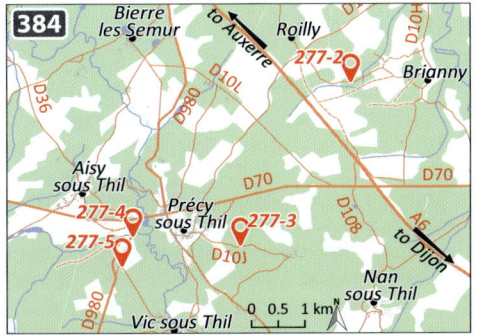

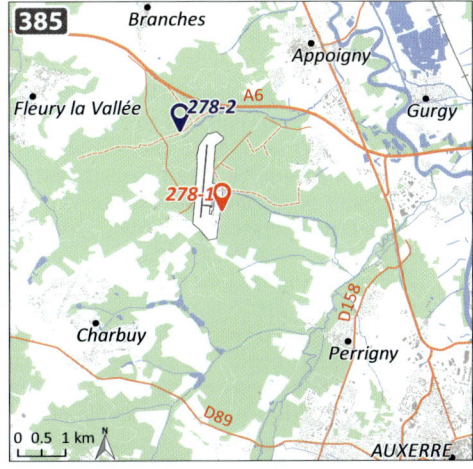

279 Ponds of northern Burgundy

2–4 HOURS; MAPS 386 & 387 The Bas-Rebourseaux nature reserve [1] is a 20 ha artificial pond embedded in a riparian mosaic of woodlands, open agricultural land and sandy banks along the River Armançon (25 km, 30 min north of Auxerre on the N77 to Troyes, turn left on the D43, bypass Vergigny and follow signs to 'Réserve ornithologique' past Bas-Rebourseaux). Most of the birding in the reserve is done from two hides (volunteers with telescopes every first and third Sunday of each month, 14.00–18.00). **The pond never freezes in winter, making it especially attractive to** wildfowl, herons **and wetland** passerines **such as** water pipit. Waders, osprey, black kite, black and whiskered terns and red-backed shrike stop over on migration in both spring and autumn. There is a bee-eater colony on the nearest loop of the Yonne to the east from late April [2]; black kite, little ringed plover and sand martin breed close to this viewpoint.

From here, drive east to Jaulges (7 km, 10 min) and take the D124 north towards Percey, stopping after 1 km. Park near a pond [3] where tundra bean geese and whooper swan can be found from time to time in winter among more common wildfowl. Waders stop over on the pond as long as there are mudflats, and a quick search of the surrounding trees and bushes is likely to yield willow tit. Further north on the D124, bypass La Chaussée and turn left on a good track after 1 km [4]. Corncrake may still be heard here in June, but hobby, wryneck, lesser whitethroat, willow tit and tree sparrow are more likely encounters.

280 The Yonne valley north of Joigny

2 HOURS – 0.5 DAY; MAP 388 Drive northwest 25 km, 30 min from Migennes to reach Saint-Julien-du-Sault gravel ponds and marshes [1] (turn right on a track 2 km past the last roundabout in Saint-Julien-du-Sault on the D3). **A ringing station operates here every autumn.** Record falls of several hundreds of passerines have been ringed. They are dominated by blackcap, reed warbler, common Phylloscopus species, but regional or national rarities are also caught every year (penduline tit is near-annual and there are several records of yellow-browed warbler). In addition to the attraction of the ringing station, single ferruginous ducks sometimes stay for a while among other wildfowl,

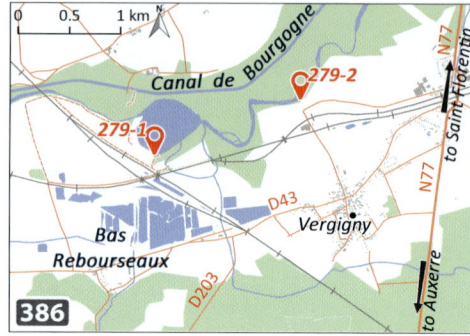

herons, Mediterranean gull, little tern, waders, black stork and osprey.

Several similar sites are ranged along a 20 km stretch of the River Yonne towards Sens: Véron [2] (little bittern, black-necked grebe, Mediterranean gull, black and whiskered terns, great reed warbler; also crested lark near Jardiland store at the northern end of Sens [3]), Villemanoche [4] (also good in winter for ferruginous duck, waders, golden oriole) and Vinneuf [5] (night-heron, stone-curlew, waders, Mediterranean gull, little tern). From here, it is a short drive to connect to the Bassée (Region 1, site 14, 11 km), a logical follow-up in late April or May if you wish to keep in with waders and breeding wetland species.

281 Bourdon reservoir and Puisaye ponds

2–5 HOURS; NO DETAILED MAP Bourdon reservoir [1] is isolated from all other spots, and mostly worth a visit for birders driving on the A77 between Paris and Nevers on the way to southern France (53 km, 55 min southwest of Auxerre, follow signs on the D965 from Saint-Fargeau). **This reservoir is embedded in a forested mosaic** where middle spotted and black woodpeckers and wood warbler are abundant. Osprey breeds nearby and can occasionally be seen flying over the pond from March to September. Black stork could also occur on migration (unreliable), and white-tailed eagle has wintered in the area. **The Bourdon is surrounded by a wide band of grass** where shanks, sandpipers and snipe sometimes stop over.

282 Rochers du Saussois

1–2 HOURS; MAP 389 **The Rochers du Saussois [1] are limestone cliffs, best known for hosting the most reliable breeding pair of** peregrine in the area (33 km, 32 min south of Auxerre; exit the D606 right, bypass Mailly-la-Ville and park 570 m past the bridge on the Yonne in Merry-sur-Yonne). Dipper occurs year-round and breeds on the nearby river. A visit at dusk means a reasonable chance of seing the local eagle owl,

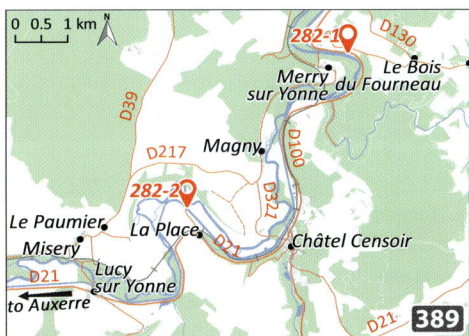

and the cliffs are a regular wintering site for wallcreeper. Bee-eater breeds on the river bank and can be seen on wires around La Place, 2 km west of Châtel-Censoir [2].

Vesoul area (sites 283–284, map 390)

HIGHLIGHTS

» The southernmost locations for collared flycatcher
» Reliable sites for grey-headed woodpecker and icterine warbler
» Old quarries and ponds with migrants and breeding marsh and great reed warbler

| J | F | M | **A** | M | J | J | **A** | **S** | O | N | D |

KEY SPECIES

YEAR-ROUND peregrine, little and eagle owls, black, middle spotted and **grey-headed woodpecker**s, wryneck, bee-eater, woodlark, cirl bunting

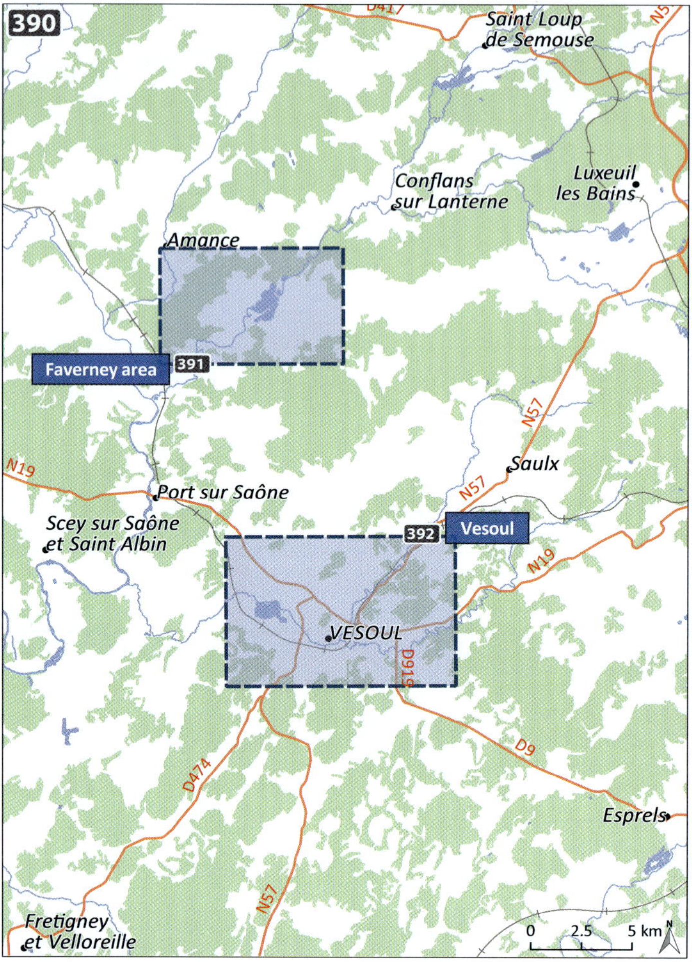

283 FAVERNEY AREA | BURGUNDY

BREEDING honey-buzzard, red kite, little ringed plover, nightjar, whinchat, melodious, **icterine**, great reed and marsh warblers, **collared flycatcher**, woodchat shrike

MIGRATION whiskered and black terns

VISIT DURATION 1 day

PLAN YOUR VISIT

The Vesoul area is mainly known for hosting three of France's most-wanted breeders: collared flycatcher, grey-headed woodpecker and icterine warbler. All three are sufficiently common and well localised to ensure sightings within one or two mornings in mid-May and June.

283 Faverney area

0.5 DAY; MAP 391 **Start your day at dawn seeking out** collared flycatcher **in the old oak–beech stands of the Bois des Balières [1] and surrounding woods** (20 km, 20 min north of Vesoul, reach Faverney and drive west for 3 km towards Amance, park for example near the Chapelle de la Dame Blanche at the junction between the D434 and the D57 and explore the surroundings). The species is relatively common in this area, but its secretive habits make it tricky to find. Learn its song and call to increase your chances. There are also a few pairs of grey-headed woodpecker, while black and middle spotted woodpeckers are common. Icterine warbler **breeds in hedgerows all around; there is no specific place to look for them, but a bit of search in late May or June should be successful (but be aware that** melodious warbler **also occurs).** The common suite of woodland mosaic breeders is present, among which honey-buzzard, little owl, bee-eater, wryneck and marsh warbler should fill your day.

In May or September, check Breurey ponds [2] for migrating stints, whiskered and black terns and breeding little ringed plover and great reed warbler (from Faverney drive 5 km

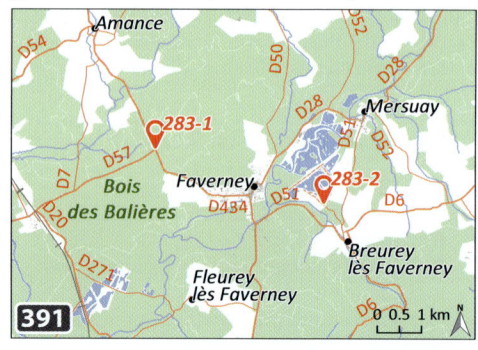

southeast to Breurey; once on the village centre drive north towards Equevilley and take the first road on left, then park after 800 m).

284 Vesoul

0.5 DAY; MAP 392   Two sites deserve a visit in the immediate vicinity of Vesoul. **East of the city, the Sabot de Frotey reserve lies at the end of the airport runway** [1] (6 km, 20 min from Vesoul centre, head to Frotey lès Vesoul, head northeast on the Rue Haute and follow signs to 'Aérodrome' and park 1 km after crossing the N19). **It hosts a typical, yet regionally uncommon community of dry open habitats**, with breeding woodchat shrike, nightjar (from May), woodlark and cirl bunting (year-round). Peregrine and eagle owl breed in the vicinity and can occasionally be seen flying over the reserve. **To the west, the Lac de Vaivre** [2] **is a highly**

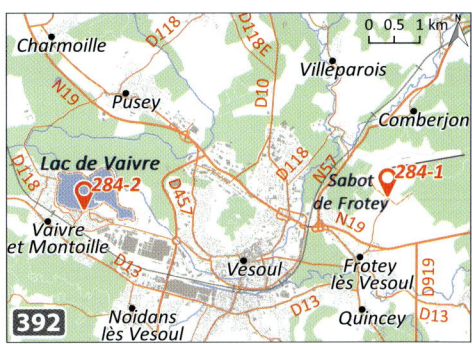

urbanised pond where terns sometimes stop over (4 km, 8 min from Vesoul, signposted). Check bushes for marsh warbler, stopping-over icterine warbler, and open areas for red kite, curlew, lapwing and whinchat (former breeder).

Dijon area (sites 285–288, map 393)

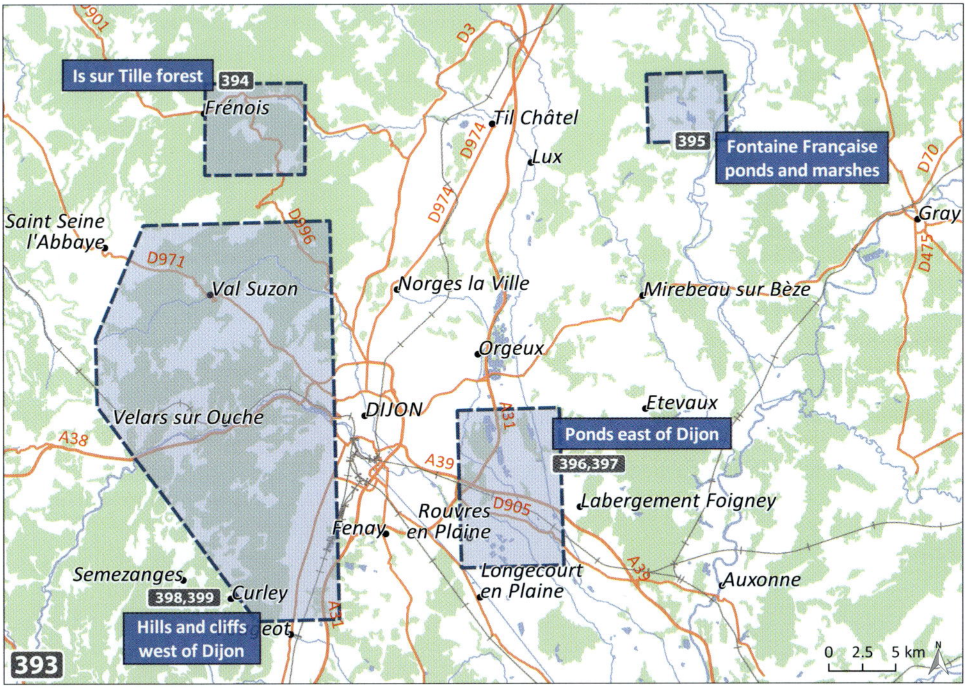

HIGHLIGHTS

» Ponds and reservoirs with reedbeds for breeding birds, migrants and wintering species
» Forests with the rare grey-headed woodpecker in the immediate vicinity of Dijon
» Cliffs and plateaus with species typical of dry, warm regions

J F **M A M J J A S O** N D

KEY SPECIES
YEAR-ROUND red-legged and grey partridges, goshawk, marsh harrier, woodcock, barn and eagle owls, **grey-headed** and middle spotted woodpeckers, **dipper**, woodlark, Eurasian treecreeper, hawfinch, cirl bunting

BREEDING red-crested pochard, short-toed eagle, Montagu's harrier, **black stork**, **little bittern**, little ringed plover, nightjar, sand martin, great reed, melodious, western Bonelli's, wood and fan-tailed warblers, common whitethroat, red-backed shrike

MIGRATION griffon vulture, **red-footed falcon**, **spotted and little crakes**, wood sandpiper, **Temminck's stint**, whiskered, black and white-winged terns, **ortolan bunting**

WINTER ferruginous duck, goosander, goldeneye, **white-tailed eagle**, **Eurasian bittern**, crag martin, Alpine accentor, **penduline tit**, **wallcreeper**, raven

VISIT DURATION   1 day

PLAN YOUR VISIT
The sites are organised in a one-day clockwise itinerary around Dijon which can be fruitful at any time of the year. They will give their best in mid-winter, and again once the late breeding birds are on their territories in mid-May.

285 Is-sur-Tille forest
0.5 DAY; MAP 394   Between March and June, start your day in Is-sur-Tille forest. **This beech and oak woodland has all the target breeding species one can expect in the region**, starring goshawk, black stork and grey-headed woodpecker as well as more common species such as middle spotted woodpecker, wood warbler and hawfinch. All these birds can be found at [1] (25 km, 35 min north of Dijon, follow signposts to Troyes on the D903 and D996, turn left towards Francheville just past the turning to Vernot; drive 700 m and park on a forest track to the right). Black stork is easier to find along the Ignon river stretch along the D901 from Lamargelle to Diénay, 12 km north on the D996 [2]. Other regional rarities breed along the river: look for grey wagtail, dipper and fieldfare. Come back at dusk for woodcock (autumn) and barn owl.

286 Fontaine Française ponds and marshes
2 HOURS; MAP 395   Fontaine-Française ponds and marshes (40 km, 40 min northeast of Dijon on the D28 through Lux) are **a regional hotspot for wetland species, worth a visit in any season**. Park on the track running alongside the northernmost pond [1] ('étang Pagosse') or in the village itself [2] to scan open waters where red-crested pochard breeds and single ferruginous duck, goosander and goldeneye winter each year with more common wildfowl. Take your time in the hope of a white-tailed eagle crossing the pond (best chances in mid-winter).

An alternative viewpoint for wildfowl is the southern end of the 'étang du Fourneau' [3], also suitable for migrating whiskered, black and the odd white-winged tern. **Boards at the northern end of the pond [4] give access to the reedbeds.** They host teal, marsh harrier, bittern and penduline tit in winter. Great reed warbler and little bittern breed, and migrating spotted crakes are regularly seen at this point (a spotted sandpiper occurred here in 2016). **Surrounding woodlands host the usual forest bird community and a few rare or uncommon breeding species:** Eurasian treecreeper, middle spotted and grey-headed woodpeckers, hobby and goshawk.

287 Ponds east of Dijon
3–5 HOURS; MAPS 396 & 397   Near Dijon, **two gravel ponds deserve a bit of time during the**

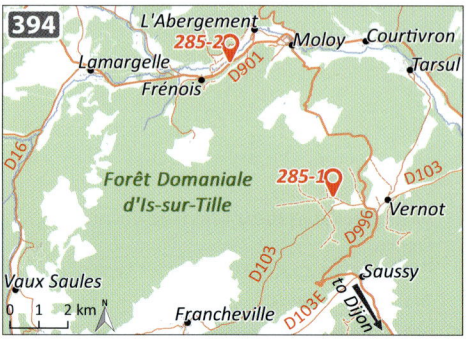

BURGUNDY | **288 HILLS AND CLIFFS WEST OF DIJON** | 279

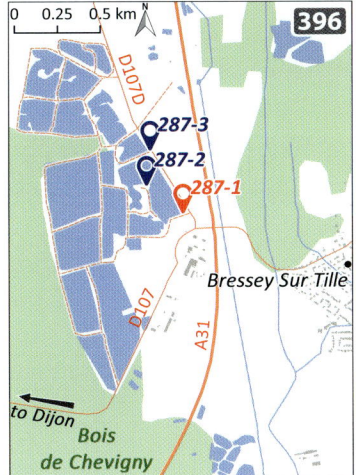

migration seasons. Start near Bressey-sur-Tille ponds (15 km, 25 min east of Dijon between Bressey-sur-Tille and Magny-sur-Tille, along the A31). Park at [1] and walk to [2] or [3] for little bittern and great reed warbler (from May). Eurasian bittern and penduline tit may occur from October to winter.

The nearby Rouvres-en-Plaine gravel ponds are a regional hotspot which has accumulated an impressive list of regional rarity records including red-necked grebe, little crake, red-footed falcon, white-winged and Caspian terns in recent years (10 km, 15 min south of Bressey-sur-Tille, park at [4] and walk to a hide 200 m northeast). More regular migrants include Temminck's stint (a few individuals per year) and wood sandpiper, while red-crested pochard, tufted duck, little ringed plover and sand martin breed on the ponds.

The two ponds 1.4 km south [5] are even better for migrants and wintering ducks; red-necked grebe regularly shows up here. A smaller pond 1.9 km from the main site [6] is also worth a visit for waders and wagtails (thunbergi yellow wagtail passes through in May). **Check the willows and fields 1.3 km to the east [7]: they are a magnet for spring migrants**. Siberian chiffchaff and ortolan bunting have been recorded here, and it is the place to search for red-footed falcon in mid-May. Red-legged and grey partdridge, quail and Montagu's harrier breed in surrounding fields, for instance on the D25 between Varanges and Genlis [8]. From here, you can go back to Dijon, connect to the Saône valley (site 293), or bypass Dijon to the south and head for the western hills.

288 Hills and cliffs west of Dijon

0.5 DAY; MAPS 398 & 399 **Just west of Dijon is a small chain of dry hills orientated roughly on a south–north axis**, home to southern species

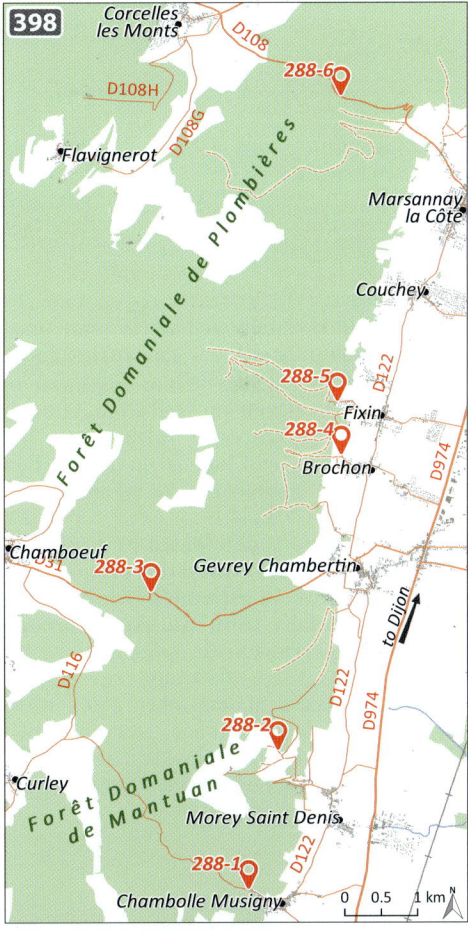

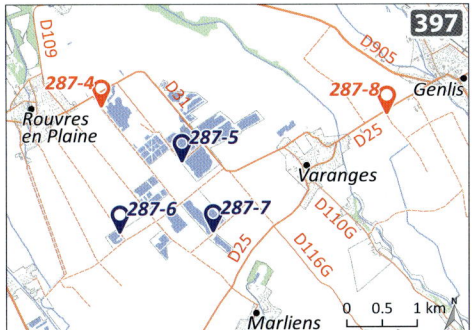

that are hard to find elsewhere in the region, not to mention wintering wallcreeper. Starting from the south (24 km, 30 min southwest of Dijon via the A311, exit 48), bird around Chambolle-Musigny [1] (an eagle owl breeds in the quarry at [2]), Gevrey-Chambertin [3], Brochon [4], Fixin [5], Marsannay-la-Côte [6] and end in Val-Suzon (from [7] to [8]).

The landscape is dominated by a pleasant mosaic of woodlands, vineyards, pastures and cliffs which offer an extremely rich breeding avifauna especially from late April through May. Common whitethroat, red-backed shrike (from mid-May), woodlark and cirl bunting (year-round) breed in semi-open areas, while the woodlands host hawfinch, western Bonelli's and wood warblers (from May) as well as a few pairs of grey-headed woodpecker (there are better sites for this species to the north of Dijon). A night drive from mid-May to July should yield nightjar without much effort. Alpine accentor, crag martin, raven and wallcreeper winter every year on cliffs and are most reliable in early March (locations change from year to year, check the news).

Baulme-La-Roche cliff is especially good for all these species (including grey-headed woodpecker**) and offers an excellent viewpoint for** raptors (park at [9] and walk to the cliff [10]): expect red kite, goshawk, short-toed eagle and maybe wandering griffon vultures (May and June). On your way, check the Suzon river for dipper (year-round). From there, it is a 30 km, 30 min drive on the A38 back to Dijon, and the same distance/time to end your day at the Auxois reservoirs (site 289).

West of Dijon and the Morvan (sites 289–291, map 400)

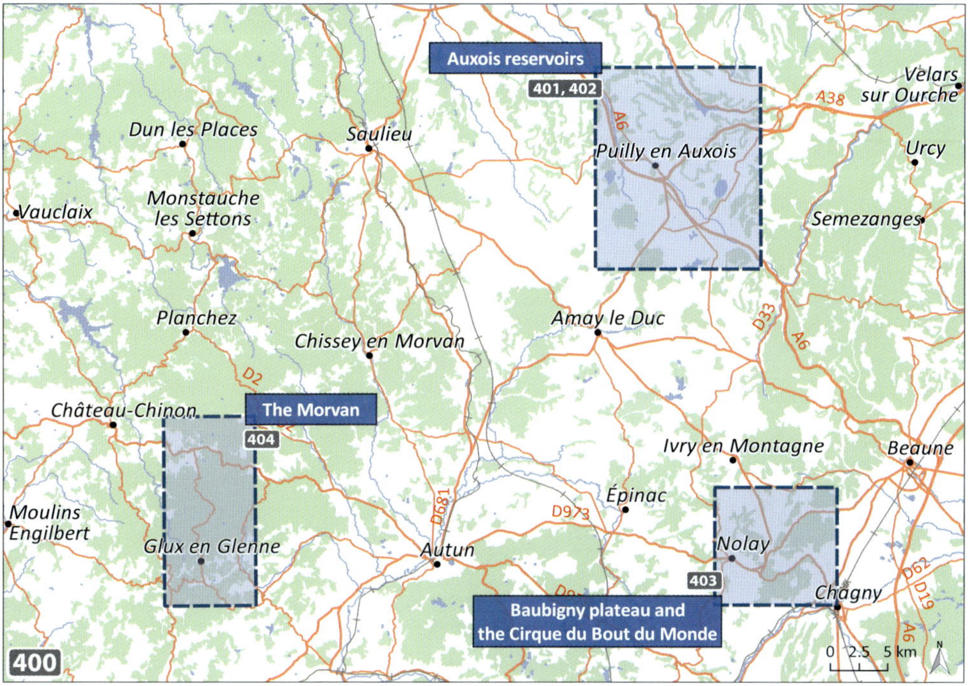

HIGHLIGHTS
» Access to mountain forest species in the Morvan
» Migration stopover sites on large reservoirs
» Search for regional or national rarities on plateaus and ponds

KEY SPECIES
YEAR-ROUND peregrine, woodcock, **pygmy, Tengmalm's, eagle and little owls**, dipper, crested tit, nutcracker, cirl bunting

BREEDING red kite, short-toed and booted eagles, quail, stone-curlew, hoopoe, nightjar, whinchat, **ring ouzel**, common whitethroat, grasshopper warbler, woodchat and red-backed shrikes

MIGRATION griffon vulture, **dotterel**, jack snipe, red-necked phalarope, white-winged tern, **Alpine swift**

WINTER Bewick's swan, **white-tailed eagle**, short-eared owl, **Alpine accentor, wallcreeper**, great grey shrike

VISIT DURATION 2 days

PLAN YOUR VISIT
Allocate a minimum of one day to the Auxois reservoirs and the Baubigny area, and one day to the Morvan mountains; do not try both in one day. **Sites are seasonal: the reservoirs are at their best at the end of summer for** waders **and from November for** wildfowl. **Higher altitudes in the Morvan are better from March to July, especially for** owls. Set out early, as the road to the Morvan can be long and mountain forests become quiet soon after sunrise.

289 Auxois reservoirs
0.5–1 DAY; MAPS 401 & 402 **A network of reservoirs west of Dijon serves as a stopover area for** waders **and** wildfowl **during autumn migration and in winter** (from Dijon, take the A38 west for 30 km, 30 min to exit 27, then continue for 4 km on the D905 towards Grosbois-en-Montagne). **Try to visit all of them, for you never know what might turn up and where.** A logical start would be Grosbois [1] (look for jack snipe in early spring and late autumn). Then visit Cercey [2], Panthier [3], Tillot [4] and Chazilly [5] (total itinerary

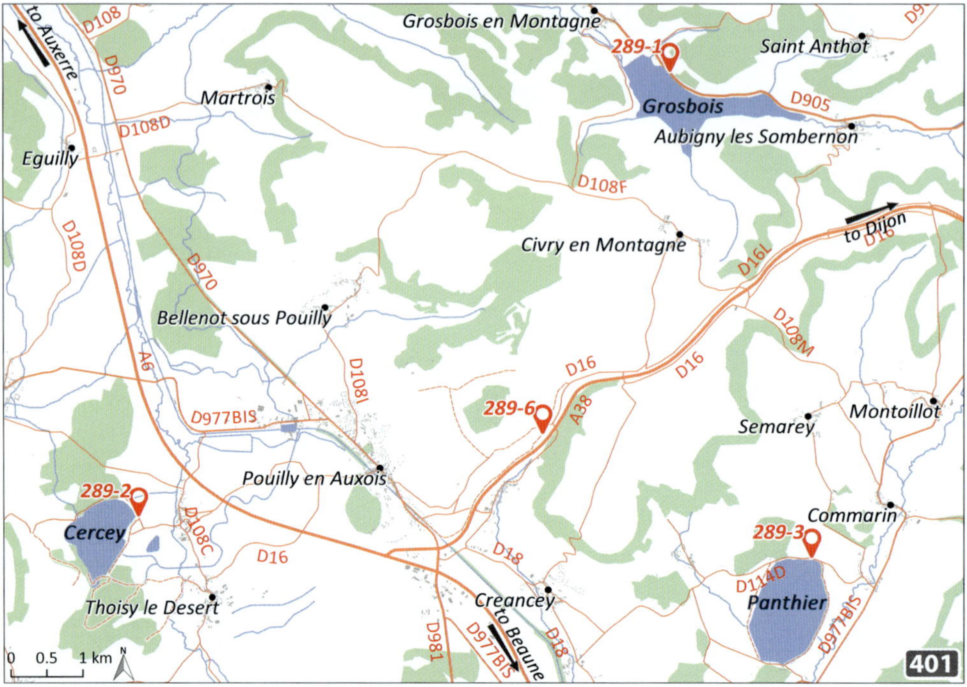

290 BAUBIGNY PLATEAU AND THE CIRQUE DU BOUT DU MONDE | BURGUNDY

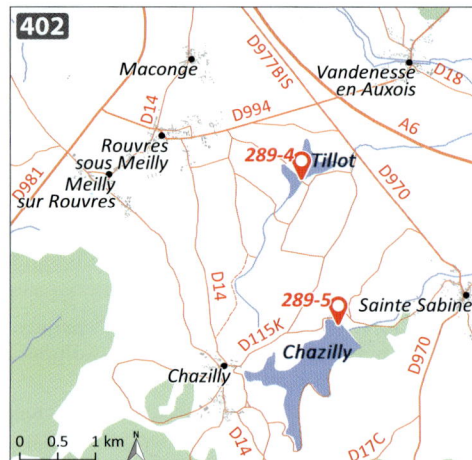

50 km). White-winged tern, red-necked phalarope, marsh sandpiper and even scarcer species such as sociable plover and pallid harrier have occured in the past. In winter, single divers, Bewick's swan, velvet scoter and white-tailed eagle sometimes show up with commoner wildfowl and herons. **One good reason to end at Chazilly reservoir at dusk is the** short-eared owl **roost which can be found on the grassy banks from October to March.**

Water levels tend to be too high from May to July in all the reservoirs, but the surrounding semi-open agricultural land hosts woodchat shrike (rare), hoopoe, little owl, red kite and booted eagle. Check fields, especially between Chazilly and Tillot, as roller, pallid harrier and Lapland bunting (all regionally scarce) have been recorded in the area along with large numbers of lapwing, golden plover and black-headed gull. The surroundings of Pouilly-Maconge airfield are also worth a check for great grey shrike. Before leaving the area, check the cliffs west of the A38 between Beaume and Civry-en-Montagne [6] (park on a track heading south near the bridge of the D16 above the motorway). Eagle owl is especially easy here, and wallcreeper sometimes winters.

290 Baubigny plateau and the Cirque du Bout du Monde

2 HOURS – 0.5 DAY; MAP 403 **The dry and stony Baubigny plateau** is an annual stopover site for dotterel (August to September) and a breeding site for stone-curlew (60 km, 50 min south of Dijon; leave the A31 at exit 24.1 south to Beaune and follow Autun until you find signposts to Baubigny). National or regional rarities have been recorded on these stony fields

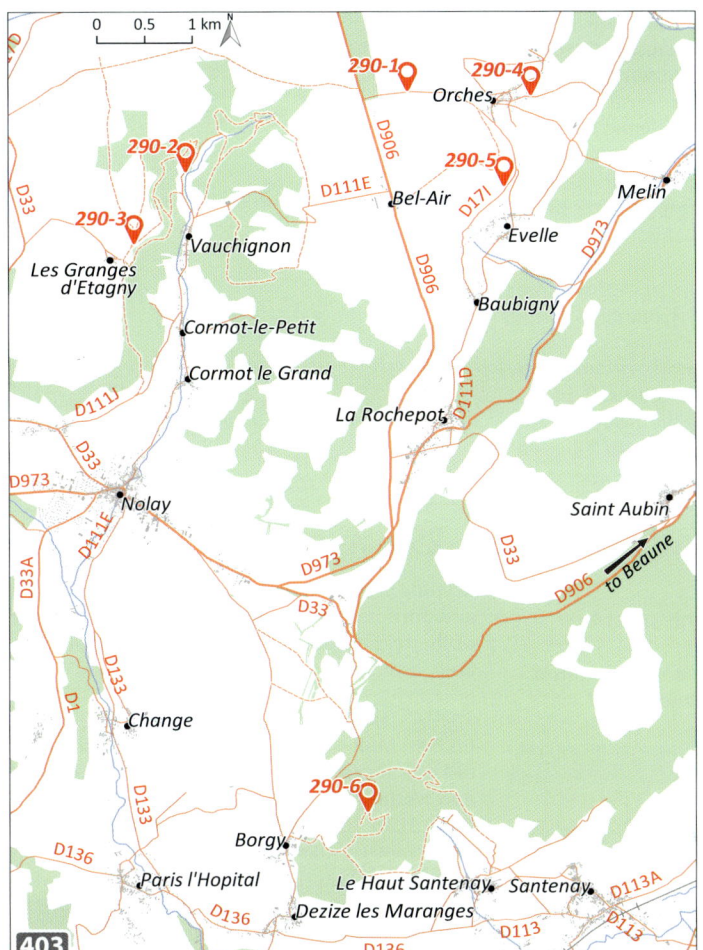

bushes interspersed within vineyards and meadows (e.g. around [4] or [5]). Commoner species include woodlark, common whitethroat (from April) and cirl bunting. Griffon vulture, short-toed eagle and Alpine accentor (early spring) show up occasionally in the Cirque. Orphean warbler could still breed, but it has not been recored since 2013.

Further south, cross the D906 towards Borgy to reach the Montagne des Trois Croix [6], an ideal viewpoint for raptors from April to August. Woodlark and cirl bunting are abundant in this area and nightjar should be easy to find during a night drive. Ring ouzel can be expected especially in the early morning in April and October.

291 The Morvan

1–2 DAYS; MAP 404 For birders visiting Burgundy between March and June, **the forested Morvan mountains offer access to a community of cold-**(rough-legged buzzard, red-footed falcon, great snipe, Pacific golden plover, roller).

A good start would be the small road between Orches and the D906 [1], but try to visit the plateau extensively and take your time, especially if you are looking for dotterel. **Nearby, the scenic Cirque du Bout du Monde** is worth a couple of hours in late afternoon as Alpine swift, peregrine and eagle owl breed and wallcreeper winters (follow signs to Vauchignon from the D906 to the west, there is an old signpost to the Cirque at the entrance of the village). Park at the bottom [2] or at the start of an easy three-hour walk up the cliffs [3]. Short toed and booted (rare) eagles may turn up in spring, as well as Alpine accentor in March.

Below the eastern end of the plateau, a few rock buntings **winter on old walls and** related species reminiscent of central European forests, including a few mountain specialities. The high point and best area for local mountain species is the Haut-Folin [1] (altitude 901 m; 30 km, 40 min from Autun, drive towards Château-Chinon on the D978 and follow signposts from Arleuf). Expect it to be hard work; concentrate on sunrise and one hour before/after sunset. Stop-and-go along the D500 between Arleuf [2] and Glux-en-Glenne [3] or park somewhere and walk along tracks and trails.

Most of the area is covered with an old mixed forest where coal and crested tits are abundant, as well as goldcrest and firecrest, crossbill and bullfinch. Woodcock occurs at higher altitudes, and there are a few pairs of

nutcracker which are usually more often heard than seen (best chance from August to October). **If the weather is clear with no wind, stay late and drive along the road several times at sunset** in the hope of a singing pygmy owl (best chance from March to May) or Tengmalm's owl (February to April, song peaks after 22.00 or even midnight). The use of playback is highly inappropriate, as populations of both owls in the area are extremely weak.

Use the quiet midday hours to explore lower altitudes around Arleuf and Glux-sur-Glenne, where quail, cirl bunting, red-backed shrike, whinchat and grasshopper warbler can all be found from mid-May. Dipper can be found on the rivers, for example along the nearby Yonne (stop-and-go on the D300 west from Glux-en-Glenne to check the banks and gravel islands).

Southern Burgundy (sites 292–295, map 405)

HIGHLIGHTS
» Among France's most reliable spots for singing corncrake
» Riparian meadows with year-round interest
» A transition between continental and Mediterranean influences

KEY SPECIES
YEAR-ROUND goshawk, marsh harrier, peregrine, **grey-headed** and middle spotted woodpeckers, Sardinian warbler, firecrest

BREEDING short-toed and booted eagles, Montagu's harrier, **night-heron**, **purple and squacco herons**, cattle egret, **corncrake**, stone-curlew, little ringed plover, **bee-eater**, **Alpine swift**, sand martin, bluethroat, whinchat, lesser whitethroat, grasshopper, willow, wood and fan-tailed warblers, woodchat and red-backed shrikes, **ortolan bunting**

MIGRATION spotted and little crakes, waders including Temminck's stint and wood sandpiper

WINTER whooper swan, goosander, goldeneye, **white-tailed eagle**, **Caspian gull**, **great grey shrike**

VISIT DURATION 2–3 days

PLAN YOUR VISIT
Start a spring birding day in the early morning in Citeaux forest, then from mid-morning explore the Saône valley starting at Maillys preserve and descending slowly southwards to Châlon. If you have some time left, also visit the Doubs valleys, keeping time at dusk to listen for spotted crake and corncrake in riverside meadows. The south of the region gives a chance for ortolan bunting and Sardinian warbler.

292 Citeaux forest
0.5 DAY; MAP 406   Citeaux forest is a protected **old oak forest on wet soil with numerous bogs and wetlands** (30 km, 40 min from Dijon: leave the A31 in Nuits-Saint-Georges following signs to Gerland). Enter the forest by turning left 4 km past Gerland on the D35 and park at [1] to walk 1.4 km around a square route. **First head north-east; stands on the left are under reserve status and host good densities of** middle spotted woodpecker, wood warbler **and** firecrest. Goshawk should not be too hard to find in this area. After 700 m turn right [2] and walk for 1 km east [3]; these stands are reliable for grey-headed woodpecker. Turn right again after 700 m [4] to head back to your car. Along the way, younger stands host willow and grasshopper warblers, together with all the commoner forest birds.

293 The Saône valley
0.5 DAY; MAPS 407–413   Several good spots for wetland species are spread along the River Saône from Dijon to Châlon sur Saône. Most of them are ponds, river meanders and oxbow lakes surrounded by meadows and arable land. These sites are at their best when spring rains change meadows into extensive marshes. Purple heron, night-heron and sometimes squacco heron join

BURGUNDY | 293 THE SAÔNE VALLEY | 285

numbers of wigeon and great white egret (36 km, 30 min southeast of Dijon, leave the A39 at exit 5 to Auxonne, drive to Champdôtre, and turn left on the D31b towards Tillenay to reach the track to the reserve). Scarcer species such as Slavonian grebe, divers or white-tailed eagle might show up during cold spells. The site is not as rich in spring but it hosts a mixed colony of cattle egret, little egret and night-heron, and curlew still breeds here in small numbers. Look for breeding lesser whitethroat and red-backed shrike along surrounding hedgerows.

Drive 15 km, 20 min south to Aillon pond [2] (access from rue Vernier, at the eastern end of Saint-Symphorien-sur-Saône), the best spot in the Dijon area for migrating waders in spring (late April to May) and autumn (as soon as mid-July through October). Concentrate on the mudflat that emerges on the northern bank. Rarities recorded in the recent past include pectoral sandpiper, long-billed dowitcher and regional rarities such as gull-billed tern, little crake and glossy ibis. Temminck's stint and spotted crakes are more reliable (annual sightings), and wood sandpiper,

passage waders from April. Corncrake and bluethroat breed in small numbers. In winter, geese and gulls roost if the meadows are not too flooded.

For a winter visit, try Les Maillys [1] reserve for wildfowl: expect goosander, goldeneye, good

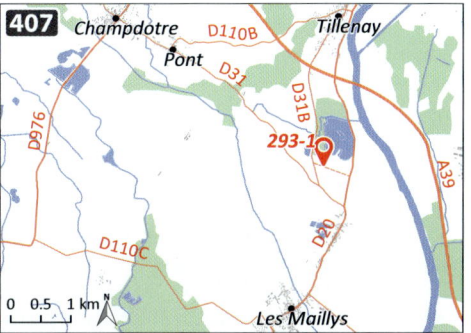

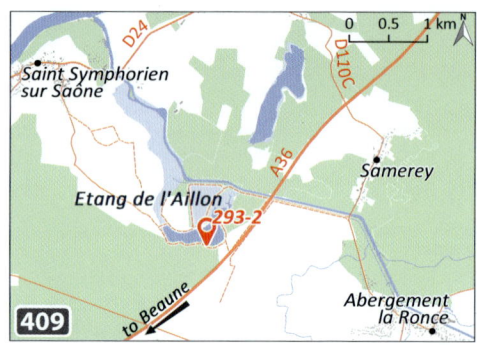
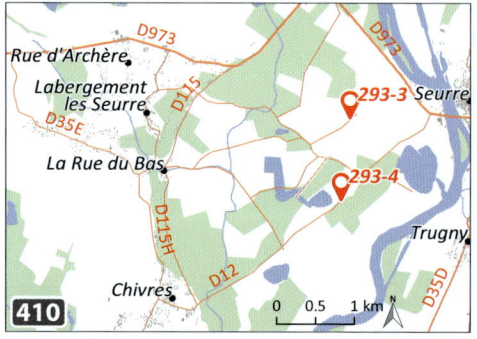
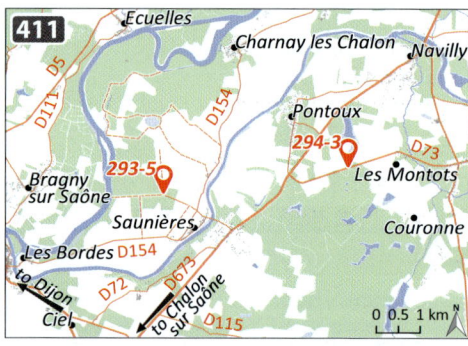

little ringed plover and commoner species occur commonly during migration peaks. Willow tit breeds in the willows surrounding the pond sluice.

Heading south from here, there are extensive meadows along the Saône river from Saint-Jean-de-Losne to Châlon-sur-Saône (roughly a 50 km section) in which red-backed shrike and lesser whitethroat breed from late April among a rich agropastoral bird community. Start in Labergement-les-Seurre [3,4], where a gravel quarry hosts a few breeding bee-eaters, bluethroats, fan-tailed warblers and tree sparrows. Waders can show up here on both spring and autumn migration. In winter, expect great grey shrike and peregrine and possibly geese.

Drive on south to Saunières and check the plain between the Doubs and the Saône [5] where rough-legged buzzard, whooper swan, marsh sandpiper and other rarities have recently been sighted. This should lead you to Saint-Maurice-en-Rivière, where Baillon's crake bred twice during recent rainy springs. Then, head to Châlon-sur-Saône to stop at the Darse Saint-Marcel [6] (walk southwards from the waypoint), a meander where Caspian gulls can reliably be found within wintering yellow-legged gulls.

South of Châlon, the flooded meadows between Varennes and Marnay [7,8] are the most productive stretch of this itinerary during migration and in the breeding season. Highlights have included breeding curlew, whinchat, corncrake and scarce migrants such as glossy ibis, collared pratincole, slender-billed gull and sociable plover.

and geese can also occur in winter. On the way, stop at Pontoux ponds [3] for wintering wildfowl or for breeding cattle egret, night-heron, purple heron and marsh harrier. From Fretterans, go back to Dijon (75 km, 1 h 15 via Dole and the A39) or connect with the Jura (region 11).

295 Southern Burgundy near Châlon

3 HOURS – 0.5 DAY; MAPS 415 & 416 South of Châlon-sur-Saône, **a hilly landscape with fields lies between the Rhône, Saône and Loire valleys. It marks a transition between woodland continental birds to the north and dry warm-climate species in the south.** Mont Péjus [1] is typical of this transition, featuring breeding stone-curlew, woodchat shrike and ortolan bunting (30 km, 40 min south of Châlon on the D49, drive towards Cluny, turn right in Saint-Gengoux-le-National, park on tracks along the D84 towards Curtil-sous-Burnand). The Roche de Solutré (signposted from Mâcon, 10 km, 20 min) is a limestone hill hosting the only pairs of Sardinian warblers in the region (year-round, search around [2] or [3]). Alpine swifts sometimes hunt around the hill, along with passing short-toed or booted eagles (which also breed in the area).

294 The Doubs valley

3–5 HOURS; MAPS 408, 411 & 414   A visit to the meadows between Fretterans [1] and Lays-sur-le-Doubs [1] along the River Doubs, a tributary of the Saône running southwest, can be a nice continuation of a spring birding day between Dijon and Châlon-sur-Saône (40 km, 40 min towards Dole, leave the N73 in Pourlans and bird along the river 5 km south). **Stop randomly wherever the habitat looks good, especially after spring floods in late April and early May** to search for bluethroat (rare), stone-curlew, little ringed plover, bee-eater and sand martin. Montagu's harrier breeds in the area and pallid harrier is seen annually on migration. Goosander

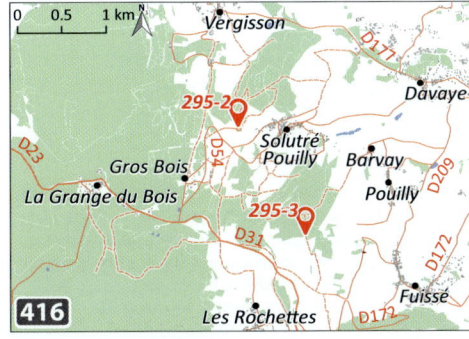

REGION 14
NORTHEAST

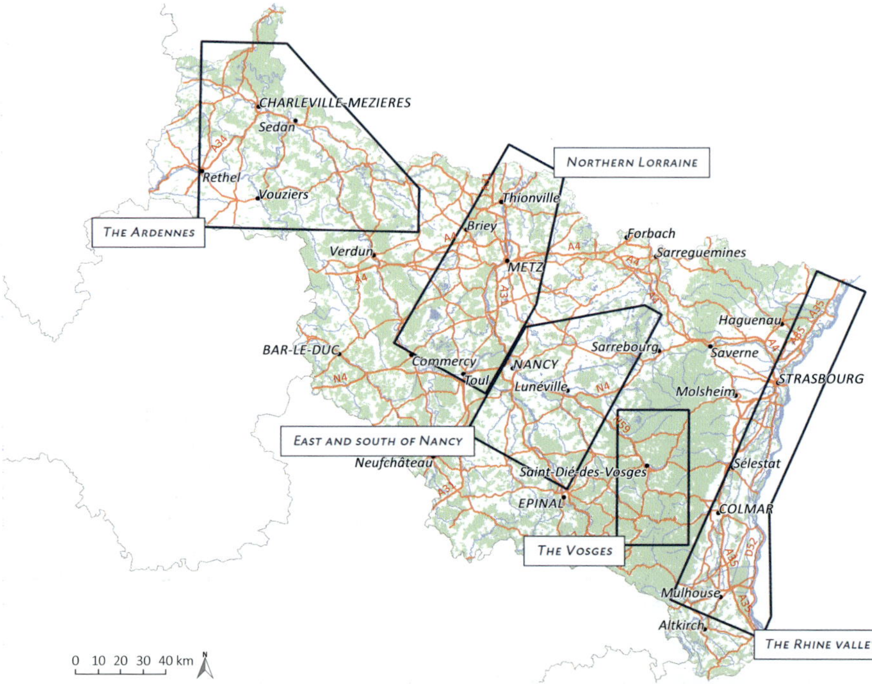

HIGHLIGHTS
» The western edge of the European collared flycatcher population
» Large waterbodies thronged with wildfowl in winter
» Mountain forests with owls, capercaillie and other specialities

The ghost of capercaillie haunts the Ballons des Vosges (sites 305–307). Even though an encounter with this species is unlikely these days, a rich accompanying suite of mountain species and scenic landscapes will fill your birding days year-round.

Oak and beech forests host the tiny collared flycatcher, giving Alsace and Lorraine a central European atmosphere.

NORTHEAST | **NORTHERN LORRAINE** | 289

REVIEWERS Jean François, Vincent Palomares, Claude Parent

USEFUL WEBSITES
www.faune-lorraine.org
www.faune-alsace.org
www.faune-champagne-ardenne.org

Northern Lorraine (sites 296–299, map 417)

HIGHLIGHTS
» Large reservoirs and ponds with wildfowl, white-tailed eagle and a rich breeding bird community
» Forests with collared flycatcher and grey-headed woodpecker

KEY SPECIES
YEAR-ROUND goshawk, marsh harrier, peregrine, stock dove, black, **grey-headed** and middle spotted woodpeckers, woodlark, **dipper**, willow tit, raven, cirl bunting

BREEDING white stork, purple heron, night-heron, honey-buzzard, red and black kites, **little bittern**, wryneck, whinchat, bluethroat, sedge, grasshopper, **Savi's** and marsh warblers, pied and **collared flycatchers**, red-backed shrike

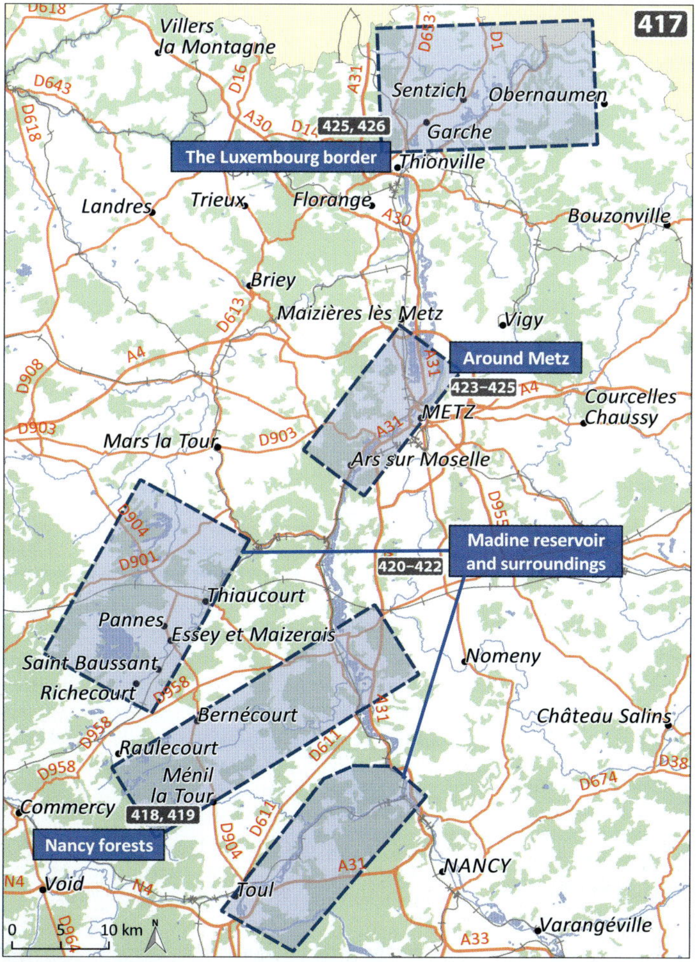

MIGRATION black stork, osprey, little crake, little ringed plover, dotterel

WINTER tundra bean goose, goosander, smew, white-tailed eagle, Caspian and yellow-legged gulls, penduline tit, great grey shrike

VISIT DURATION 2 days

PLAN YOUR VISIT
The main targets in this forested area are grey-headed woodpecker and collared flycatcher. Both live in old deciduous forests which you should visit in March (woodpecker) and May (flycatcher). **Bird early in the morning, as forests quickly become quiet after sunrise.** Most of the sites can be visited in a couple of hours, but keep time for birding in the surrounding open habitats and riparian woods in search of raptors, little owl, chats, warblers, shrikes and an overall rich community. Winter should lead you to Madine reservoir, with its wildfowl and white-tailed eagle. Regional rarities, including divers and diving

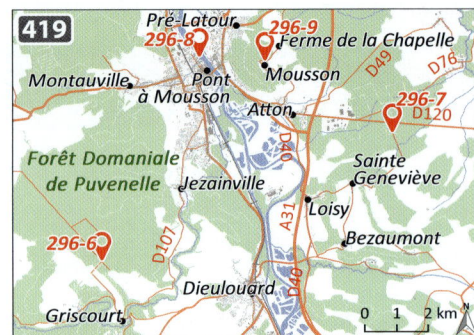

ducks, are seen there every year. A few great grey shrikes winter in the surrounding bocages and can be found perching on fences and hedgerows.

296 Nancy forests

0.5 DAY; MAPS 418 & 419 In mid-May, schedule a day trip from Nancy to start at first light in one of three forests northwest of Nancy: La Reine, Puvenelle and Facq-et-Juré. You will not manage to visit all three in the 2–3 hours of bird activity of a spring morning, so focus on one of them.

La Reine is the most obvious choice (from Nancy: 60 km, 1 h; take the A31 and leave it at exit 14 in Toul, then head north to Ménil-la-Tour, where the forest is signposted on the D10, left before the church). Leave your car on a track and walk towards the forest, for instance from a clearing on the D10, 2 km past Sanzey [1] or on a track 1 km further towards Boucq [2]. Stands are dominated by old oak and beech, with locally dense understorey, bogs and ponds. **Your main target should be** collared flycatcher, **which is locally common in old oak stands with large trees and light undergrowth.** Although it is not particularly rare, it can be hard to detect because of its secretive habits, preference for higher tree strata, dense foliage and its inconspicuous song. Catch a singing male by ear; it will likely fly away as you approach, but will soon come back if you wait quietly at some distance. **Several ponds are scattered through the forest.** Little (and sometimes Eurasian) bitterns, marsh harrier and sedge warbler breed in the surrounding reedbeds and little crake has been heard in the past. **Visit the most extensive of them, Romé pond, as a priority** (park on a track heading west at the northern edge of the forest on the D904 [3] and walk for 4 km to [4]). You can also make a stop at Gérard Sas pond, along the D147 towards Raulecourt [5]. Several other ponds hosting the same species are within walking distance from this latter.

Closer to Pont-à-Mousson, Puvenelle beech–oak forest [6] **hosts** grey-headed woodpecker, **while Facq-et-Juré beech forest [7] is a better choice for** collared flycatcher **and** red kite. In addition to these target species, the three forests host hobby, honey buzzard (from late May) and grasshopper warbler (in young stands with tall grassy vegetation), plus all the common forest breeding birds. Red-backed shrike, wryneck **and** whinchat **(spring migration) occur in fields and hedgerows around the edges of the forests.** These species remain active all day, so keep them for the heat of the day once the forests have become quiet.

Then head back towards Pont-à-Mousson and **stop on the Moselle's bank at the end of Boulevard Ney [8] to look for the** peregrines **that breed on the Abbaye des Prémontrés.** Cross the Moselle and follow the 'Metz–Nancy' signs to reach the D910. Leave it near the dump towards Mousson. Park at the base of the Butte de Mousson [9] and climb to the top, **a good place to see** white storks **and** red **(year-round) and** black **(April–July)** kites that frequent the dump.

297 Madine reservoir and surroundings

3 HOURS – 0.5 DAY; MAPS 420–422 Madine reservoir (Lac de Madine) is a large and shallow artifical lake used for watersports in summer (70 km, 1 h from Nancy, signposted from Toul and Pont-à-Mousson). **Its main interest lies in large numbers of wintering** wildfowl **from October to March, which usually involve a few** goosander, smew **and the three** divers. Red-necked and Slavonian grebes unpredictably show up from time to time, especially after cold weather or on migration. One or more white-tailed eagles winter in the surrounding woods and are often seen flying over the reservoir.

The main viewpoints are Nonsard dyke and harbour [1], Heudicourt harbour [2] and

hide [3] (follow signposts to 'observatoire ornithologique' from the village on the 'rue du lac' heading southeast, opposite the church) **and the wooded southern banks from the D119 near Montsec** (follow signs to 'tour du lac' from the village and park at [4] or [5], then walk through the woods). **On a sunny day, stay for a while in the Butte de Montsec [6], which offers a panoramic view over the reservoir** (400 m southwest of Montsec, signposted from the main street). This is an ideal viewpoint for gulls (including Caspian and yellow-legged, scarce), terns, raptors around midday and roosting duck and goose flocks in the evening. **On your way to, from and around the reservoir, keep an eye open for the few great grey shrikes that winter in the surrounding bocages: you will usually find them perched on fences and hedgerows.**

In spring and autumn, Madine is a good location for migrating osprey, and penduline tit occasionally stops over in reedbed patches, but it has become rarer in recent years. The reservoir is less birdy in spring, but still worth a visit for Eurasian and little bitterns (the latter from May), common tern, Savi's and marsh warblers. **Walk the footpath that circles along the reservoir banks for an introduction to the local diversity of forest, wetland and meadow breeding birds.**

The same species are easier to find at Lachaussée pond (Étang de Lachaussée), 20 km, 20 min north of Madine (drive north towards Verdun from Pannes or Heudicourt; the pond is signposted from the roundabout in Saint-Benoît-en-Woëvre). Here, you can also expect purple heron, Eurasian and little bitterns and bluethroat. Little crake used to breed and still occurs as a passage migrant in both spring and autumn. **There are viewpoints all around the pond, some of which give access to smaller peripheral wetlands**, for instance [7] (walk on tracks from Haumont-lès-Lachaussée), along the D131 [8], on tracks west of Lachaussée [9] or on the western bank [10] (access from the D904, turn right 1 km past a large farm 2.5 km north of Saint-Benoît-en-Woëvre).

From Madine or Lachaussée, head back to Nancy (or Metz, if you have time for a detour) along the banks of the Moselle, where you can expect wildfowl **(winter) or breeding** marsh warblers (spring), for instance near Toul [11], Gondreville [12], Liverdun [13] and Frouard (on the access road to the 'parc Effeil Energia') [14].

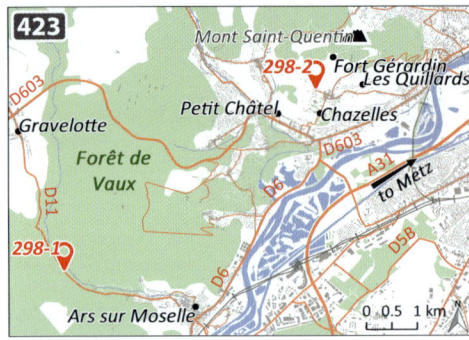 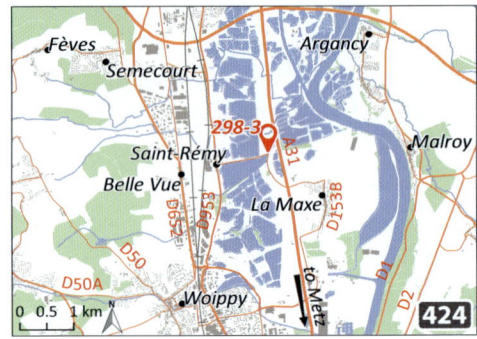

298 Around Metz

0.5–1 DAY; MAPS 423–425 A few sites around Metz make a logical follow-up to a morning at Madine or in the forests. **The Mance river valley (west of Ars-sur-Moselle, 13 km, 16 min south of Metz) runs through woodland** where dipper, goshawk, black woodpecker, stock dove and willow tit can be found year-round (bird along the D11 towards Jarny [1]). In spring, a wise follow-up is Mont Saint-Quentin, 10 km, 15 min north [2] (from Plappeville in the western suburbs of Metz, follow signs to 'col de Lessy' and Scy-Chazelles on the D103G). **This dry hill has breeding** woodlark, red-backed shrike **and** cirl bunting **among a suite of commoner grassland and forest species.**

Head to the northern edge of the Metz suburbs along the railway to visit the ponds between Woippy and Maizières [3] (turn east on Rue de l'Étang from the D953 in Saint-Rémy and park at the eastern end of the ponds). **They host some** wildfowl **in winter, but are at their best during both migrations for** waders, **and from mid-April for breeding** little ringed plover, little bittern, night-heron, marsh harrier **and** marsh warbler.

Further north again, the River Moselle and artificial ponds north of Thionville between Manom and Gavisse, especially in Garche [4] and south of Sentzich [5], are an annual wintering site for tundra bean goose.

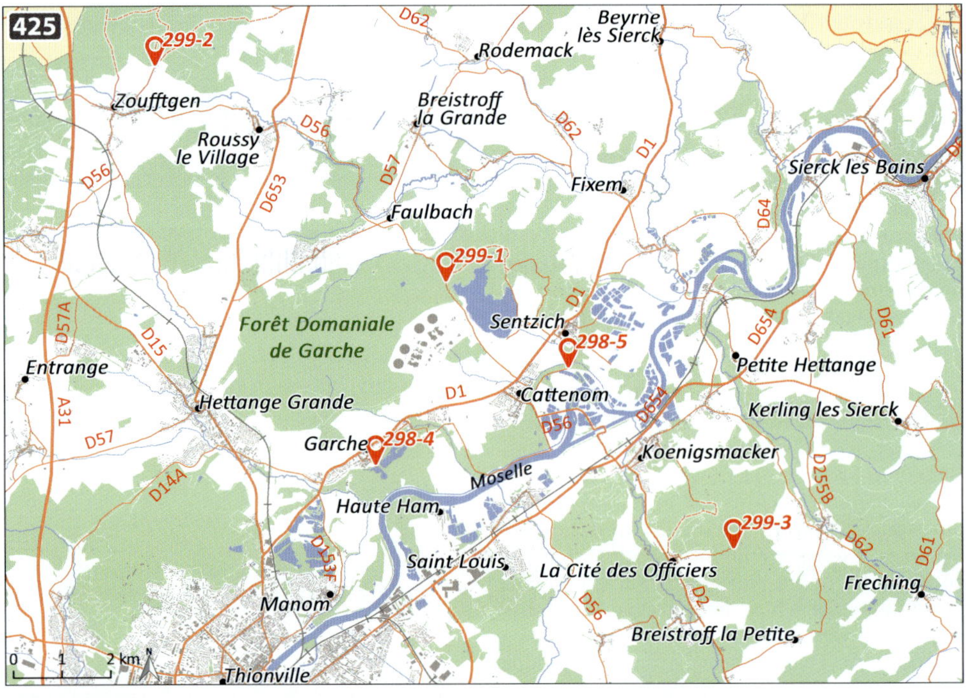

299 The Luxembourg border

0.5 DAY; MAPS 425 & 426 The mixed beech and oak forests that spread along the Luxembourg border host breeding grey-headed woodpecker, collared and pied flycatchers. They deserve a visit in May as an alternative to the Nancy forests if you are staying in Metz or Thionville. The two most extensive woodlands are known as Cattenom forest (11 km, 30 min north of Thionville, turn left 200 m after the northern limit of the nuclear power plant on a narrow road to enter the forest [1]) and Zoufftgen forest (take the first street to the right when entering the village from the south, park at the 'Maison Forestière d'Unterwald' [2]). **Both host good numbers of** collared **and a few** pied flycatchers, **both** treecreepers **and a few remaining** grey-headed woodpeckers, plus regionally common species such as black and middle spotted woodpeckers, goshawk (easier early in season) and stock dove.

Further southeast, Koenigsmacker has the same species plus breeding raven (cross the

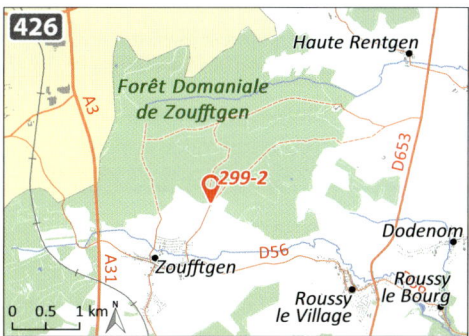

Moselle in Koenigsmacker and follow signs to Kédange; 2.5 km after joining the D2, turn left in La Cité des Officiers to get into the forest [3]).

Besides forests, **the higher parts of Merschweiller** [4 – see GPS file] **along the German border are stopover places for migrating** black stork, raptors **and** dotterel from mid-August to late September (30 km, 30 min northeast of Thionville on the D654 along the Moselle then east towards Kitzing on the D64).

East and south of Nancy (sites 300–304, map 427)

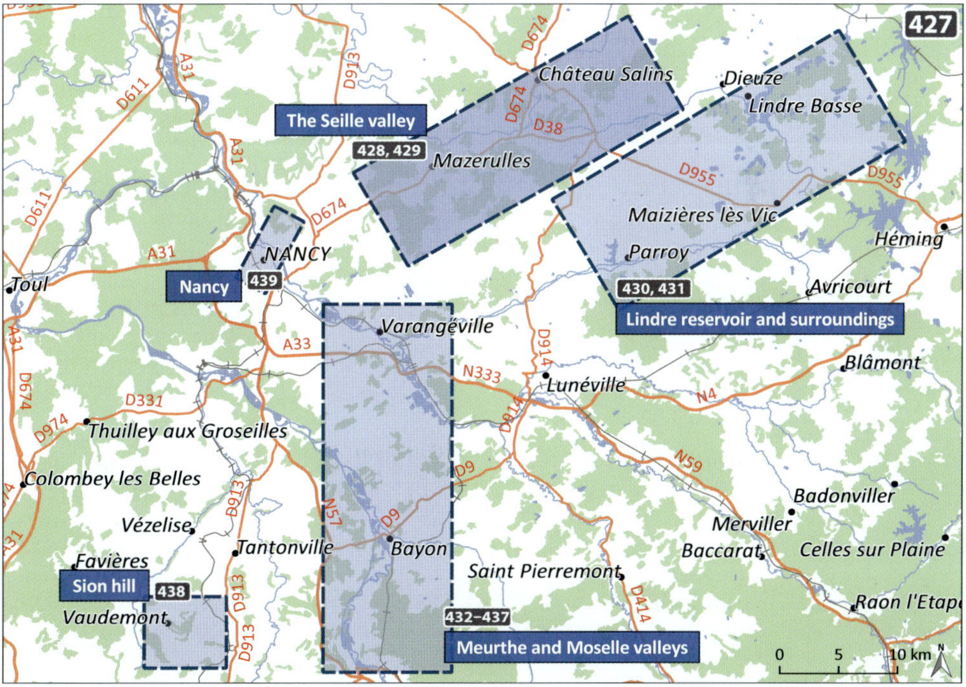

300 THE SEILLE VALLEY | NORTHEAST

HIGHLIGHTS
» Lindre reservoir is a must in winter and during migration periods for cranes, waders and wildfowl
» Several large mixed beech–oak forests with collared flycatcher, middle spotted and grey-headed woodpeckers and other forest targets
» Rare breeding species in well-preserved forests and meadows: osprey, white-tailed eagle and corncrake
» Warm-climate species including bee-eater and woodchat shrike to the south of Nancy

KEY SPECIES
YEAR-ROUND marsh harrier, goshawk, peregrine, stock dove, **grey-headed** and middle spotted woodpeckers, woodlark, cirl bunting

BREEDING red-crested pochard, garganey, black-necked grebe, **black stork**, purple heron, night-heron, **little bittern**, **white-tailed eagle**, honey-buzzard, osprey, black kite, **corncrake**, black-winged stilt, little ringed plover, bee-eater, hoopoe, sand martin, tawny pipit, bluethroat, **Savi's**, marsh, great reed, melodious and wood warblers, **collared flycatcher**, red-backed and woodchat shrikes

MIGRATION Spotted and little crakes, dotterel

WINTER wildfowl, merlin, common crane, **bearded and penduline tits**

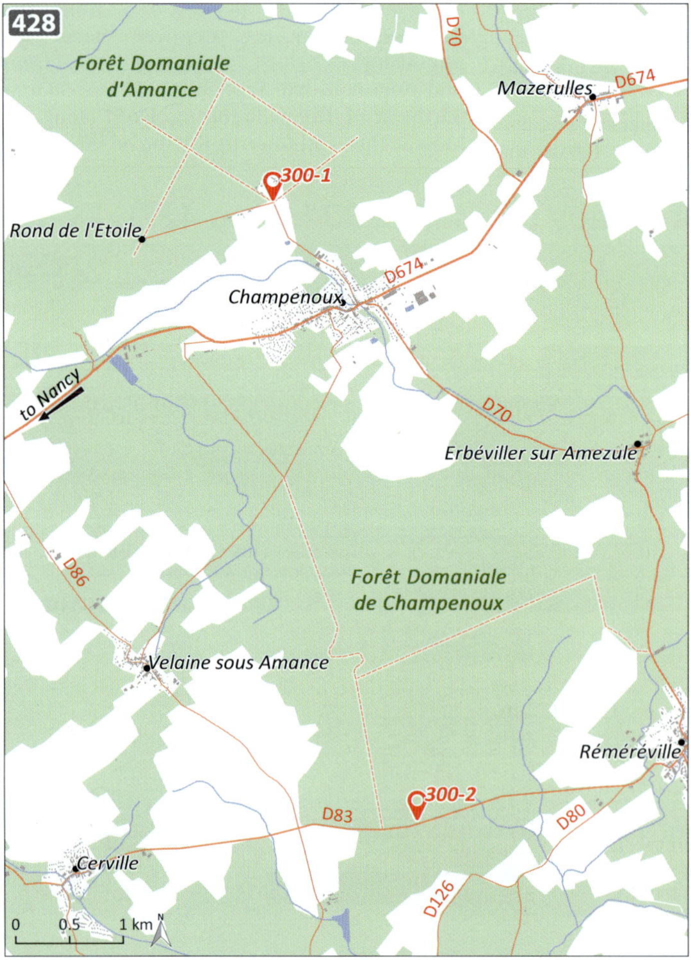

VISIT DURATION 2 days
PLAN YOUR VISIT
The sites are organised as a round circuit from and to Nancy, which is best undertaken in May in search of collared flycatcher and wetland breeding birds. Concentrate on Lindre in winter and during wader migration, especially when the pond is being drained in November and December, which happens every two years.

300 The Seille valley

0.5 DAY; MAPS 428 & 429 The Seille valley runs east of Nancy across a mosaic of farmland and forest, interspersed with arable fields, pasture, oak and beech woods and wetlands. **Champenoux forest is one of the region's most reliable spots for** collared flycatcher (from early May) just 16 km, 30 min east of Nancy on the D674 towards Sarreguemines. Try the track leading northeast behind the INRA campus [1] (signposted from the

D674) and the area south of Velaine-sous-Amance [2] (south on the D86, 4 km west of Champenoux, then park along forest tracks on the D83). In addition to the flycatcher, a couple of morning hours in this area should yield the two treecreepers, middle spotted woodpecker and possibly goshawk and honey-buzzard, plus all the common forest breeding species.

If you are short of time once out of the forest, bird along the River Seille between Pettoncourt and Chambrey (13 km, 15 min east of Champenoux, turn north in Moncel-sur-Seille): meadows and artificial ponds [3] host regionally rare breeders such as curlew, marsh and great reed warblers, marsh harrier plus wildfowl and migrating waders. Black stork is regular, albeit unpredictable, in the area. **If you can spend a full day in the area, head 17 km, 20 min further east**: corncrake still breeds in very low numbers some years between Marsal and Mulcey [4] and can be heard singing at dusk and dawn within a rich community of wetland species. Curlew also occurs in the area.

301 Lindre reservoir and surroundings

0.5 DAY; MAPS 430 & 431 **Lindre reservoir (600 ha) is the largest of a vast network of lakes and ponds embedded in an essentially forested region,** 50 km, 1 h east of Nancy towards Sarreguemines on the D674 (signposts from Dieuze). **Birdwise, it is probably the richest spot of the region year-round, with several rarities breeding or wintering in the vicinity,** starring white-tailed eagle, which occurs at all seasons around the reservoir.

Park in Lindre-Basse [1] and walk to the main dam [2] for a first overview of the site. Most wildfowl, geese and cranes will be quite distant, but water depth makes this spot suitable for diving

ducks, divers and gulls. When the water level falls in autumn, mudflats can host a good diversity of waders and gulls. The dam is also good in spring for breeding purple heron, osprey and goshawk. **Walk to a hide 500 m east of the dam** [3] (the access track is marked by a gate and sign at the base of the dam) for close views of black-necked grebe in spring and all wildfowl species in winter; white-tailed eagle and peregrine sometimes perch on nearby tall trees. The access trail crosses a wood in which both treecreepers and middle spotted woodpeckers can occur.

Head back to Dieuze and drive south on the D999, then turn left onto the D199F towards Tarquimpol. **On your way, stop in front of the 'île de la Folie'** [4]. **It is a fine viewpoint for** wildfowl, **and good for breeding** red-crested pochard. Garganey, little bittern, spotted **and** little crakes **are more likely encountered in remote bays on the southeastern side of the reservoir around Guermange, but these are either hard to access or closed to the public.** In winter, bearded tit shows up from time to time all around the reservoir (penduline tit is also possible on migration although much scarcer). Reed bunting, water and meadow pipits are common in open areas, meadows and mudflats. **Make sure you have enough time to walk along the path that starts from a car park at the end of the street to the left of Tarquimpol church** [5] (6 km, 10 min from the roundabout south of Dieuze). This is the best place to scan roosting crane and goose flocks before dusk, and it offers an excellent view over parts of the main waterbody that cannot be seen from the dam. The final area open to the public in the immediate vicinity of Lindre reservoir is the Cornée d'Assenoncourt, a wide open waterbody which is best scanned with a scope from Alteville castle, in the south [6]. The reservoir is a bit distant, but it is a good spot for singing Savi's warbler and bluethroat in spring and for roosting wildfowl in winter.

Other ponds (Gelucourt [7 – see GPS file]**, Donnelay** [8 – see GPS file]**, Ommeray** [9 – see GPS file] **and Parroy** [10]**) are worth a visit** in spring for breeding red-crested pochard, black-necked grebe, purple heron, little bittern, bluethroat and great reed and Savi's warblers.

A 25 km, 30 min drive southeast from Lindre leads to the Parc Animalier de Sainte-Croix [11 – see GPS file] (signposted from the D955 to Strasbourg), which harbours a breeding colony of night-herons. **Do not neglect the large**

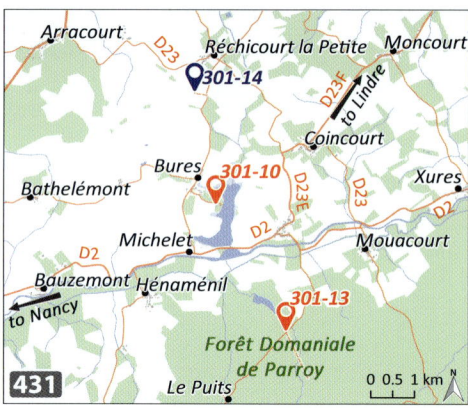

expanses of surrounding oak and beech forests between April and June.

Just east of Lindre reservoir, good numbers of collared flycatcher **breed in Romersberg forest** [12] **(turn right to Guermange, 5.5 km east of Dieuze), together with** osprey, goshawk, stock dove, **all the** woodpeckers, **both** treecreepers **and** wood warbler. **For** grey-headed woodpecker, **prefer Parroy forest** [13] (access through a narrow road from the D2 just south of the village), where all the above-mentioned species also occur. Still near Parroy, dotterel is regular from mid-August to early September in fields close to Réchicourt-la-Petite [14].

302 Meurthe and Moselle valleys

1 DAY; MAPS 432–437   The Meurthe and Moselle rivers run along valleys that justify a spring day trip from Nancy **for** waders **and breeding wetland species.** Although the Meurthe valley is quite heavily urbanised, ponds and quarries alongside the river maintain patches of suitable habitats that are worth visiting between March and June. Start at Embanie pond in Art-sur-Meurthe (8 km, 20 min southeast from Nancy on the D2 towards Lunéville), a fenced site which can be viewed from its northwestern gate [1]. Black-necked grebe, black-winged stilt, little ringed plover, little bittern and great reed warbler breed here and the site can be excellent for waders when mudflats are exposed in late August.

Two former quarry ponds, 14 km, 25 min south between Rosières-aux-Salines and Damelevières, are an alternative for the same species (park at [2] and walk among the ponds, including the one west of the road). Little bittern **is easier here than anywhere else in the region, especially at** [3]. There is a mixed black-headed/

NORTHEAST | 302 MEURTHE AND MOSELLE VALLEYS | 297

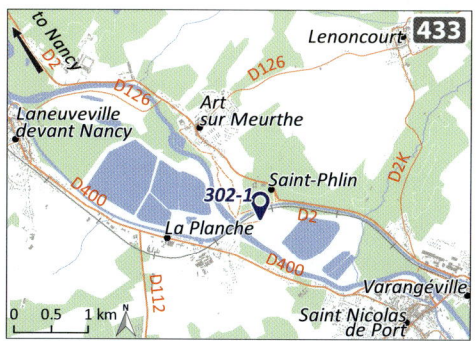

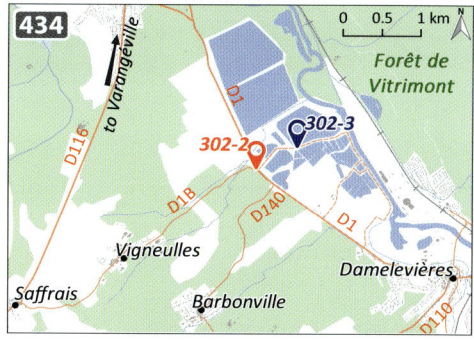

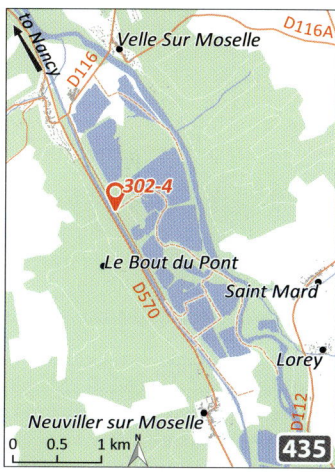

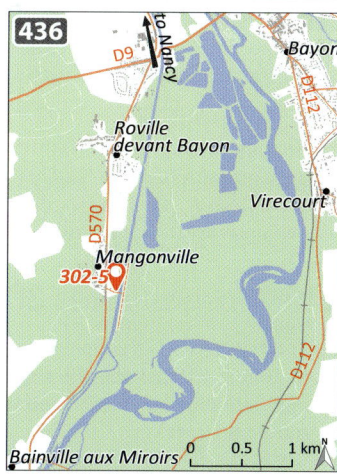

Mediterranean gull colony with some pairs of common tern in the active quarry, which has recently been joined by several pairs of yellow-legged gull. Although the colony itself cannot be seen, the birds roost and feed in surrounding fields with lapwing and larks.

Shift to the Moselle valley and head southwards, stopping here and there along the small roads that run parallel to the D570, for instance in the vicinities of Crévéchamps [4] (30 km, 30 min from Rosières-aux-Salines; follow signposts

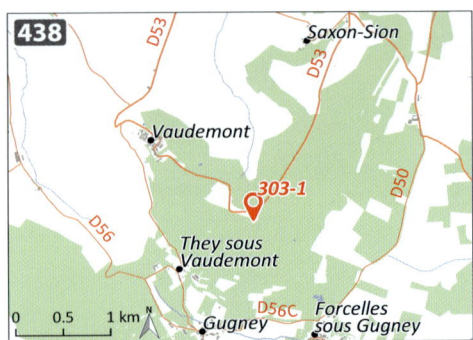

to the 'Plan d'eau du Grand Rozot' and turn right just after the canal), Mangonville [5] and Charmes [6]. **Bee-eater and** sand martin **breed on sandy banks on these sites** along with black kite, honey-buzzard, hobby, little ringed plover, common tern and kingfisher. On the passerine side, expect melodious and grasshopper warblers, red-backed shrike and reed bunting.

303 Sion hill
1–3 HOURS; MAP 438   Sion hill [1] is the closest autumn migration site to Nancy (36 km, 40 min south of Nancy: D913 towards Mirecourt and follow signs from Tantonville). Do not expect high counts, but passage can still be significant under clear skies with light winds. Hirundines, tree pipit, woodpigeon and fieldfare, finches and larks later in season dominate the passage. Small numbers of raptors (both kites, honey-buzzard, osprey, hobby, merlin) pass through in late summer, sometimes joined by black stork. Woodland edges and fields should yield hoopoe (April to August), while woodlark and a few woodchat shrike breed from May in surrounding groves.

304 Nancy
1–2 HOURS; MAP 439   Nancy inner city has a few good species to offer, the most obvious being **a huge roost of** ring-necked parakeets **in the Parc de la Pépinière** [1]. Peregrines breed on the Basilique Notre-Dame de Lourdes [2] and on two other buildings. **In the immediate suburbs, visit the open meadows and fields of the Malzéville plateau** [3], where you should at least find woodlarks and cirl buntings (mainly on the western side). Among other local breeders, focus on tree pipit, the common warblers and red-backed shrike. Tawny pipit and whinchat are regular additions on both spring and autumn migration, together with commoner migrants.

The Vosges (sites 305–307, map 440)

HIGHLIGHTS
» Mountain species, starring a relictual population of capercaillie and the two mountain owls
» Celles-sur-Plaine reservoir and its surrounding forests for a winter or spring visit

KEY SPECIES
YEAR-ROUND goshawk, peregrine, **capercaillie**, **hazel grouse**, pygmy and **Tengmalm's owls**, black, **grey-headed** and middle spotted woodpeckers, **dipper**, willow and marsh tits, Eurasian and short-toed treecreepers, nutcracker, raven, **citril finch**, rock bunting

BREEDING honey buzzard, red and black kites, **Alpine accentor**, ring ouzel, wood warbler, pied flycatcher, red-backed shrike, siskin

| J | F | M | A | M | J | J | A | S | O | N | D |

NORTHEAST | **305 CELLES-SUR-PLAINE** | 299

MIGRATION osprey, red-footed falcon, **black stork**, dotterel
WINTER goldeneye, goosander, **wildfowl**, merlin, great white egret
VISIT DURATION 2–3 days
PLAN YOUR VISIT
The Vosges are the northeastern counterpart of the Jura in terms of both landscapes and birds. These low and rounded mountains (max. altitude 1424 m) are largely covered by beech and conifer forests, with small amounts of Alpine meadow on the highest plateaus. **You will find most mid-mountain specialities, including** Tengmalm's **and** pygmy owls **and** citril finch. A few capercaillie remain; forget about looking for them, as all locations are confidential. In winter, weather can turn cold and changeable, with snow restricting access to many minor roads: you will do better to concentrate on the lower parts where reservoirs hold large numbers of wildfowl. Rather than visiting many sites, it is better to concentrate on just a few and cover them intensively. If you have only one day, spend it on the Hohneck.

305 Celles-sur-Plaine
2 HOURS – 0.5 DAY; MAP 441
Celles-sur-Plaine reservoir [1] (altitude 300 m) is the most obvious choice in winter when access to higher altitudes is impaired by snow (from Nancy: 78 km, 1 h east towards Saint-Dié-des-Vosges, follow 'Pierre-Percée' from Raon centre; from Strasbourg: 80 km, 1 h 10 west through Schirmeck and the Col du Donon). This wide artificial lake closed by a dam is surrounded by wide expanses of old mixed forest where lowland species mix with mid-altitude specialists. The area is at its best in late winter as wildfowl stop on the reservoir on their way northwards while early-spring forest breeding species start to sing. Great white egret, goldeneye and goosander frequent the reservoir from October to March. Peregrine and goshawk breed in the vicinity and are most easily spotted when flying over open water.

Ascend into the forest west of the reservoir from Raon to Neufmaisons on the D8 and park at the start of the last wide track before leaving the forest [2] (look for the sign 'Forêt Domaniale des Reclos' facing a small clearing). Pygmy owl breeds around here [3] (on foot only, 2 km southwest on forest tracks). Grey-headed woodpecker still occurs in good densities in surrounding deciduous stands, together with the more common middle spotted and black woodpeckers. The usual passerine community of cool mixed forest is of course present, with willow and marsh tits (year-round), wood warbler (from mid-April), Eurasian and short-toed treecreepers and dipper on rivers.

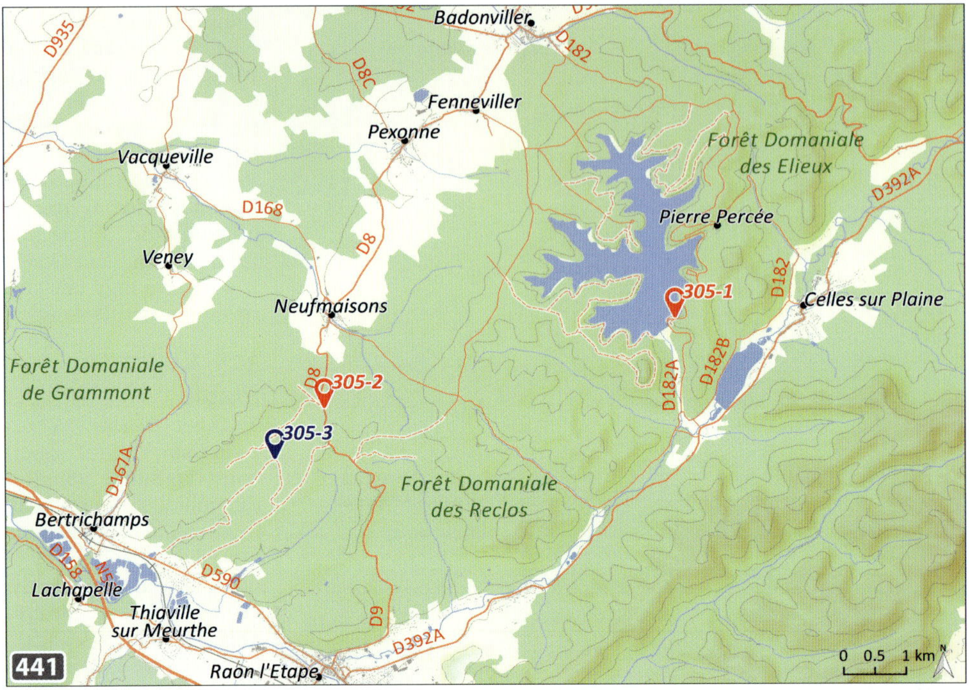

306 Tanet-Gazon du Faing area

0.5 DAY; MAPS 442 & 443 **Tanet-Gazon du Faing reserve protects old undisturbed forests and high-altitude meadows along some of the highest ridges of the Vosges** (from Saint-Dié-des-Vosges, follow 'Colmar par le col' for 26 km until you reach the Col du Bonhomme, then turn right on the 'Route des crêtes'). **Drive on a north–south axis between the Col du Louschbach [1] and the Hohneck (site 307) or walk along the ridges from the Col du Calvaire [2] towards [3], preferably in the early morning**. This is the only reliable area for capercaillie and hazel grouse in the Vosges, but encountering any of them will require lots of luck. **Be aware that the area is highly protected, with strict access restrictions in some places year-round and permanent rangers.**

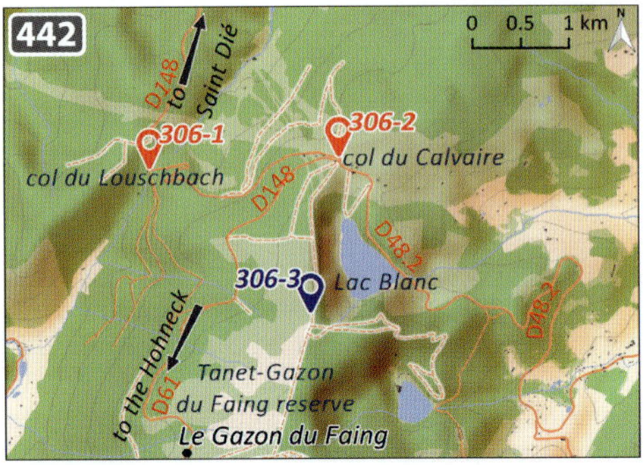

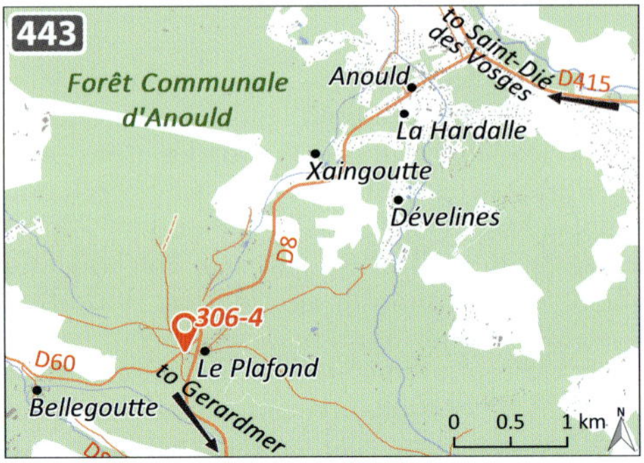

Nutcracker is reasonably common and easier to find in autumn when feeding in hazel trees. **A night drive in March** can yield pygmy owl (before sunset) and Tengmalm's owl (best chances one hour after dusk or in complete darkness). All forest passerines breed in the reserve. Higher altitudes add species that are hard to find elsewhere in the region: citril finch and ring ouzel (both in pines at the tree limit), red-backed shrike and rock bunting.

From July to November, the Col du Plafond [4] **is among the best migration spots of the Vosges, especially in southwesterly winds** (22 km, 25 min north of the Col du Calvaire on the D8 through Fraize and Anould). Settle along the GR533 either north or south of the pass depending on the wind (news on www.migraction.net). Suitable weather should lead raptors to fly over the pass: honey-buzzard and black kite until September; hobby, red kite, sparrowhawk and merlin later in season, plus osprey or rarities such as red-footed falcon), with the possibility of black stork and loads of passerines (hirundines, larks, pipits and finches). Citril finch, raven and peregrine occur on this site year-round.

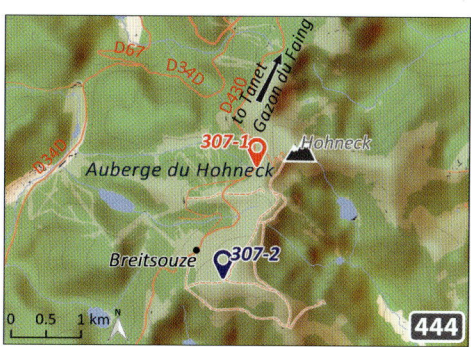

307 The Hohneck

1 DAY; MAP 444   **The Hohneck (1360 m) is covered by old-growth forest and Alpine meadows typical of the Vosges mountains. It consequently holds one of the richest bird communities of the region** (20 km, 20 min south of Tanet-Gazon du Faing through the 'Route des crêtes'). Allocate a day-long visit to the mountain in May or June, possibly in combination with a visit to Tanet-Gazon du Faing (site 306) as a complement.

Park at the Auberge du Honeck [1], then walk along the ridge towards [2] to cover the whole range of possible habitats from old beech–fir forests to wide expanses of Alpine meadow. **Start at the bottom and bird while climbing, as forests get quiet early after dawn.** Middle spotted woodpecker, wood warbler, pied flycatcher, Eurasian and short-toed treecreepers are confined to old beech stands, while nutcracker, crossbill, siskin and citril finch are easier to find in conifer areas or at the upper edge of the forest, where they are joined by rock bunting and ring ouzel. **A post-midnight drive could yield singing** Tengmalm's owl, **most likely from February to April.** Dipper breeds on all the rivers running down from the Hohneck. You should easily find stonechat and water pipit along the ridges, where dotterel sometimes stops over in early May and late August. Alpine accentor could still breed on the eastern rocky slopes. There are winter or migration records of shore lark, desert wheatear, snowfinch and snow bunting (all vagrants).

The Rhine valley (sites 308–309, map 445)

HIGHLIGHTS
» Wintering wildfowl on the Rhine, with sometimes a grebe or a diver mixed in
» Riparian forests with good populations of pied flycatcher
» Wader and raptor migration along the Rhine in both spring and autumn

| J | F | M | A | M | J | J | A | S | O | N | D |

KEY SPECIES
YEAR-ROUND goshawk, middle spotted and **grey-headed woodpeckers**, willow and marsh tits
BREEDING night-heron, purple heron, **little bittern**, black and red kites, bluethroat, reed, marsh, great reed, grasshopper, **Savi's** and icterine **warblers**, pied flycatcher, red-backed shrike
MIGRATION osprey
WINTER whooper swan, **tundra bean goose**, wildfowl, smew, goldeneye, goosander, scaup, black-throated and red-throated divers, **white-tailed eagle**, **Caspian** and yellow-legged gulls

VISIT DURATION   1 day

PLAN YOUR VISIT
Each site described here is organised as an independent itinerary suitable for both spring and winter with good results. The sites are similar, so choose one, cover it intensively and use the other as a backup if you missed one of your targets.

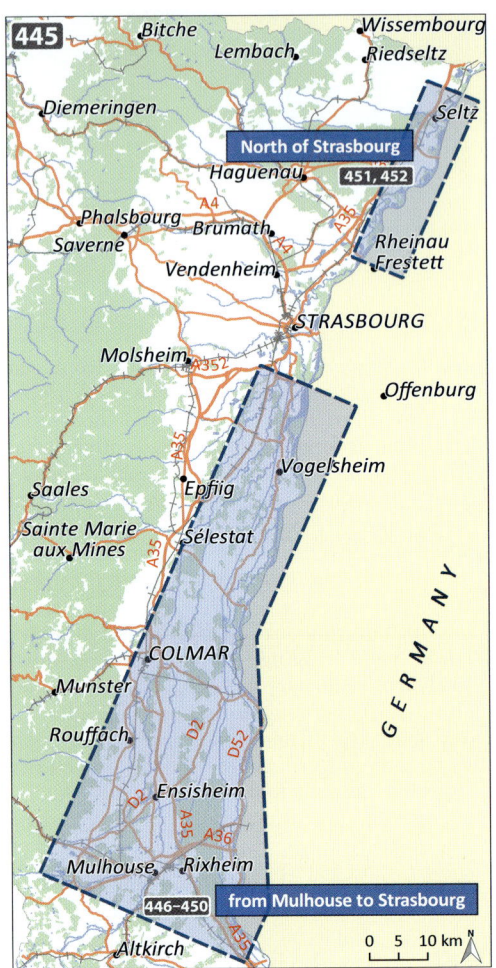
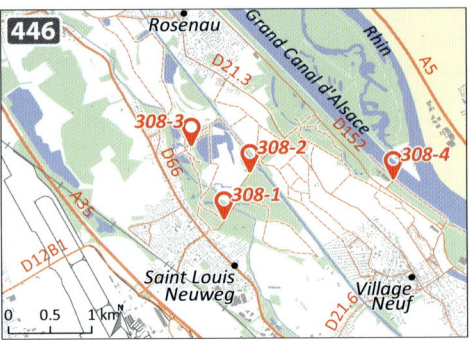

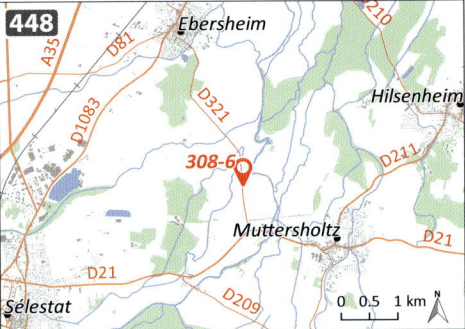

308 From Mulhouse to Strasbourg

1–2 DAYS; MAPS 446–450 **Start a spring morning in the Petite Camargue Alsacienne**, a wetland surrounded by deciduous woods where regionally rare breeding species such as little bittern, night-heron, purple heron, great reed and Savi's warblers should be easy to find (130 km, 1 h 20 south of Strasbourg; leave the A35 towards Bâle-Mulhouse airport and drive to Saint-Louis Neuweg. There are signposts from the village: paths start from Neuweg stadium [1], 120 m south of the canal lock [2] or from the old fish farm [3]). Surrounding forests host the two treecreepers, middle spotted woodpecker and most other forest species.

In winter, forget this first site and spend your time stopping here and there along the Rhine between Kembs and Huningue (for instance near the water treatment pond [4]) for wildfowl. Caspian (rare) and yellow-legged gulls winter on this section of the river.

Michelbach reservoir [5], 50 km, 30 min west of Mulhouse, is worth a quick stop on your way to the Vosges in winter or during wader migration (exit 15 on the A36, drive towards Cernay, exit in Aspach-le-Bas and further west to Aspach-le-Haut, then follow signposts). **Alternatively, you could decide to visit meadows around Muttersholtz** [6] (6 km, 10 min east of Sélestat on the D21, turn left on the last road before the village): if partly flooded, they can also host waders and wildfowl during migrations.

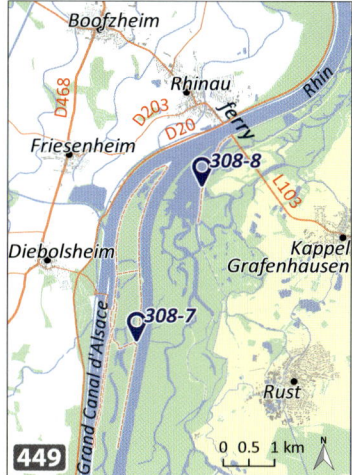
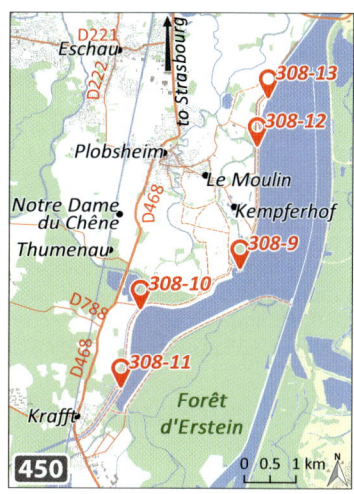

The Île de Rhinau reserve [7] is a long narrow island in the Rhine (27 km, 30 min east of Sélestat or 45 km, 40 min south of Strasbourg near Rhinau, park at the power plant and cross the canal locks). It is entirely protected, with hunting prohibited and habitat management aimed at preserving old wood stands, meadows and wetlands. Red-necked grebe, goldeneye, scaup, smew and rarer species such as velvet scoter (near-yearly), whooper swan or black-throated and great northern divers winter in the reserve. Among the most notable breeders are all woodpecker species (including grey-headed), black kite, goshawk, red-backed shrike, marsh and reed warblers.

Further north, Rhinau-Kappel pond, an oxbow lake connected to the Rhine, is a renowned viewpoint for wildfowl and white-tailed eagle (easiest access from the hide on the German side [8]: cross the Rhine on the ferry from Rhinau and park as soon as you can, then walk south for 1 km on the dyke). The surrounding forest hosts all the regional specialities.

End your itinerary south of Strasbourg at Krafft pond, near Plobsheim [9] (follow signposts to 'Golf Kempferhof' from the village). This is arguably the best spot in Alsace for wildfowl, in terms of both diversity and quantity, and is the one place you should not miss if you are short of time. The pond is large and has viewpoints both south [10,11] and north [12,13].

309 North of Strasbourg

0.5–1.5 DAYS; MAPS 451 & 452 Although they are somewhat isolated from other sites, **the ponds around Munchhausen [1] are worth a**

winter visit for whooper swan and tundra bean goose, plus the near-certainty of finding Caspian gulls in good numbers (55 km, 40 min north of Strasbourg: leave the A35 at exit 57 and drive through Seltz eastwards). These rare species are accompanied by more usual ones, among them goosander, smew and great white egret. Willersinn pond [2] is especially suitable for these species (from the previous point, follow the Rhine for 2 km until you find a track on the right with a small signpost). Whooper swans may be easier on the flooded meadows of the Sauer delta in Munchhausen village [3] (6 km north of Seltz). **In spring, you should find a nice combination of wetland, woodland and meadow breeding species**, some of which are regionally rare: bluethroat, marsh, grasshopper and Savi's warblers, red-backed shrike. A few pairs of icterine warbler also breed in the area from late May. Pied flycatcher, willow and marsh tits and middle spotted woodpecker inhabit woods in the immediate vicinity. Wander along edges

or in open areas for hobby (from mid-April), goshawk, osprey (on stopover) and red kite, the latter being easier at Wintzenbach dump [4] (2 km west of the roundabout between Seltz and Munchhausen), which is also a roosting site for gulls, storks and peregrine.

If you are around during wader **peak passage (early May and July to mid-September) or if you are specifically looking for** icterine warbler, go back halfway to Strasbourg and check mudflats on the Rhine around Gambsheim dam [5] (30 km, 30 min from Seltz or Strasbourg, leave the A35 at exit 52 towards Achern).

The Ardennes (sites 310–312, map 453)

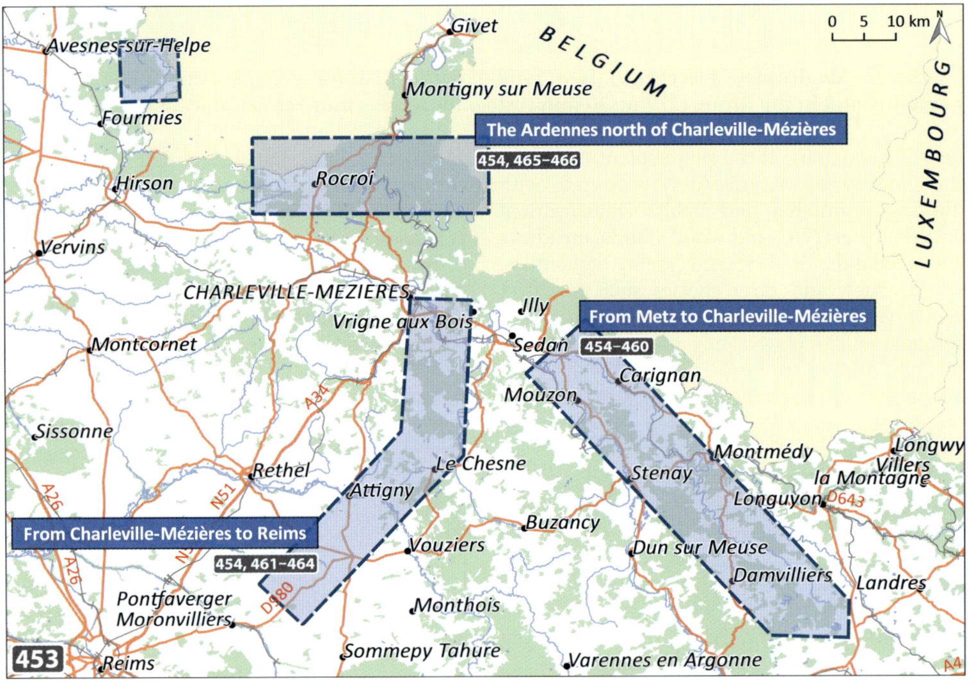

HIGHLIGHTS

- Some of the last reliable places for corncrake in the whole of France, in the Meuse and Chiers valleys
- Wildfowl and wader stopover areas
- Large expanses of well-preserved forest in low mountains with marshes and small rivers
- An underwatched region with a high diversity of habitats

KEY SPECIES

YEAR-ROUND **whooper swan**, grey partridge, **Eurasian bittern**, great white egret, goshawk, marsh harrier, **black and hazel grouse**, **Tengmalm's** and little owls, middle spotted and black woodpeckers, **dipper**, crested, willow and marsh tits, Eurasian treecreeper, **great grey shrike**, hawfinch, redpoll

BREEDING **black stork**, **little bittern**, night-heron, black and red kites, Montagu's harrier, quail, **corncrake**, stone-curlew, little ringed

NORTHEAST | **310 FROM METZ TO CHARLEVILLE-MÉZIÈRES** | 305

plover, wryneck, nightjar, whinchat, bluethroat, lesser and common whitethroats, sedge, great reed, marsh, melodious, **icterine** and wood warblers, **collared** and pied flycatchers, red-backed shrike, siskin

MIGRATION short-eared owl, **aquatic warbler (scarce)**, **bearded** and **penduline tits**

WINTER wildfowl, smew, goldeneye, scaup, common crane, **Caspian gull**

VISIT DURATION   3 days

PLAN YOUR VISIT

From Reims to the Belgian border, the landscape changes from agricultural plains to forests interspersed with rivers running along well-preserved riparian meadows and low mountains (below 500 m) at the extreme north (the Ardennes per se). We describe three one-day itineraries, all at their best from April to June. The region is largely underwatched, so do not refrain from stopping wherever habitat looks suitable, and find your own birds.

310 From Metz to Charleville-Mézières

1 DAY; MAPS 454–460   Start your day early at Amel pond to try for wetland breeding birds (80 km, 1 h west of Metz, leave the A4 on exit 32 and follow signs towards Longwy on the D908 for 20 km until you find a left turn towards 'Réserve Naturelle Régionale' and Amel sur l'étang; park at the reserve entrance [1]). **Two hides** can be accessed by a short walk [2,3]. Expect great white egret, marsh harrier, black kite, sedge warbler and with some patience great reed warbler, little bittern (both from May), night-heron and Eurasian bittern (year-round). Grey partridge, quail, lesser and common whitethroats are rather common from early May in the surrounding plain, and little owl and red-backed shrike breed in the hedgerows.

From October to March, cranes forage in fields around Billy-sous-Mangiennes, 10 km

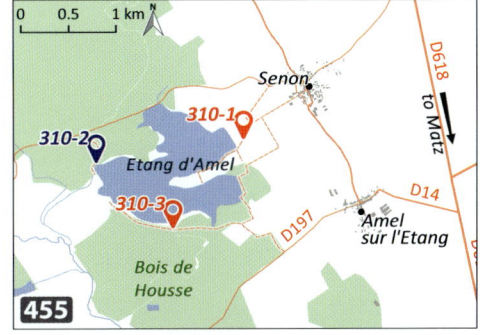

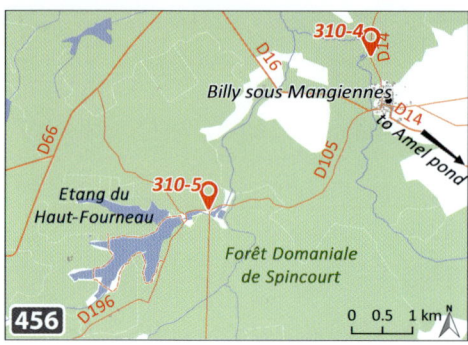

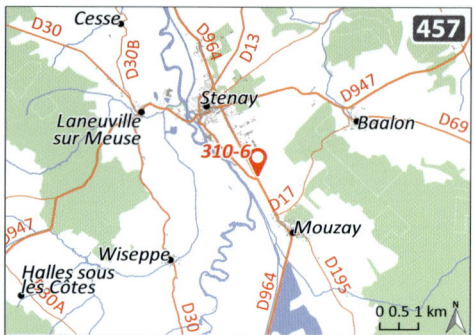

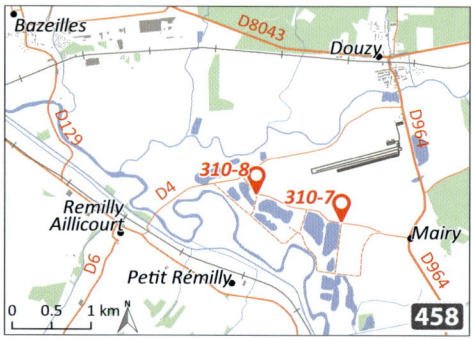

north [4]. Check wires, fences and hedgerows as a few great grey shrike could also winter around here. In late May, collared flycatcher breeds in Spincourt forest, which starts at the western end of Amel pond. Search for it in old beech and oak stands, where middle spotted and black woodpeckers, wood warbler, willow and marsh tits, hawfinch and all common forest species also occur. Look for wryneck (rare) in younger stands or along forest edges. Similar species can be found a bit further south around the Haut Fourneau pond [5] (head to Azannes from Billy-sous-Mangiennes on the D105 and turn towards Gremilly). **The whole forest and its surroundings are generally bird-rich because of the habitat diversity, so do not refrain from spending some time in this area.**

Leave the forest through Azannes, then turn left on the D905 to the Meuse valley, which you will join in Vacherauville. Follow it downstream (north) to Sedan (80 km, 1 h 20). **A few** corncrakes **still breed in well-preserved wet meadows around Stenay** [6] (40 km from Vacherauville), but there is no specific viewpoint: **your best chance will be to find suitable areas during the day and stop-and-go from place to place at dawn or dusk.** Hobby, curlew, wryneck, red-backed shrike, sedge, marsh and melodious warblers, lesser whitethroat, golden oriole and a generally rich community of meadow and hedgerow species occurs all along the river in suitable habitats, so **stop frequently on the way.**

The junction of the Meuse and Chiers rivers between Mairy and Rémilly can be especially nice **if spring has been sufficiently wet to flood marshes and meadows, making them prime stopover grounds** for wildfowl, geese, swans (including whooper), cranes and waders (turn left on the D4 2 km from Mairy just past Douzy airfield and park carefully at the start of quarry tracks to walk among ponds [7,8]). Red-throated pipit and Lapland bunting have been recorded in this area. Corncrake and curlew, plus dipper,

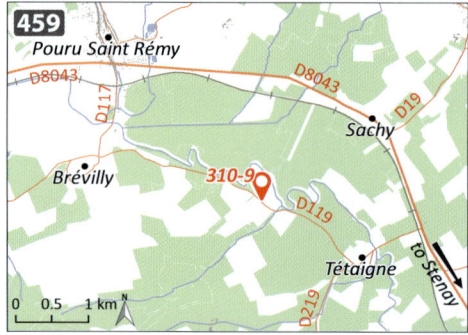

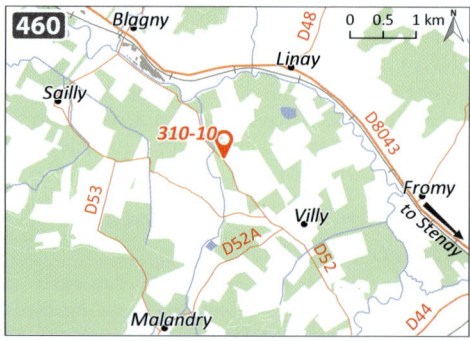

Montagu's harrier, great grey shrike (scarce) and whinchat also breed along the Chiers between Rémilly-Aillicourt and La Ferté-Sur-Chiers. Follow this 25 km stretch of the river, **preferably along small roads on the southern bank**, for instance between Brévilly and Tétaigne [9] or between Blagny and Villy [10]. This will lead you back to Stenay, ideally at dusk to try for corncrake.

311 From Charleville-Mézières to Reims

0.5 DAY; MAPS 454 & 461–464   Les Ayvelles pond [1], 10 km, 15 min south of Charleville-Mézières, hosts **good numbers of** wildfowl **and** waders **in winter and during migration** (follow signs to Villers-Semeuse or Flize, once in Les Ayvelles turn left at the bakery). Otherwise rare species like scaup, smew and goldeneye occur regularly, and breeding birds include common tern and white stork.

Drive south for 30 km past Flize, Vendresse and Sauville to visit Bairon pond [2], another good spot for wildfowl and waders. In spring, **walk along the extensive reedbed which covers the pond west of the dyke** [3] to look for bluethroat, sedge and reed warblers. Scarcer species like bearded and penduline tits or aquatic warbler may stop over on migration, but finding them is essentially a matter of chance and being there at the right time.

A logical follow-up would be to drive south again for 20 km past Le Chesne to **stop-and-go along the 10 km stretch of the Aisne valley between Voncq and Vandy** [4]. **A few** corncrakes **still breed in wet meadows along this stretch of the river and are best sought as they sing at dawn and dusk in May.** More reliable are curlew, white stork, night-heron, little ringed plover, whinchat, red-backed shrike and marsh warbler. In March and April, cranes, geese, wildfowl and waders may stop over if meadows are flooded.

The landscape dramatically changes a few kilometres southeast as it becomes a flat cultivated plain with sparse woods. Visit the area between Cauroy, Fuy and Juniville (follow the D315 to Neuville en Tourne [5], or any other road in the vicinity, parking carefully to walk on tracks). Hen and Montagu's harriers and stone-curlew breed in the fields together with other open-country species such as yellow wagtail, common whitethroat and corn bunting. In winter, short-eared owls may roost in

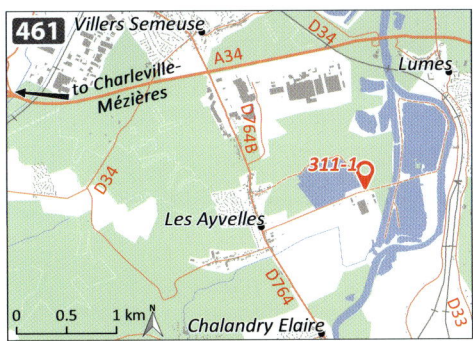

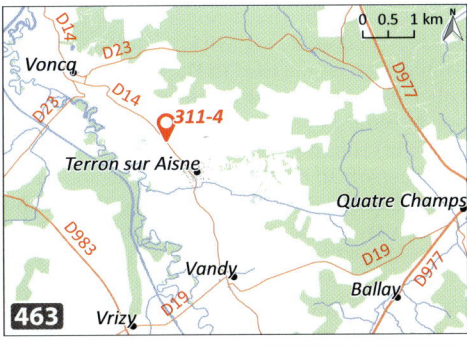

uncultivated plots, and rough-legged buzzards have been recorded several times.

312 The Ardennes north of Charleville-Mézières

1 DAY; MAPS 454, 465 & 466 The hills of the Ardennes north of Charleville-Mézières peak at 500 m on the Plateau de la Croix-Scaille, **a wide expanse of beech–spruce mixed forest with bogs** spreading between Monthermé (20 km, 30 min from Charleville, follow the River Meuse on the D1 past Nouzonville), Thilay and Hargnies at the Belgian border. **Several marshy clearings** in Les Hauts-Buttés (park near the abbey [1] and walk to [2], 10 km, 10 min north of Monthermé on the road to Hargnies), La Croix-Gillet [3] (5 km further north on the D989, turn left towards Fumay and park immediately on the right) and the road up to the Belgian border can be good starting points. **On the southeast part of the plateau, a drive along the Vallon de l'Ours offers more opportunities** (drive north for 4 km from Les Hautes Rivières to Lingchamps [4] and turn left on a narrow road towards La Neuville aux Haies [5], then left again after 1.5 km to cross the Ours stream [6] and join the D989 in Les Hauts Buttés). **Overall, your best bet will be to stop at any suitable place for your target**; checking the news on www.faune-champagne-ardenne.org can be a good idea to save time.

Be patient – the forest is usually quiet and birds occur at low densities. Focusing on clearings and edges will make things easier, especially for pied flycatcher (in beech stands), siskin and redpoll. The local highlights are black stork (spring and summer; try to locate soaring individuals as the day warms), goshawk, hazel grouse and Tengmalm's owl (singing activity peaks in clear nights of February and March). Black grouse **used to breed on the plateau but is now probably extirpated.** Other species of interest include middle spotted and black woodpeckers, red-backed shrike and nightjar, as well as forest passerines such as Eurasian treecreeper, crested tit, wood warbler, hawfinch and crossbill.

The Plateau de Rocroi is more open and can be a good follow-up for the warm hours of the day in spring (30 km, 30 min from Charleville-Mézières, follow Charleroi until you find signposts to Rocroi). There is no specific location, and the whole area between Rocroi, Regniowez and Petite-Chapelle can be good, for instance around Taillette [7]. Search for icterine warbler (from late May), great grey and red-backed shrikes, little owl and soaring red kites or storks. In winter, check gull roosts in the meadows around Eteignières dump [8] (10 km southwest of Rocroi) in search of Caspian gull.

BIRDFINDER

This section provides details of 30 species that are frequently high on the wishlist of birders visiting France. Phenological bars display the likelihood of seeing the species at a suitable location, month by month:

- ▭ absent
- ▭ present but hard to find
- ▭ present and easy to find

A non-exhaustive list of the most suitable sites to see the species is also provided. This does not mean that the species occurs only at these sites, but that it is easier here than anywhere else owing to local commonness, ease of access, close monitoring or other reasons. We also provide recommendations based on our own experience or that of relevant experts. However, your best chance to encounter your dream-bird will always be to study its identification and ecology in the specialised literature. **There is no excuse for disturbance**: avoid the use of playback during the breeding season and/or for sensitive species, do not enter restricted areas, do not harass birds in any way. Finally, remember that for virtually all birds, an early-morning start, discreet behaviour and patience are the keys to success.

Hazel Grouse – Gélinotte des bois – *Tetrastes bonasia*

J	F	M	A	M	J	J	A	S	O	N	D

BEST SITES Risoux forest (224), Vallorcine (230), Vercors (234).

Arguably the hardest of all grouse, typically offering just a frustrating glimpse as it flies over dense vegetation. Setting up a hide in a known location will be your only chance to achieve good views or photos. It is slightly easier to see in protected areas, especially from April to early June and in September, when family groups sometimes wander conspicuously along deserted forest tracks. Hazel grouse favours chaotic rocky terrains under mixed beech–fir or hazel stands with a dense understorey and berries.

Capercaillie – Grand Tétras – *Tetrao urogallus*

J	F	M	A	M	J	J	A	S	O	N	D

BEST SITES Orlu forest (165), Col de Mantet (187), Pic de Madrès (188), Risoux forest (224).

Capercaillie has undergone a marked decline, and its French populations are now facing extinction. The eastern stronghold is located in the Jura, where it is closely watched and under strong protection measures. For these reasons, we do not advise targeted searches for this species, which can be found more easily and less intrusively elsewhere in Europe. There are still a few individuals in the Vosges, but whether they constitute a viable population is unclear. A few individuals occur in the Cévennes (Mont Aigoual, site 179), but they are of captive origin. The Pyrenean aquitanicus subspecies is slightly less threatened and still occurs in mid-altitude conifer forests throughout the eastern Pyrenees. As elsewhere in Europe, capercaillie live in old beech or conifer forests above an altitude of 1000 m (up to 2500 m in the Pyrenees) as long as the understorey is sufficiently clear for a cover of herbaceous undergrowth to develop (especially, but not only, blueberries).

Black Grouse – Tétras lyre – *Lyrurus tetrix*

J	F	M	A	M	J	J	A	S	O	N	D

BEST SITES Mercantour (204), Col de la Colombière (228), Argentière (230), Glières plateau (231), Vercors (234, 235), Chaillol (240).

Search for black grouse on grassy slopes with willows at an altitude of 1000–2000 m. Singing activity reaches its peak from May to mid-June and is largely constrained by disturbance from hikers, so be on site before first light. In places such as the Col de la Colombière, leks can be watched from the roadside with a scope; elsewhere you may need to walk a bit on tracks – but in general, with some patience, finding birds should not be too complicated once you have detected them by ear.

Rock Partridge – Perdrix bartavelle – *Alectoris graeca*

J	F	M	A	M	J	J	A	S	O	N	D

BEST SITES Mercantour (204), Col de Floray (227), Col de la Colombière (228), Termignon (233), Prapic and the Ecrins (240)

Rock partridge has significantly declined in recent decades and is confined to the Alps, but the sites listed here are still reliable. The best areas are rocky slopes with little, if any vegetation, at an altitude of 1000–2500 m. Be on site before first light, as birds will become secretive and hide high on mountain slopes as soon as the first hikers show up. Catch singing birds by ear (mainly May–June) and search for them with a scope along ridges or on the tops of large boulders.

Black Stork – Cigogne noire – *Ciconia nigra*

J	F	M	A	M	J	J	A	S	O	N	D

BEST SITES Orient, Temple, Amance and Der lakes (1–4), Val-Joly (40), Loire valley (89, 95, 100), Saint-Denis-du-Payré (114), Le Teich (145), southern Aquitaine (149), Organbidexka (151), Gruissan (182), Canet-en-Roussillon (185), Eyne (189), Camargue (191), Dombes (221), Défilé de l'Écluse (223), Le Hucel (227), Châtillon-sur-Seine forest (275), Yonne valley (280), Is-sur-Tille (285), Ardennes (312)

The best way to see this iconic species is to spot from a migration watchpoint, Gruissan (April–May and August–September), the Camargue or Organbidexka (July–September) being probably the most reliable for this species. It is also regular along large rivers and on ponds or marshes all across the country, specifically along the Loire, Seine and Rhône valleys, as well as in the coastal marshes of Aquitaine and the Mediterranean. Finding a breeding black stork is a near-impossible task unless you can spare a whole season to wander among the oldest oak and beech forests of the central regions. A few birds winter, most of them in the Camargue.

Little Bittern – Blongios nain – *Ixobrychus minutus*

J	F	M	A	M	J	J	A	S	O	N	D

BEST SITES Paris (8, 10, 11), Marne and Bassée valleys (13, 14), Saint-Quentin-en-Yvelines (18), the Brenne (96), southern Médoc and Aquitaine (143, 148, 149), Aude coast (180, 184), Camargue (190, 191), Dombes (221), Lac du Bourget (232), Madine (297), Meurthe valley (302), Petite Camargue Alsacienne (308)

This small reedbed specialist is not especially hard to see with a bit of patience. Sit in a hide or at some distance along the edge of a reedbed and catch it as it flies over. The first migrants arrive on the breeding grounds in mid-April, but the species will not be reliable before the first week of May. Birds become especially conspicuous as they make frequent journeys to their nests to feed their nestlings in late June and July. They remain around, although more secretive, until late August.

Black-winged Kite – Élanion blanc – *Elanus caeruleus*

J	F	M	A	M	J	J	A	S	O	N	D

BEST SITES Médoc (142), Adour plain (152, 173), Dordogne (154), Saint-Gaudens plain (159)

The range of this most-wanted species has been steadily increasing for over a decade, and finding it is no longer such a big deal. Once confined to southern Aquitaine, it now breeds regularly as far north as Bordeaux, and sightings or even breeding attempts in other regions (including Vendée, Loire valley, Brittany or along the Rhône valley) are now annual. Look for it along hedgerows or on isolated trees in cultivated fields. Birds are usually found hunting from conspicuous lookout posts or in flight. Over 100 individuals may roost together in winter, especially in fields in the Toulouse region.

Lammergeier – Gypaète barbu – *Gypaetus barbatus*

| J | F | M | A | M | J | J | A | S | O | N | D |

BEST SITES Organbidexka and surroundings (151, 152), central Pyrenees (165–171), Col de Mantet (187), Pic de Madrès (188), Eyne (189), Asco and Bavella (214, 218), Col de la Colombière (228), Termignon (233), Lautaret (237), Lauzon lake (240)

The species has undergone a good recovery following the combined efforts of a reintroduction programme and close monitoring and protection of a few remaining wild nests in the last decades of the twentieth century. It is now relatively easy to find in the northern Alps and the Pyrenees even without targeted searches, especially when birds soar on warm, sunny days.

Black Vulture – Vautour moine – *Aegypius monachus*

| J | F | M | A | M | J | J | A | S | O | N | D |

BEST SITES Verdon gorges (200), Rémuzat (206), Larzac (245), Jonte gorges (246)

Most of the French black vultures have been released from captivity or are offspring of reintroduced individuals. However, Spanish birds sometimes cross the border in the Pyrenees. Owing to its low numbers the species remains hard to find: head to known viewing points such as the Gorges de la Jonte and Rémuzat and spend time there, preferably on a sunny day.

Bonelli's Eagle – Aigle de Bonelli – *Aquila fasciata*

| J | F | M | A | M | J | J | A | S | O | N | D |

BEST SITES Causse d'Aumelas (176), Saint-Hyppolyte-du-Fort (179), La Clape (182), Alpilles (196)

Bonelli's eagle is a flagship of French bird conservation. It is rare and local, most pairs occurring along limestone cliffs in Mediterranean plains from the Aude to the Alpilles. The sites listed do not indicate nests or even known territories, but locations where the species is seen more frequently than elsewhere. Nest locations are kept confidential, as they remain under high illegal pressure from falconry interests. Should you find one, keep it for an official contact at the LPO or another institution. Your best chance of seeing the species will be to wait on a high point in the hope of catching a soaring bird. It is inconspicuous and does not fly as much as other raptors, so be patient.

Western Swamphen – Talève sultane – *Porphyrio porphyrio*

| J | F | M | A | M | J | J | A | S | O | N | D |

BEST SITES Montpellier marshes (174, 176), Aude coastal marshes (180, 181, 182), Canet-en-Roussillon (185), Camargue Gardoise (190), Route de Cacharel (191), Vigueirat marshes (192)

A common bird of Mediterranean reedbeds. Search along mudflats and reedbed edges in the early morning, and learn its call. Swamphens sometimes roost in parties of over 10 or 20 individuals after reed harvest in the western Camargue (February), and then becomes ridiculously easy to find – however, it is trickier after cold winters. The most famous site where the species is near-guaranteed is the Pont des Tourradons in the Camargue, but many of the other places listed above are just as good.

Slender-billed Gull – Goéland railleur – *Chroicocephalus genei*

| J | F | M | A | M | J | J | A | S | O | N | D |

BEST SITES Lansargues (174), Aude valley (181, 184), Camargue (191), Hyères (197)

This most sought-after of all French gulls breeds in large colonies near Montpellier, in the Camargue and in Hyères saltmarshes. It is uncommon outside the breeding and wintering sites but remains easy to find year-round, most often in small parties of up to 10 birds feeding in shallow waters or roosting at dusk. Learn its typical flight and feeding postures.

Roseate Tern – Sterne de Dougall – *Sterna dougallii*

| J | F | M | **A** | **M** | **J** | **J** | **A** | S | O | N | D |

BEST SITES Saint-Jacut-de-la-Mer (55), Morlaix bay (58), Larmor-Baden (late August, 78)

Roseate terns are rare and mostly restricted to a few colonies in northern Brittany and post-breeding roost sites in southern Brittany. Your best chance will probably be a late-August visit to Larmor-Baden, where the birds usually perch near the shore, allowing much closer views than at their colonies, which are located on distant, inaccessible rocks.

Pin-tailed Sandgrouse – Ganga cata – *Pterocles alchata*

| **J** | **F** | **M** | **A** | **M** | **J** | **J** | **A** | **S** | **O** | **N** | **D** |

BEST SITE the Crau (193)

The French sandgrouse population is confined to the Crau plain between the Rhône and Marseille, with very few records outside this area. Although small flocks are sometimes seen flying over random places on the plain, the best way to find the species is to search for it at dawn or dusk around the places described in the text. It is gregarious year-round, but flocks are possibly easier to find in winter, especially around Eyguières airfield, at the northern edge of the Crau.

Pygmy Owl – Chevêchette d'Europe – *Glaucidium passerinum*

| J | **F** | **M** | **A** | **M** | **J** | **J** | **A** | **S** | **O** | **N** | D |

BEST SITES Pic de Madrès (188), Saint-Etienne de Tinée (204), Jura (222, 224), south of Lake Geneva (227), upper Chamonix valley (230), Glières plateau (231), Vercors (234, 235), Chaillol (240), Dévoluy (241), Hautes Chaumes (266), Morvan (291), Celles-sur-Plaine (305), Tanet-Gazon du Faing (306)

The French population of pygmy owl is expanding, and it has become relatively easy to find in the Jura and the Alps as well as at a few localities in the eastern quarter of the country. It was recently reported on the Pic de Madrès in the eastern Pyrenees, but it is far from having established a population yet. Search it along edges or clearings of old spruce or other conifer stands with numerous dead trees, at an altitude of 1200–2000 m, sometimes along with Tengmalm's owl – although the latter has somewhat different habitat requirements. Pygmy owl is also more crepuscular and rarely sings after full dark; it can even be heard during the day, especially in summer once nestlings have fledged.

Tengmalm's Owl – Chouette de Tengmalm – *Aegolius funereus*

| **J** | **F** | **M** | **A** | **M** | **J** | **J** | **A** | **S** | **O** | **N** | **D** |

BEST SITES Mont Aigoual (179), Col de Mantet (187), Pic de Madrès (188), Risoux forest (224), Vallorcine (230), Glières plateau (231), Vercors (234, 235), La Serre (260), Puy de Dôme (261), Morvan (291), Vosges (306, 307)

This is the commoner of the two mountain owls, but also harder to see, and there are strong population fluctuations from year to year even in the species' core area. It generally breeds in old beech or spruce stands at an altitude of 800–1800 m, sometimes at relatively high densities, and always close to breeding black woodpecker, whose old holes it reuses. Song (mid-January to early April) peaks in early March, with some low-level activity in early summer and from November. The species is quite demanding: expect it only under optimum conditions (clear sky with no wind) and do not hesitate to stay late at night. In our experience, song starts at full dark but peaks around 23.00 or even later. Breeding birds (around April–June) are extremely secretive and are usually only seen at the nest hole. It would be highly inappropriate to cause any kind of disturbance in pursuit of a sighting.

Three-toed Woodpecker – Pic tridactyle – *Picoides tridactylus*

| J | F | M | A | M | J | J | A | S | O | N | D |

BEST SITES Risoux forest (224), upper Chamonix valley (230), Glières plateau (231)

While this species was once a ghost of old Alpine forests, it has now become a classic of trips in the Jura and the northern Alps owing to population increases and better knowledge, although finding it requires some patience. As for other woodpeckers, the best periods are March, when birds are drumming, and early July, when family groups wander over restricted areas around the nest hole. Look for the species along the edges and clearings of old spruce or birch forests, where individuals especially favour dead or dying trees. Be on site at the very first daylight as birds tend to become particularly secretive as the day warms up. Although the drummings are quite different, separating its calls from those of great spotted woodpecker (which is always commoner, even at high altitudes) requires experience; all audio records should therefore be confirmed by a sonogram or a visual sighting.

White-backed Woodpecker – Pic à dos blanc – *Dendrocopos leucotos lilfordi*

| J | F | M | A | M | J | J | A | S | O | N | D |

BEST SITES Iraty and Hayra forests (151)

The most reliable place is Iraty forest, especially between the barrage and Organbidexka pass. The species is not especially hard to find and tends to be confident or even curious, but it can require from several hours to several days to get a first contact. Walk along forest tracks behind Iraty camp site and search for birds inside old beech stands and in clearings with dead trees. For those who want a greater thrill, the species is present in beech forests from the Atlantic shore to inner Pyrenees, maybe as far as western Ariège. Learn to separate its call from that of great spotted woodpecker, which breeds alongside it. There are sightings year-round, but in our experience the species is easier to find before the onset of the breeding season (March) and once the nestlings have fledged (between mid-June and mid-July).

Grey-headed Woodpecker – Pic cendré – *Picus canus*

| J | F | M | A | M | J | J | A | S | O | N | D |

BEST SITES Orient forest (2), the Perche (47), Loches (94), Orléans forest (99), Allogny forest (102), Tronçais (259), Burgundy (Region 13), Nancy forests (296), Parroy forest (301), Celles-sur-Plaine (305)

Once common, this species now ranks among the hardest French breeding birds to find, being restricted to large expanses of oak and beech forests in the centre of France. Population strongholds are now located in Burgundy and Lorraine, although a few still breed in the west. It is especially secretive, and even hearing a call on known sites can be difficult. Vocal activity peaks in March to April, and birds virtually disappear once the nestlings have hatched. Spend time in old, clear deciduous stands (even clearings with large isolated trees can be suitable, but the species seems to avoid dense understorey) during the first morning hours. Avoid using playback – any disturbance to such a sensitive species would be highly inappropriate.

Lesser Kestrel – Faucon crécerellette – *Falco naumanni*

| J | F | M | A | M | J | J | A | S | O | N | D |

BEST SITES Vendémian and Saint-Pont-de-Mauchiens (176), Aude valley (181), Eyne (189), the Crau (193), Barcaggio (217)

Once almost extirpated from the country, the lesser kestrel has made a remarkable recovery and is now a classic sighting on any day tour in the Crau, Aude and southern Montpellier areas from late March to September, when birds flock in sometimes large parties before leaving on migration. Experience and caution are required to separate females and immatures from common kestrel, which always breeds nearby.

Try to take photos, and make careful records of the jizz, behaviour, head pattern, streaks on the underparts and the colour of the talons. Besides the Crau, where sightings are guaranteed in spring, the best spots are in the vicinity of Vendémian (30 min southwest of Montpellier) and Fleury d'Aude, where a colony breeds in a purpose-built structure.

Southern Grey Shrike – Pie-grièche méridionale – *Lanius meridionalis*

| J | F | M | A | M | J | J | A | S | O | N | D |

BEST SITES Causse d'Aumelas (176), Causse de Blandas (178), the Crau (193), the Calissanne (194), Alpilles (196), Valensole (201)

This is a bird of the garrigues and maquis of the Mediterranean backcountry. Once common, it has undergone a sharp decline and has now become hard to find outside the Crau plain and the listed areas around Montpellier. Birds are usually seen perching on bushes but they tend to be secretive and silent. Be persistent when looking for this species in the spots described, as they may be hard to find if hiding inside dense shrub cover.

Penduline Tit – Rémiz penduline – *Remiz pendulinus*

| J | F | M | A | M | J | J | A | S | O | N | D |

BEST SITES Saint-Quentin-en-Yvelines (18), Brouage marshes (122), Braud-et-Saint-Louis (140), Villeneuve-lès-Maguelone (176), Scamandre, Pont des Tourradons (190), Cacharel, Verdier marshes (191), Vigueirat marshes (192), the Dombes (221), south of Lake Geneva (226), southern Lac du Bourget (232), Marcenay reservoir (276), Yonne valley (280)

Penduline tit is mainly a winter bird and is easily found in the reedbeds of Camargue from October to March. It also occurs more or less reliably in coastal marshes of Aquitaine, along the Rhône valley and in Lorraine; it is irregular elsewhere. It primarily frequents large expanses of reedbed, but a small cattail patch even within highly urbanised areas can be enough for stop-over migrants. Learn the call, especially to differentiate it from that of reed bunting.

Iberian Chiffchaff – Pouillot ibérique – *Phylloscopus ibericus*

| J | F | M | A | M | J | J | A | S | O | N | D |

BEST SITE Iraty area (151)

Another tricky breeding species, for two reasons: first, there is no regular place where it is found (singing birds are recorded on the local news website); second, it is a true identification challenge. It is normally unwise to identify an Iberian chiffchaff outside the song season, as there is currently no known stable and definitive identification criterion. There are many intermediate birds (and song patterns), so it is always preferable to record the birds and check sonograms before assessing an identification as certain. The species is restricted to the Pyrenean foothills of western Aquitaine, roughly in a triangle between Saint-Jean-Pied-de-Port, Hendaye and Dax, and it is easiest to find from April to June. Most of the singing birds are recorded on mountain slopes, usually in low deciduous (willow or beech) stands near waterbodies or rivers.

Moustached Warbler – Lusciniole à moustaches – *Acrocephalus melanopogon*

| J | F | M | A | M | J | J | A | S | O | N | D |

BEST SITES Villeneuve-lès-Maguelone (176), Vendres (180), the Aude coast (181, 182, 184), Canet-en-Roussillon (185), Camargue Gardoise (190), Verdier marsh (191), Saint-Chamas (194)

Unlike other *Acrocephalus* warblers, moustached warbler is sedentary and relatively easy to find in reedbeds at the recommended locations even in mid-winter. Being confident, it comes easily to pishing and moderate playback (do not insist – it will either come at the first play or it won't come at all). Pick a windless and sunny day, and wait for the reeds to warm up, especially in cold periods. The species is most conspicuous when singing in March, sometimes in high numbers. However, it gets trickier to identify as soon as the first sedge warblers show up in April, and it will become near-silent by late April. Your best chance then is to spot feeding adults or family flocks as they wander along reedbed edges.

Spectacled Warbler – Fauvette à lunettes – *Sylvia conspicillata*

J F M **A M J** J A S O N D

BEST SITES La Gacholle and Cacharel (191), Valensole (201)

The easiest places to see this species are in the saltmarshes around La Gacholle lighthouse and Route de Cacharel in Camargue. The lavender fields of the Valensole plateau are another stronghold, closer to the Alps. Singing birds or breeding pairs settle in low garrigues south of Montpellier and along the Aude coast (for instance in Lapalme), but their occurrence is unpredictable.

Corsican Nuthatch – Sittelle corse – *Sitta whiteheadi*

J F M **A M J J A S O N D**

BEST SITES Aïtone (212), Restonica valley (213), Haut-Asco (214), Col de Bavella (218)

This species is a strict endemic of Corsican mountains, where it breeds in old pine stands with dead trees. It is resident and relatively easy to find year-round, but will be easier when displaying in April. Playback works well but is usually unnecessary.

Wallcreeper – Tichodrome échelette – *Tichodroma muraria*

J F M A M J J A S **O N D**

BEST SITES Iparla (151), Lapiaz de la Pierre Saint-Martin (152), Hortus cliffs and the Cévennes (177, 178, 179), Les Baux-de-Provence (196), Vallée des Merveilles (204), Haut-Asco (214), Alps (Region 11), Massif Central (in winter, Region 12)

A most-wanted of birders visiting the Alps, but also a true challenge as the species lives on high, shadowy, damp cliffs usually above 2000 m and requires approach hikes and patient searches. It is present all through the Alps wherever the habitat is suitable but its secretive habits and discreet song make it difficult to catch; chances will be higher at the sites listed, owing to easy access or good viewpoints. Locate a bird by its characteristic flight when feeding along ridges in May or June. The species can be easier to find in winter as it settles on old buildings and small cliffs on the plains, especially along the Rhône, in the Massif Central or in the Cévennes foothills, but even sometimes as far west as Paris. Wintering birds are usually located on regional birding databases, so check the news.

Collared Flycatcher – Gobemouche à collier – *Ficedula albicollis*

J F M A **M J** J A S O N D

BEST SITES Faverney (283), Nancy forests (296), Champenoux (300), Lindre reservoir (301), Amel pond (310)

The French population marks the western end of the species' distribution. Core sites are located in the old beech and oak forests of Lorraine, but they breed as far west as around the Lac du Der and south of Vesoul. Search for singing birds early in the morning in May along forest tracks (learn to separate the call from that of pied flycatcher, as the two species sometimes breed close together). Although usually easy to hear, catching a sight of a bird may require a bit of patience as they often sing from close to the trunk under the crown of a large tree, above dense understorey. Single migrants are recorded annually along the Mediterranean coast in early May and September: beware of grey pied flycatcher females and young, and take careful note of the pattern of white on wings, coloration features and the call (photos and sonograms are preferable).

Common Rock Thrush – Monticole de roche – *Monticola saxatilis*

J	F	M	A	M	J	J	A	S	O	N	D

BEST SITES western Pyrenees (152), Mont Valier (167), Pic de Cagire (168), Col du Tourmalet (170), Causse de Blandas (178), Pic Carlit (187), Mercantour (204), Mont Ventoux (205), Alps (Region 11), Causse Méjean (247), Plomb du Cantal (250), Mont Mézenc (256), Chaîne des Puys (261, 262), Hautes Chaumes (266)

This is a common breeding species of Alpine rocky slopes from 1500 to 2500 m (lower in the Cévennes), although it does not occur everywhere. Being usually on breeding grounds by mid-April, and departing in mid-August, it is easy to find during the main Alpine birding season. Odd individuals are regularly recorded at migration watchpoints or even on small cliffs in Mediterranean garrigues.

Snowfinch – Niverolle Alpine – *Montifringilla nivalis*

J	F	M	A	M	J	J	A	S	O	N	D

BEST SITES Mont Valier (167, Pic de Bigorre (170), Cime de la Bonette (204), Mont Ventoux (winter, 205), Col de la Colombière (228), upper Chamonix (230), Grand Veymont (235), Lautaret and Galibier (237), the Ecrins (240)

Except in winter, when birds concentrate around ski resorts, the only way to connect with snowfinch is to climb up to the snow limit, which can be as high as 2200 m in June. It does not occur everywhere – stick to the spots we have mentioned to avoid a long and strenuous hike for no reward. When present, however, it is usually very confident towards hikers, relatively numerous and thus easy to find.

CHECKLIST

This checklist lists all the species and subspecies of annual occurrence in France in systematic order, with their English, scientific and French names. The 'Category' column follows the British Ornithologists' Union (BOU) categories (**A**: species recorded in an apparently natural state at least once since 1 January 1950; **B**: species recorded in an apparently natural state at least once between 1 January 1800 and 31 December 1949, but not recorded subsequently; **C**: species that, although introduced, now derive from the resulting self-sustaining populations). We have not listed species in categories D, E and F. The column 'Regions' indicates the regions in which the rarest and most local species can reasonably be expected – they can still occur elsewhere, but will be less reliable, highly local or too rare to be expected. The columns on the right show each species' average phenology and abundance in its main regions of occurrence.

Key for phenological bars

- abundant
- common
- locally common
- uncommon
- rare/local
- scarce
- absent
- feral, common
- feral, rare or local

English name	Scientific name	French name	Category	Regions	J	F	M	A	M	J	J	A	S	O	N	D
Brent goose	*Branta bernicla*	Bernache cravant	A													
Pale-bellied brent goose	*B. b. hrota*	Bernache à ventre pâle	A	4												
Black brant	*B. b. nigricans*	Bernache du Pacifique	A	4,6												
Canada goose	*Branta canadensis*	Bernache du Canada	C													
Barnacle goose	*Branta leucopsis*	Bernache nonnette	AC	2,3												
Greylag goose	*Anser anser*	Oie cendrée	AC													
Pink-footed goose	*Anser brachyrhynchus*	Oie à bec court	A	2												
Tundra bean goose	*Anser serrirostris*	Oie des moissons	A	1,14												
White-fronted goose	*Anser albifrons*	Oie rieuse	A	1,2												
Mute swan	*Cygnus olor*	Cygne tuberculé	AC													
Bewick's swan	*Cygnus columbianus*	Cygne de Bewick	A	1, 10												
Whooper swan	*Cygnus cygnus*	Cygne chanteur	A	1,2,11												
Egyptian goose	*Alopochen aegyptiaca*	Ouette d'Égypte	BC													
Shelduck	*Tadorna tadorna*	Tadorne de Belon	A													
Ruddy shelduck	*Tadorna ferruginea*	Tadorne casarca	BC													

CHECKLIST

English name	Scientific name	French name	Category	Regions	J	F	M	A	M	J	J	A	S	O	N	D
Mandarin duck	Aix galericulata	Canard mandarin	C													
Garganey	Spatula querquedula	Sarcelle d'été	A													
Shoveler	Spatula clypeata	Canard souchet	A														
Gadwall	Mareca strepera	Canard chipeau	A														
Wigeon	Mareca penelope	Canard siffleur	A														
American wigeon	Mareca americana	Canard à front blanc	A	4												
Mallard	Anas platyrhynchos	Canard colvert	AC													
Pintail	Anas acuta	Canard pilet	A														
Teal	Anas crecca	Sarcelle d'hiver	A														
Red-crested pochard	Netta rufina	Nette rousse	A														
Pochard	Aythya ferina	Fuligule milouin	A														
Ferruginous duck	Aythya nyroca	Fuligule nyroca	A	1,9,10,11,13,14												
Ring-necked duck	Aythya collaris	Fuligule à bec cerclé	A	1,4,6,11												
Tufted duck	Aythya fuligula	Fuligule morillon	A														
Scaup	Aythya marila	Fuligule milouinan	A	1,2,4,13,14												
Eider	Somateria mollissima	Eider à duvet	A	2,3,4,6												
Velvet scoter	Melanitta fusca	Macreuse brune	A	3,4												
Common scoter	Melanitta nigra	Macreuse noire	A														
Long-tailed duck	Clangula hyemalis	Harelde boréale	A	4,6												
Goldeneye	Bucephala clangula	Garrot à œil d'or	A	1,4												
Smew	Mergellus albellus	Harle piette	A	1												
Goosander	Mergus merganser	Harle bièvre	A														
Red-breasted merganser	Mergus serrator	Harle huppé	A														
Ruddy duck	Oxyura jamaicensis	Erismature rousse	C														
Hazel grouse	Tetrastes bonasia	Gélinotte des bois	A	11												
Capercaillie	Tetrao urogallus	Grand Tétras	A	8,10,11												
Black grouse	Lyrurus tetrix	Tétras lyre	A	10,11												
Ptarmigan	Lagopus muta	Lagopède alpin	A	8,9,10,11												
Rock partridge	Alectoris graeca	Perdrix bartavelle	A	10,11												

English name	Scientific name	French name	Category	Regions	J	F	M	A	M	J	J	A	S	O	N	D
Red-legged partridge	Alectoris rufa	Perdrix rouge	AC													
Grey partridge	Perdix perdix	Perdrix grise	AC													
Pyrenean grey partridge	P. p. hispaniensis	Perdrix grise des Pyrénées		8,9												
California quail	Callipepla californica	Colin de Californie	C													
Northern bobwhite	Colinus virginianus	Colin de Virginie	C	10												
Quail	Coturnix coturnix	Caille des blés	A													
Reeves's pheasant	Syrmaticus reevesii	Faisan vénéré	C													
Pheasant	Phasianus colchicus	Faisan de Colchide	C													
Red-throated diver	Gavia stellata	Plongeon catmarin	A													
Black-throated diver	Gavia arctica	Plongeon arctique	A													
Great northern diver	Gavia immer	Plongeon imbrin	A													
Storm petrel	Hydrobates pelagicus	Océanite tempête	A													
Leach's petrel	Oceanodroma leucorhoa	Océanite culblanc	A	2,3,4												
Fulmar	Fulmarus glacialis	Fulmar boréal	A													
Scopoli's shearwater	Calonectris diomedea	Puffin cendré	A	9, 10												
Cory's shearwater	Calonectris borealis		A	4,6												
Sooty shearwater	Ardenna grisea	Puffin fuligineux	A													
Great shearwater	Ardenna gravis	Puffin majeur	A	4												
Manx shearwater	Puffinus puffinus	Puffin des Anglais	A													
Yelkouan shearwater	Puffinus yelkouan	Puffin yelkouan	A													
Balearic shearwater	Puffinus mauretanicus	Puffin des Baléares	A													
Little grebe	Tachybaptus ruficollis	Grèbe castagneux	A													
Red-necked grebe	Podiceps grisegena	Grèbe jougris	A	1,2,3,4												
Great crested grebe	Podiceps cristatus	Grèbe huppé	A													
Slavonian grebe	Podiceps auritus	Grèbe esclavon	A													
Black-necked grebe	Podiceps nigricollis	Grèbe à cou noir	A													
Black stork	Ciconia nigra	Cigogne noire	A	5,7,9,10,13,14												
White stork	Ciconia ciconia	Cigogne blanche	A													
Glossy ibis	Plegadis falcinellus	Ibis falcinelle	A	9, 10												

CHECKLIST

English name	Scientific name	French name	Category	Regions	J	F	M	A	M	J	J	A	S	O	N	D
African sacred ibis	Threskiornis aethiopicus	Ibis sacré	C													
Great white pelican	Pelecanus onocrotalus	Pélican blanc	A													
Spoonbill	Platalea leucorodia	Spatule blanche	A													
Bittern	Botaurus stellaris	Butor étoilé	A	1,3,9,10,11,14												
Greater flamingo	Phoenicopterus roseus	Flamant rose	A	9, 10												
Little bittern	Ixobrychus minutus	Blongios nain	A	1,5,9,10,11,13,14												
Night-heron	Nycticorax nycticorax	Bihoreau gris	A													
Squacco heron	Ardeola ralloides	Crabier chevelu	A	9, 10,11												
Cattle egret	Bubulcus ibis	Héron garde-bœufs	A													
Grey heron	Ardea cinerea	Héron cendré	A													
Purple heron	Ardea purpurea	Héron pourpré	A													
Great white egret	Ardea alba	Grande Aigrette	A													
Western reef heron	Egretta gularis	Aigrette des récifs	A	10												
Little egret	Egretta garzetta	Aigrette garzette	A													
Gannet	Morus bassanus	Fou de Bassan	A													
Shag	Phalacrocorax aristotelis	Cormoran huppé	A													
Mediterranean shag	P. a. desmarestii	Cormoran de Desmarest		9, 10												
Cormorant	Phalacrocorax carbo	Grand Cormoran	A													
Osprey	Pandion haliaetus	Balbuzard pêcheur	A													
Black-winged kite	Elanus caeruleus	Elanion blanc	A	7,8												
Bearded vulture	Gypaetus barbatus	Gypaète barbu	A	7,8,9,10,11												
Egyptian vulture	Neophron percnopterus	Vautour percnoptère	A	8,9,10,12												
Honey-buzzard	Pernis apivorus	Bondrée apivore	A													
Griffon vulture	Gyps fulvus	Vautour fauve	AC													
Cinereous (black) vulture	Aegypius monachus	Vautour moine	AC	12												
Short-toed eagle	Circaetus gallicus	Circaète Jean-le-blanc	A													
Lesser spotted eagle	Clanga pomarina	Aigle pomarin	A	10,11												
Spotted eagle	Clanga clanga	Aigle criard	A	7, 10												
Booted eagle	Hieraaetus pennatus	Aigle botté	A													

CHECKLIST | 321

English name	Scientific name	French name	Category	Regions	J	F	M	A	M	J	J	A	S	O	N	D
Golden eagle	Aquila chrysaetos	Aigle royal	A													
Bonelli's eagle	Aquila fasciata	Aigle de Bonelli	A	9, 10												
Sparrowhawk	Accipiter nisus	Epervier d'Europe	A													
Goshawk	Accipiter gentilis	Autour des palombes	A													
Marsh harrier	Circus aeruginosus	Busard des roseaux	A													
Hen harrier	Circus cyaneus	Busard Saint-Martin	A													
Pallid harrier	Circus macrourus	Busard pâle	A	9,10												
Montagu's harrier	Circus pygargus	Busard cendré	A													
Red kite	Milvus milvus	Milan royal	A													
Black kite	Milvus migrans	Milan noir	A													
White-tailed eagle	Haliaeetus albicilla	Pygargue à queue blanche	A													
Rough-legged buzzard	Buteo lagopus	Buse pattue	A	1,2,13												
Long-legged buzzard	Buteo rufinus	Buse féroce	A	9, 10												
Buzzard	Buteo buteo	Buse variable	A													
Little bustard	Tetrax tetrax	Outarde canepetière	A	5,6,9,10												
Water rail	Rallus aquaticus	Râle d'eau	A													
Corncrake	Crex crex	Râle des genêts	A	5,14												
Little crake	Porzana parva	Marouette poussin	A	9, 10,11												
Baillon's crake	Porzana pusilla	Marouette de Baillon	A	9, 10												
Spotted crake	Porzana porzana	Marouette ponctuée	A	4,9, 10,11,14												
Western swamphen	Porphyrio porphyrio	Talève sultane	AC	9, 10												
Moorhen	Gallinula chloropus	Gallinule poule-d'eau	A													
Coot	Fulica atra	Foulque macroule	A													
Crane	Grus grus	Grue cendrée	A													
Stone-curlew	Burhinus oedicnemus	Œdicnème criard	A													
Oystercatcher	Haematopus ostralegus	Huîtrier pie	A													
Black-winged stilt	Himantopus himantopus	Echasse blanche	A													
Avocet	Recurvirostra avosetta	Avocette élégante	A													
Lapwing	Vanellus vanellus	Vanneau huppé	A													

CHECKLIST

English name	Scientific name	French name	Category	Regions	J	F	M	A	M	J	J	A	S	O	N	D
Sociable plover	Vanellus gregarius	Vanneau sociable	A	1,5												
Golden plover	Pluvialis apricaria	Pluvier doré	A													
Grey plover	Pluvialis squatarola	Pluvier argenté	A													
Ringed plover	Charadrius hiaticula	Grand Gravelot	A													
Little ringed plover	Charadrius dubius	Petit Gravelot	A													
Kentish plover	Charadrius alexandrinus	Gravelot à collier interrompu	A													
Dotterel	Charadrius morinellus	Pluvier guignard	A	2,4,10,11,12,14												
Whimbrel	Numenius phaeopus	Courlis corlieu	A													
Curlew	Numenius arquata	Courlis cendré	A													
Bar-tailed godwit	Limosa lapponica	Barge rousse	A													
Black-tailed godwit	Limosa limosa	Barge à queue noire	A													
Turnstone	Arenaria interpres	Tournepierre à collier	A													
Knot	Calidris canutus	Bécasseau maubèche	A													
Ruff	Calidris pugnax	Combattant varié	A													
Broad-billed sandpiper	Calidris falcinellus	Bécasseau falcinelle	A	10												
Curlew sandpiper	Calidris ferruginea	Bécasseau cocorli	A													
Temminck's stint	Calidris temminckii	Bécasseau de Temminck	A	1,5,6,9,10,13												
Sanderling	Calidris alba	Bécasseau sanderling	A													
Dunlin	Calidris alpina	Bécasseau variable	A													
Purple sandpiper	Calidris maritima	Bécasseau violet	A													
Baird's sandpiper	Calidris bairdii	Bécasseau de Baird	A	4												
Little stint	Calidris minuta	Bécasseau minute	A													
White-rumped sandpiper	Calidris fuscicollis	Bécasseau de Bonaparte	A	4												
Buff-breasted sandpiper	Calidris subruficollis	Bécasseau rousset	A	4												
Pectoral sandpiper	Calidris melanotos	Bécasseau tacheté	A	4,9,10												
Woodcock	Scolopax rusticola	Bécasse des bois	A													
Jack snipe	Lymnocryptes minimus	Bécassine sourde	A													
Great snipe	Gallinago media	Bécassine double	A	9, 10												
Snipe	Gallinago gallinago	Bécassine des marais	A													

English name	Scientific name	French name	Category	Regions	J	F	M	A	M	J	J	A	S	O	N	D
Terek sandpiper	Xenus cinereus	Chevalier bargette	A	10												
Red-necked phalarope	Phalaropus lobatus	Phalarope à bec étroit	A	9, 10												
Grey phalarope	Phalaropus fulicarius	Phalarope à bec large	A	2,3,4												
Common sandpiper	Actitis hypoleucos	Chevalier guignette	A													
Green sandpiper	Tringa ochropus	Chevalier culblanc	A													
Lesser yellowlegs	Tringa flavipes	Chevalier à pattes jaunes	A	4												
Redshank	Tringa totanus	Chevalier gambette	A													
Marsh sandpiper	Tringa stagnatilis	Chevalier stagnatile	A	9, 10												
Wood sandpiper	Tringa glareola	Chevalier sylvain	A													
Spotted redshank	Tringa erythropus	Chevalier arlequin	A													
Greenshank	Tringa nebularia	Chevalier aboyeur	A													
Collared pratincole	Glareola pratincola	Glaréole à collier	A	9, 10												
Kittiwake	Rissa tridactyla	Mouette tridactyle	A													
Sabine's gull	Xema sabini	Mouette de Sabine	A													
Slender-billed gull	Chroicocephalus genei	Goéland railleur	A	9, 10												
Black-headed gull	Chroicocephalus ridibundus	Mouette rieuse	A													
Little gull	Hydrocoloeus minutus	Mouette pygmée	A													
Audouin's gull	Ichthyaetus audouinii	Goéland d'Audouin	A													
Mediterranean gull	Ichthyaetus melanocephalus	Mouette mélanocéphale	A													
Common gull	Larus canus	Goéland cendré	A													
Ring-billed gull	Larus delawarensis	Goéland à bec cerclé	A	4,7												
Great black-backed gull	Larus marinus	Goéland marin	A													
Glaucous gull	Larus hyperboreus	Goéland bourgmestre	A	2,4,7												
Iceland gull	Larus glaucoides	Goéland à ailes blanches	A	2,4,7												
Herring gull	Larus argentatus	Goéland argenté	A													
Caspian gull	Larus cachinnans	Goéland pontique	A	1,2,5,11,14												
Yellow-legged gull	Larus michahellis	Goéland leucophée	A													
Lesser black-backed gull	Larus fuscus	Goéland brun	A													
Gull-billed tern	Gelochelidon nilotica	Sterne hansel	A	9, 10												

CHECKLIST

English name	Scientific name	French name	Category	Regions	J	F	M	A	M	J	J	A	S	O	N	D
Caspian tern	Hydroprogne caspia	Sterne caspienne	A	6,9,10												
Sandwich tern	Thalasseus sandvicensis	Sterne caugek	A													
Elegant tern	Thalasseus elegans	Sterne élégante	A	6												
Little tern	Sternula albifrons	Sterne naine	A													
Roseate tern	Sterna dougallii	Sterne de Dougall	A	4												
Common tern	Sterna hirundo	Sterne pierregarin	A													
Arctic tern	Sterna paradisaea	Sterne arctique	A													
Whiskered tern	Chlidonias hybrida	Guifette moustac	A													
White-winged black tern	Chlidonias leucopterus	Guifette leucoptère	A	6,9,10												
Black tern	Chlidonias niger	Guifette noire	A													
Great skua	Stercorarius skua	Grand Labbe	A													
Pomarine skua	Stercorarius pomarinus	Labbe pomarin	A													
Arctic skua	Stercorarius parasiticus	Labbe parasite	A													
Long-tailed skua	Stercorarius longicaudus	Labbe à longue queue	A	2												
Little auk	Alle alle	Mergule nain	A	2,4												
Guillemot	Uria aalge	Guillemot de Troïl	A													
Razorbill	Alca torda	Pingouin torda	A													
Puffin	Fratercula arctica	Macareux moine	A	4												
Pin-tailed sandgrouse	Pterocles alchata	Ganga cata	A	10												
Rock dove/feral pigeon	Columba livia	Pigeon biset	A													
Stock dove	Columba oenas	Pigeon colombin	A													
Woodpigeon	Columba palumbus	Pigeon ramier	A													
Turtle dove	Streptopelia turtur	Tourterelle des bois	A													
Collared dove	Streptopelia decaocto	Tourterelle turque	A													
Great spotted cuckoo	Clamator glandarius	Coucou geai	A													
Cuckoo	Cuculus canorus	Coucou gris	A													
Barn owl	Tyto alba	Effraie des clochers	A													
Scops owl	Otus scops	Petit-duc scops	A													
Eagle owl	Bubo bubo	Grand-duc d'Europe	A													

CHECKLIST | 325

English name	Scientific name	French name	Category	Regions	J	F	M	A	M	J	J	A	S	O	N	D
Tawny owl	Strix aluco	Chouette hulotte	A													
Little owl	Athene noctua	Chevêche d'Athéna	A													
Pygmy owl	Glaucidium passerinum	Chevêchette d'Europe	A	11												
Tengmalm's owl	Aegolius funereus	Chouette de Tengmalm	A	9,11,12												
Long-eared owl	Asio otus	Hibou moyen-duc	A													
Short-eared owl	Asio flammeus	Hibou des marais	A	1,3,4,6												
Nightjar	Caprimulgus europaeus	Engoulevent d'Europe	A													
Alpine swift	Tachymarptis melba	Martinet à ventre blanc	A													
Swift	Apus apus	Martinet noir	A													
Pallid swift	Apus pallidus	Martinet pâle	A	9, 10												
Roller	Coracias garrulus	Rollier d'Europe	A													
Kingfisher	Alcedo atthis	Martin-pêcheur d'Europe	A													
Bee-eater	Merops apiaster	Guêpier d'Europe	A													
Hoopoe	Upupa epops	Huppe fasciée	A													
Wryneck	Jynx torquilla	Torcol fourmilier	A													
Three-toed woodpecker	Picoides tridactylus	Pic tridactyle	A	11												
Middle spotted woodpecker	Dendrocoptes medius	Pic mar	A													
Lesser spotted woodpecker	Dryobates minor	Pic épeichette	A													
Great spotted woodpecker	Dendrocopos major	Pic épeiche	A													
White-backed woodpecker	Dendrocopos leucotos lilfordi	Pic à dos blanc	A	7												
Black woodpecker	Dryocopus martius	Pic noir	A													
Green woodpecker	Picus viridis	Pic vert	A													
Iberian green woodpecker	P. v. sharpei	Pic de Sharpe		9												
Grey-headed woodpecker	Picus canus	Pic cendré														
Lesser kestrel	Falco naumanni	Faucon crécerellette	A	9, 10												
Kestrel	Falco tinnunculus	Faucon crécerelle	A													
Red-footed falcon	Falco vespertinus	Faucon kobez	A	9, 10,11,12												
Eleonora's falcon	Falco eleonorae	Faucon d'Éléonore	A	7,9,10												
Merlin	Falco columbarius	Faucon émerillon	A													

CHECKLIST

English name	Scientific name	French name	Category	Regions
Hobby	*Falco subbuteo*	Faucon hobereau	A	
Peregrine	*Falco peregrinus*	Faucon pèlerin	A	
Ring-necked parakeet	*Psittacula krameri*	Perruche à collier	C	
Fischer's lovebird	*Agapornis fischeri*	Inséparable de Fischer	C	
Red-backed shrike	*Lanius collurio*	Pie-grièche écorcheur	A	
Lesser grey shrike	*Lanius minor*	Pie-grièche à poitrine rose	A	9
Great grey shrike	*Lanius excubitor*	Pie-grièche grise	A	12,14
Southern grey shrike	*Lanius meridionalis*	Pie-grièche méridionale	A	9, 10
Woodchat shrike	*Lanius senator*	Pie-grièche à tête rousse	A	
Balearic woodchat shrike	*L. s. badius*			
Golden oriole	*Oriolus oriolus*	Loriot d'Europe	A	
Jay	*Garrulus glandarius*	Geai des chênes	A	
Magpie	*Pica pica*	Pie bavarde	A	
Nutcracker	*Nucifraga caryocatactes*	Cassenoix moucheté	A	
Alpine chough	*Pyrrhocorax graculus*	Chocard à bec jaune	A	
Chough	*Pyrrhocorax pyrrhocorax*	Crave à bec rouge	A	
Jackdaw	*Coloeus monedula*	Choucas des tours	A	
Rook	*Corvus frugilegus*	Corbeau freux	A	
Carrion crow	*Corvus corone*	Corneille noire	A	
Hooded crow	*Corvus cornix*	Corneille mantelée	A	
Hooded crow (Corsica)	*C. c. sharpii*	Corneille de Corse		10
Raven	*Corvus corax*	Grand Corbeau	A	
Waxwing	*Bombycilla garrulus*	Jaseur boréal	A	2
Coal tit	*Periparus ater*	Mésange noire	A	
Crested tit	*Lophophanes cristatus*	Mésange huppée	A	
Marsh tit	*Poecile palustris*	Mésange nonnette	A	
Willow tit	*Poecile montanus*	Mésange boréale	A	
Blue tit	*Cyanistes caeruleus*	Mésange bleue	A	
Great tit	*Parus major*	Mésange charbonnière	A	

English name	Scientific name	French name	Category	Regions	J	F	M	A	M	J	J	A	S	O	N	D
Penduline tit	*Remiz pendulinus*	Rémiz penduline	A													
Bearded tit	*Panurus biarmicus*	Panure à moustaches	A	4,9,10,11												
Woodlark	*Lullula arborea*	Alouette lulu	A													
Skylark	*Alauda arvensis*	Alouette des champs	A													
Crested lark	*Galerida cristata*	Cochevis huppé	A													
Thekla lark	*Galerida theklae*	Cochevis de Thekla	A	9												
Shore lark	*Eremophila alpestris*	Alouette haussecol	A	2												
Greater short-toed lark	*Calandrella brachydactyla*	Alouette calandrelle	A	9, 10												
Calandra lark	*Melanocorypha calandra*	Alouette calandre	A	10												
Sand martin	*Riparia riparia*	Hirondelle de rivage	A													
Swallow	*Hirundo rustica*	Hirondelle rustique	A													
Crag martin	*Ptyonoprogne rupestris*	Hirondelle de rochers	A													
House martin	*Delichon urbicum*	Hirondelle de fenêtre	A													
Red-rumped swallow	*Cecropis daurica*	Hirondelle rousseline	A	9, 10												
Cetti's warbler	*Cettia cetti*	Bouscarle de Cetti	A													
Long-tailed tit	*Aegithalos caudatus*	Mésange à longue queue	A													
Willow warbler	*Phylloscopus trochilus*	Pouillot fitis	A													
Chiffchaff	*Phylloscopus collybita*	Pouillot véloce	A													
Siberian chiffchaff	*P. c. tristis*		A	4												
Iberian chiffchaff	*Phylloscopus ibericus*	Pouillot ibérique	A	7												
Western Bonelli's warbler	*Phylloscopus bonelli*	Pouillot de Bonelli	A													
Wood warbler	*Phylloscopus sibilatrix*	Pouillot siffleur	A													
Yellow-browed warbler	*Phylloscopus inornatus*	Pouillot à grands sourcils	A	2,3,4,6,7												
Great reed warbler	*Acrocephalus arundinaceus*	Rousserolle turdoïde	A													
Moustached warbler	*Acrocephalus melanopogon*	Lusciniole à moustaches	A	9, 10												
Aquatic warbler	*Acrocephalus paludicola*	Phragmite aquatique	A	4,6,7,9												
Sedge warbler	*Acrocephalus schoenobaenus*	Phragmite des joncs	A													
Reed warbler	*Acrocephalus scirpaceus*	Rousserolle effarvatte	A													
Marsh warbler	*Acrocephalus palustris*	Rousserolle verderolle	A													

CHECKLIST

English name	Scientific name	French name	Category	Regions	J	F	M	A	M	J	J	A	S	O	N	D
Melodious warbler	*Hippolais polyglotta*	Hypolaïs polyglotte	A													
Icterine warbler	*Hippolais icterina*	Hypolaïs ictérine	A													
Grasshopper warbler	*Locustella naevia*	Locustelle tachetée	A													
Savi's warbler	*Locustella luscinioides*	Locustelle luscinioïde	A													
Fan-tailed warbler	*Cisticola juncidis*	Cisticole des joncs	A													
Red-billed leiothrix	*Leiothrix lutea*	Léiothrix jaune	C													
Blackcap	*Sylvia atricapilla*	Fauvette à tête noire	A													
Garden warbler	*Sylvia borin*	Fauvette des jardins	A													
Barred warbler	*Sylvia nisoria*	Fauvette épervière	A													
Lesser whitethroat	*Sylvia curruca*	Fauvette babillarde	A													
Western orphean warbler	*Sylvia hortensis*	Fauvette orphée	A													
Whitethroat	*Sylvia communis*	Fauvette grisette	A													
Dartford warbler	*Sylvia undata*	Fauvette pitchou	A													
Marmora's warbler	*Sylvia sarda*	Fauvette sarde	A	10												
Spectacled warbler	*Sylvia conspicillata*	Fauvette à lunettes	A	9, 10												
Subalpine warbler	*Sylvia cantillans*	Fauvette passerinette	A													
Eastern subalpine warbler	*S. c. albistriata*		A	9, 10												
Moltoni's warbler	*Sylvia subalpina*	Fauvette moltonée	A													
Sardinian warbler	*Sylvia melanocephala*	Fauvette mélanocéphale	A													
Firecrest	*Regulus ignicapilla*	Roitelet à triple bandeau	A													
Goldcrest	*Regulus regulus*	Roitelet huppé	A													
Wren	*Troglodytes troglodytes*	Troglodyte mignon	A													
Corsican nuthatch	*Sitta whiteheadi*	Sittelle corse	A	10												
Nuthatch	*Sitta europaea*	Sittelle torchepot	A													
Wallcreeper	*Tichodroma muraria*	Tichodrome échelette	A	8,9,11												
Treecreeper	*Certhia familiaris*	Grimpereau des bois	A													
Short-toed treecreeper	*Certhia brachydactyla*	Grimpereau des jardins	A													
Rose-coloured starling	*Pastor roseus*	Étourneau roselin	A	4												
Starling	*Sturnus vulgaris*	Étourneau sansonnet	A													

CHECKLIST

English name	Scientific name	French name	Category	Regions
Spotless starling	*Sturnus unicolor*	Étourneau unicolore	A	
Ring ouzel	*Turdus torquatus*	Merle à plastron	A	
Blackbird	*Turdus merula*	Merle noir	A	
Fieldfare	*Turdus pilaris*	Grive litorne	A	
Redwing	*Turdus iliacus*	Grive mauvis	A	
Song thrush	*Turdus philomelos*	Grive musicienne	A	
Mistle thrush	*Turdus viscivorus*	Grive draine	A	
Spotted flycatcher	*Muscicapa striata*	Gobemouche gris	A	
Mediterranean flycatcher	*M. s. tyrrhenica*			
Robin	*Erithacus rubecula*	Rougegorge familier	A	
Bluethroat	*Luscinia svecica*	Gorgebleue à miroir	A	
Nightingale	*Luscinia megarhynchos*	Rossignol philomèle	A	
Pied flycatcher	*Ficedula hypoleuca*	Gobemouche noir	A	
Collared flycatcher	*Ficedula albicollis*	Gobemouche à collier	A	14
Red-breasted flycatcher	*Ficedula parva*	Gobemouche nain	A	4
Black redstart	*Phoenicurus ochruros*	Rougequeue noir	A	
Redstart	*Phoenicurus phoenicurus*	Rougequeue à front blanc	A	
Rock thrush	*Monticola saxatilis*	Monticole de roche	A	8,9,10,11,12
Blue rock thrush	*Monticola solitarius*	Monticole bleu	A	9,10
Whinchat	*Saxicola rubetra*	Tarier des prés	A	
Stonechat	*Saxicola rubicola*	Tarier pâtre	A	
Wheatear	*Oenanthe oenanthe*	Traquet motteux	A	
Black-eared wheatear	*Oenanthe hispanica*	Traquet oreillard	A	9,10
Dipper	*Cinclus cinclus*	Cincle plongeur	A	
House sparrow	*Passer domesticus*	Moineau domestique	A	
Italian sparrow	*Passer italiae*	Moineau cisalpin		
Spanish sparrow	*Passer hispaniolensis*	Moineau espagnol	A	10
Tree sparrow	*Passer montanus*	Moineau friquet	A	
Rock sparrow	*Petronia petronia*	Moineau soulcie	A	

English name	Scientific name	French name	Category	Regions	J	F	M	A	M	J	J	A	S	O	N	D
White-winged snowfinch	Montifringilla nivalis	Niverolle alpine	A													
Indian silverbill	Euodice malabarica	Capucin bec-de-plomb	C													
Alpine accentor	Prunella collaris	Accenteur alpin	A													
Dunnock	Prunella modularis	Accenteur mouchet	A													
Yellow wagtail	Motacilla flava	Bergeronnette printanière	A													
	M. f. thunbergi	Bergeronnette nordique														
	M. f. feldegg	Bergeronnette des Balkans		10												
Citrine wagtail	Motacilla citreola	Bergeronnette citrine	A	4, 9, 10,11												
Grey wagtail	Motacilla cinerea	Bergeronnette des ruisseaux	A													
Pied wagtail	Motacilla alba	Bergeronnette grise	A													
	M. a. yarrellii	Bergeronnette de Yarrell														
Richard's pipit	Anthus richardi	Pipit de Richard	A	4, 8,10												
Tawny pipit	Anthus campestris	Pipit rousseline	A													
Meadow pipit	Anthus pratensis	Pipit farlouse	A													
Tree pipit	Anthus trivialis	Pipit des arbres	A													
Red-throated pipit	Anthus cervinus	Pipit à gorge rousse	A	9, 10												
Water pipit	Anthus spinoletta	Pipit spioncelle	A													
Rock pipit	Anthus petrosus	Pipit maritime	A													
Chaffinch	Fringilla coelebs	Pinson des arbres	A													
Brambling	Fringilla montifringilla	Pinson du Nord	A													
Hawfinch	Coccothraustes coccothraustes	Grosbec casse-noyaux	A													
Bullfinch	Pyrrhula pyrrhula	Bouvreuil pivoine	A													
Common rosefinch	Carpodacus erythrinus	Roselin cramoisi	A	2,4												
Greenfinch	Chloris chloris	Verdier d'Europe	A													
Twite	Linaria flavirostris	Linotte à bec jaune	A													
Linnet	Linaria cannabina	Linotte mélodieuse	A													
Common redpoll	Acanthis flammea	Sizerin boréal														
Lesser redpoll	Acanthis cabaret	Sizerin flammé cabaret	A													
Crossbill	Loxia curvirostra	Bec-croisé des sapins	A													

CHECKLIST | 331

English name	Scientific name	French name	Category	Regions	J	F	M	A	M	J	J	A	S	O	N	D
Goldfinch	*Carduelis carduelis*	Chardonneret élégant	A													
Citril finch	*Carduelis citrinella*	Venturon montagnard	A													
Corsican finch	*Carduelis corsicana*	Venturon corse	A													
Serin	*Serinus serinus*	Serin cini	A													
Siskin	*Spinus spinus*	Tarin des aulnes	A													
Corn bunting	*Emberiza calandra*	Bruant proyer	A													
Yellowhammer	*Emberiza citrinella*	Bruant jaune	A													
Rock bunting	*Emberiza cia*	Bruant fou	A													
Ortolan bunting	*Emberiza hortulana*	Bruant ortolan	A													
Cirl bunting	*Emberiza cirlus*	Bruant zizi	A													
Little bunting	*Emberiza pusilla*	Bruant nain	A	4												
Black-headed bunting	*Emberiza melanocephala*	Bruant mélanocéphale	A	10												
Reed bunting	*Emberiza schoeniclus*	Bruant des roseaux	A													
	E. s. witherbyi															
Lapland bunting	*Calcarius lapponicus*	Bruant lapon	A	2,3,4												
Snow bunting	*Plectrophenax nivalis*	Bruant des neiges	A	2,3,4												

Key for phenological bars

- abundant
- common
- locally common
- uncommon
- rare/local
- scarce
- absent
- feral, common
- feral, rare or local

SITE INDEX

A

Adour valley, The 173
Aiguillon bay, The 117
Aïtone forest 218
Allier gorges, The 253
Allogny forest 109
Alpilles, The 205
Alpine foothills south of Lake Geneva (Lac Léman), The 230
Alps, the 228–244
Amance Lake *see* Lac d'Amance
Angers 93, 95
Aquitaine 134–60
Arcachon Bay, Northern 141
   Southern 143
Ardèche gorges 264
Ardennes 52, 304, 308
Ariège–Garonne confluence, The 162
Aude valley, The 183
Audierne bay 83
Auxerre 273
Auxois bocages 273
Auxois reservoirs 281

B

Baie d'Authie 47
Baie de Canche 47
Barcaggio 220
Basque coastline, The 147
Basses Vallées Angevines 94
Baubigny plateau 282
Béarn 151
Beauce 34
Beaufour pond 132
Belle Henriette lagoon, The 119
Bellebranche forest 93
Biguglia 219
Biscarosse ponds 144
Bonifacio 221
Bonne Anse bay 127
Boulogne-sur-Mer 45
Bourbonnais bocage, The 257
Bourdon reservoir and Puisaye ponds 274

Bourget Lake *see* Lac du Bourget
Brenne, The 102
Brest 80
Brignogan semaphore 78
Brittany 73–91
Brotonne forest 59
Brouage marshes 125
Bugey, The 225
Burgundy 270–87
   Southern near Châlon 287

C

Cabane de Moins, The 123
Calais harbour 43
Camargue Gardoise, The 197
Camargue, The 199
Cambounet-sur-le-Sor 165
Canet-en-Roussillon 189
Cap Blanc-Nez 43
Cap Fréhel 75
Cap Gris-Nez 44
Cap-Sizun 82
Capu Rossu 218
Careil pond 86
Caucalières plateau (Causse de Caucalières) 165
Causse Comtal 247
Causse de Blandas 180
Causse Méjean 249
Cébron reservoir (Lac du Cébron) 130
Celles-sur-Plaine 299
Cerisy forest 64
Cévennes 181
Ceyssat/Chaîne des Puys 261
Chamonix 232
Chamonix valley 232
Champeigne Tourangelle 99
Channel, The 58
Charente estuary, The 124
Charleville-Mézières 305, 307, 308
Chassezac 262
Châtillon-sur-Seine forest 272
Chausey islands 69
Chinon forest 98

Cirque de Navacelles 180
Cirque du Bout du Monde 282
Citeaux forest 284
Col de Bavella 220
Col de la Colombière 231
Col de Mantet 190
Col de Vergio 218
Col du Tourmalet 171
Combourg pond 133
Contres marshes 110
Corbières, The 185
Corsica 217–21
Corte 219
Cotentin, Northern 66
   Western 68
Crau, The (Plaine de la Crau) 202
Cré-sur-Loir 93
Créteil Lake *see* Lac de Créteil
Cuers/Pierrefeu-du-Var airfield 209
Curécy pond 93

D
Dain lagoon, The 113
Défilé de l'Écluse 226
Der-Chantecoq Lake *see* Lac du Der-Chantecoq
Dévoluy, The 244
Dijon 278, 279
Dives marshes 64
Dombes wetlands, The 224
Dordogne, Northern 157
   Southern 155
Dordogne gorges 269
Douarnenez bay 82
Doubs valley, The 287
Dune du Pilat 144
Dunkerque 40
   Parc du Vent 40

E
Ecopôle du Forez 266
Ecrins, The 242
Étang de Berre 204
Étang de la Rincerie 93
Étang du Puits 108
Eyne 194

F
Faverney area 276
Fontaine Française ponds and marshes 278
Forêt de Fontainebleau (Fontainebleau forest) 33
Forez 264
Fort de la Revère 217
Frasne 228
Fréjus – Villepey lagoon 216
Frontignan 177

G
Galens lake 247
Galibier pass 239
Garonne 162
Gâvres bay 87
Gironde, Northern 136
Glières plateau 234
Golfe de Fos 205
Golfe du Morbihan *see* Morbihan gulf
Goulven bay 77
Grand-Lieu Lake *see* Lac de Grand-Lieu
Grée marshes 91
Grésigne forest 165
Gruissan 185
Guérande marshes 89
Guissény 78

H
Haut-Asco 219
Hautes Chaumes 264
Hemmes de Marck 42
Hérault gorges 181
Héron Lake *see* Lac du Héron
Heurteauville marshes and meadows 60
Hoëdic island 88
Hyères 207

I
Île d'Ambès 136
Île de Ré 120
Iraty 149, 151
Is-sur-Tille forest 278

SITE INDEX

J
Joigny 274
Jonte gorges, The 248

L
L'Anglade marshes 129
La Bassée 32
La Flèche marshes 93
La Margeride 254
La Marque marshes 54
La Rochelle 117, 118, 120, 123
La Serre 259
La Terrière 119
La Vacherie nature reserve 119
Lac d'Amance 25
Lac de Créteil 29
Lac de Grand-Lieu 91
Lac de Rillé *see* Rillé reservoir
Lac du Bourget 234
Lac du Der-Chantecoq 25
Lac du Héron 54
Lac du Temple 25
Lac Léman *see* Lake Geneva
Lairoux-Curzon marshes 119
Lake Geneva (Lac Léman) 230
Lamartine ponds (Lac Lamartine) 162
Lampaul-Ploudalmézeau 79
Landes pond (Étang des Landes), The 268
Lansargues 176
Lapalme 186
Larmor-Baden 89
Larzac, The 248
Lautaret pass 239
Lavaud reservoir 133
Le Barcarès 186, 187
Le Carladez 251
Le Conquet 79
Le Havre 58
Le Hohneck 301
Le Hucel 230
Le Pibeste 170
Le Puy-Sainte-Réparade 212
Le Teich nature reserve 142
Les Gras plateau 263
Leucate 187

Liamone estuary 217
Lindre reservoir 295
Loches forest 100
Loire – Langeais 97
Loire gorges 265
Loiré-sur-Nie 133
Loire Valley, The 92–110
Longeyroux peat bog 268
Lorraine 289
Lot valley, The, and hills 154
Louroux pond (Étang du Louroux) 100
Luxembourg border, The 293
Lyon 223

M
Macinaggio 220
Madine reservoir 290
Maine reservoir (Lac de Maine) 95
Marais Breton, The 113
Marcenay reservoir 272
Mare à Goriaux 54
Marne valley, The 30
Martres-Tolosane gravel ponds 163
Massif Central 245–69
Mauny loop 60
Mazères 163
Médoc 137
    Inner 139
    Lakes of Southern 140
Mellois plateau, The 133
Mercantour National Park 212
Metz 292, 305
Meurthe 296
Mirebalais plain, The 130
Moëze-Oléron reserve 126
Moissons-Lavacourt 36
Mont Aigoual 181
Mont Blanc, northern slopes 232
Mont Lozère 250
Mont Mézenc 255
Mont Pilat 266
Mont Saint-Michel 69
    North of 70
    Southern bay 72
Mont Valier 167

SITE INDEX | 335

Mont Ventoux 214
Mont-Dore 262
Montdauphin 241
Montlouis 99
Montpellier 176, 180
Morbihan gulf (Golfe du Morbihan) 88
Mormal forest 55
Morvan, The 283
Moselle valleys 296
Moulière forest 131
Moulin-Neuf pond 83
Mulhouse 302
Murat pond 268

N
Nancy 290, 298
Nantes 91
Narse de Lascols 252
Néons-sur-Creuse 104
Néouvielle 171
Névache 240
Nîmes 177
Noirmoutier island 113
Normandy 56–72

O
Oléron island 127
Olonne marshes (Marais d'Olonne) 116
Ondes gravel ponds 162
Organbidexka pass 150
Orient Lake *see* Paris: Lac d'Orient
Orléans 108
Orléans forest 107
Orlu reserve 166
Orne 64
Orx marshes 146
Ouessant *see* Ushant island

P
Paimpont forest 85
Païolive 262
Pamiers airfield 163
Paris:
    Bois de Vincennes 28
    Buttes Chaumont 27
    Jardin des Plantes 27
    La Courneuve park 27
    Lac d'Orient 24
    Parc de Sausset 28
    Parc Montsouris 27
    Père-Lachaise cemetery 27
    Vincennes 28
Penmarc'h 84
Penn Mané wetlands 87
Perche forests, The 61
Perpignan 189
Pic Carlit, The 190
Pic de Cagire 169
Pic de Madrès 191
Pic du Midi de Bigorre 171
Picardie, Inner 51
Pinail nature reserve 131
Pithiviers 34
Plateau des Chaux – Gigors et Lozéron 239
Platier d'Oye, The 41
Plomb du Cantal 251
Pointe de Grave, The 138
Pointe de La Torche 84
Poitiers 130
Porquerolles island 209
Port-Saint-Louis-du-Rhône 205
Portiragnes 183
Poses 61
Prades hill 254
Prat de Bouc 251
Prés du Hem 53
Puy de Dôme 261
Puy de Sancy 262
Puydarrieux reservoir 173
Pyrenees, The 151, 160–73
    Atlantic 149

Q
Quiberon 87

R
Reims 307
Rémuzat 215
Rennes 85
Restonica valley 219

SITE INDEX

Rillé reservoir (Lac de Rillé) 97
Risoux forest 226
Roanne area 266
Rochefort water treatment plant 124
Rochers du Saussois 275
Rodez airport 247
Romelaëre ponds (Étangs de Romelaëre) 53
Ronce-les-Bains 126
Roumare meadows 60
Ruan 34

S

Saclay 35
Saint-Brieuc bay 75
Saint-Cyr 130
Saint-Denis-du-Payré 119
Saint-Hubert ponds (Étangs de Saint-Hubert) 36
Saint-Jacut-de-la-Mer 75
Saint-Jean-Pied-de-Port 150
Saint-Quentin-en-Yvelines (Étang de Saint Quentin) 36
Saintes 129
Salses 186
Sanguinet 144
Saône valley, The 284
Sault plateau 191
Seille valley 294
Sein island 83
Seine estuary 59
Seine river 36
Sénart forest 29
Sept-Îles, The 76
Serre-Ponçon reservoir 241
Seudre estuary, The 126
Sion hill 298
Sion-sur-l'Océan 115
Sologne, The 105
Somme bay, Northern 49
Strasbourg 302
  North of 303

T

Tanet-Gazon du Faing 300

Tarascon 167
Temple Lake *see* Lac du Temple
Termignon 235
Thionville 292–93
Tours 97–100
Trélon forest 55
Tronçais forest 257
Truyère gorges, The 252

U

Ushant island (Île d'Ouessant) 79

V

Val d'Allier 256
Val d'Auron 110
Val-Joly ponds (Étang du Val-Joly) 55
Valensole plateau, The 211
Vallorcine 233
Vanoise, The 235
Var estuary, The 217
Vendée 88, 111, 113
Vendres 183
Vercors, Northern 237
  Southern 238
Verdon gorges, The 211
Verneuil-sur-Seine 36
Vesoul 277
Vicdessos 167
Vigueirat marsh (Marais du Vigueirat) 202
Villeneuve-lès-Maguelone 177
Vincennes 28
Vinon-sur-Verdon airfield 212
Vire 64
Virelay 110
Vosges mountains 298–301

W

Wissant marshes 44

Y

Yeu island (Île d'Yeu) 115
Yonne valley 274
Yves bay 123